AUDEL®

Refrigeration: Home and Commercial

by Edwin P. Anderson
revised by Rex Miller

An Audel® Book

Macmillan Publishing Company

New York

Collier Macmillan Publishers

London

FOURTH EDITION

Macmillan Publishing Company
866 Third Avenue, New York, NY 10022
Collier Macmillan Canada, Inc.

Production services by the Walsh Group, Yarmouth, ME

Library of Congress Cataloging-in-Publication Data

Anderson, Edwin P., 1895-
      Refrigeration, home and commercial / by Edwin P. Anderson; revised by Rex Miller.–4th ed.
            p.   cm.
      At head of title: Audel.
      ISBN 0-02-584875-5
      1. Refrigeration and refrigerating machinery.   I. Miller, Rex, 1929- . II. Title.
TP492.A583   1990
621.56–dc20   89-13875
                                                                        CIP

Foreword

This book has been prepared to assist personnel, who are involved with the installation, servicing, and operating of refrigeration equipment. Fundamentally, the principles and physics of refrigeration have not changed to any degree during the many years that we have relied on refrigeration as a means of keeping foods from spoiling.

Household as well as commercial types of refrigeration are covered. Some of the items in this book involve both household and commercial refrigeration, as the principles are basically the same.

It is not the intent of this book to cover large industrial installations, as this is a separate subject, involving refrigerating machines of very large capacity.

The personnel involved in installing, servicing, and operating refrigeration equipment must have a firm foundation and knowledge of the principles of refrigeration, including the fundamentals of refrigeration and the laws of physics that govern these fundamentals.

Troubleshooting guides are included as necessary so that one may refer to them for troubles that present themselves in servicing and operating refrigeration equipment.

It is most important to have a thorough understanding of the working parts of both electrically driven and absorption-type refrigeration units. This will include the properties of the various refrigerants used, including some of the obsolete refrigerants, as it is possible that you may come across some of these obsolete refrigerants still in use. Once you have the theory of refrigeration well organized in your mind, troubles in the systems will be much easier to locate.

Some of the information that appears dated has been left for the benefit of those who must still operate and repair these pieces of equipment. As you know, refrigeration equipment is well known for its long life.

Sincere thanks goes to Carl W. McPhee, editor of the ASHRAE *Handbooks and Product Directory*. ASHRAE, the American Society of Heating, Refrigeration and Air Conditioning Engineers, Inc., is the source of information on refrigeration, heating, and air conditioning and has issued many books on these subjects. Credit is given for information used from their *Handbooks*.

REX MILLER
EDWIN P. ANDERSON

Contents

CHAPTER 1

FUNDAMENTALS OF REFRIGERATION 11
  Applications of refrigeration—physical units—atmospheric, absolute, and gage
  pressures—Boyle's law for perfect gases—Charles's law—power and work—
  units of power—energy—theory of heat—thermometers—absolute zero—
  units of heat—specific heat—latent heat—change of state—latent heat of fusion—latent heat of evaporation—steam—refrigerants—use of ice—ton of
  refrigeration—evaporation—pressure—pressure-temperature relations—effect of low pressures—heat transmission—condensation—altitudes and refrigeration capacity—refrigeration by vaporization—basic systems—laws of gases—
  the refrigeration cycle—pressure—summary—review questions

CHAPTER 2

REFRIGERANTS 71
  Desirable properties—Freon refrigerants—leak detection—pressure-temperature chart—care in handling refrigerants—handling refrigerant cylinders—a
  simple refrigeration system—summary—review questions

CHAPTER 3

COMPRESSORS 93
  How the compressor works—reciprocating compressors—rotary compressors—hermetic compressors—centrifugal compressors—system precautions—
  compressor maintenance—compressor maintenance and service—compressor removal—compressor replacement—belt alignment—running-time
  check—compression ratio—energy consumption—refrigeration-system lubricants—compressor calculations—safety precautions—summary—review
  questions

CHAPTER 4

DOMESTIC REFRIGERATION 135

The refrigeration cycle—compressors—condensers—drier-strainer—capillary tubes—evaporators—accumulators—control methods—summary—review questions

CHAPTER 5

ABSORPTION SYSTEM FOR DOMESTIC REFRIGERATION 153

Principle of operation—absorption-refrigeration cycle—the flue system—gas-control devices—electrical accessories—installation and service—burner-orifice cleaning—burner-air adjustment—thermostat-valve cleaning—gas-pressure adjustment—flame stability—refrigerator-unit removal and replacement—refrigerator maintenance—**absorption-refrigeration troubleshooting guide**—summary—review questions

CHAPTER 6

THERMOELECTRIC COOLING 191

Performance of thermoelectric couples—advantages and applications of thermoelectric cooling—summary—review questions

CHAPTER 7

REFRIGERATION SERVICE EQUIPMENT AND TOOLS 201

Service valves—combination gage set—sealed systems—expansion-valve units—leak testing—service tools—making a hand bend—making a relieved bend—use of reseating tools—copper tubing—silver brazing—summary—review questions

CHAPTER 8

DOMESTIC REFRIGERATION OPERATION AND SERVICE 249

Sealed-system operation—electrical testing—compressor testing with direct power—compressor motor relays—compressor terminals—cabinet light and switch—condenser cooling fan—replacing system components—evaporator cooling fan—temperature-control service—**refrigeration troubleshooting guide**—summary—review questions

CHAPTER 9

HOUSEHOLD-CABINET DEFROSTING SYSTEMS 289
Manual defrosting—automatic defrosting—defrost-thermostat testing—summary—review questions

CHAPTER 10

CABINET MAINTENANCE AND REPAIRS 315
Door replacement—summary—review questions

CHAPTER 11

HOUSEHOLD FREEZERS 325
Cycle of operation—characteristics of a capillary-tube system—electrical system—upright freezers—**household freezer troubleshooting guide**—summary—review questions

CHAPTER 12

STYLES OF DOMESTIC REFRIGERATORS AND
FOOD ARRANGEMENT 349
Food arrangement—door openings—warm-weather effect—cooling-coil frosting—cleaning—odor prevention—summary—review questions

CHAPTER 13

INSTALLATION METHODS 359
Installation of absorption-type refrigerators—summary—review questions

CHAPTER 14

COMPRESSOR LUBRICATION SYSTEM 369
Motor lubrication—compressor lubrication—servicing data—summary—review questions

CHAPTER 15

REFRIGERATION CONTROL DEVICES 383
    Automatic expansion values—thermostatic expansion valves—capillary-tube
    systems—domestic electrical systems—temperature control by suction pres-
    sure—servicing valves—summary—review questions

CHAPTER 16

THE ELECTRICAL SYSTEM 411
    Units of electricity—the electric circuit—direct current—single- and three-
    phase current—energy—motor efficiency—pulley speed and sizes—alternat-
    ing-current motors—single-phase motors—motor-compressor controls—oil-
    pressure controls—temperature controls—defrost controls—electrical-sys-
    tem maintenance—compressor-motor replacement—**electrical-system trou-
    bleshooting guide**—summary—review questions

CHAPTER 17

COMMERCIAL REFRIGERATION PRINCIPLES 465
    Effect of pressure on refrigerants—essential parts of a refrigeration plant—
    condensers, fittings, and accessories—cooling-water requirements—cooling-
    system components—refrigeration-system operation—**compressor trou-
    bleshooting guide**—summary—review questions

CHAPTER 18

BRINE SYSTEMS 497
    Indirect refrigeration system—brine—brine-system corrosion—preparing new
    brine—summary—review questions

CHAPTER 19

ICE-MAKING SYSTEMS 509
    Air piping—ice cans—inferior ice—freezing tanks—can dumps—plate-ice
    systems—automatic ice makers—dry ice—ice-making calculations—ice stor-
    age—summary—review questions

CHAPTER 20

SUPERMARKET AND GROCERY REFRIGERATION 527

Display-case arrangement—open display cases—display-case lighting—radiant-heat effect—close meat-display cases—multiple connection—display-case maintenance—installation—mechanical centers—combined systems—display-case defrosting—drainage requirements—walk-in coolers—meat preparation—summary—review questions

CHAPTER 21

LOCKER PLANTS 555

Locker-plant construction—design considerations—insulation requirements—equipment and controls—summary—review questions

CHAPTER 22

SPECIAL REFRIGERATION APPLICATIONS 569

Water coolers—ice cream refrigeration—bottle and can beverage coolers—milk coolers—servicing and repair—milk pasteurization—summary—review questions

CHAPTER 23

COLD-STORAGE PRACTICE 593

Quick freezing—food preparation for quick freezing—summary—review questions

CHAPTER 24

FANS AND BLOWERS 603

Propeller fans—centrifugal fans—fan characteristics—power requirements—motor types—fan selection—summary—review questions

CHAPTER 25

REFRIGERATION PIPING 611

Pressure-drop considerations—liquid refrigerant lines—interconnection of suction lines—discharge lines—water valves—multiple-unit installation—pipe connections—piping insulation—summary—review questions

CHAPTER 26

COMMERCIAL ABSORPTION SYSTEMS 633
Absorption-refrigeration cycles—basic absorption-refrigeration cycle—characteristics of the refrigerant-absorption pair—water-lithium bromide machine—ammonia-water cycle—solar cooling—summary—review questions

CHAPTER 27

CIRCULATING PUMPS 657
Rotary pumps—centrifugal pumps—reciprocating pumps—power requirements—pump maintenance—summary—review questions

CHAPTER 28

INSTALLATION AND OPERATION 669
Foundations—installing belt-driven compressors—installing direct-connected, motor-driven compressors—leak testing—dehydration and evacuation—charging the system—summary—review questions

CHAPTER 29

HEAT LEAKAGE THROUGH WALLS 685
Summary—heat leakage tables—review questions

CHAPTER 30

REFRIGERATION-LOAD CALCULATIONS 709
Calculating heat-transfer coefficient—calculating heat leakage—capacity of machines—load calculation for a walk-in refrigerator—evaporator capacity—load calculations for a dairy plant—load calculations for small freezers—summary—review questions

APPENDIX 725

GLOSSARY 737

INDEX 761

CHAPTER 1

Fundamentals of Refrigeration

Refrigeration, regardless of the means by which it is obtained, may be defined as *a process of the removal of heat from a substance or from a space.* Thus, if heat is removed from any substance or space, that substance or space becomes cool. In making the statements *warm* or *cool,* one must remember that these are relative statements. What is warm or cool is relative to surrounding temperatures. Thus, we must consider temperatures as measured by means of thermometers to arrive at the relationship between heat and cool.

From the foregoing statement, it follows that the problem of refrigeration concerns itself with heat removal from a substance or a space. Heat is a form of *energy* like electrical, chemical, and mechanical energy. It must be stated that energy can be neither created nor destroyed, but can be converted from one form to another, with temperature indicating its relative intensity.

The first method of refrigeration in recent history was by means of cutting and storing ice in cabinets. Later, considerable progress was made in the commercial manufacture of artificial ice. Probably the most widely used form of refrigeration was the manufacture of ice by means of an electrically-driven compressor, with ammonia used as the refrigerant, from which 100-lb cakes of ice were produced commercially. The development of household refrigeration followed, and since 1923 this type of refrigeration machine has been manufactured in large quantities and has become commercially successful. Today, this type of machine is no longer considered a luxury; it is considered a necessity by all of us. From this effort, we further expanded into the commercial refrigeration and the frozen foods field.

APPLICATIONS OF REFRIGERATION

Food preservation is the largest application of mechanical refrigeration today. Without modern refrigerating machinery, the packing-house industry, transportation of perishable foods, and preparation of many other edibles would not only be difficult, but in many cases almost impossible. Refrigeration has not only saved quantities of meat, fish, eggs, milk, and cream from spoilage; it has also played an important part in the diet revision of the world. No longer are the inhabitants of one hemisphere, country, or locality dependent upon local foods, but may draw upon the entire world as a resource. This has made possible great developments in agriculture and livestock raising in countries far distant from potential markets. It has permitted the utilization of the great productivity of the tropics in supplying fruits and foods for other parts of the world.

Nearly all of the industries involved in the preparation of foods and drinks make extensive use of artificial ice or mechanical refrigeration. The dairy industry, for example, finds that for pre-cooling milk and cream and in the manufacture of butter and ice cream, refrigeration is indispensable. In many industries, cooling and conditioning of air is an important phase of the manufacturing process. Among these processes are the manufacture of photographic films, explosives, machine production of cigars and cigarettes, candy and chewing gum, and rayon. They are important users of refrigeration, not only because of their size, but also because of their dependence on conditioned air for successful operation.

The cooling of liquids plays an important part in the mechanical and chemical industries. Oil-tempering baths are kept at constant low temperatures through refrigeration; and in the jackets of nitrators and mixing machines for the celluloid, smokeless powder, and rubber industries, cooled brines are used to keep the mixtures at correct temperatures. In the chemical industry, use of refrigeration for drying air and gases and for purifying solutions by means of crystallization at definite temperatures is steadily increasing.

PHYSICAL UNITS

In order to obtain a clear conception of the functioning of a mechanical refrigeration system, it is imperative to consider and understand the physical and thermal properties underlying the production of the artificial cold.

Practical Dimensions

We are users of the English system of measurement. However, we must also consider the metric system in our discussions. You will notice that many items purchased in stores give weights and measures in both the English and the metric systems. A comparison will now be made between the two systems, but this book will use primarily the English system (Fig. 1-1) From the following information, you may readily convert from one system to another.

The following prefixes are used in the metric system:

Micro = one-millionth
Milli = one-thousandth
Centi = one-hundredth
Deci = one-tenth
Deca = ten
Hecto = one hundred
Kilo = one thousand
Mega = one million

10 millimeters = 1 centimeter
100 centimeters = 1 meter
1000 millimeters = 1 meter
25.4 millimeters = 1 inch
2.54 centimeters = 1 inch
30.48 centimeters = 1 foot
304.8 millimeters = 1 foot
3.28083 feet = 1 meter
39.36996 inches = 1 meter
0.62137 miles = 1 kilometer
1.60935 kilometers = 1 mile
0.9144 meters = 1 yard

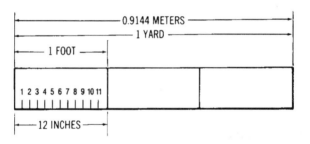

Fig. 1-1. Relationship of centimeters to feet and meters to yards.

Square Measure

Common units of area, covering both the English and the metric systems, are (Fig. 1-2):

1 square foot = 144 square inches
1 square yard = 9 square feet
1 square foot = 929.03 square centimeters
1 square yard = 0.836 square meters
1 foot = 30.48 centimeters
1 yard = 0.9144 meters

Cubic Measure or Measure of Solids

Common measurements of three-dimensional spaces of volume, in both the English and the metric systems, are (Fig.1- 3):

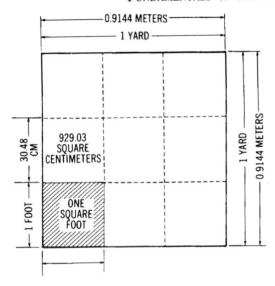

Fig. 1-2. Relationship between various metric and English square units.

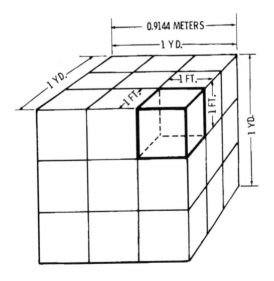

Fig. 1-3. Relative amount of space occupied by a cubic foot as compared to a cubic yard.

$$1 \text{ cubic foot} = 1728 \text{ cubic inches}$$
$$1 \text{ cubic yard} = 27 \text{ cubic feet}$$
$$1 \text{ cubic foot} = 28{,}316.84659 \text{ cubic}$$
$$\text{centimeters}$$
$$1 \text{ cubic foot} = 0.02832 \text{ cubic meters}$$
$$1 \text{ cubic yard} = 0.7646 \text{ cubic meters}$$

Mass

The quantity of matter that a body contains is called its *mass*. The space a body occupies is called its *volume*. The relative quantity of matter contained in a given volume is called its *density*. The relation of volume to mass is expressed by the term density and is written:

$$\text{Density} = \frac{\text{mass}}{\text{volume}}$$

$$\text{Mass} = \text{density} \times \text{volume}$$

$$\text{Volume} = \frac{\text{mass}}{\text{density}}$$

The relative density of a substance is the ratio of its density to the density of pure water at a temperature of approximately 39.2°F (Fahrenheit) or 4°C (Celsius, formerly centigrade).

Mass is a property of all matter. Gas, air, water, and metals have mass. The weight of a substance is due to the earth's attraction on a substance (gravity). The only condition in which a substance has no weight is when it is falling in a perfect vacuum under the influence of gravity. At all other times it has weight; therefore, weight divided by acceleration due to gravity is mass:

$$g = \text{acceleration due to gravity} = 32.2 \text{ ft/s}^2$$

The unit of mass is the slug. The definition of a slug is: 32.74 pounds or 14.505 kilograms. Our units of weight are the ounce, the pound, and the ton:

$$16 \text{ ounces} = 1 \text{ pound}$$
$$2000 \text{ pounds} = 1 \text{ ton}$$

Example: What is the mass of a 10-lb substance?

$$\text{Mass} = \frac{W}{g}$$

where W = weight of substance
g = acceleration due to gravity (32.2 ft/s^2)

$$\text{Mass} = \frac{10}{32.2} = 3.22 \text{ slugs}$$

The metric units of weight and their conversion to the English system are as follows:

$$
\begin{aligned}
1 \text{ grain} &= 0.6480 \text{ grams} \\
1 \text{ ounce} &= 28.3495 \text{ grams} \\
1 \text{ pound} &= 0.45359 \text{ kilograms} \\
1 \text{ short ton} &= 0.9718 \text{ metric tons} \\
1 \text{ long ton} &= 1.01605 \text{ metric tons} \\
1 \text{ gram} &= 15.4324 \text{ grains} \\
1 \text{ gram} &= 0.03527 \text{ ounces} \\
1 \text{ kilogram} &= 2.20462 \text{ pounds} \\
1 \text{ metric ton} &= 1.10231 \text{ short tons} \\
1 \text{ metric ton} &= 0.98421 \text{ long tons}
\end{aligned}
$$

Before defining specific gravity, it should be explained that there must be a certain temperature used in arriving at specific gravity because liquids, solids, and gases will expand when heated and contract when cooled. In the preceding discussion on density of water, 39.2°F (4°C) was used.

Specific Gravity

Specific gravity is the ratio between the weight of a given volume of any substance and the weight of the same volume of some other substance taken as a standard. For solids and liquids, this standard is distilled water at a temperature of 39.2°F. Figure 1-4 gives the volume and weight of 1 gallon of water at 62°F. In Fig. 1-5, the weight of 1 cu in. of water is compared with 1 cu ft.

For gases and vapors, some substance must be taken that is itself a gas, and so the standard for air is *hydrogen*. For solids and liquids, therefore:

$$\text{Specific gravity} \quad = \quad \frac{\text{weight of body in air}}{\text{weight of same volume of water}}$$

$$= \quad \frac{\text{weight of body in air}}{\text{loss of weight in water}}$$

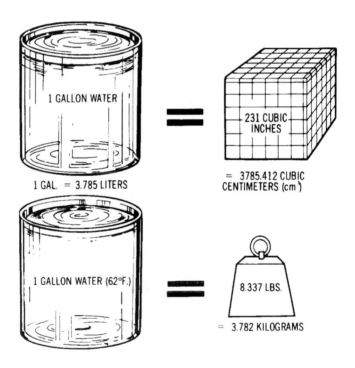

Fig. 1-4. Difference between volume and weight as shown by a 1–gallon unit of water at 62°F.

Figure 1-6 shows how to determine the specific gravity of a piece of cast iron. It will be noted that its weight in air equals 15

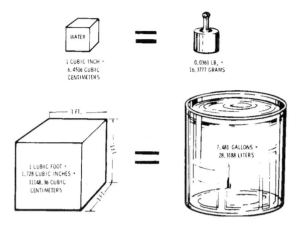

Fig. 1-5. Comparison of the weight of 1 cu in. of water and the volume of 1 cu ft and their metric equivalents.

lb, and the weight registered by the same scale when the pieces are totally submerged in water is 12.9 lb. Substituting in the formula,

$$\text{Specific gravity } = \frac{W}{W - w} = \frac{15}{15 - 12.9} = 7.14$$

ATMOSPHERIC, ABSOLUTE, AND GAGE PRESSURES

It is of the utmost important that refrigeration students understand the meaning of the various kinds of pressure as related to refrigeration.

Atmospheric Pressure

Atmospheric pressure is that pressure which is exerted by the atmosphere in all directions, as indicated by a barometer. Standard atmospheric pressure is considered to be 14.695 pounds per

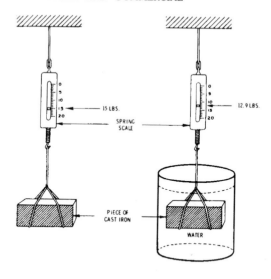

Fig. 1-6. Comparison of the specific gravity of a piece of cast iron in air and in water.

square inch (usually written 14.7 psi), which is equivalent to 29.92 inches of mercury (in. Hg).

When using 14.7 psi or 29.92 in. Hg as atmospheric pressure, bear in mind that this is the perfect atmospheric pressure at sea level and will vary, plus or minus, with altitude and local atmospheric pressures. The temperature used for determining standard atmospheric pressure is 62°F.

As the altitude increases above sea level, the barometric pressure decreases. This is mentioned since in working with refrigeration, you will find later that vacuum measurements must be made. Altitude also affects the boiling point of the refrigerants. To show the results, pressures are given in Table 1-1 for various altitudes.

Barometers should always be corrected to the sea-level reading, but from the preceding figures the actual barometric pressure can be found for the altitude at which you are working. This is a standard reading and does not take into account high and low atmospheric pressures that occur with weather changes.

Absolute Pressure

Absolute pressure is the sum of gage pressure and atmospheric pressure at any particular time. For example, if the pressure gage at one particular time reads 53.7 lb, the absolute pressure would be 53.7 + 14.7, or 68.4 psi.

The aforementioned definitions may be written as follows:

Absolute pressure = gage pressure + 14.7

where 14.7 is the normal atmospheric pressure. Then

Gage pressure = absolute pressure − 14.7

BOYLE'S LAW FOR PERFECT GASES

Boyle's law refers to the relationship between the pressure and volume of a gas and may be stated as follows: *With temperatures constant, the volume of a given weight of gas varies inversely as its absolute pressure.* This is illustrated in Fig. 1-7. Boyle's law is written mathematically as

$$P_1 V_1 = P_2 V_2$$

or
$$\frac{V_1}{V_2} = \frac{P_2}{P_1}$$

where P_1 = absolute pressure of a quantity of perfect gas before a pressure change
P_2 = absolute pressure after pressure change
V_1 = volume of gas at pressure P_1
V_2 = volume of gas at pressure P_2

Since $P_1 V_1$ for any given case is a definite constant quantity, it follows that the product of the absolute pressure and volume of a gas is constant:

$PV = C$ (when the temperature is kept constant)

In this connection it should be mentioned that any change in

Table 1-1. Effects of Altitude on Atmospheric Pressure

Altitude in Feet	Psi	Inches of Mercury
Sea Level = 0	14.72	29.92
1,000	14.17	28.80
2,000	13.64	27.73
3,000	13.13	26.69
4,000	12.64	25.69
5,000	12.17	24.74
6,000	11.71	23.80
7,000	11.27	22.91
8,000	10.85	22.05
9,000	10.45	21.24
10,000	10.06	20.45
11,000	9.69	19.70
12,000	9.33	18.96
13,000	8.98	18.25
14,000	8.64	17.56
15,000	8.32	16.91

the pressure and volume of a gas at constant temperature is called an *isothermal change*.

CHARLES'S LAW

Charles's law refers to the relationship between the pressure, volume, and temperature of a gas and may be stated as follows:

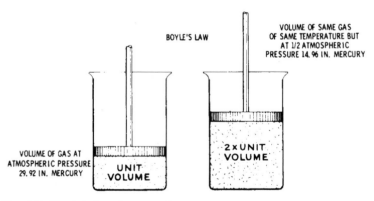

Fig. 1-7. Boyle's law for a perfect gas.

*At a constant pressure, the volume of a gas varies directly as the absolute temperature; at a constant volume, the pressure varies directly as the absolute temperature.*This is illustrated in Fig. 1-8.

When heat is added to a constant volume, the relation is written:

$$P_1T_2 = P_2T_1$$

or

$$\frac{P_1}{P_2} = \frac{T_1}{T_2}$$

For the same temperature range at a constant pressure:

$$V_1T_2 = V_2T_1$$

or

$$\frac{V_1}{V_2} = \frac{T_1}{T_2}$$

Combined, these laws read

$$P_1V_1T_2 = P_2V_2T_1$$

or

$$\frac{P_1V_1}{V_1} = \frac{P_2V_2}{T_2}$$

Since volume is proportional to weight, the relationship between P, V, and T for any given weight of gas W is

$$PV = WRT$$

where P = absolute pressure of gas in points
  V = volume of gas
  W = weight of gas
  R = constant, depending on gas under consideration
  T = absolute temperature, °F

Figure 1-9 shows the four standard temperature scales. A comparison of the absolute Celsius and Celsius scales shows the absolute scale has 0° at 273° below the freezing point of water, which is 0° on the Celsius scale; thus, 0° Celsius would be 273°C absolute, and 10° Celsius would be 283°C absolute.

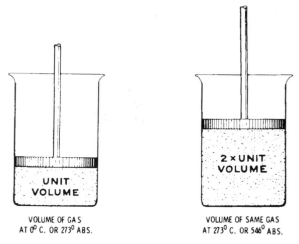

VOLUME OF GAS
AT 0° C. OR 273° ABS.

VOLUME OF SAME GAS
AT 273° C. OR 546° ABS.

PRESSURE UNCHANGED

Fig. 1-8. Charles's law for a perfect gas.

ABSOLUTE CELSIUS   CELSIUS   FAHRENHEIT   ABS. FAHRENHEIT

ABSOLUTE CELSIUS	CELSIUS	FAHRENHEIT	ABS. FAHRENHEIT	
373 ABS. C°	100 C°	F° 212	672 ABS. F°	
363	90	194	654	
353	80	176	636	BOILING POINT
343	70	158	618	OF WATER
333	60	140	600	AT SEA LEVEL
323	50	122	582	
313	40	104	564	
303	30	86	546	AVERAGE ROOM
293	20	68	528	TEMPERATURE
283	10	50	510	
273	0	32		FREEZING TEMPERATURE
263	-10	14	492	OF WATER
253	-20	-4	474	AT SEA LEVEL
243	-30	-22	456	
233	-40	-40	438	
223	-50	-58	420	
33	-240	-400	402	
23	-250	-418	60	
13	-260	-436	42	
3	-270	-454	24	
0	-273	-460	6 0	

Fig. 1-9. Four standard temperature scales.

Figure 1-9 also shows the Fahrenheit and absolute Fahrenheit scales. Freezing of water on the Fahrenheit scale would be 32°; on the absolute scale it would be 492°.

The methods and scales used to measure temperature have been chosen by scientists, and the following standards have been established. The common American scales are the Fahrenheit scale and the Fahrenheit absolute scale (Rankine scale), while the metric system embodies the Celsius and Celsius absolute scales (Kelvin scale).

The Fahrenheit scale is so fixed that it divides the heat level from the melting temperature of ice to the boiling point of water into 180 equal divisions and sets the melting point of ice at 32 divisions above the zero scale. Therefore, ice melts at 32°F, and water boils at 212°F (180° + 32°F), assuming the standard atmospheric pressure of 14.7 psi at sea level. The Fahrenheit absolute (F_a) scale uses the same divisions as the Fahrenheit scale but sets zero at the temperature where molecular action of all substances ceases (absolute zero), where no heat exists in the body and the temperature cannot be lowered any further. This temperature corresponds to –460°F, and water boils at 672°F_a, assuming standard atmospheric pressure.

The Celsius scale has coarser divisions than the Fahrenheit scale, and the zero (0°) of this scale is set at the melting temperature of ice. The boiling point of water is fixed 100 divisions above that point, or 100°C, assuming atmospheric pressure to be 14.7 psi at sea level.

Occasionally you will have to convert Celsius reading to Fahrenheit reading and vice versa. To convert Celsius degrees into Fahrenheit degrees:

$$°F = \tfrac{9}{5} \, (°C + 32)$$

or
$$°F = 1.8 \, (°C + 32)$$

To convert Fahrenheit degrees into Celsius degrees:

$$°C = \tfrac{5}{9} \, (°F - 32)$$

or
$$°C = 0.55555555 \, (°F - 32)$$

To convert Fahrenheit degrees to Fahrenheit absolute:

$$°F_a = °F + 460$$

To convert Celsius degrees into Celsius absolute:

$$°C_a = °C + 273$$

POWER AND WORK

Power may be defined as the capability of performing mechanical work as measured by the rate at which it is or can be done. Stated in simple language, power is *the rate of doing work.* If a horse pulls a weight of 220 lb against gravity at the rate of 2.5 ft/s, as shown in Fig. 1-10, and if the friction in the pulling arrangement is neglected, the horse will develop one horsepower (1 hp). The work done in a given time divided by the units of time gives *the average rate of doing work,* or the *power,* which may be written

$$\text{Power} = \frac{\text{work}}{\text{time}}$$

UNITS OF POWER

The most commonly used units of power are the foot-pounds per unit time and horsepower. The horsepower is the rate of doing work at 550 foot pounds per second (ft-lb/s), or 33,000 foot-pounds per minute (ft-lb/min). It is equivalent to the power used to raise a weight of 33,000 pounds against gravity at a rate of one foot per minute:

$$\text{hp} = \frac{\text{foot - pounds}}{33,000 \times \text{time (minutes)}}$$

Another way of illustrating one horsepower is shown in Fig. 1–11. The motor is said to develop 1 hp when the 550-lb weight is lifted against gravity at a rate of 1 ft/s. If a weight of 3000 lb is lifted through a distance of 35 feet in 2 minutes, the required number of horsepower would be:

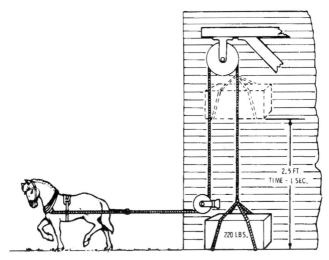

Fig. 1-10. One horsepower, using a horse to lift.

$$hp = \frac{3000 \times 35}{33,000 \times 2} = 1.59$$

ENERGY

A body is said to possess energy when it can do work. Energy exists in various forms, such as:

1. Mechanical energy
2. Electrical energy
3. Heat energy (thermal)

Other well-known forms of energy are *kinetic* and *potential energy.*

Energy may be transmitted from one form to another by various processes, such as mechanical, thermal, chemical, or electrical, but the energy lost or gained is either kinetic or potential. If the process is mechanical, such as impact, compression, or the application of a mechanical force, work is done. If it is thermal,

27

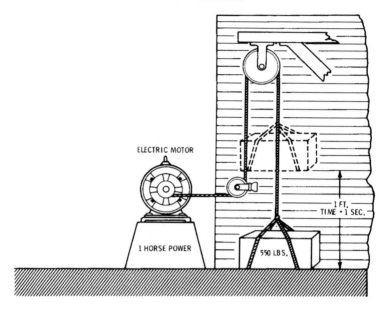

Fig. 1-11. One horsepower, using a motor to lift.

radiation, or conduction, then heat is added or withdrawn. Very simple relations exist between mechanical, electrical, and heat energy, and they may be written as follows:

$$778 \text{ foot-pounds (ft-lb)} = 1 \text{ British thermal unit (Btu)}$$
$$2546 \text{ British thermal units (Btu)} = 1 \text{ horsepower-hour (hp-hr)}$$
$$39,685 \text{ British thermal units (Btu)} = 1 \text{ kilogram-calorie (kg-cal)}$$
$$746 \text{ watts (W)} = 1 \text{ horsepower (hp)}$$

Other equivalents are:

1 Btu = 0.001036 lb water evaporated at 212°F
  = 0.0000688 lb carbon oxidized
  = 0.000393 hp-hr
  = 107.6 kg-m
  = 778 ft.lb
  = 1055 W-s

From these equivalents, it may immediately be apparent that heat is an entity, a real something, for it has a unit of measure and each unit is convertible to other forms of energy at a constant rate of exchange. Since refrigeration is an art that is concerned with the heat problem, those who desire a thorough understanding of the refrigeration process must know the simple facts concerning heat.

THEORY OF HEAT

The technical nature of heat was not understood for a long time, and several different theories were advanced to explain it. The concept that is generally accepted is the so-called molecular theory because it is based on the theory that all matter is composed of innumerable, separate, and minute particles called *molecules.* The molecule is composed of two or more atoms and is so small that an extremely powerful microscope is required to see it.

Molecular Theory

The molecular theory is based on the supposition that the molecules of a substance are not attached to each other by any bond or cement but are held together by a force known *as cohesion,* or *mutual attraction,* a phenomenon somewhat similar to the attraction offered by a magnet for steel particles. The molecules, however, are not physically bound or in contact with each other, like the iron filings, but are actually separated to such an extent in some instances (and usually so at more elevated temperatures) that the space separating two adjoining molecules is larger than either particle. Furthermore, the molecules are not fixed or stationary but revolve and vibrate within the orbits or limits of their allotted space.

Each substance on earth is composed of different ingredients or various combinations of molecules, and therefore each particular kind and mixture, together with additional peculiarities in physical assembly, has a structure differing from other materials. It is the rapidity of the motion of the molecules composing a body or mass that determines the intensity of its heat. As a specific example, to point out the connection between the theory of heat and the molecular theory, a bar of lead may be used for illustration. When cold, the molecules undergo comparatively slow motion. But if the bar is heated, the molecular activity becomes more rapid until its temperature has been increased to a point where molecular motion becomes so rapid and the individual particles are so far separated from each other that, with further assimilation of heat, they become so weakly cohesive they can no longer hold the body in a rigid or solid mass and it devolves into a liquid.

Further application of heat forces the molecules to greater separation and speeds up their motion to such an extent that the liquid becomes more mobile and volatilization finally results, so that a gas or vapor is produced. The vapor thus formed no longer has a definite volume, such as it had in either the solid or molten (liquid) form, but will expand and completely fill any space that is provided for it. The vapor, of course, contains the same number of molecules that were in the original solid, but molecular action is extremely active in the vapor phase, comparatively slow in the liquid, and slowest in the solid form. The motion of the molecule is somewhat like a bicycle rider: at a slow speed, a very small orbit can be adhered to, but with each advance in the rate of travel, larger orbits are required.

To one learning the theory of heat and matter for the first time, it may appear extremely complex. Whether the molecular theory herein described (which is the one accepted by leading scientists) or any other theory is used as a basis of explanation, the facts regarding heat as we know it are in agreement with the molecular theory and from all tangible proof substantiate the theory in all respects. One of the simplest indications of value in support of the molecular theory is the age-old knowledge concerning expansion and contraction, caused by heat and cold respectively, or, as we look at it from the molecular theory, the rate of molecular activity.

Temperature

Heat and temperature are closely allied, but it must be remembered that heat is energy itself representing the kinetic activity of the molecules composing a substance, while temperature is but a measure of the condition of a mass or body as it affects our ideas of warmth or cold. Heat is convertible into electrical, chemical, or mechanical energy.

Our ideas of cold and hot are merely relative ranges of

temperature that affect our sense of feeling, which we term hot and cold. In a way, heat itself may be likened to water, for, of its own accord, it will flow only downhill; that is, it will pass from the hotter (or higher) range of temperature to the colder (or lower) plane.

Two miscible liquids poured together will find their own temperature level, as will a solid and a liquid, or even two solids in intimate contact. For instance, if an ice cube is dropped into a tumbler of tepid water, it will be found that the colder substance will be able to assimilate a certain quantity of heat from the warmer substance. Thus, through this extraction, we find that the water has been cooled or reduced in temperature. Two substances of different temperatures in intimate contact tend to reach an equilibrium or balance by the dispersion of heat by one body and assimilation of heat by the colder body. If a thermometer is placed in contact with the substance, the degree or level of its temperature, or the measure of the intensity of its heat, will be indicated.

THERMOMETERS

The instrument in common use for measuring temperature, known as the *thermometer,* operates on the principle of the expansion and contraction of liquids (and solids) under varying intensities of heat. The ordinary mercury thermometer operates with a fair degree of accuracy over a wide range. It becomes useless, however, where temperatures below –38°F are to be indicated because mercury freezes at that point. Some other liquid, such as alcohol (usually colored for easy observation), must be substituted. The upper range for mercurial thermometers is quite high, about 900°F, and so it is apparent that for ordinary service and general use the mercury thermometer is usually applicable.

The operation of the thermometer depends on the effect of heat on the main body of mercury or alcohol contained in a bulb or reservoir. The liquid will expand or contract (rise or drop) in the capillary tube, which is inscribed with the various increments of an arbitrary scale, as shown in Fig. 1-12. Several temperature scales are in existence and are used in various countries. The

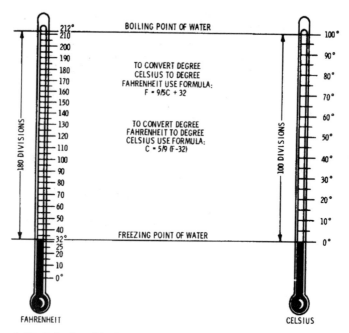

Fig. 1-12. Relationship between Fahrenheit and Celsius scales used on thermometers.

English, or Fahrenheit, scale is commonly used in the United States. Since the Celsius scale is so widely used in scientific work in all countries, an illustration of the comparison of thermometers is presented so that any one scale may be converted to another.

Freezing on the Fahrenheit scale is fixed at 32°; on the Celsius scale, freezing is placed at 0°. On the Fahrenheit scale, the boiling point of pure water under the normal atmospheric pressure encountered at sea level is 212°; on the Celsius scale, under the same conditions, the boiling point is 100°.

In Fig. 1-9, all four temperatures are shown; only the Fahrenheit and Celsius scales are shown in Fig. 1-12 because these two scales are used most often.

As previously stated, temperature is a measurement of the intensity of the heat contained in a body, accomplished by means of a thermometer, just as a yardstick is used to measure the length of a body. However, a thermometer will not indicate the quantity or amount of heat contained in a body.

ABSOLUTE ZERO

In the study of thermometer scales, the question of thermal limits is naturally considered. From the molecular theory we are led to believe that with the removal of heat, molecular action is slowed down accordingly. It must naturally follow that, at some point, all heat will be removed and molecular activity will cease entirely.

The calculation in Fig. 1-13 indicates that absolute zero is attained at a temperature of 460° below zero on the Fahrenheit scale. Temperatures as low as this have never been reached, although in some instances it has been approached within a few

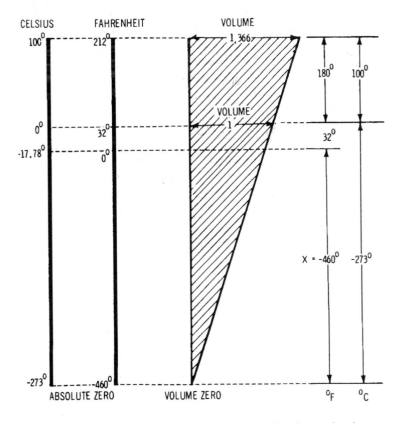

Fig. 1-13. How absolute-zero temperature may be determined.

tenths of one degree. Bodies subjected to extremely low temperatures take on characteristics entirely different from those exhibited under normal conditions.

This calculation in Fig. 1-13 is based upon the fact that atmospheric air expands 1.366 of its volume on being heated from 32° to 212°F. Therefore, it can be assumed inversely that if all possible heat were withdrawn, the volume of air would shrink to zero. With reference to our diagram, the volume of air at 32°F = 1. It follows that the volume of air at 212°F = 1.366. Constructing our triangles as shown, the following relations are obtained:

$$\frac{32 - X}{212 - X} = \frac{1}{1.366}$$

$$X = \frac{1.366 \times 32 \ - \ 212}{0.366}$$

$$= 460°F$$

(−459.62°F Rounded to −460° here)

UNITS OF HEAT

The quantity of heat contained in a body is not measurable by a thermometer, which indicates only the temperature or intensity. For example, a gallon of water and a pint of water may have the same temperature, but we are certainly aware that the larger body must contain more heat as energy than the smaller body.

The unit of heat measure employed in this country is called the *British thermal unit* (more commonly referred to as the Btu), which is that quantity of heat required to be added to one pound of pure water initially at a temperature of its greatest density (that is, 39°F) to raise its temperature one degree on the Fahrenheit thermometer (in this case, to 40°F). Roughly, a Btu may be said to be the quantity of heat required to raise the temperature of one pound of water one degree on the Fahrenheit scale, as shown in Fig. 1-14.

Just as the thermometer is used as a measure of intensity of heat of a body, the Btu is used to represent the quantitative energy. For example, one body at 50°F may contain twice as many Btu as another body at the same temperature because the

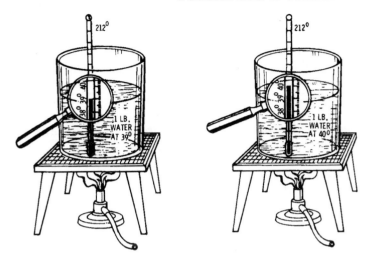

Fig. 1-14. British thermal unit (Btu).

bodies may be different in size or weight and, of great importance, may have different capacities for absorbing heat.

SPECIFIC HEAT

It could be said that each and every substance on the earth has a different capacity for absorbing heat. Some identical materials, especially natural formations, even give different values for samples secured in different localities. As an illustration of the difference in the heat capacities of materials, let us consider a pound of iron and a pound of water, both at 80°F. If the heat is removed from these bodies and the number of Btu extracted in cooling each mass to the same temperature is recorded, it will be found that each will give up a different amount. If each body had been at a temperature of 80°F and was cooled to 60° (a reduction of 20° on the Fahrenheit scale), it would be found that 1 lb of water at 80° cooled to 60° would evolve 20 heat units or 20 Btu.

If the iron (of the same weight) is cooled over the same range, only 2.6 Btu will be extracted (approximately ⅛ of that taken from the same weight of water under identical conditions). We there-

35

fore come to the conclusion that all materials absorb heat in different capacities; and by comparing the heat-absorbing qualities with a standard, we have a standard of measure, a gage to compare all substances. This measure is the amount of heat, expressed in Btu, required by one pound of a substance to change its temperature on degree Fahrenheit. Since water has a very large heat capacity, it has been taken as a standard; and since one pound of water requires one Btu to raise its temperature one degree, its rating on the specific-heat scale is 1.00. Iron has a lower specific heat, its average rating being 0.130; that of ice, 0.504; of air, 0.238; and of wood, 0.330. The more water an object contains, as in the case of fresh food or air, the higher the specific heat. Materials usually stored in a refrigerator have a high specific heat, averaging about 0.80.

Table 1-2 lists the average specific heats of common substances, comparing their relative heat capacities with water. Observe that metals have limited heat-storing powers as compared with water. This is one of the reasons why scalds from hot water burn so deeply, for the water contains so much heat energy that a considerable amount is released and causes a worse burn than molten metal at a much higher temperature.

In Table 1-2 various substances are listed with their average specific values. By finding this factor, the amount of heat in Btu to be added or taken from a substance of known weight to bring about a change of one degree in its temperature may be calculated. Specific heat problems may be calculated easily by use of the following formula:

$$Btu = Sp.\ heat \times W(t_2 - t_1)$$

where W = weight of the substance, in lb
Sp. heat = specific heat of the substance to be heated
$t_2 - t_1$ = temperature change, °F

Where very accurate scientific work is done, certain allowances are made for the fact that the specific heat of a substance does not remain constant throughout the entire temperature range. Since the difference is not appreciable (except over a wide range), the refrigeration engineer usually regards the specific heat factor as a constant. Other materials, such as liquids and

Table 1-2. Average Specific Heats

Water	1.000	Pine	0.650
Copper	0.900	Strong brine	0.700
Vinegar	0.920	Oak	0.570
Alcohol	0.659	Ice	0.504
Air	0.238	Glass	0.194
Mercury	0.333	Iron	0.130
Coal	0.241	Sulfur	0.202
Brass	0.094	Zinc	0.095

gases, also have specific heats. However, the calculations concerning the heat capacities of gases are further complicated by the pressures as well as the varying temperatures imposed. Despite the tendency to ignore the variance in specific heat and employ a constant, the values of water are given in Table 1-3.

Table 1-3. Specific Heats of Water
(Value at 55°F, Taken as Unity)

Temp., °F	Specific Heat	Temp., °F	Specific Heat
20	1.0168	140	0.9986
30	1.0098	160	1.0002
40	1.0045	180	1.0019
50	1.0012	200	1.0039
60	0.9990	220	1.0070
70	0.9977	240	1.0120
80	0.9970	260	1.0180
90	0.9967	280	1.0230
100	0.9967	300	1.0290

The specific heat of various foods and their containers are of interest to the refrigeration engineer in estimating the amount of heat to be extracted in cooling a refrigerator load. The heat that can be felt and detected is termed the *sensible heat*; its name denotes the heat we can sense or feel, preventing confusion with other heats. It is the sensible heat which forms the preponderance of the heat load the refrigerating machine is called on to remove in cooling most edibles. Of course, ice manufacturing plants and other storage warehouses that freeze edibles are additionally concerned with other heat problems.

Most foods have a high water content and therefore a large heat capacity, averaging about 0.80. A few foods are listed in Table 1-4 so that the specific heats of various edibles can be ascertained and the truth of the statement regarding the water content verified. The specific heats listed are those of fresh foods before freezing.

An estimate of the heat quantities that must be removed from the food and its containers can be easily accomplished. For instance, if we have 1000 pounds of cider contained in glass bottles having a weight of 75 lb, which in turn have been packed in pine boxes totaling 50 lb, the whole shipment being at a temperature of 80°F, the heat in Btu that would have to be extracted to cool these materials to 50°F can be calculated as follows:

$$80° - 50° = 30° \text{ difference in each case}$$

1. 1000 lb of cider having a specific heat of 0.90, cooled over a 30° range = $1000 \times 0.90 \times 30 = 27,000$ Btu.
2. 75 lb of glass with a specific heat of 0.194, cooled 30° = $75 \times 0.194 \times 30 = 436.5$ Btu.
3. 50 lb of pine wood having a specific heat of 0.650 cooled 30° = $50 \times 0.650 \times 30 = 975$ Btu, or a total of $27,000 + 436.5 + 975 = 28,411.5$ Btu.

By observing a thermometer immersed in the cider and taking an average reading, we could calculate at any time during the cooling process just how much of the sensible heat had been removed. In fact, we could calculate any heat load of any material by finding the specific heat value and the cooling range; and if we desired to freeze the product (in this specific case, the cider), an additional factor would have to be considered, namely, the *latent heat*.

Table 1-4. Specific Heats of Foods

Food	Specific Heat	Food	Specific Heat
Apples	0.92	Eggs	0.76
Beans	0.91	Fish	0.80
Beef	0.75	Grapes	0.92
Butter	0.60	Milk	0.90
Cabbage	0.93	Peaches	0.92
Cheese	0.64	Pork	0.50
Chicken	0.80	Potatoes	0.80
Celery	0.91	Veal	0.70
Cider	0.90	Watermelon	0.92

LATENT HEAT

One of the most mystifying laws to the layman is that of *latent heat*. The word *latent* expresses it aptly enough, for it means hidden or not apparent. In order to indicate clearly just what latent heat is, an example will be cited, and an illustration shown in Fig. 1-15.

We are aware that, under proper conditions, most substances are capable of assuming two or more physical states. For instance, lead, when cold, is a solid; when heated and molten, it is a liquid. Water is an outstanding example, for it can assume three states— solid, liquid, and vapor—within a relatively short temperature range. Ice, of course, represents the solid state; water, the liquid; and steam, the vaporous, or gaseous, state.

CHANGE OF STATE

Temperature and heat play important parts in effecting changes from one state to another. For instance, the only quantity

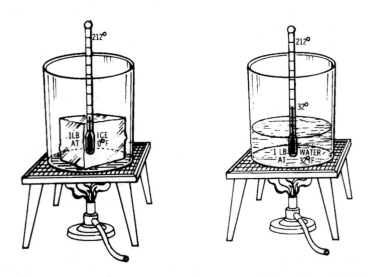

Fig. 1-15. Apparatus needed for conversion of 1 lb of ice at 0°F to water at 32°F.

39

that would convert a tumbler of water to either of the other states (that is, a solid or a vapor) would be the addition or extraction of heat.

To prove the function that heat plays in effecting a change of state, let us take a block of ice and perform an experiment with it. In order to simplify the calculations, we will utilize a piece of ice that weighs exactly 1 lb with a thermometer frozen in the center of it, and we will assume that the block of ice is at a temperature of 0°F. We will not discuss the apparatus or computations necessary to measure the heat values applied but will assume the use of an imaginary Btu meter just as though such a device existed. Thus, every time a Btu of heat energy is expended, it will be registered on our meter.

Prior to beginning the experiment, a graph or a squared section of paper is obtained so that the results can be plotted on it. In the lower right-hand section of the paper (a part that will not be required for our log), we will put down the various findings. Setting up the apparatus is a simple matter; we require only a vessel and a source of heat, the latter being measurable by our "meter." Place the piece of ice, weighing exactly 1 lb, in the vessel and indicate the start of the experiment by marking an *A* at the lower left-hand corner of the chart, as shown in Fig. 1-16, to denote that the ice is at 0°F, and that no heat has been added as yet. The specific heat of ice is not as great as that of water; in fact, it is just about half the value, 0.504 to be exact. For this experiment let us assume it is 50 percent, or 0.5.

Let us now begin to add heat. In accordance with its specific heat capacity, we will find that every time 0.5 of a Btu is added, the temperature of the ice will increase 1°F. Continuing to add heat and marking off the progress on the chart, we find that up to 32°F, the number of Btu required to bring the ice to this degree of sensible heat would be

1 (lb of ice) × 0.5 (specific heat) × 32 (degrees rise) = 16 Btu

Thus, from *A* to *B* (a range of 32°) 16 Btu were required.

At 32°F, an interesting stage is reached. We find that the further addition of heat does not warm up the ice and the thermometer frozen in the test block of ice continues to indicate a sensible heat of 32°. Even providing the ice with a greater quan-

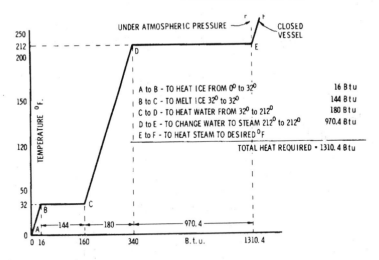

Fig. 1-16. Heat units required to change 1 lb of ice at sea level and 0°F to steam at 212°F.

tity of heat causes no observable change in temperature. The only noticeable feature is the melting of ice or conversion of a solid (ice) to a liquid (water). This important transformation is called the *change of state* due to the fact that a solid is converted to a liquid. If we continued to add heat to the vessel containing the ice and water, we would find that when a total of 144 Btu has been added, the entire block of ice would be converted to water and the water would have a temperature of 32°F. In other words, to change 1 lb of ice at a temperature of 32°F to water at the same temperature, 144 Btu are required to effect this change in state. We will plot this on our chart as line *B* and *C*.

LATENT HEAT OF FUSION

We have had to apply 144 Btu, which was taken up by the ice at 32° and caused it to melt and assume its liquid state without a single indication of any rise or increase in its sensible heat. This masked assimilation of heat is termed *latent heat* and may be said to be the amount of heat units that must be supplied to a solid in order to change it to a liquid without an increase in temperature. It is readily understood that a reversal of the process, that is, for

41

the liquid to assume its solid state (ice), heat must be extracted from water.

It requires the extraction of 144 Btu to cause 1 lb of water at a temperature of 32°F to freeze into a solid block of ice at 32°F. Every solid substance, in varying degrees, has a latent-heat value, and that amount required to convert it or bring about a change of state is termed the *latent heat of fusion*. This heat, assimilated or extracted, as the case may be, is not measurable with the thermometer because the heat units are absorbed or expended in intermolecular work. Latent heat of fusion separates the molecules from their attractive forces so that a change of state is effected.

LATENT HEAT OF EVAPORATION

Now let us refer again to the experiment we have under way. We have converted the solid to a liquid, both at 32°F, by adding 144 Btu. Having to deal with water, and knowing that it has a specific-heat factor of 1.00, we may look for a rise of 1° in temperature for each Btu added. This will hold until 212°F is indicated on the thermometer, assuming that the experiment is made at atmospheric pressure existing at sea level. Over the 180° range of 32 to 212°, we will be obliged to add 180 Btu in order to raise the temperature of 1 lb of water from the former to the latter temperature. This may be plotted as line C and D on the chart, with the number of heat units recorded on the lower edge.

The further addition of heat would serve to bring out the fact that while the phenomenon of boiling would occur at 212°F, the temperature or sensible heat would not be increased. Just as ice requires a certain quantity of heat units (144 Btu/per lb) to melt or convert it from a solid to a liquid without a rise in temperature, we find a similar condition existing. This time a liquid is being converted to a vapor, or steam in this case.

Careful measurements have determined that the conversion of 1 lb of pure water at 212°F to steam at 212°F requires exactly 970.4 Btu when carried out at the normal pressure of the atmosphere encountered at sea level. If we carefully add heat and keep count of the Btu expended, we will find that when all the water has been changed to steam, 970.4 heat units will have been used. Thus, line *D* and *E* may be plotted on the chart and a note of the number of Btu expended jotted down on the lower right-hand corner. The further addition of heat would serve only to heat the steam, such as would be possible if it had been trapped or the experiment had been performed in a closed vessel so that heat could be applied to it.

STEAM

Steam is the hot, invisible vapor state of water at its boiling point. The visible white vapor is really a collection of fine watery particles formed from true steam by condensation. Steam acts like all trite vapors or gases in that it:

1. Has fluidity
2. Has mobility
3. Has elasticity
4. Exerts equal pressure in all directions

The difference in volume between water and steam at atmospheric pressure is 1646 to 1, and this wide ratio is manifested by nearly all gases and vapors. The heating of the steam generated in our experiment could be represented by the line *E* and *F* and would be extended over a considerable range. By referring to the log of our experiment, we can trace the heat quantities required to convert ice to water and steam and the action of heat on both liquid and solid. A tabulated form in the lower right-hand portion of the log represents the quantities of heat energy required to effect a change of state, rise in temperature, and other values.

From this experiment it is apparent that heat added to a substance either increases its temperature or changes its state. The heat that brings about the change of state from a solid to a liquid is known as the *latent heat of fusion*, while the heat required to convert a liquid to a gas is termed the *latent heat of evaporation or vaporization*. One of the interesting facts brought out previously is that a liquid, once brought to the boiling point, does not increase in temperature but utilizes the heat energy it takes up in converting more liquid to a gas or vapor.

If we were creatures dwelling on another planet and accustomed to a normal atmospheric temperature of 250° or 300°F, we would be able to utilize water to cool substances by immersing them in a bath of water. Since heat and solid are only relative to our senses, materials immersed in a pan of water will lose heat to it, and the heat extracted will be taken up by the water and used to generate steam. Through the extraction or flow of heat from objects so immersed, the water (assumed to be under the same normal sea-level pressures) would remain at 212°F, which would be cooler than the atmospheric temperature of our hypothetical planet. Fortunately for earth dwellers, there are other liquids that boil at temperatures much below that of water.

REFRIGERANTS

Among common liquids that boil at a temperature below that of water is alcohol, with a boiling point of 173°F, while ether boils at ordinary summer heat, which is 94°F. Other substances boil at still lower temperatures. Carbon dioxide boils at −110°F, ammonia at −20°F, sulfur dioxide at 14°F, methyl chloride at −11.6°F, ethyl chloride at 54°F, Freon-12 at 21.6°, and Freon-22 at 41.4°F—all at atmospheric pressure encountered at sea level.

Those materials, solid or liquid, which vaporize or liquefy at comparatively low temperatures and are suitable for use in refrigeration work are termed *refrigerants*. The refrigerants are employed in specially designed apparatus so that the extraction of heat from rooms or perishables can be accomplished as inexpensively as possible.

The latent heat of vaporization of liquid refrigerants varies with the material and pressure at which vaporization is allowed to proceed. The latent heat of vaporization in Btu per pound of various refrigerants in common use and volatilizing at 0°F is presented as follows. There will be notes accompanying these refrigerants because some of them are now practically obsolete, but you may run across some older equipment which uses them:

Ammonia—572.2

Ammonia is considered a hazardous material, and you will

encounter it only in very large commercial installations, where the hazards have been allowed for.

Methyl chloride—176.0

Methyl chloride mixed in the right proportions with air will burn or explode. It has not been used much in home refrigeration and is not being used today to any degree.

Ethyl chloride—173.4

Ethyl chloride has been practically eliminated from the refrigeration field.

Sulfur dioxide (SO_2)—171.8

Sulfur dioxide was once used to a large degree, but it is very toxic and the only places that you will encounter it are in very old domestic refrigerators.

Carbon dioxide (CO_2)—117.5

Carbon dioxide is not used in domestic refrigerators.

Freon-12—71.0

Freon-12 is very popular at the present time.

Freon-22—100.7

Freon-22 is used in many domestic freezers.

Freon-502—74.8

Freon-502 is used in large air conditioning systems.

Freon-21—119.4

From this list, it is observed that ammonia is an excellent refrigerant in that it absorbs a great quantity of heat as compared to other agents. However, it is toxic and classified as a hazardous material so that its use is confined mostly to domestic absorption-type refrigeration and large industrial manufacturers of ice. Methyl chloride is combustible under the right mixtures with air, and so it is seldom used any more. Sulfur dioxide (SO_2), while quite popular in the past, is extremely toxic and no longer used. It will be covered to some extent in this book, as you may come across an SO_2 system that needs repair. Carbon dioxide (CO_2) in solid form is used in transporting ice cream, fish, and similar products and where a great quantity of escaping gas fumes would not endanger human life. This substance is also known as dry ice.

It is very easy to understand that a vessel containing a refrigerant, which is allowed to absorb heat, will cause cooling or a refrigerating effect. Since heat (as energy) cannot be created or destroyed, it follows that heat removed from one body must show up in some form of energy in another body. Essentially, the removal of heat from one body and its transference to and dissipation by another body is refrigeration.

USE OF ICE

Where ice is employed, the latent heat of fusion results in the cooling of other materials suffering such loss of heat energy that the heat is taken up by the ice and it reverts to water. The heatladen liquid (water) is drained away, carrying its heat load out and away from the refrigerator. A graphical illustration of this cycle is shown in Fig. 1-17.

TON OF REFRIGERATION

Due to the fact that refrigeration was first produced by ice, the rate of removal of heat in a cooling operation is expressed in terms of pounds or tons of ice required per unit time, usually per day. It has been found that 1 lb of ice absorbs 144 Btu when it melts; 1 ton of ice consequently absorbs 2000×144, or 288,000 Btu. When 1 ton of ice melts in 24 hours, the rate is $288,000/24$ or 12,000 Btu/hr, or $12,000/60 = 200$ Btu/min. This rate has been officially designated as 1 ton of refrigeration and is the basis for rating all refrigerating machinery.

EVAPORATION

We have previously discussed the quantities of heat required to bring about a change of state. Considerable heat energy was

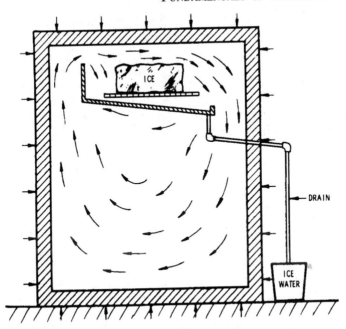

Fig. 1-17. Example of the ice-refrigeration cycle.

necessary when a change took place from the liquid to the vapor phase. The term *evaporation* is well known to all. A common example of refrigeration by means of evaporation is found in the cooling of the human body by perspiration. The beads of moisture appear on the skin and, by evaporation into the air, actually maintain the body at a lower temperature than the surrounding atmosphere.

The normal internal temperature of human beings is 98.6°F. Everyone at some time or another has encountered atmospheric temperatures that were much higher than body heat, never realizing that it was evaporation that acted as the refrigerating element to keep the body cool. The human body is extremely sensitive and is always within a very limited range of temperature, allowing a variation of only a few degrees. For example, if the inside body temperature were raised above 108°F, death is almost certain; yet men have lived in atmospheres where temperatures were as high as 140°F and even higher. Their bodies were maintained at normal body heat by evaporation of moisture from their skins.

Another example of cooling by evaporation found in nature is the reduction in temperature which usually follows a summer rain. The required heat to bring about the evaporation of the rain water is drawn from the atmosphere and the earth and thus effects the coolness we note after such precipitation.

PRESSURE

In the previous instances of boiling temperatures, the wording "at pressures found at sea level" was used. For instance, water boils at 212°F when it is heated in a vessel open to the atmosphere and provided the experiment takes place at an altitude at or near sea level or under equivalent pressure. Regardless of how rapidly heat may be added to the open vessel of water, no rise in boiling temperature can be secured, the only result achieved being that steam is formed with greater rapidity commensurate with the amount of heat. If, however, pressure is allowed to enter as a factor in the conversion of a liquid to a vapor, or vice versa, it will be found that the boiling point is altered. We are all more or less familiar with the story of the man who tried to prepare his eggs on the top of a high mountain where atmospheric pressure, due to altitude, was so low that the water boiled at such a low temperature that it would not cook the eggs. From this simple example, it is apparent that the lower the pressure exerted on the liquid, the lower will be the boiling point.

By way of demonstrating the effect of pressure, let us take a sturdy flask, one capable of withstanding both heat and pressure, and pour in some water. We will assume it to be a glass vessel so that we can view the process and make certain when boiling actually occurs. A stopper provided with three perforations, so that a pressure gage, thermometer, and a valve can be inserted in the individual openings, is forced tightly into the neck of the flask. The completely assembled apparatus is then placed over a source of heat, such as a gas flame, as depicted in Fig. 1-18. The exit valve is allowed to remain wide open. Soon the water will

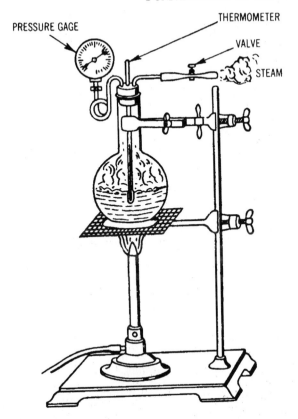

Fig. 1-18. Apparatus needed for evolution of pressure-temperature relations of water.

boil, and if the experiment is performed at or near sea level, the pressure gage will register 0 lb of pressure while the thermometer will indicate a temperature of 212°F. As long as the valve orifice is large enough to offer a free and unobstructed exit to all of the steam formed, no amount of heat will raise the temperature.

Now let us turn the valve so that the opening is slightly restricted. The pressure will build up, and by regulating the amount of gas burned and the exit valve so that a jet of steam is released, we can obtain or impose any pressure we desire on the vessel. Let us select a pressure of about 5 lb as the pressure desired. With proper regulation of the gas and steam valves, this value can easily be obtained and maintained. When secured, let us read the thermometer. By readjusting the steam and the gas valves, let us raise the pressure to 10.3 lb gage pressure (10.3 psig). At this point we would find the water would start to boil at 240.1°F. Again we would find that no amount of heat added rapidly or otherwise to our flask would raise its temperature above that point provided the pressure remained constant.

With heat added rapidly, more steam would be formed and we would of course be obliged to open the steam exit valve in order to maintain the pressure at the desired point. If only a little heat were added, we would have to close the valve somewhat to maintain the pressure. By referring to a table giving the properties of saturated steam, various gage pressures and corresponding temperatures can be determined. For instance, if our boiler were of heavy steel, we could increase the pressure to a point where the gage would indicate 100.3 lb, at which point the boiling of the water would occur at 338.1°F. With the increase in pressure, the boiling point advances, and, again, in accordance with this pressure-temperature relationship, the boiling point is lowered as the pressure is reduced.

When the pressure decreases, the boiling temperature of a liquid will decrease also. For example, water at an atmospheric pressure equal to 4800 feet above sea level would boil at approximately 202°F. By the same token, water at 10,000 feet altitude would boil at approximately 193.2°F.

PRESSURE-TEMPERATURE RELATIONS

Most bodies expand when heated, whether of solid, liquid, or gaseous form. If the expansion is restricted, great forces are set up by the body in the effort to expand. Every gas or vapor confined in a closed vessel exerts a certain pressure against the restraining walls of the vessel. The pressure imposed depends on the amount and temperature of the gas. The natural tendency for a confined gas is to expand when the temperature is increased; if the gas is contained in a vessel with rigid walls, the pressure will increase. This is of such importance that a law has been formulated to express it, stating that the pressure exerted by a gas or vapor in a closed vessel is *directly proportional to the absolute temperature of the gas.*

Low Pressures

Before proceeding with our experiments with lower pressures, it is best that we understand the comparative terms. As pointed out previously, objects at sea level are taken as being subjected to a certain constant pressure. This pressure is due to the weight of the atmosphere, which at sea level is found to exert a pressure of 14.7 pounds per square inch (psi) of surface.

The atmosphere is like a high sea— if we dive into water similar conditions are encountered because the deeper we go, the greater the pressure becomes due to the volume above. As we rise, the pressure becomes less. Modern aircraft designers take into consideration the rarefication of the atmosphere with increasing altitude and know that at a certain height the air density will not allow an airplane to rise any higher. This they term the *ceiling*, or *density* at which further altitude for each particular design is not possible. Since pressure is so instrumental in a great many engineering calculations, a universal standard was necessary. Thus, the pressure existing at sea level was accepted as a standard, being termed 0 lb gage, or 14.7 lb absolute.

In the manufacture of gages, these instruments are usually set up to indicate zero pressure at normal atmospheric sea-level density. It must be remembered that this setting at zero is made when the actual and true pressure is in reality 14.7 psi, the pressure that the weight of the atmosphere exerts at sea level. Nearly all steam and pressure gages are set in this fashion, and the pressures indicated by these instruments are termed *gage pressures*. The usual increments are in 1-, 5-, and 10-lb readings so that only approximate pressures can be determined. For most commercial applications, these roughly calibrated gages are sufficiently accurate and quite inexpensive.

For fine work and experimental and scientific determinations, another pressure scale is employed. This scale is called the *absolute pressure* and is based on true pressures, for zero on this scale indicates no pressure at all, or, in other words, a perfect vacuum. An external view and construction details of a typical pressure gage are shown in Fig. 1-19. It consists of a metal tube of elliptical

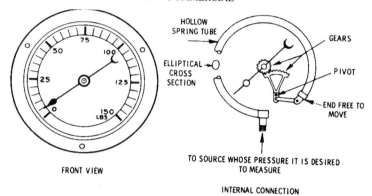

HOLLOW
SPRING TUBE

GEARS

ELLIPTICAL
CROSS
SECTION

PIVOT

END FREE TO
MOVE

FRONT VIEW

TO SOURCE WHOSE PRESSURE IT IS DESIRED
TO MEASURE

INTERNAL CONNECTION

Fig. 1-19. External view and construction details of a typical pressure gage.

cross section bent into a nearly complete ring and closed at one end. The flatter sides of the tube form the inner and outer sides of the ring. The open end of the tube is connected to the pipe through which the liquid under pressure is admitted. The closed end of the tube is free to move. As the pressure increases, the tube tends to straighten out, moving a pointer through a lever-and-gear arrangement. The scale is graduated directly in pounds per square inch (psi).

On the pressure-gage scale, the normal atmospheric pressure found at sea level (14.7 psi) is indicated as 0 lb pressure, and below that point the term of pressure is converted to vacuum, or inches of mercury (in. Hg). A long glass tube, sealed at its upper extremity, is exhausted of air and filled with mercury, as shown in Fig. 1-20, its lower open end dipping into a reservoir of mercury open to the atmosphere or to the pressure imposed. Such a device is called a *barometer* and is used to measure the pressure in terms of inches of mercury in place of a term such as pounds under zero gage pressure. Just as variance in temperature will cause a rise or fall of the mercury level in a thermometer, so will a difference of pressure exerted on the mercury in the reservoir of the barometer bring about a certain reading in terms of inches of mercury. The pressure of the atmosphere or any other pressure exerting its force on the mercury exposed in the reservoir drives the liquid up into the evacuated end to a height corresponding with the pressure. In a way, the barometer is much like the thermometer, for it indicates a value by the height of a mercury column, one recording in terms of temperature and the other in terms of an equivalent of pressure.

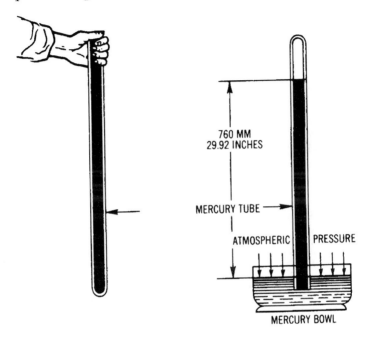

760 MM
29.92 INCHES

MERCURY TUBE

ATMOSPHERIC   PRESSURE

MERCURY BOWL

Fig. 1-20. Balancing the pressure of air with a mercury column.

On the absolute scale, it will be observed that pressures begin from an absolute zero or perfect vacuum, and that a gage calibrated in this manner will indicate 14.7 lb at normal sea-level pressures. On the other hand, a gage calibrated in the fashion employed for the commercial field of refrigeration and air conditioning will indicate this latter pressure as 0 lb where normal atmospheric conditions prevail.

EFFECT OF LOW PRESSURES

In the previous experiments, we determined that whenever pressure was increased, the boiling point was raised. Making use of the same apparatus, let us investigate the effect of a reduced pressure on the evaporating temperature. To do so, we will again place a quantity of water in the flask, force the stopper containing a pressure gage, thermometer, and control valve tightly in place so that no leakage can occur to mar the accuracy of our experiment, and place the apparatus over a gas flame. This time, however, instead of using the exit valve for holding the steam within the vessel to cause a pressure to build up, we will connect a vacuum line to it and use it to prevent the steam from being drawn out or evacuated so that a certain low pressure, or inches of vacuum, can be maintained within the vessel.

Referring to the pressure scale (Fig. 1-20), we find that 30 in. Hg is roughly equivalent to 15 lb on the gage-pressure scale, so that a 2-inch vacuum means a pressure of 1 lb less than the sea-level atmospheric pressure. For our first experiment, let us take a pressure of 14 lb absolute, or 0.7 lb less pressure than normally exerted. From a table on the properties of saturated steam, we will find that this corresponds to 1.42 in. Hg on the gage-pressure scale.

By regulating the exit valve so that the proper vacuum or low pressure is obtained, we can easily arrive at a point where boiling occurs at the constant pressure we desire. At this point of 14 lb absolute pressure, we would find that the water boils at 209.55°F. Then, we can open the exit valve a trifle more and regulate it so that a pressure of 12 lb absolute (5.49 in. Hg) is maintained. This pressure results in a boiling point of 201.96°F.

The fact is thus brought out that the pressure to which water is subjected has just as much to do with its boiling as does the temperature. This holds true for any liquid, for evaporation can be made to occur at any temperature above its freezing point if the pressure to which it is subjected is made low enough.

As you progress in your studies, you will find that sometimes you will have to put a vacuum pump on a refrigeration system to get rid of all unwanted gases and water. At sea level theoretically you should get the vacuum pump to pull down to 29.92 in. Hg. This will not be possible because the pump you will be using will not pull a perfect vacuum. Depending on the atmospheric pressure and efficiency of your pump, you will probably be able to pump down to between 27 and 28 in. Hg. At a 5000-foot altitude you may reach approximately 23 inches of vacuum.

HEAT TRANSMISSION

In refrigeration, we are interested in getting the heat contained in a room or refrigerator to a medium that will effect its removal. Ice is a simple method and formerly was widely used. The more modern refrigerator is equipped with a cooling apparatus that supplants ice and provides for more constant and cooler temperatures. Before we can study how heat is taken up and removed mechanically, it is imperative to learn the behavior of heat.

One of the most important laws has already been mentioned—the flow of heat from a body of higher temperature to one having a lower sensible heat. Never of its own accord will heat or water flow uphill or in the opposite direction. Therefore, it follows that heat in a refrigerator or room will flow to the cooler object, such as ice or the cooling device. The transmission of heat may be accomplished in three ways: by *conduction, convection, and radiation.*

Conduction

Conduction is the transference of heat by molecular impact from one particle to another in contact. For instance, if the end of a bar of iron is heated in the fire, some of the heat will pass through the bar to the cooler portion. Heat traveling in a body or from one body to another where the two are in intimate contact is termed *conduction*. This transfer of heat is shown in Fig. 1-21.

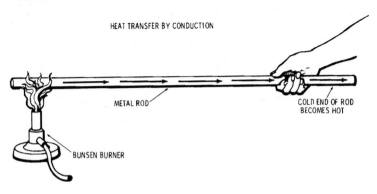

HEAT TRANSFER BY CONDUCTION

METAL ROD

BUNSEN BURNER

COLD END OF ROD BECOMES HOT

Fig. 1-21. Transfer of heat by conduction.

Metals are usually splendid heat conductors. Each and every material has a conduction value—some good, like the metals, others mediocre, and a few very poor. For instance, heat will quickly pass through a piece of copper but will have considerable difficulty in passing through a piece of cork. The materials that have very low heat conductivities are termed *heat insulators*. Even the very poorest conductors, or insulation materials, allow a certain amount of heat to pass through. There is no material that offers a perfect barrier or resistance to the passage of heat.

Convection

Convection is the principle used in hot-air heating. Air that is free to circulate, such as in any air body of appreciable size, will be set into motion where a difference of temperature occurs because it will absorb heat from the warmer wall, become heated, expand, and become lighter. The heated portion of the air will rise, and cooler air will move into its place, which, in turn, will become heated. The heated portion of the air eventually moves over to the colder wall, and the heat flows from the air to the colder object. Thus, any body of air capable of motion will transmit heat by convection. Hot-air and hot-water heating systems work on the the convection principle. They convey heat by bodily moving the heated substance from one place to another, as shown in Fig. 1-22.

The most efficient heat insulation known is a vacuum; but except for very small containers, it is structurally impractical to employ it commercially. The next best insulating medium is air subdivided into the smallest possible units so that it is still, or stagnant. Air that is contained in spaces of appreciable size, such as between the double walls of refrigerators, will circulate and transmit heat by convection. Cork is an insulation material of a high order and has great resistance to the passage or transmission of heat because of its air content. The air cells in cork are of such minute size that the air trapped in them is so restricted that only a little circulation is possible; for all practical purposes, it is still air, and little or no convection occurs.

Radiation

Heat energy transmitted through the air in the same way light is sent out by a lighted lamp, a radiant heater, or the sun is called

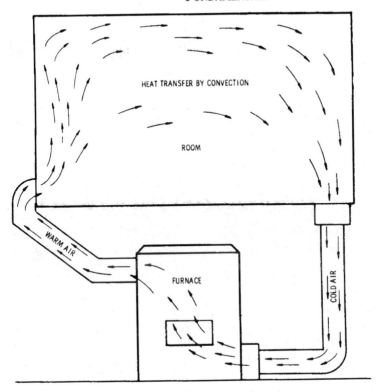

Fig. 1-22. Transfer of heat by convection.

radiated energy, as shown in Fig. 1-23. Large cold-storage warehouses, auditoriums, theaters, and homes are built with consideration of the heat evolved through radiant energy of the sun. Small household appliances rarely require the consideration of any radiant-heat factor, for they are used in existing structures without any change in building design and are sheltered from direct heat.

CONDENSATION

In a previous experiment we illustrated that a liquid, heated to the boiling point corresponding to the pressure imposed, will assimilate heat and produce a vapor, or gas. The heat taken up by the liquid is used to speed up molecular activity until a vapor is evolved. It stands to reason, then, that any vapor, or gas, contains a considerable quantity of heat.

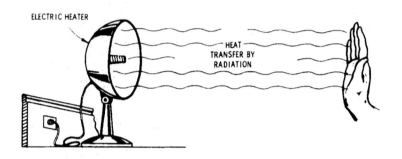

Fig. 1-23. Transfer of heat by radiation.

In accordance with the foregoing experiments, we found that heat itself always flows from the warmer to the colder body. To prove this again and illustrate just what condensation is, let us take a vessel, fill it about half full of water, and bring it to the boiling point over a gas flame, as shown in Fig. 1-24. As soon as steam is generated, let us take a dry, cold plate and hold it at an angle over the jet of steam as it issues from our crude boiler. It will be observed that the steam impinges on the plate and is converted again into water, the droplets forming and dripping off the edge of the plate. This was caused by the hot steam giving up its heat load (latent heat of vaporization) to the cold plate, the heat flowing from the steam to the plate.

To make sure that this is really the case, let us take the vessel we used in a previous experiment and again fill it with water, secure the stopper in place, and apply heat. This time let us set the exhaust valve so that we can maintain a few pounds of steam pressure, perhaps 5 lb. While steam is generating and we are regulating the valve to get this constant pressure, let us build a condenser. This is easily done; all we need is about 12 feet of copper tubing. Any tubing or material will do, but copper is so easily bent and formed that it is not a difficult task to wind it in a spiral form, such as shown in Fig. 1-25.

COLD PLATE

Fig. 1-24. Example of condensation.

The formed coil is submerged in a pail of cold water with its open end at the bottom. The top end is supplied with a piece of rubber tubing. When the gas and steam valves have been regulated so that there is a constant steam pressure of 5 lb within the boiler and a fair amount of steam being exhausted from the exit, quickly connect the rubber tube on the condenser to the steam pipe. It will be observed that the pressure gage has dropped somewhat; and after the steam has blown out the air contained in our condenser, we will see that no steam is issuing from the open end of the copper coil submerged in the water.

If we weighed the steam-generating apparatus and the condenser equipment before and after the experiment, we would find that the boiler lost weight through evaporation of water while the condenser gained through condensation exactly what had been lost. Also, if a thermometer has been placed in the condenser water and a reading taken at the start and end of our experiment, it would have been found that the water increased in temperature. In fact, if we ignore various losses, the amount of steam condensed can be estimated from the increase in temperature of the water.

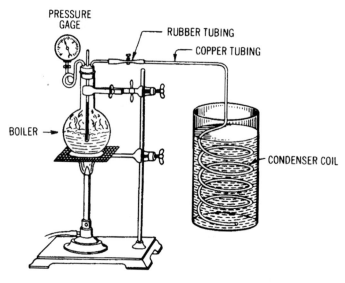

Fig. 1-25. Example of condensation by steam vapors.

It is apparent that the steam is easily convertible to its liquid form if the heat of vaporization is extracted. Then, too, our old rule would apply, not only to steam, but to all vapors and gases; that is, if we increase the pressure we will find that the gas will condense at the lower temperature found in the condenser, for the temperature of the gas will be raised far above the condenser temperature. Thus, heat will flow from the hot gas to the cold condenser; the greater the temperature difference, the faster will be the heat exchange.

In refrigeration and air conditioning applications, the gases employed as a refrigerating medium, for any given pressure, have a corresponding temperature at which they condense, or liquefy. Where both gas and liquid are present in the same vessel and the closed container is heated to cause boiling, the temperature of both gas and liquid will be the same at the boiling point. If pressure is increased, the boiling point will be raised. Above a certain point the gas will cease to have any latent heat of vaporization, and it will remain a gas regardless of the intensity of the pressure imposed.

ALTITUDES AND REFRIGERATION CAPACITY

As altitude increases, there will be a decrease in the efficiency of refrigeration due to the lower air pressures; thus the cooling power of the condenser of the refrigerator and the horsepower of the motor naturally decrease. Do not be alarmed about this on home refrigerators, as they are oversized for sea level and work well at higher altitudes but with less efficiency.

Atmospheric Factors in Refrigeration

Dry-bulb temperature is the temperature that the plain thermometer will register. Wet-bulb temperature is that temperature observed if an ordinary thermometer bulb is covered with material such as linen, wet with distilled water, and exposed to atmospheric evaporation. Table 1-5 shows dry- and wet-bulb temperatures as related to summer air conditioning design conditions.

The amount of refrigeration required is affected, not only by dry-bulb temperature, but also by humidity, indicated by wet-bulb temperature, due to the latent heat required to condense the water vapor, which will collect as frost on the evaporator of the air conditioner.

Refrigerant pressure versus temperature is valuable information to have when repairing or installing refrigeration equipment. Table 1-6 is based on gage pressure in psi at sea level and may be used in home refrigeration and air conditioning to determine the

Table 1-5. Typical Outdoor Design Conditions for Air Conditioning

City	Dry-Bulb Temp., °F	Wet-Bulb Temp., °F
Birmingham, Alabama	95	78
Phoenix, Arizona	105	76
Los Angeles, California	90	70
Denver, Colorado	95	64
Washington, D.C.	95	78
Tampa, Florida	95	78
Atlanta, Georgia	95	76
Chicago, Illinois	95	75
New Orleans, Louisiana	95	80
Boston, Massachusetts	92	75
St. Louis, Missouri	95	78
New York City, New York	95	75
Cincinnati, Ohio	95	78
Dallas, Texas	100	78
Seattle, Washington	85	65

relative temperatures of evaporators. In commercial refrigeration, where pressure switches instead of thermostats are used to control the temperature desired, the table can be an extremely valuable tool in setting the desired temperatures. On present-day home refrigeration and air conditioning, R-12 and R-22 are the most commonly used refrigerants. In this text you will notice other refrigerants have been included, as you no doubt will come across some of these in older machines that you may be called upon to service.

REFRIGERATION BY VAPORIZATION

The temperature at which a liquid boils or vaporizes is called its boiling point. For liquids with very low boiling points, it is not necessary to supply heat by fire or other artificial means because the heat in surrounding objects may be sufficient to cause boiling or vaporization. This is true, for example, with anhydrous ammonia, the boiling point of which is –28°F, which is sufficiently low to cause it to boil violently when placed in an open vessel at ordinary temperatures. The absorption of heat by the vaporization of the ammonia will cause the outside of the container to

Table 1-6. Refrigerant Pressure vs. Temperature*

Temp.,°F	REFRIGERANT			
	R-12	R-22	R-502	R-717 NH$_3$
-40	11.0 †	0.5	4.1	8.7 †
-35	8.4 †	2.6	6.5	5.4 †
-30	5.5 †	4.9	9.2	1.6 †
-25	2.3 †	7.4	12.1	1.3
-20	0.6 †	10.1	15.3	3.6
-15	2.4	13.2	18.8	6.2
-10	4.5	16.5	22.6	9.0
-5	6.7	20.1	26.7	12.2
0	9.2	24.0	31.1	15.7
5	11.8	28.2	35.9	19.6
10	14.6	32.8	41.0	23.8
15	17.7	37.7	46.5	28.4
20	21.0	43.0	52.5	33.5
25	24.6	48.8	58.8	39.0
30	28.5	54.9	65.6	45.0
35	32.6	61.5	72.8	51.6
40	37.0	68.5	80.5	58.6
45	41.7	76.0	88.7	66.3
50	46.7	84.0	97.4	74.5
55	52.0	92.6	106.6	83.4
60	57.7	101.6	116.4	92.9
65	63.8	111.2	126.7	103.1
70	70.2	121.4	137.6	114.1
75	77.0	132.2	149.1	125.8
80	84.2	143.6	161.2	138.3
85	91.8	155.7	174.0	151.7
90	99.8	168.4	187.4	165.9
95	108.2	181.8	201.4	181.1
100	117.2	195.9	216.2	197.2
105	126.6	210.8	231.7	214.2
110	136.4	226.4	247.9	232.3
115	146.8	242.7	264.9	251.5
120	157.6	259.9	282.7	271.7
125	169.1	277.9	301.4	293.1
130	181.0	296.8	320.8	
135	193.5	316.6	341.3	
140	206.6	337.2	362.6	

*Gage pressure in psi at sea level.
†In. Hg below standard atmosphere.

become heavily frosted by moisture condensed and frozen from the air immediately surrounding the container.

Since the boiling temperature of any liquid may be changed by the pressure exerted upon it, it is easy to cause a liquid refrigerant to boil at any desired temperature by placing it in a vessel where the required pressure is maintained. The process of boiling or vaporization by which a liquid is changed to a vapor can be reversed; that is, the vapor can be reconverted into a liquid by the removal of beat. This is called condensation. An increase in pressure, by raising the boiling point, will assist in the condensation of the vapor. Liquids used as refrigerants must be recovered because of their initial cost, and the process of condensation is usually employed for this purpose.

Although some liquids boil at temperatures suitable for refrigeration, comparatively few possess all the requirements of a practical refrigerant. Those used commercially are Freon-12 (R-12), Freon-22 (R-22), Freon-502 (R-502), and ammonia (R-717, NH_3). Others used previously are sulfur dioxide, methyl chloride, and carbon dioxide. It is very questionable that you will run into any of the latter refrigerants now. For large installations, anhydrous ammonia will be used as well as lithium bromide. The first four refrigerants mentioned, listed in Table 1-6, are the refrigerants which you will come in contact with today in household and industrial refrigeration.

BASIC SYSTEMS

Refrigeration is distributed by several methods. In household and commercial refrigeration, the one you will find in most cases is the *direct expansion system*. The volatile refrigerant is allowed to expand in a pipe placed in the room or household refrigerator to be cooled, where the refrigerant absorbs its latent heat of evaporation from the material to be cooled. This method is used in small cold-storage rooms, constant-temperature rooms, freezer rooms, household refrigerators, and where possible losses due to the leakage of refrigerant would be low.

In the *indirect system*, a refrigerant medium (such as brine) is cooled down by the direct expansion of the refrigerant and is then pumped through the material or space to be cooled, where it absorbs its sensible heat. Brine systems are used to advantage in large installations where the danger of the large amount of refrigerant is important and the rooms or series of rooms have fluctuating temperatures. It will also be found in homes that are heated with hot-water systems, except that instead of brine, ordinary water is used, with temperature controls for keeping the water from freezing. You will find these systems primarily used in air conditioning.

For small refrigerated rooms or spaces where it is desired to operate the refrigerating machine only part of each day, the brine coils are supplemented by holdover, or congealing, tanks. A *hold-over tank* is a steel tank containing strong brine in which direct expansion coils are immersed. During the period of operation of the refrigeration machine, the brine is cooled down and is capable of absorbing heat during the shutdown period, the amount depending upon the quantity of brine, its specific heat, and the temperature head. *Congealing tanks* serve the same general purpose but operate on a different principle. Instead of a strong brine, they contain a comparatively weak brine solution, which freezes or congeals to a slushy mass of crystals during the period of operation. In addition to its sensible heat, this mass of congealed brine is capable of absorbing heat equivalent to its latent heat of fusion.

LAWS OF GASES

Extensive investigations of the behavior of gases have shown that a given weight of gas expands or contracts uniformly ⅟₄₅₉ of its original volume for each degree it is raised or lowered above or below 0°F, provided the pressure on the gas remains constant. This fact is known as the *law of Charles*. Following this same reasoning, we find that at −459.62°F, a gas would cease to exist; this assumption, therefore, establishes −459.62°F as *absolute zero*. Actually, this temperature or condition has never been attained. The *law of conservation of matter* states that matter can be changed from one form into another. Temperatures within a few degrees of absolute zero have been reached when liquefying oxygen, nitrogen, and hydrogen, but these (like other gases) change their physical state from gas into liquid and fail to disappear entirely at these low temperatures.

The fact that absolute zero has never been reached is also explained by another law, known as the *law of conservation of energy*. It has already been explained that heat is a form of energy. This law is stated as follows: Energy can be neither created nor destroyed, though it can be changed from one form into another. Table 1-7 gives factors for converting energy from one form to another.

Table 1-7. Energy Conversion Factors

1 watt hour (Wh)	= 3.411 British thermal units (Btu)
1 British thermal unit (Btu)	= 0.252 calorie (cal)
1 calorie (cal)	= 3.968 British thermal units (Btu)
1 pound melting ice equivalent (MIE)	= 144 British thermal units (Btu)
1 British thermal unit (Btu)	= 0.00695 pound melting ice equivalent (MIE)
1 pound melting ice equivalent (MIE)	= 36.3 calories (cal)
1 calorie (cal)	= 0.0276 melting ice equivalent (MIE)

Having considered the effect of temperature on a gas, the effect of pressure on gases to aid the study of refrigeration must be considered next. In 1662, Robert Boyle announced a simple relationship existing between the volume of a gas and the pressure applied to it, which has since become known to scientists as *Boyle's law* and may be stated as follows: At a constant temperature, the volume of a given weight of gas varies inversely as the pressure to which it is subjected. The more pressure applied to a gas, the smaller its volume becomes if the temperature remains the same; likewise, if the pressure is released or reduced, the volume of the gas increases. Mathematically, this is expressed as

$$PV = p \times v$$

where P = pressure on the gas at volume V, and p = pressure on the same weight of gas at volume v.

Boyle's law has been found to be only approximately true, especially for refrigerant gases, which are more easily liquefied. The variations from the law are greater approaching the point of liquefaction, or condensing point, of any gas, although the material movement of air is determined by this law.

It will be found that if the temperature is held constant and sufficient pressure is applied to a given weight of gas, it will change from the gaseous state into the liquid state. The point at which this change of state takes place is known as the *point of liquefaction, or condensing point.*

It should now be evident that a definite relationship exists between the pressures, temperatures, and volumes at which a given weight of gas may exist. This relationship is used extensively in scientific work. It is known as the *combined law of Boyle and Charles* and may be expressed mathematically as

$$\frac{PV}{T} = \frac{pv}{t}$$

where pressures P and p are expressed in the absolute pressure scale in pounds per square foot; volumes V and v are expressed in cubic feet; and temperatures T and t are expressed in degrees on the absolute temperature scale.

When the pressure, temperature, or volume of a gas is varied, a new set of conditions is created under which a given weight of gas exists in accordance with the preceding mathematical equation. If a gas is raised to a certain temperature (which varies with each individual gas, no matter how much pressure is applied to it), it will be found impossible to condense it. This temperature is known as the *critical temperature*. The pressure corresponding to the critical temperature is termed the *critical pressure*. Above the critical points it is impossible to vaporize or condense a substance.

When a liquid is evaporated to a gas, the change of physical state is always accompanied by the absorption of heat. Evaporation has a cooling effect on the surroundings of the liquid since the liquid obtains the necessary heat from its surroundings to change the molecular structure. This action takes place in the evaporator of a refrigeration system. Any liquid tends to saturate the surrounding space with its vapor. This property of liquids is an important element in all refrigeration work. On the other hand, when a gas is condensed into a liquid, the change of physical state is always accompanied by the giving up of heat. This action takes place in the condenser or the condensing unit of the refrigeration system due to the mechanical work exerted on the gas by the compressor. If gas or liquid is placed in a closed container and the temperature of the container changed, it will be found that the pressure exerted by that gas in the container is directly proportional to the absolute temperature. Thus, if the temperature is raised, the pressure increases; if the temperature is lowered, the pressure decreases.

THE REFRIGERATION CYCLE

The refrigeration cycle is simply a means of heat extraction. In a compression system four distinct parts are required:

1. Compressor (pump)
2. Condenser-receiver
3. Evaporator (cooling coils)
4. Expansion valve (pressure-reducing device)

Figure 1-26 shows schematically the parts that make up the compression system. In this process, a refrigerant is used that can be alternately vaporized and liquefied. The heat energy required to change the liquid refrigerant into a gas is obtained from the air space surrounding the evaporator. This low pressure is then drawn through the suction line and into the compressor. In passing through the compressor, the heat-laden gas is raised from the low pressure in the suction line to a higher pressure, thereby raising its temperature. It is then forced from the compressor through the discharge line into the condenser, where the heat is removed from the vapor by means of natural air circulation. The removal of heat from the vapor causes it to liquefy and then flow from the bottom of the condenser and into the receiver.

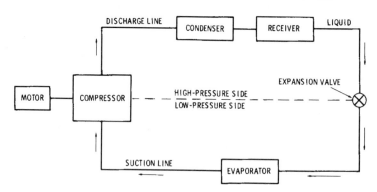

Fig. 1-26. Basic refrigeration circuit.

The expansion valve is adjusted to control the flow of refrigerant into the evaporator at a rate that is sufficient to maintain a desired temperature. In this connection, it should be noted that once the refrigerant has returned to a liquid state, it is again ready to be admitted through the expansion valve (or other pressure-reducing device) to the evaporator. In the evaporator, the pressure is reduced, the boiling point is lowered, and vaporization takes place, resulting in extraction of heat. This action is repeated continually as long as the compressor is running.

PRESSURE

It is of the utmost importance that the refrigeration student understand the meaning of the various kinds of pressure as related to refrigeration.

Atmospheric Pressure

Atmospheric pressure is pressure that is exerted by the atmosphere in all directions, as indicated by a barometer. Standard atmospheric pressure is considered to be 14.695 psi (usually written 14.7 psi), which is equivalent to 29.92 in. Hg.

Absolute Pressure

Absolute pressure is the sum, at any particular time, of gage pressure and atmospheric pressure. Thus, for example, if the pressure gage at one particular time reads 53.7 lb, the absolute pressure will be 53.7 + 14.7, or 68.4 psi. The preceding definitions may be written as follows:

$$\text{Absolute pressure} = \text{gage pressure} + 14.7$$
$$\text{Gage pressure} = \text{absolute pressure} - 14.7$$

where 14.7 is the normal atmospheric pressure.

SUMMARY

In this chapter the fundamental principles of refrigeration have been presented. Since refrigeration deals with the removal of heat from space or material substances, it is important that the student or serviceman clearly understand the method used in refrigeration as well as the nature of heat, heat of vaporization, latent heat, and change of state.

It should be noted that modern refrigeration is accomplished simply by a change of state from liquid to gas in various types of refrigerants, the change of its physical state being accomplished by the absorption of heat. The laws of gases have been fully treated in addition to the energy sources utilized for pumping the refrigerant through the refrigeration-system components.

The measurement of heat and a simple calculation for removal of heat from various material substances using specific heats and Btu values will further assist the reader in understanding the nature of refrigeration, its laws, and, finally, its utilization.

REVIEW QUESTIONS

1. What is meant by units of heat?
2. What is meant by conduction, convection, and radiation?
3. How does specific heat affect various substances?
4. How does latent heat affect the change of state in various fluids?
5. What is meant by heat of vaporization?
6. Name various commercial refrigerants.
7. How is refrigeration accomplished?
8. What is meant by an indirect system of refrigeration?
9. Define the law of conservation of matter.
10. What is the absolute zero reading on the Fahrenheit scale?
11. How many calories are represented by 1 Btu?
12. Define Boyle's law.
13. State the law of conservation of energy.
14. State the relation between pressure, temperature, and volume in a given weight of gas.
15. Explain how the refrigeration cycle is accomplished.
16. What will be the temperature reading on the Fahrenheit scale when the centigrade thermometer reads –10°?
17. Give the relation between absolute and gage pressure.

CHAPTER 2

Refrigerants

Refrigerants are heat-carrying mediums, which, during their cycle, absorb heat at a low temperature level and are compressed by a heat pump to a higher temperature where they are able to discharge the absorbed heat. The ideal refrigerant is one that can discharge to the condenser all the heat that the refrigerant is capable of absorbing in the evaporator or cooler. All refrigerating mediums, however, carry a certain portion of heat from the condenser back to the evaporator, which reduces the heat-absorbing capacity of the medium on the low side of the system. Tables showing the physical and chemical properties of refrigerants can be found in the Appendix.

DESIRABLE PROPERTIES

The requirements of a good refrigerant for commercial use are:

1. Low boiling point
2. Safe and nontoxic
3. Ease of liquefication action at moderate pressure and temperature
4. High latent-heat value
5. Ability to operate on a positive pressure
6. No effect on moisture
7. Mix as well with oil
8. Noncorrosive to metal

Classifications

Refrigerants may be divided into three classes, according to their manner of absorption or extraction of heat from the substances to be refrigerated.

Class 1—This class includes those refrigerants that cool materials by the absorption of the latent heat. The temperature and pressure properties of these refrigerants are shown in Table 2-1.

Table 2-1. Temperature and Pressure Properties of Refrigerants

Refrigerant	Boiling Point, °F	Freezing Point, °F	Critical Temperature, °F	Critical Pressure, p.s.i.a
Freon-14	−198.2	−312.0	−49.9	542.0
Freon-13	−114.5	−296.0	83.8	579.0
Carbon dioxide	−108.4	−69.9 triple	87.8	1,071.0
Freon-22	−41.4	−256.0	204.8	716.0
Ammonia	−28.0	−107.9	271.2	1,651.0
Freon-12	−21.6	−252.4	232.7	582.0
Methyl chloride	−10.76	−143.7	289.6	969.2
Sulfur dioxide	14.0	−98.9	314.8	1,141.5
Freon-114	38.4	−137.0	294.3	474.0
Freon-21	48.0	−211.0	353.3	750.0
Freon-11	74.7	−168.0	388.4	635.0
Methylene chloride	103.7	−143.0	421.0	640.0
Freon-113	117.6	−31.0	417.4	495.0

Class 2—The refrigerants in this class cool substances by absorbing their sensible heats. They are: air, calcium chloride brine, sodium chloride (salt) brine, alcohol, and similar nonfreezing solutions.

Class 3—This group consists of solutions that contain absorbed vapors of liquefiable agents or refrigerating media. These solutions function by the nature of their ability to carry the liquefiable vapors that produce a cooling effect by the absorption of latent heat. An example of this group is aqua ammonia, which is a solution composed of distilled water and pure ammonia.

The refrigerants in class 1 are employed in the standard *compression* type of refrigerating systems. The refrigerants in class 2 are employed as *immediate cooling agents* between class 1 and the substance to be refrigerated and do the same work for class 3. The latter is employed in the standard *absorption* type of refrigerating systems.

FREON REFRIGERANTS

The Freon family of refrigerants is presently used almost universally in household-type refrigerators. In the past, refrigerants were selected for use principally for their boiling points and pressures and their stability within the system or unit regardless of other important necessary properties, such as nonflammability and nontoxicity. Of course, there are many factors that must be taken into account when selecting a chemical compound for use as a refrigerant other than boiling point, pressure, stability, toxicity, and flammability. They must include molecular weight, density, compression ratio, heat value, temperature of compression, compressor displacement, design or type of compressor, etc., to mention only a few of the major considerations.

Chemical Properties

The Freon refrigerants are colorless, almost odorless substances, the boiling points of which vary over a wide range of temperatures. Freon refrigerants are nontoxic, noncorrosive, nonirritating, and nonflammable under all conditions of usage. They are generally prepared by replacing chlorine or hydrogen with fluorine. Chemically they are inert and thermally stable up to temperatures far beyond conditions found in actual operation.

73

Physical Properties

Pressures required to liquefy the refrigerant vapor affect the design of the system. The refrigerating effect and specific volume of the refrigerant vapor determine the compressor displacement; and the heat of vaporization and specific volume of liquid refrigerant affect the quantity of refrigerant to be circulated through the pressure-regulating valve or other device. Table 2-1 covers the boiling point at 1 atmospheric pressure (atm), freezing point, critical temperature, and critical pressures of not only the Freon refrigerants, but other commonly used refrigerants as well.

Operating Pressures

Pressures of saturated vapor at various temperatures of typical refrigerants are listed in Table 2-1. Table 2-2 lists the pressure of saturated vapor under standard ton conditions. Operating pressures will vary with the temperature of the condensing medium, amount of condenser surface, whether an air- or water-cooled condenser is used, operating back pressure, presence of noncondensable or foul gas in the condenser, circulation of the condensing medium through the condenser, condition of the condenser surface, extent of superheating of the refrigerant gas, and other factors.

Table 2-2 Operating Pressures of Refrigerants
(standard ton)

Refrigerant	Pressure, psig, 86°F	Pressure, psig, 5°F	Compression Ratio
Carbon dioxide	1024.30	319.70	3.110
Freon-22	159.80	28.33	4.045
Ammonia	154.50	19.57	4.940
Freon-12	93.20	11.81	4.075
Methyl chloride	80.00	6.46	4.480
Sulfur dioxide	51.75	5.87*	5.630
Freon-114	21.99	16.14*	5.420
Freon-21	16.53	19.25*	5.960
Freon-11	3.58	23.95*	6.240
Methylene chloride	9.44	27.53*	8.570
Freon-113	13.93*	27.92*	8.010

*In. Hg below 1 atm.

Freon-12

Freon-12, CCl_2F_2, has a boiling point of $-21.6°F$ and is used extensively as a refrigerant in both direct and indirect industrial, commercial, and household air-conditioning systems, as well as in household refrigerators, ice cream cabinets, frozen food cabinets, food locker plants, water coolers, etc., employing reciprocating-type compressors ranging in size from fractional to 800 hp. Freon-12 is also used in household refrigerating systems, ice cream and frozen food cabinets employing rotary-vane-type compressors. Freon-12 is used in industrial process water and brine cooling to $-110°F$, employing multistage centrifugal-type compressors in cascade of 100-ton refrigeration capacity and larger.

The health hazards resulting from exposure to Freon when used as a refrigerant are remote. Freon belongs to a class of special, nontoxic gases. Vapor in any proportion will not irritate the skin, eyes, nose, or throat; and, being odorless and nonirritating, it will eliminate all possibilities of panic hazards should it escape from a refrigerating system. At $68°F$ room temperature, 1 lb of Freon-12 liquid expands to 3.8 cu ft of vapor. Freon-12 is a stable compound capable of undergoing, without decomposition, the physical change to which it is commonly subjected in service, such as freezing, vaporization, and compression.

The low boiling point permits low temperatures to be reached without the compressor operating on a vacuum. This low but positive pressure prevents moisture-laden air from accidentally entering the system and also permits detection and location of the source of leaks, which is difficult if the refrigeration system operates at a negative back pressure.

Under ordinary conditions, when no moisture is present, Freon does not corrode the metals commonly used in refrigeration systems. In the presence of water, it will discolor brass, steel, and copper; but there is little or no evidence of any serious corrosive action. It is only slightly soluble in water, and the solution formed will not corrode any of the common metals used in refrigeration construction. It is nonflammable and noncombustible under fire conditions or where appreciable quantities come in contact with flame or hot-metal surfaces. It requires an open flame at 1382°F to decompose the vapor, and then the vapor decomposes to form only hydrogen chloride and hydrogen fluoride, both of which are irritating but readily dissolved in water. Air mixtures are not capable of burning and contain no elements that will support combustion. Therefore, Freon is considered nonflammable. It is so safe and nontoxic that the only possible way to cause death with Freon is to get concentrations so great as to exclude the oxygen. However, Freon can damage the ozone layer that protects the earth from ultraviolet rays. Any unnecessary release of Freon into the air should be avoided.

Freon-21

Freon-21, $CHCl_2F$, has a boiling point of 48°F and may be used as a refrigerant in industrial and commercial air-conditioning systems. It is also used in industrial process water and brine cooling to –50°F, employing single- or multistage centrifugal-type compressors of 100-ton refrigeration capacity and larger. Freon-21 has been used in fractional-horsepower household refrigerating systems and drinking water coolers employing rotary-vane-type compressors; it has also been used in comfort-cooling air conditioning systems of the absorption type where dimethyl ether or tetraethylene glycol is used as the absorbent.

Freon-22

Freon-22, $CHClF_2$, has a boiling point of –41.4°F and is used as a refrigerant in industrial and commercial low-temperature refrigerating systems to –150°F; it is also used in window-type and unit package room coolers and air conditioning units. Freon-22 is used in many installations where more efficient operation is desired in providing the necessary lower temperatures for low-temperature locker plants, resulting in quicker freezing of foods, greater volume of products handled by the quick-freezing units, home or farm freezers, and countless numbers of low-temperature industrial applications.

Freon-113

Freon-113, $CCl_2F-CClF_2$, has boiling point of 117.6°F, and is used as a refrigerant in most all industrial and commercial air conditioning systems. Freon-113 is also used in industrial process water and brine cooling to 0°F, employing four-stage (or more) centrifugal-type compressors of 25-ton refrigeration capacity and larger.

Freon-11

Freon-11, CCl_3F, has a boiling point of 74.7°F, and has a wide usage as a refrigerant in indirect industrial and commercial air conditioning systems. It is also used in industrial process water and brine cooling to –40°F, employing single- or multistage centrifugal-type compressors of 100-ton refrigeration capacity and larger.

Freon-114

Freon-114, $C_2Cl_2F_4$, has a boiling point of 38.4°F, and is used as a refrigerant in fractional-horsepower household refrigerating systems and drinking-water coolers employing rotary-vane-type compressors. It is also used in indirect industrial and commercial air conditioning systems and in industrial process water and brine cooling to –70°F, employing multistage centrifugal-type compressors in cascade of 100-ton refrigeration capacity and larger.

Freon-13 and Freon-14

Freon-13, $CClF_3$, and Freon-14, CF_4, are two compounds recently added to the Freon family, having boiling points of –114.5 and –198.2°F, respectively. These refrigerants will undoubtedly find usage in extremely low-temperature industrial refrigerating systems approximating liquid-air temperatures, and they will be employed in cascaded reciprocating-type compressors. The use of these two Freon refrigerants will require thorough investigation and research as to the behavior of metals at low temperatures while under stress, lubrication of mechanical compressors, and possible development of more efficient insulating materials.

Sulfur Dioxide

Sulfur dioxide, SO_2, was formerly one of the most common refrigerants employed in household refrigerators. It is a colorless gas or liquid. It is toxic, has a very pungent odor, and is obtained by burning sulfur in air. It is not considered a safe refrigerant, especially in quantities. It combines with water and forms sulfurous and sulfuric acids, which are corrosive to metal. Sulfur dioxide has an adverse effect on almost anything with which it comes in contact. It boils at about 14°F (standard conditions) and has a latent-heat value of 166 Btu/lb.

Sulfur dioxide has the disadvantage that it must operate in a vacuum to give temperatures required in most refrigeration work. Should a leak occur, moisture-laden air will be drawn into the system, eventually corroding the metal parts and ruining the compressor. Also, in relation to Freon or methyl chloride, approximately one-third more vapor must be pumped in order to get the same amount of refrigeration. This means that either the condensing unit must be speeded up to give the desired capacity, or the size of the cylinders must be increased proportionately.

Sulfur dioxide does not mix well with oil. The suction line must be on a steady slant to the machine; otherwise the oil will trap out, making a constriction in the suction line. On many installations, it is not possible to avoid traps, and on these jobs sulfur dioxide is not satisfactory. Because of its characteristic pungent odor, comparatively small leaks are readily detected. Even the smallest leaks are readily located by means of an ammonia swab. A small piece of cloth or sponge may be secured to a wire and dipped into strong aqua ammonia or household ammonia and then passed over points where leaks may be present. A dense white smoke forms where the sulfur dioxide and ammonia fumes come in contact. When no ammonia is available, leaks may be located by the usual soap-bubble or oil test. The soap solution or oil is put on the tube joints and points where bubbles would be noted. *Note: liquid sulfur dioxide on any part of the body produces freezing. In any such case a physician should be consulted immediately.*

Methyl Chloride

Methyl chloride has been used mostly in commercial refrigerator units; its use in household refrigerators has been very limited. It is a good refrigerant, but because it will burn under some conditions and is slightly toxic, it does not conform to some of the strict city codes now enforced. Roughly speaking, the average relative concentration by weight of different refrigerant vapors in a room of a given size that produces the same effect on a person breathing the air thus contaminated can be specified approximately as follows:

Carbon dioxide	100
Methyl chloride	70
Ammonia	2
Sulfur dioxide	1

In other words, methyl chloride is 35 times safer than ammonia and 70 times safer than sulfur dioxide. To produce any serious effects from breathing methyl chloride, a considerable quantity is required. For example, in a room $20 \times 20 \times 10$ feet, it would be necessary to liberate about 60 lb of methyl chloride to produce any serious effects.

Methyl chloride has a low boiling point. Under standard atmospheric pressure, it boils at $-10.8°F$. It is easy to liquefy and has a comparatively high latent value—approximately 176 Btu/lb under standard conditions. It will operate on a positive pressure as low as $-10°F$ and mixes well with oil. In its dry state it has no corrosive effect on metal, but in the presence of moisture, copper plating of the compressor parts results; and in severe cases of moisture, a sticky black sludge is formed, which is detrimental to the working parts of the system.

Methyl chloride is not irritating and, consequently, does not serve as its own warning agent in case of leaks as does sulfur dioxide. In some cases, a warning agent is added, such as a small percentage of acrolein (1%). Many consider the addition of 5% sulfur dioxide a dependable warning agent, but there is some controversy as to the desirability of doing this.

Ammonia

Ammonia is a refrigerant employed in refrigerators operating on the absorption principle. It is also used in large machines for industrial and other purposes. It is a colorless gas with a pungent characteristic odor. Its boiling temperature at normal atmospheric pressure is –28°F, and its freezing temperature is –107.86°F. It is very soluble in water, one volume of water absorbing 1.148 volumes of ammonia at 32°F.

Ammonia is combustible or explosive when mixed with air in certain proportions (about one volume of ammonia to two volumes of air), and much more so when mixed with oxygen. Because of its high latent-heat value (555 Btu at 18°F) large refrigeration effects are possible with relatively small-sized machinery. It is very toxic and requires heavy steel fittings. Pressures of 125 to 200 psi are not uncommon, and water-cooled units are essential.

Carbon Dioxide

At ordinary temperatures carbon dioxide is a colorless gas with a slightly pungent odor and acid taste. It is harmless to breathe except in extremely large concentrations, when the lack of oxygen would cause suffocation. It is nonexplosive, non-flammable, and does not support combustion. The boiling point of carbon dioxide is so extremely low that at 5°F a pressure of well over 300 psi is required to prevent its evaporation. At a condenser temperature of 80°F, a pressure of approximately 1000 psi is required to solidify the gas. Its critical temperature is 87.8°F, and –69.9°F is the triple point.

Because of its high operating pressure, the compressor of the carbon dioxide refrigerator unit is very small, even for a comparatively large refrigerating capacity. Because of its low efficiency as compared to other common refrigerants, carbon dioxide is seldom used in household units, but is used in some industrial applications and aboard ships. Leakage of carbon dioxide gas can be tested by making sure that there is pressure on the part to be tested and then using soap solutions at the suspected points. Leakage into condenser water can be tested with the use of bromthymol blue. The water entering and leaving the condenser should be tested at the same time because of the sensitivity of the test. When carbon dioxide is present, the normal blue color changes to yellow.

LEAK DETECTION*

The detection of leaks in refrigeration equiptment is a major problem for both manufacturers and service engineers. Several methods of leak detection will be described.

Electronic Detector

The electronic detector is widely used in the manufacture and assembly of refrigeration equipment. The operation of the instrument depends on the variation of current flow due to ionization of decomposed refrigerant between two oppositely charged platinum electrodes. This instrument can be used to detect any of the halogenated refrigerants except refrigerant 14. It is not recommended for use in atmospheres containing explosive or flammable vapors. Some other vapors, such as alcohol and carbon monoxide, may interfere with the test.

The electronic detector is the most sensitive of the various leak-detector methods, reportedly capable of sensing a leak of $\frac{1}{100}$ oz/year of refrigerant 12, or approximately 1×10^{-6}cc.

A portable model is available with an automatic balancing system which corrects for refrigerant vapors that might be present in the atmosphere around the test area.

Halide Torch

The halide torch has been used for many years as a fast and reliable method of detecting leaks of halogenated refrigerants. Air is drawn over a copper element heated by a methyl alcohol or hydrocarbon flame. If halogenated vapors are present, they will be decomposed and the color of the flame will change to bluish-green. Although not as sensitive as the electronic detector, this method is suitable for most purposes. *Note: Do not breathe the fumes from the halide torch.*

---

*The methods of leak detection discussed in this section are reprinted by permission from ASHRAE *Handbook of Fundamentals.*

Bubble Method

In the bubble method of leak detection, the object to be tested is pressurized with air or nitrogen. A pressure corresponding to operating conditions is generally used. The object can be immersed in a water bath and any leaks detected by observing the formation of bubbles in the liquid. Addition of a detergent to the water will decrease the surface tension, prevent escaping gas from clinging to the side of the object, and promote the formation of a regular stream of small bubbles. Kerosene or other organic liquids are sometimes used for the same reason. A solution of soap or detergent can be brushed or poured onto joints and any bubbles that form can be readily detected.

Note: If air or nitrogen is used for the test, the system must be thoroughly purged of these gases by means of a good vacuum pump for quite a period of time to remove all of the air or nitrogen and any moisture before charging the unit with refrigerant.

Detection of Ammonia and Sulfur Dioxide Leaks

Ammonia can be detected by burning a sulfur candle in the vicinity of the suspected leak or by bringing a solution of hydrochloric acid near the object. If ammonia vapor is present, a white cloud of ammonium sulfite or ammonium chloride will be formed. Ammonia can also be detected with any indicating paper that changes color in the presence of a base.

Sulfur dioxide can be detected by the appearance of a white smoke when aqueous ammonia is brought near the leak.

The presence of leaks can also be determined by pressurizing or evacuating the system and observing the change in pressure or vacuum over a period of time. This is good practice in checking on the "tightness" of a system but is of little help in locating the point of leakage.

PRESSURE-TEMPERATURE CHART

A logarithmic-scaled chart giving the relationship between pressure and corresponding temperatures in degrees Fahrenheit of common refrigerants is shown in Fig. 2-1. The axis of the abscissa shows the temperature, and the axis of the ordinate shows the pressure in pounds per square inch gage (psig) and absolute (psia), respectively. To ascertain the pressure of a refrigerant at any particular temperature, follow the desired

PRESSURE -LB./ IN.² GAGE

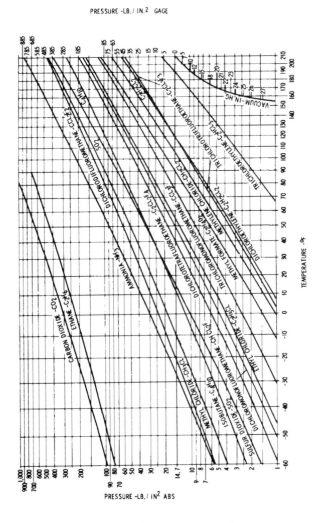

Fig. 2-1. Vapor pressure of refrigerants at various temperatures.

temperature until the curve of the particular refrigerant is reached. The corresponding pressure is then found on the pressure axis.

For example, the temperature corresponding to a pressure of 40 psig for sulfur dioxide is approximately 60°F, and the corresponding temperature at the same pressure for dichlorodifluoromethane (Freon-12) is approximately 27°F, etc.

CARE IN HANDLING REFRIGERANTS

In the preceding discussion it was observed that one of the requirements of an ideal refrigerant is that it must be nontoxic. In reality, however, all gases (with the exception of pure air) are more or less toxic or asphyxiating. It is therefore important that wherever gases or highly volatile liquids are used, adequate ventilation should be provided because even nontoxic gases in air produce a suffocating effect.

Vaporized refrigerants, especially ammonia and sulfur dioxide, bring about irritation and congestion of the lungs and bronchial organs, accompanied by violent coughing and vomiting; and when breathed in sufficient quantity, they cause suffocation. It is therefore of the utmost importance that the serviceman subjected to a refrigerant gas should find access to fresh air at frequent intervals to clear his lungs. When engaged in the repair of ammonia and sulfur dioxide machines, approved gas masks and goggles should be used. Carrene, Freon-12, and carbon dioxide fumes are not irritating and can be inhaled in considerable concentrations for a short period without serious consequences. It should be remembered that liquid refrigerants will refrigerate or remove heat from anything with which they come in contact when released from a container, as in the case of an accident.

HANDLING REFRIGERANT CYLINDERS

It is of the utmost importance to handle cylinders of compressed gas with care (Fig. 2-2) and observe the following precautions:

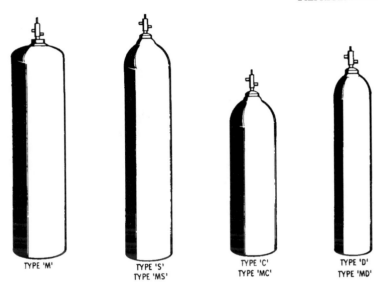

TYPE 'M'
TYPE 'S'
TYPE 'MS'
TYPE 'C'
TYPE 'MC'
TYPE 'D'
TYPE 'MD'

Fig. 2-2. Various sizes of refrigerant cylinders.

1. Never drop cylinders nor permit them to strike each other violently.
2. Never use a lifting magnet or a sling (rope or chain) when handling cylinders. A crane may be used when a safe cradle or platform is provided to hold the cylinders.
3. Where caps are provided for valve protection, such caps should be kept on the cylinders except when the cylinders are in use.
4. Never overfill cylinders. Whenever the refrigerant is discharged from or into a cylinder, immediately thereafter weigh the cylinder and record the weight of refrigerant remaining in the cylinder.
5. Never attempt to mix gases in a cylinder.
6. Never use cylinders for rollers, supports, or any purpose other than to carry gas.
7. Never tamper with the safety devices in valves or cylinders.
8. Open cylinder valves slowly. Never use wrenches or tools except those provided or approved by the gas manufacturer.

9. Make sure that the threads on regulators or other unions are the same as those on cylinder-valve outlets. Never force connections that do not fit.
10. Regulators and pressure gages proved for use with a particular gas must not be used on cylinders containing different gases.
11. Never attempt to repair or alter cylinders or valves.
12. Never store cylinders near highly flammable substances, such as oil, gasoline, waste, etc.
13. Cylinders should not be exposed to continuous dampness, salt water, or salt spray.
14. Store full and empty cylinders apart to avoid confusion.
15. Protect cylinders from any object that will produce a cut or other abrasion in the surface of the metal.

Refrigerant Containers

Refrigerants may be purchased in containers from 1 to 150 lb. If you use the 150-lb containers, it is well for domestic use to also purchase 10-lb containers, which are easier to handle. The 10-lb containers may be refilled from the 150-lb container by lowering the temperature of the 10-lb container with ice or cold water and connecting the two tanks with a flexible hose. Be certain that the air is purged out of the hose before connecting to the 10-lb tank. This is accomplished by cracking the valve on the 150-lb tank and allowing some gas to escape; then connect the hose to the small tank. Place the small tank on a scale, turn the large tank on its side or end, and watch the scale, stopping at the point that you reach 10 lb plus the weight of the tank. *Do not overfill. If anything, underfill* the small tank.

A SIMPLE REFRIGERATION SYSTEM

The principle of using the latent heat of vaporization of a liquid, such as sulfur dioxide, for producing refrigeration can be illustrated very easily by thinking of a refrigerator of very simple design, similar to the one shown in Fig. 2-3. The refrigerator is made up of a box that is completely insulated on all six sides to prevent the entrance of heat by conduction, convec-

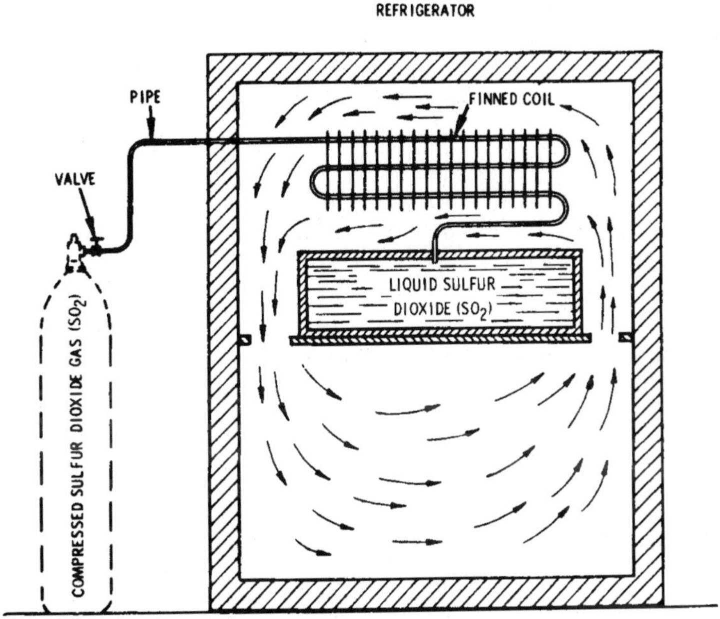

REFRIGERATOR

Fig. 2-3. Simple refrigeration cycle.

tion, and radiation. We then place in the top of the cabinet a series of finned coils, with one end connected to the cylinder charged with sulfur dioxide. Through this end we will charge, for example, 2 lb of sulfur dioxide into the coil, after which the compressed cylinder will again be sealed and disconnected from the line, with the charging end of the pipe open to the atmosphere.

Since the liquid sulfur dioxide is exposed to the air, the only pressure to which the liquid is subjected is atmospheric pressure, which is approximately 14.7 psi absolute or 0 psig. At this pressure, as previously explained, sulfur dioxide liquid will boil or vaporize at a temperature of 14°F or at any higher temperature. We will say, just for example, that the temperature of the room in which the refrigerator is located is 70°F. If this is the case, the temperature of the cabinet at the time of the addition of sulfur dioxide will also be 70°F. The liquid sulfur dioxide in the coils will therefore immediately start boiling and vaporizing because the surrounding temperature is above the boiling point (14°F) of the liquid. As the liquid boils away, it will absorb heat from the cabinet because for every pound of sulfur dioxide vaporized, 168 Btu of heat will be extracted from the cabinet. As soon as the temperature of the cooling coil is reduced to a point lower than the cabinet temperature, the air in the cabinet will start circulating in the direction shown by the arrows in Fig. 2-3 because heat always flows from the warmer to the colder object.

With this method, however, the 2 lb of sulfur dioxide liquid would soon be vaporized and the gas given off to the air outside the cabinet; refrigeration would then stop until a new charge was placed in the cooling coil. Sulfur dioxide is expensive and difficult to handle, and therefore some means must be used to reclaim the vapor in order to use the original charge continuously. The inconvenience of recharging the coil must also be prevented, and the refrigerator must be built so that it will automatically maintain proper food-preservation temperatures at all times with absolutely no inconvenience to the customer. This is accomplished by the compressor pulling the warm sulfur dioxide gas from the cooling unit and pumping it into the condenser, where it is changed to a liquid ready to return to the cooling unit.

SUMMARY

Refrigerants are heat-carrying mediums that absorb heat at a low temperature level and are compressed by a heat pump to a higher temperature where they are able to discharge the absorbed heat. The ideal refrigerant is one that can discharge to the condenser all the heat that the refrigerant is capable of absorbing in the evaporator.

Commercial requirements for a refrigerant are low boiling point, safety and nontoxicity, ease of liquefaction at moderate pressure and temperature, high latent-heat value, ability to operate on a positive pressure, no effect on moisture, that it mix well with oil and be noncorrosive to metal.

Refrigerants are divided into three classes: Class 1, Class 2 and Class 3. Each class has its own properties and uses.

Freon refrigerants are used almost universally in household refrigerators. Freon refrigerants are colorless, almost odorless and boiling points vary over a wide range of temperatures. They are generally prepared by replacing chlorine or hydrogen with fluorine. Chemically they are inert and thermally stable up to temperatures far beyond conditions found in actual operation.

Operating pressures of saturated vapor will vary with the temperature of the condensing medium, amount of condenser surface, whether an air- or water-cooled condenser is used, operating back pressure, presence of noncondensable or foul gas in the condenser, circulation of the condenser medium through the condenser, condition of the condenser surface, extent of superheating of the refrigerant gas and other factors.

Freon-12 is CCl_2 and has a boiling point of $-21.6°F$.
Freon-21 is $CHCl_2F$ and has a boiling point of $48°F$.
Freon-22 is $CHClF_2$ and has a boiling point of $-41.4°F$.
Freon-113 is CCl_2F-$CClF_2$ and has a boiling point of $117.6°F$.
Freon-11 is CCl_3F and has a boiling point of $74.7°F$.
Freon-114 is $C_2Cl_2F_4$ and has a boiling point of $38.4°F$
Freon-13 is $CClF_3$ and has a boiling point of $-114.5°F$.
Freon-14 is CF_4 and has a boiling point of $-198.2°F$.
Sulfur dioxide is SO_2 and has a boiling point of $14°F$.
Methyl chloride has a boiling point of $-10.8°F$.
Ammonia has a boiling point of $-28°F$.

Ammonia is used in commercial refrigeration systems. It is combustible or explosive when mixed with air in certain proportions and much more so when mixed with oxygen. Methyl chloride is non-irritating and does not serve as its own warning agent in case of leaks. Sulfur dioxide, however, has a very noticeable odor. Carbon dioxide, CO_2, at ordinary temperatures is a colorless gas with a slightly pungent odor and acid taste. It is harmless to breathe except in extremely large concentrations. It is nonexplosive, nonflammable, and does not support combustion. The boiling point of carbon dioxide is $5°F$.

Leaks in refrigeration units can be detected by a number of methods. The electronic detector is widely used in the manufacture and assembly of refrigeration equipment. The instrument is used to detect leaks in refrigerants except refrigerant 14. It is not recommended for use in atmospheres containing explosive or flammable vapors.

The halide torch has been used for many years as a fast and reliable method of detecting leaks of halogenated refrigerants. The flame turns a bluish-green in the presence of a refrigerant. And, of course, the bubble method can be used to detect leaks where the pressure is such as to cause leakage.

Ammonia leaks can be detected by using a burning sulfur candle. It produces a white cloud near the leaking area. Litmus paper can also be used to detect a leak. Sulfur dioxide can be detected by the appearance of a white smoke when aqueous ammonia is brought near the leak.

A logarithmic-scaled chart is used for the relationship between pressure and temperatures in degrees Fahrenheit for common refrigerants for vapor pressure at various temperatures.

Refrigerant cylinders are of four types, M, MS or S, C or MC, and D or MD. Containers are available with from 1 to 150 pounds of refrigerant. Ten-pound containers are usually the easiest to handle for domestic work.

REVIEW QUESTIONS

1. Name the desirable properties of a good refrigerant for commercial use.
2. How are commercial refrigerants classed?
3. What is the purpose of calcium chloride brine as used in commercial refrigeration plants?
4. Why is ammonia employed in preference to other refrigerants in commercial plants?
5. State the boiling point and heat of vaporization for Freon-12.
6. In what respect does the Freon family of refrigerants differ from such refrigerants as methyl or ethyl chloride?
7. Name the chemical and physical properties of Freon.
8. Why is Freon-12 preferred in both commercial and house-

hold refrigeration units?

9. What is meant by the critical temperature of a refrigerant?
10. Why does a high latent heat of evaporation per unit of weight affect the desirability of a refrigerant?
11. Describe the methods of leak detection in a refrigeration system.
12. What precautions should be observed in the storage and handling of refrigerant cylinders?
13. What type of industrial processes use refrigeration?
14. What are the basic units of measurement in the metric system?
15. What is the difference between a square foot and a cubic foot?
16. What is a slug?
17. What is the difference between a short ton and a long ton?
18. What is specific gravity?
19. What is the atmospheric pressure? Why is it important to know when working with refrigerants?
20. How is Boyle's law used in reference to refrigerants?
21. What relationships does Charles' law deal with?
22. What is isothermal change?
23. How do you convert °C to °F?
24. How do you convert °F to °C?
25. What is the difference between kinetic and potential energy?

CHAPTER 3

Compressors

Compressors may be divided into several types, depending on the size of the plant, refrigerant used, and other factors: They may be *single-cylinder* or *multicylinder*. With respect to the method of compression, they are classified as: *reciprocating*, *rotary*, or *centrifugal*. They may employ either direct drive or belt drive. With respect to the location of the prime mover, they are classified as either *independent* (belt drive) or *semihermetic* (direct drive, motor and compressor in separate housings).

HOW THE COMPRESSOR WORKS

The passage of the refrigerant to and from the compressor is controlled by means of a discharge and suction valve located on a specially designed valve plate, which forms the lower part of the cylinder head. The flapping action of the valve permits the flow of refrigerant out only through the discharge valve port and flow in only through the suction valve port. Thus, when the piston moves away from the valve plate (suction stroke), a pressure reduction takes place. Because the pressure in the cylinder is now below that in the suction line, a flow of refrigerant occurs that pushes open the suction valve and permits a certain quantity of refrigerant to enter the compressor.

As the motion of the piston reverses and moves toward the valve plate (compression stroke), it increases the pressure, which forces the suction valve to close. A further compression as the piston moves close to the valve plate opens the discharge valve and forces the refrigerant into the discharge line, thus causing what is known as the *high-side pressure*, or *positive working pressure*, of the refrigerant.

RECIPROCATING COMPRESSORS

Reciprocating compressors derive their name from the recip- rocating (back-and-forth) action of the piston. The length of the piston movement within its cylinder is known as the stroke. The capacity of the compressor, as previously noted, depends on such factors as the number of cylinders, stroke, revolutions per minute of the crankshaft. Figure 3-1 shows a typical reciprocating com- pressor.

Figure 3-2 shows a heavy-duty, two-cylinder industrial com- pressor of vertical design and low-speed operation. The motion of the piston is transferred from the crankshaft, which, in turn, receives its motion from a belt wheel, as shown. The prime mover for this type of compressor is usually an electric motor, although there is no reason why other types of prime movers cannot be used.

The piston motion is effected by the crankshaft arm (or arms), to each of which is fastened one end of a connecting rod; the other end of the connecting rod is fastened to the piston by means of a piston pin. In this manner, the connecting rod and crank arm transfer the rotation of the crankshaft to the reciprocating or back-and-forth motion of the piston.

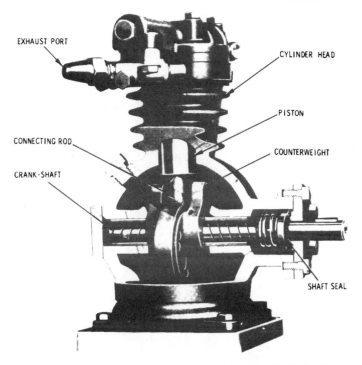

EXHAUST PORT

CYLINDER HEAD

PISTON

CONNECTING ROD

COUNTERWEIGHT

CRANK-SHAFT

SHAFT SEAL

Multiplex Manufacturing Company

Fig. 3-1. Cutaway view of a single-cylinder, belt-driven reciprocating refrigeration compressor.

Lubrication

Force-feed lubrication is usually provided in a reciprocating compressor by a reversible oil pump directly driven from the crankshaft. The pump provides lubrication for the main and connecting-rod bearings, wrist-pin bushings, shaft seal, and all other movable compressor parts. Certain large commercial compressors are equipped with a mechanical sight-feed system for supplying oil under pressure to the suction inlet and to the thrust side of the cylinder walls through special rod passages.

Shaft Seal

The function of the shaft seal in a refrigeration compressor is to prevent gas from escaping from the compressor at the point

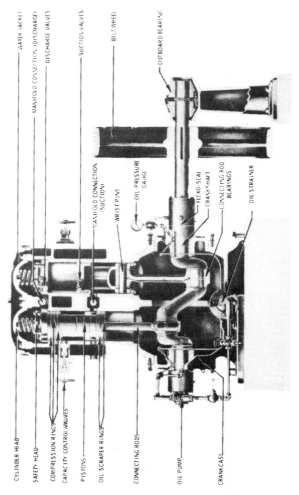

WATER JACKET

MANIFOLD CONNECTION (DISCHARGE)

DISCHARGE VALVES

SUCTION VALVES

BELT WHEEL

OUTBOARD BEARING

MANIFOLD CONNECTION (SUCTION)

WRIST PINS

OIL PRESSURE GAUGE

FLEXO-SEAL

CRANKSHAFT

CONNECTING ROD BEARINGS

OIL STRAINER

CYLINDER HEAD

SAFETY HEAD

COMPRESSION RINGS

CAPACITY CONTROL VALVES

PISTONS

OIL SCRAPER RINGS

CONNECTING RODS

OIL PUMP

CRANKCASE

The Frick Company

Fig. 3-2. Sectional view of a heavy-duty vertical industrial reciprocating compressor, belt driven.

where the shaft leaves the compressor housing. Depending on the size of the machine, a variety of leakfree shaft seals are used. A bellows-type or flexible seal usually provides protection against leaks for either pressure or vacuum. Figure 3-3 shows the seal arrangement of a typical heavy-duty compressor.

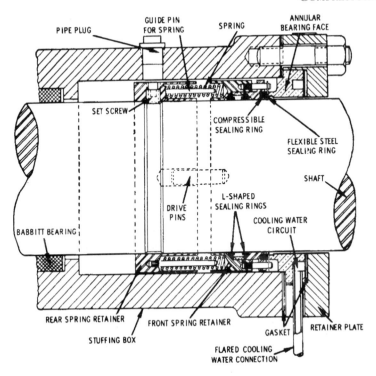

The Frick Company

Fig. 3-3. Sectional view of a heavy-duty compressor shaft seal.

Pistons and Rings

Pistons employed in the majority of compressors are made of the best grade of cast iron. They are carefully machined, polished, and fitted to the cylinders at close tolerances. There are several different designs for pistons. Some early models were not equipped with piston rings but were lapped to the cylinder in a very close fit. At present, however, two or more piston rings are employed, their function being to ensure proper lubrication and sealing of the cylinder walls. Pistons should fit the cylinder walls so that, when inserted, they do not drop of their own weight but must be urged through with the fingers. This fit is usually obtained by allowing 0.0003 inch (three ten-thousandths of an inch) of clearance per inch of cylinder diameter. Thus, a 1½-inch cylinder would have a 0.00045-inch clearance between the piston and the cylinder wall.

If cylinders are scored, they must be honed after reboring to ensure a high degree of wall finish, which maintains cylinder proportions over long periods. After rings have been fitted, the compressor should be operated overnight to permit the rings to wear in. The compressor should then be drained and flushed with petroleum spirits or dry-cleaning fluid and dehydrated before being returned to service. Typical piston rings are shown in Fig. 3-4.

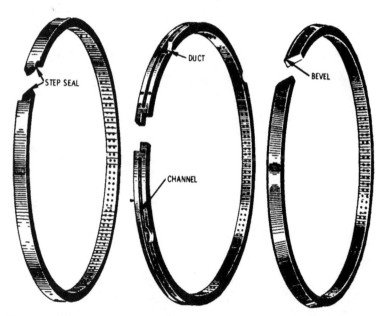

Fig. 3-4. Compressor piston rings.

Connecting Rods and Wrist Pins

The connecting rod forms the link between the piston and the crankshaft. One end is connected to the piston by means of a hardened, ground, and highly polished steel wrist pin. As shown in Fig. 3-5, compressor connecting rods are very similar to those employed in automobile engines except for their size.

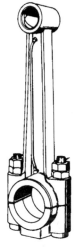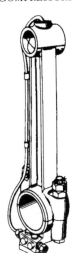

Fig. 3-5. Compressor connecting-rod construction.

Several different methods are used to secure the piston wrist pin and connecting rod. In some designs, the wrist pin is tightly clamped to the connecting rod, with the moving bearing surface in the piston, while others have a bushing in the connecting rod that allows the pin to turn freely in both piston and connecting rod. Again, others have the wrist pin held solidly in the position. Connecting rods are usually made of a high grade of cast iron or drop-forged steel.

Crankshafts

The crankshafts in refrigeration compressors are commonly made of steel forgings, which are machined to the proper tolerance for the main and connecting-rod bearing. The double crank is employed in all two-cylinder compressors, while the single crank is used on the single-cylinder type. Crankshafts are equipped with counterweights and are carefully balanced to ensure smooth and vibrationless compressor operation.

Some types of compressors are equipped with what is known as the *eccentric shaft*, which is a different application of the crankshaft principle. It employs an assembly consisting of the main shaft (over which an eccentric is fitted) and the outer eccentric strap.

Valves

The function of the valves in a compressor is to direct the flow of refrigerants through the unit. These valves are named according to the function they perform, such as *suction* and *discharge* valves. The so-called poppet type of valve has largely been replaced by the ring-plate valve. The ring-plate valve (Fig. 3-6) is designed for quick, positive, leakfree operation and is usually precision machined from specially heat-treated alloy steel. Because of their location in the cylinder safety head, valves can easily be removed and replaced as an assembly. Discharge valves consist of two spring-loaded ring plates, and the suction valve consists of one ring plate.

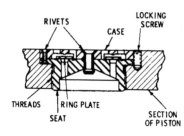

Fig. 3-6. Sectional view of a compressor ring valve.

ROTARY COMPRESSORS

The rotary compressor, although not as common as the reciprocating type, employs a slightly different method in compression of the refrigerant. The rotary compressor compresses gas by the movement of blades in relation to a pump chamber. In one type, the blades slide on an eccentric on the drive shaft inside a stationary concentric ring known as the *pump chamber*. In another design, the blades revolve concentrically with the shaft inside an off-center ring (Fig. 3-7). Hermetically-sealed rotary compressors are used in some makes of domestic refrigerators and freezers.

A representative type of a rotary booster compressor is shown in Fig. 3-8. During rotation, each gas space between the sliding blades is at maximum volume as it passes the cylinder suction port. As the rotor turns, this space is gradually reduced and the

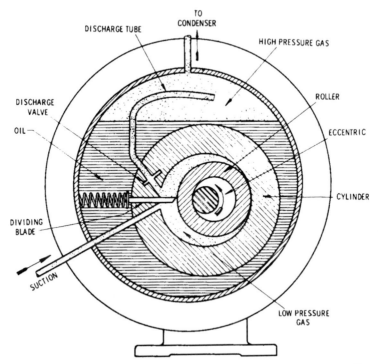

Fig. 3-7. Interior of a rotary compressor (seal and cover removed).

trapped gas is compressed. When the space reaches the minimum volume point (the cylinder discharge port), the gas is forced into the discharge line.

HERMETIC COMPRESSORS

Hermetic, or sealed-type, compressors are directly connected to an electric motor; the motor and compressor operate on the same shaft and are enclosed in a common casing. Condensing units of this type are used almost exclusively in domestic refrigerators and also in locker and home cold-storage plants, drinking fountains, ice cream and food display cabinets, soda fountains, and the like. They are made to operate on either the reciprocating or rotary principle and may be mounted with the shaft in either

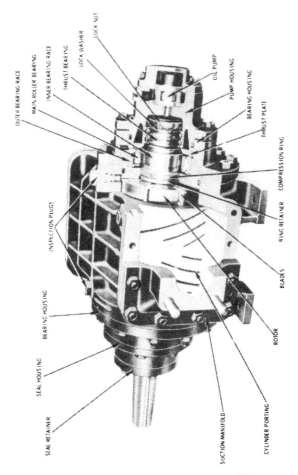

The Frick Company

Fig. 3-8. Cutaway view of a rotary booster compressor.

the vertical or horizontal position. In a unit of this type, the revolutions per minute (rpm) obviously are the same for both compressor and motor. This factor has a very important bearing on the size and design of the unit since it determines the type of refrigerant, the type of controls to be used, etc.

A representative hermetic horizontal compressor is shown in Fig. 3-9. In a unit of this type, the one-piece housing provides for

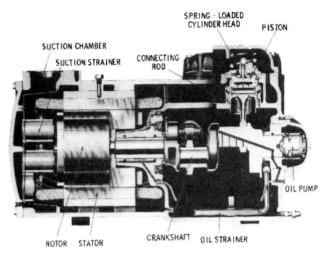

SPRING - LOADED
CYLINDER HEAD   PISTON

SUCTION CHAMBER

SUCTION STRAINER   CONNECTING
ROD

OIL PUMP

ROTOR   STATOR   CRANKSHAFT   OIL STRAINER

Trane Company

Fig. 3-9. Sectional view of a horizontal hermetic compressor.

quietness and a minimum of vibration. In addition, the seal and coupling (always a maintenance problem in open compressors) are eliminated. Still another dependable feature is the fact that the motor operates in an ideal atmosphere. Because of the fact that it is entirely enclosed, no airborne dust or dirt can reach it. Suction gas at 50° to 60°F cools the motor and shell. Together, the foregoing factors ensure long, troublefree motor operation.

A different type of compressor is shown in Fig. 3-10. This particular unit is internally spring-mounted. The motor, located above the compressor, operates in a vertical position, whereas the compressor is horizontal. This construction permits operation of the compressor in oil, simplifying the lubricating problem. The suction intake is placed so that the suction vapor must travel through the holes in the motor rotor in order to get to the top of the shell and then to the intake tube.

Certain models of this type of compressor are provided with internal thermostats, which are inserted in the motor windings and therefore measure motor temperature exactly, without allowing for the air gap between the motor and the top of the shell where overloads are normally located. This is particularly impor-

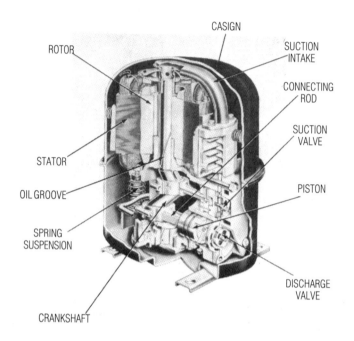

CASIGN

ROTOR

SUCTION INTAKE

CONNECTING ROD

SUCTION VALVE

STATOR

OIL GROOVE

PISTON

SPRING SUSPENSION

DISCHARGE VALVE

CRANKSHAFT

Tecumseh Products Company

Fig. 3-10. Sectional view of a vertical hermetic compressor.

tant in heat-pump applications where the ambient temperature may have considerable influence on the protection system. Because the thermostat is located at the most critical point, it gives instantaneous and accurate sensing of the motor temperature and therefore can remove the compressor from the line at a safe temperature level. It is always operative when the compressor is running.

Another important feature is an antislug device, consisting basically of two assemblies. One is the centrifuge, which is press fit on the crankshaft and therefore rotates at the speed of the compressor. The refrigerant is drawn in through the holes in the top. Any liquid or oil is expelled through the slots on the side by centrifugal force, and the gas (being lighter) is drawn through the slots in the hub. The second assembly collects the gas and directs it to the cylinder heads. This system always operates when the compressor is running and functions under all conditions that may affect slugging. It is not dependent upon any external component, which may fail, and so is practically foolproof.

CENTRIFUGAL COMPRESSORS

The centrifugal compressor is a relatively high-speed machine in which a continuous stream of gas is compressed by centrifugal force. The centrifugal compressor is, by its very nature, a large-volume machine, which means that a large amount of refrigeration can be handled within a relatively small floor area.

From the foregoing, it follows that the centrifugal compressor is well suited when a large amount of refrigeration is required but is not practical for *small* applications. To be specific in defining the term "small," it should be noted that, at the present time, the centrifugal machine has a practical lower-capacity limit of approximately 700 tons when using ammonia as a refrigerant. Eventually, as smaller, higher-speed compressors are designed, this lower limit will be reduced.

Centrifugal compressors are available in a single- or multistage arrangement to meet specific requirements in refrigeration applications. In cases of staged systems, however, a centrifugal compressor should be used as the low-stage unit together with a reciprocating high stage. In such a case, the centrifugal compressor could be used for capacities down to about 200 tons, depending on the temperature level. Centrifugal compressors have also been added as booster units for existing reciprocating systems to provide lower evaporating temperatures and increased capacity. A cutaway view of a centrifugal compressor is shown in Fig. 3-11.

Multistage centrifugal compressors are readily adapted to cycles involving intercooling, including interstage liquid-flash cooling. According to Carrier Corporation engineers, a line of centrifugal compressors designed to be universally applicable to all gases and vapors must take into account the chemical properties of these fluids. Ammonia and many other gases are peculiar in their reaction to certain metals. For instance, the copper and copper

Carrier Corporation

Fig. 3-11. Cutaway of an open centrifugal compressor with an electric-motor drive.

alloys that are used in Freon machines have no place in ammonia compressors. Therefore, all parts of the centrifugal compressor that come into direct contact with the ammonia are made of materials other than copper and copper alloys. Inter-stage and balancing piston labyrinths are made of aluminum, and the rotating part of the shaft labyrinth is "free-machining" stainless steel. The stationary part of the shaft labyrinth is cast iron with a pure lead lining. Because of its inert property, a lead coating is applied to the wheels, which are basically steel forgings. A special contact-type floating carbon-ring seal, from which all copper parts have been eliminated, is used on ammonia applications. This same seal is used for practically all other gases, except atmospheric air.

SYSTEM PRECAUTIONS

As with all machines, centrifugal or reciprocating, there are certain precautions to be observed during installation and initial operation. All piping should be carefully cleaned of rust, scale, sand, and other foreign materials. If water has been permitted to accumulate in the piping, it should be removed and the system dried out as thoroughly as possible. Piping should be properly designed so that there are no liquid traps that might cause flooding over into the compressor. A knockout drum in the suction piping at the compressor inlet is helpful in this regard.

Although a centrifugal compressor will take a slug of liquid with much less detrimental effect than will a reciprocating machine, it is well to keep in mind that the centrifugal type is *not* designed for this service. It is particularly important to note that only a little liquid entering the compressor suction will increase the weight flow tremendously. This can increase the horsepower requirement to the extent that it may cause the unit to bog down in the case of turbine drive or to trip out on overload in the case of motor drive. Appreciable amounts of liquid flowing into the suction of the compressor can cause excessive thrust, resulting in damage to the thrust bearing and possible damage to other internal parts if the rotor shifts axially as a result.

Piping should be designed so that strains due to weight and thermal changes are not imposed on the compressor and turbine equipment. Close shaft alignment is necessary on rotating equipment, which is impossible if piping forces are imposed on the units.

COMPRESSOR MAINTENANCE

Normal and correct operation can be observed and determined by the temperature and sound of the compressor. The normal compressor noise consists of a slight valve ticking. Normal compressor temperature at the crankcase should not be more than 25 to 30°F above the suction-inlet-gas temperature. The head of the machine and the discharge pipe should not be more than 30° warmer than the temperature corresponding to the discharge pressure. This discharge temperature may vary up or down with the temperature of the suction gas. The foregoing approximation is correct for most ordinary conditions. Heating of the compressor cylinder or of the heads higher than mentioned indicates trouble in the form of broken or leaky valves and should be investigated promptly.

Oil Level

The oil level should be a part of the daily observation by the operators. Once the system is charged and the proper oil return effected, normally there should be no cause for adding additional oil. If the compressor keeps losing oil, then look for a trap in the suction piping or a place of high lift through a vertical pipe from an evaporator from which the oil is not returning. Also check for leaks. Be sure that the oil level is at least halfway up on the compressor-gage glass but not over the top. A higher oil level is harmful because it will cause an oil bump in the compressor.

Oil Pressure

When the compressor is used on booster service at low temperatures and is handling either halocarbons or ammonia, the oil-pressure gage should read at least 10 lb above atmospheric pressure. To set the pressure, close the suction-gas stop valve until a vacuum of 27 inches is maintained while the compressor is running, then adjust the oil-relief valve at the seal end but keep this valve open at least five turns. If necessary, also adjust the relief valve above the pump until a pressure of 10 lb is obtained.

Suction Strainer

The compressor will have either one or two suction traps, and inside each one is a metal strainer and cloth filter. When installed, the filter bag must be pushed into the strainer screen until it makes contact with the bottom at the closed or spring end, and the top must then be folded back over the edge of the open end. The filter and screen are then inserted in the suction trap with the closed end next to the flanged cover, which provides the spring tension for holding the screen and filter in place. The hem on the filter at the open end will prevent it from being pulled through by the suction gas when properly positioned in the bottom groove of the trap.

Valves

The plate valves inside the machine will very rarely require grinding. A visual inspection of these valves is recommended about once every six months. The stop valves are usually double-seated and can be repacked under pressure by opening the valve wide until the back of the valve button sets against the bonnet. Use first-grade braided packing.

Compressor Care

Compressors, especially those connected to an iron or steel pipe system, should be opened and thoroughly cleaned every two weeks of operation. To pump out the crankcase before opening a machine, the suction stop valve B (Fig. 3-12) is closed, the discharge valve A is kept open, and a vacuum or near vacuum is pumped on the compressor crankcase. Usually, running it several minutes will remove all of the refrigerant mixed with the oil from the crankcase. Stop the compressor, and close valve A; the oil-filling plug is removed next to relieve the pressure. Open slowly, making sure the pressure is first relieved. The crankcase cover plate can then be removed and the drain plug opened to drain all remaining oil. The inside of the compressor should be thoroughly cleaned, using lintfree rags.

After the crankcase has been cleaned, all of the old oil, if dirty, should be replaced with a fresh charge. It is also an excellent precaution to remove the cover plate from the scale trap and thoroughly clean it. Never use gasoline for cleaning purposes. It is recommended that the wiping rags be saturated with good clean oil such as that used in the compressor. A really thorough cleaning is recommended; halfway measures are never satisfactory and will only lead to eventual trouble.

At the time the machine is opened to be cleaned, it is well to check the nuts on the connecting-rod bolts to be sure that they are tight. Clean the oil screen in a new machine at the end of the first month of operation. Cleaning once a year will be sufficient thereafter. Before restarting the compressor, remove the oil plug in the discharge side of the cylinder block; then start the unit with the suction and discharge valves closed. The air in the compressor

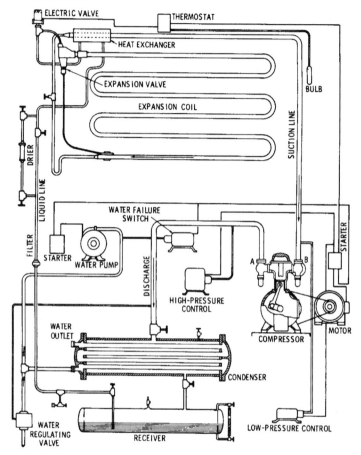

Fig. 3-12. Piping diagram of a typical refrigeration system.

will then be discharged. Just as the machine is shutting down, replace the plug, after which the compressor may be put in operation in the normal manner.

COMPRESSOR MAINTENANCE AND SERVICE

Numerous refrigeration units are of the open type with a belt-driven compressor mounted separately and in which the compressor bearings are provided with shaft seals. The following information will assist in servicing this type of unit.

Shaft Seal

The function of the seal in an open-type refrigerating compressor system is to prevent the gas from escaping the compressor at the point where the shaft leaves the compressor housing. In the early stages of the refrigeration industry, this was a difficult problem to solve. The shaft, of necessity, must revolve, and yet the refrigerant must not be allowed to escape when there is a pressure within the crankcase. Neither must air be allowed to enter when the pressure in the crankcase is below 0 lb pressure. The solution to this problem of crankshaft leaks was found in the development of a bellows-type seal, which is now an important part of most compressors.

The seal is an assembly of parts consisting of a seal ring fastened to a bellows, which, in turn, is fastened to a seal flange, as shown in Fig. 3-13. A spring surrounds the bellows, with one end resting against the seal ring and the other against the seal flange. The face of the seal ring is lapped flat and smooth. When assembled in the

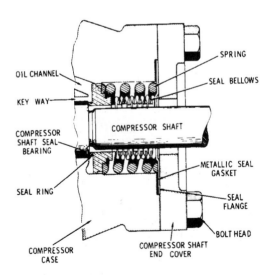

Fig. 3-13. Typical shaft-seal assembly.

111

compressor, the seal ring presses against a shoulder on the shaft that has also been lapped, forming a gas- and oil-tight joint at this point. The bellows provides the degree of flexibility necessary to keep the seal ring in perfect contact with the shaft. The seal flange is clamped against a gasket around an opening in the side of the compressor. The compressor shown in Fig. 3-2 is of the hermetic type and needs no shaft seal.

To determine if a compressor is leaking at the seal, proceed as follows: Close both shutoff valves by turning the stem to the right as far as possible. To ensure adequate refrigerant pressure in the compressor crankcase and on the seal bearing face, attach a refrigerant drum containing the correct refrigerant to the suction shutoff-valve outlet port. When making this connection, there should be a gage in the line from the drum to the compressor to accurately determine the pressure of the refrigerant in the compressor.

Test pressures for this purpose should be approximately 70 to 80 lb. If the compressor is located in a cool location, it may be necessary to raise the pressure in the drum by adding heat. In this process, care must be taken not to exceed 100 lb because pressure greater than this may damage the bellows assembly of the seal. With this pressure on the crankcase of the compressor, test for leaks with a halide torch, moving the finder tube close to the seal nut, crankshaft, seal-plate gasket, and point at which the seal comes in contact with the seal plate. If this process does not disclose a leak, turn the flywheel over slowly by hand, holding the finder tube close to the aforementioned parts.

After the leak has been detected, locate the exact place where it is leaking, if possible. If the leak is around the seal-plate gasket, replace the gasket. If the leak should be at the seal or nut, replace with a new seal plate, gasket, and seal assembly.

To replace the seal, observe the following instructions:

1. Remove the compressor from the condensing unit.
2. Remove the flywheel, using a puller. Leave the flywheel nut on the crankshaft so that the wheel puller will not distort the threads.
3. Remove the seal guard, seal nut, and seal assembly.
4. Remove the seal plate and gasket.

5. When assembling the seal, put a small quantity of clean compressor oil on the seal face plate and the seal.
6. To reassemble, reverse these operations, making sure the seal plate is bolted in place and the seal guard is at the top.

Compressor Efficiency Test

An efficiency test is a check on the relative amount of useful work that the compressor will accomplish. Strictly speaking, the efficiency of any machine is taken as the ratio of the power output to the power input in the same unit. This is usually written

$$\text{Efficiency} = \frac{\text{output}}{\text{input}}$$

The factors that determine the efficiency of a compressor are:

1. The degree to which the piston valve remains closed on the up stroke.
2. The degree to which the discharge valve remains tight when the piston is on the down stroke.

To test the compressor efficiency, proceed in the following manner:

1. Stop the compressor and install a compound gage on the suction-line valve and a pressure gage on the discharge-line valve.
2. Close the suction valve and operate the compressor until about 25 inches of vacuum is obtained on the compound gage; then stop the unit and note the gage readings. If the compressor will not pull a 25-inch vacuum or better, it is probable that air is leaking by the discharge valve and piston valve.
3. If the head pressure drops and the vacuum-gage reading remains practically constant, it is probable that there is an external leak at one of the points on the head of the compressor or at the gage and shutoff-valve connection.
4. To repair either a leaky suction valve or discharge valve, remove the head of the compressor and remove the discharge-valve plate carefully. In disassembling the discharge

valve, caution should be used so as not to disturb the actual conditions prevailing during the test. It is possible for some dirt, scale, or other foreign matter to get under the valve disk and on the seat, causing poor performance. If there is no evidence of dirt or foreign matter, check the seat on both the discharge valve and piston assembly for low spots or scratches. If these are found, replace the disks or the valve plate completely.

Generally the trouble can be found at either the suction valve or the discharge valve. Badly scored seats on either side of the discharge or suction valve require replacing with new assemblies. Removal of deep scores changes the valve lift, further endangering the efficiency. After the repairs are made, thoroughly clean the parts with gasoline or other solvent and reassemble them, using new gaskets. Repeat the efficiency test to ensure that the trouble has been eliminated.

Stuck or Tight Compressor

The reason for a compressor becoming *stuck* is usually a result of moisture in the system or lack of lubrication. When this occurs, the compressor should be thoroughly cleaned. The compressor should be completely disassembled and the parts thoroughly cleaned and refitted. New oil and refrigerant should be put into the cleaned system.

A tight compressor will result when a cylinder head, seal cover, or similar part has been removed and not replaced carefully or when the screws have been tightened unevenly. This will develop a misalignment, causing a bind in the moving parts, which may cause the compressor to become stuck.

Compressor Knocks

A knock in the compressor may be caused by a loose connecting rod, eccentric strap and rod, eccentric disk, piston pin, crankshaft, or too much oil in the system. A compressor knock can be determined by placing the point of a screwdriver against the crankcase and the ear against the handle. It will not be possible to determine what causes the knock until the compressor is disassembled.

Sometimes it may be possible to determine a looseness of the aforementioned parts without completely disassembling the compressor. First, remove the cylinder head and valve plate to expose the head of the piston. Now rotate the compressor by hand and press down on the top of the piston with the finger. Any looseness can be felt at each stroke of the piston. The loose part should be replaced.

It is always well to check the compressor-oil level before analyzing and determining compressor repairs. Oil knocks are usually caused by adding too much oil in the servicing of the unit. It should never be necessary to add oil to a system unless there has been a leakage of oil. A low charge is sometimes diagnosed as a lack of oil. Always make sure that a low oil level is actually due to lack of oil rather than to a low charge before adding oil.

Lubrication

In conventional reciprocating compressors the lubrication is accomplished by the so-called splash system. Special *dippers*, or *slingers*, fastened to the crankshaft distribute the crankcase oil to the pistons, pin bearings, cylinder walls, crankshaft bearings, and seal. Perhaps the most important point for the serviceman to remember in connection with compressor service is to check the amount of oil in an open-type compressor. This is accomplished as follows.

After the unit has been operated for several minutes, attach a compound gage to the suction service valve and close the valve. The unit should be pumped down to balance the pressure (zero on the gage), and then stopped. Allow a few minutes before removing the oil-filter plug, and carefully measure the oil level. This level should be checked with the manufacturer's instructions concerning the oil level required for the unit in question. Only oil specifically recommended by the manufacturer for the particular unit and refrigerant should be used when adding oil.

COMPRESSOR REMOVAL

To remove a compressor from the unit, proceed as follows. Attach a compound gage to the suction-line valve. Close the

115

suction-line shutoff valve and run the compressor until 20 to 25 inches of vacuum are obtained; then close the discharge shutoff valve. Before removing any fittings, crack the suction-line valve to bring the gage reading back to zero. Before removing service valves, loosen the pressure gage and relieve the pressure in the head of the compressor. Remove the cap screws holding the suction and discharge valves to the compressor. If the compressor is to be taken away from the premises for repairs, place service valves over both the discharge and suction-line openings. This prevents air and moisture from entering and oil from leaking out. Loosen the nuts that hold the compressor to the base, and bend the tubing away from the compressor just enough to permit the assembly to be lifted out. Care should be taken not to loosen the mounting pads and washers on which the compressor rests.

COMPRESSOR REPLACEMENT

Installation of a compressor is roughly a reversal of the process of removal. First, place the mounting pads and washers in position. Place the compressor carefully on the base in the same position it occupied before removal. Bolt the compressor in place. Replace the suction and discharge valves, using new gaskets. After the compressor and valves are bolted in place, the compressor must be evacuated to remove the air.

BELT ALIGNMENT

The importance of proper belt alignment in open units must be kept constantly in mind when service necessitates changing a compressor, motor, or motor pulley. To determine the correct belt alignment and tension, proceed as follows.

The correct belt tension for a V-type belt is to have the tension so adjusted that it is possible to depress or raise the belt ½ inch from its original position with the fingers without undue pressure. With the compressor and motor in their respective positions and in line as nearly correct as possible, loosen the setscrew on the motor pulley so that the pulley turns freely on the motor shaft. On older motors it may be necessary to dress down the motor shaft with emery cloth. A little oil should be added to ensure that the motor pulley turns freely.

At this point, attach the belt with the proper tension; turn the flywheel forward several times and then backward several times. The point at which the belt is in correct alignment will be that position where there is no in or out travel of the motor pulley on the motor shaft, whether the flywheel is turned forward or backward.

The movement of the motor pulley outward when the flywheel is turned forward, or inward when the flywheel is turned backward (or vice versa), indicates that the motor pulley is out of plane with the flywheel; the motor must be readjusted so that this condition does not exist. Close observance of the aforementioned procedure will greatly increase the length of time the belt will remain in a serviceable condition.

RUNNING-TIME CHECK

The percentage of running, or operating, time of the unit, assuming it functions normally, depends upon the amount of work being done. The amount of work needed depends upon the size of the refrigerator cabinet to be cooled, the temperature of the air that surrounds it, and the amount of heat that must be extracted from it. The amount of heat that leaks into the refrigerator depends upon the temperature of the room, the frequency and length of time of the door openings, and the amount and temperature of the food placed in the refrigerator. Climatic conditions also affect the operations of a refrigerator. Thus, a refrigerator will operate more efficiently in a dry climate than in a climate of high humidity.

Keeping the aforementioned factors in mind, the length of the "on" and "off" period on the average well-adjusted refrigerator should be somewhat as follows:

At 75°F room temperature, the "on" period should be around 2 to 4 minutes and the "off" period around 10 to 15 minutes. In many cases, where a complaint is received that the refrigerator is operating too much, a check may very easily be made by connecting a self-starting electric clock to the motor terminals. It is evident that since the clock will operate only when the motor operates, the total running time of the compressor will be registered.

COMPRESSION RATIO

The compression ratio in a refrigeration system is defined as the absolute head pressure divided by the absolute suction pressure:

$$\text{Compression ratio} = \frac{\text{absolute head pressure}}{\text{absolute suction pressure}}$$

It should be noted that the ordinary compound gage does not register atmospheric pressure but reads zero when not connected to a pressurized system. To obtain the absolute head pressure or absolute suction pressure at zero gage or above, 15 lb must be added to the gage reading.

Example—At zero gage or above, we have

$$\text{Absolute head pressure} = \text{gage reading} + 15 \text{ lb}$$

$$\text{Absolute suction pressure} = \text{gage reading} + 15 \text{ lb}$$

Example—When the low side is reading in the vacuum range, we have

$$\text{Absolute head pressure} = \text{gage reading} + 15 \text{ lb}$$

$$\text{Absolute suction pressure} = \frac{30 - \text{gage reading (in inches)}}{2}$$

The calculation of compression ratio can be shown by the following examples:

$$\text{Head pressure} = 160 \text{ lb}$$

$$\text{Suction pressure} = 10 \text{ lb}$$

$$\text{Compression ratio} = \frac{\text{absolute head pressure}}{\text{absolute suction pressure}}$$

$$= \frac{160 + 15}{10 + 15}$$

$$= \frac{175}{25}$$

$$= 7{:}1$$

Example—

$$\begin{aligned}
\text{Head pressure} &= 160\,\text{lb} \\
\text{Suction pressure} &= 10 \text{ inches of vacuum} \\
\text{Absolute head pressure} &= 160 + 15 = 175\,\text{lb} \\
\text{Absolute suction pressure} &= \frac{30 - 10}{2} = \frac{20}{2} = 10 \\
\text{Compression ratio} &= \frac{\text{absolute head pressure}}{\text{absolute suction pressure}} \\
&= \frac{175}{10} \\
&= 17.5{:}1
\end{aligned}$$

ENERGY CONSUMPTION

The electrical energy (usually referred to as the *power consumption*) used by a refrigerator depends on all the factors that influence the "on" and "off" period, in addition to the size of the motor employed. Motors used generally range in size from ¼ to about ¾ hp, depending upon the size of the unit, type of refrigerant employed, and various other factors.

Experience has shown that in the northern half of the United States, a good average range for the summer months is 25 to 30 kWh/month and 20 kWh for other months. In the southern half of the country, these figures (especially for the summer months) will be somewhat higher.

Newer models are energy efficient and have the factors relating to power consumption listed on a tag attached to all new refrigerators.

REFRIGERATION-SYSTEM LUBRICANTS

The necessity of providing lubrication to prevent frictional contact on metal surfaces is universally recognized. Correct lubrication of refrigeration machinery demands special consideration; the employment of incorrect oil or the excessive use of even the most suitable oil will create objectionable operating conditions. An excessively heavy oil will not be distributed properly to the working parts and therefore will fail to do the required job. An excessively light oil will not cling and will not satisfactorily seal the pistons and rings from the bypass of the gas back into the suction port and crankcase. Because of this fact, selection of the proper oil should be made with care. It must be viscous enough to give proper lubrication. In a rotary compressor, the oil acts as a lateral seal and efficiency depends on its proper functioning. Due to superheating at the discharge valves, the oil must be able to withstand high temperatures.

Reputable refiners who have assisted in lubrication research are now able to supply proper oils and, from experience, can advise a potential user which oil will best serve his purpose. The soundest advice on lubrication, therefore, is to use the oil that experience has shown to be the best. This information generally can be secured from the manufacturer of the machine or a refiner actively engaged in supplying oils for refrigeration work.

Lubrication of Motor and Fan Bearings

Lubrication of the motor and fan bearings usually can be accomplished by feeding the oil through a piece of wicking or felt from a reservoir directly to the bearings in continuous small quantities. It is necessary, therefore, to use an oil that is light enough in body to pass though the wick but heavy enough to form a proper film on the bearing surface at operating temperatures. It must also have a low "pour point" so that it will flow freely to the bearings when starting cold. The best oil will resist deterioration due to oxidation and polymerization, which cause stickiness and decreased lubrication. The use of properly refined oil will eliminate most of these difficulties. When one considers the small amount required and the costly damage that may result from using an inferior oil, it can be seen that the best oil is the cheapest over a long period of time.

Compressor Lubrication

There are so many types of refrigerating compressors that it is impossible to specifically determine which lubricant fits all requirements. Even when compressors and machines are quite similar in design, they are apt to require individual consideration for proper lubrication.

Speed, clearance, temperature conditions, etc., have an important bearing and the system of operation and the refrigerant used are the determining factors. Most oils are chosen for a given job as a result of actual tests. Therefore, servicemen in the field will always profit by using the oil specified by the manufacturer of the unit being serviced or an oil known to answer the same specifications. Buy oils from reputable refiners only, preferably those who have had real experience in refrigeration problems.

Types of Lubrication

Compressors are usually lubricated by splash, although a semi-forced feed is occasionally used; in certain types a separate oil pump is required. There are two systems of oil circulation—one in which the oil circulates through the low side, and the other in which it does not.

As the temperature of the gas discharged from the compressor rises, the vapor pressure of the oil carried with it will also rise and the amount of oil that can be separated will be larger. In addition, the quantity of oil carried over can be increased by the mechanical action in the crankcase of the compressor, and foaming can be induced by the release of dissolved refrigerant during the suction stroke of the piston. This latter difficulty may be quite pronounced in a machine charged with methyl chloride. If the gas and oil from the compressor are discharged into a chamber, baffles or centrifugal force can be employed to separate the oil from the gas and return it to the intake of the machine.

This system is very satisfactory when employing refrigerants that dissolve oil or in a dry system where the velocity of the gas carries the oil back to the compressor. In commercial installations, the oil usually circulates through the low-pressure side. In a dry system, the high velocity of the gas keeps the oil moving. In a flooded system, the evaporation of the refrigerant leaves the oil behind, and special arrangements are used for its return.

If sufficient care has not been exercised in removing all impurities from the oil, emulsification in the liquid refrigerant can result. The effect would be the same as excessive solubility since it can change the liquid level of the refrigerant in the evaporator.

121

It can also result in decreased free flow of the refrigerant and prevent proper boiling in the evaporator, which causes a temperature that disagrees considerably with the corresponding pressure.

COMPRESSOR CALCULATIONS

Mechanical Efficiency—Mechanical efficiency is defined as the ratio of the work to the gas (as may be obtained from an indicator diagram) to the work delivered to the compressor drive shaft.

Compression Ratio—Compression ratio is defined as the ratio of the absolute head pressure to the absolute suction pressure:

$$\text{Compression ratio} \ = \ \frac{\text{absolute head pressure}}{\text{absolute suction pressure}}$$

Volumetric Efficiency—Volumetric efficiency is defined as the volume of fresh vapor entering the cylinder per stroke divided by the piston displacement.

Compressor Capacity—Compressor capacity is the refrigeration effect that can be accomplished by a compressor and is usually measured in Btu. By definition, compressor capacity is equal to the difference in total heat content between the refrigerant liquid at a temperature corresponding to the pressure of the vapor leading to the compressor and the refrigerant vapor entering the compressor.

Performance Factor—The performance factor of a hermetic compressor is the ratio between the capacity in Btu and the power input in watts. It is written:

$$\text{Performance factor} \ = \ \frac{\text{capacity, Btu}}{\text{power input , watts}}$$

Brake Horsepower—The brake horsepower is a function of the power input of an ideal compressor and its compression, the mechanical efficiency, and volumetric efficiency.

As noted in the foregoing, the volumetric efficiency or displacement efficiency of a compressor is the ratio of the volume of the refrigerant actually removed reduced to the temperature and pressure conditions of the evaporator to the volume of the piston displacement. Several factors affect the volumetric efficiency: (1) the volume of the vapor is greater after it enters the cylinder, as it takes up heat from the valves and walls to become superheated; (2) pressure in the cylinder is lower than in the evaporator because a lower pressure is required to cause a flow that will increase the volume. Again, reexpansion of the gases left in the clearance spaces between the piston, cylinder head, and valves will partly fill the suction chamber so that a full charge of fresh vapor cannot be drawn in from the evaporator. The effect of the superheating and leakage cannot be calculated and must be determined by experiment, but the clearance effect can be calculated when the clearance ratio is known. It is only practical to determine the volumetric efficiency by a calorimeter test, which is the accepted method used by practically all manufacturers today.

Theoretical Compressor Capacities

Example—A certain condensing unit is equipped with a two-cylinder compressor having a 2½-inch bore by 3½-inch stroke, and is operating at 500 rpm. Calculate:

1. Capacity (in Btu/hr) when using Freon-12 as the refrigerant, operating at 125 lb head pressure and 35 lb suction pressure.
2. Capacity (in Btu/hr) when using Freon-12 as the refrigerant, operating at 125 lb head pressure and 11 lb suction pressure.

Solution—The cubic inch displacement per revolution will be obtained as follows:

$$\text{Displacement} = \frac{\pi D^2}{4} = \times 2 \times L$$

$$= \frac{3.1416 \times 2.5^2 \times 2 \times 3.5}{4}$$

$$= 34.4 \text{ cu in.}$$

Accordingly, the cubic feet displacement per hour at 500 rpm will be:

$$\frac{34.4 \times 500 \times 60}{1728} = 597 \text{ cu ft / hr}$$

1. From the Freon-12 table (Table 3-1), column v_g at 35 lb suction pressure, the specific volume of vapor equals 0.819 cu ft/lb (ft/lb³). Thus,

$$\frac{597}{0.819} = 729 \text{ lb Freon-12/hr}$$

From Table 3-1, we obtain further:

Total heat h_g of vapor/lb at 35 lb = 82.49 Btu
Total h_f liquid/lb at 125 lb = 32.15 Btu
Net refrigeration effect = 82.49 – 32.15
= 50.34Btu

From the foregoing, it follows that the theoretical capacity is:

$$729 \times 50.34 = 36,700 \text{ Btu / hr (approximately)}$$

2. From Table 3-1, column y_g, the specific volume at 11 lb suction pressure equals 1.514 cu ft/lb. Therefore,

$$\frac{597}{1.514} = 394 \text{ lb Freon-12/hr}$$

The total heat h_g of vapor at 11 lb (from Table 3-1) is 8.67 Btu/lb.

Total h_f liquid/lb at 125 lb = 32.15 Btu

Solution—

Net refrigeration effect = 78.67 – 32.15
46.52 Btu/lb

that is, 394×46.52 = 18,330 Btu/hr
(approximately)

If the required refrigeration (in Btu/hr) is known, the size and characteristics of the compressor can be determined as follows: First determine the refrigerating effect (RE) per pound of refrigerant under standard conditions.

Table 3-1. Freon-12 Properties of Saturated Vapors

Temp.	Pressure		Volume		Density		Heat Content from −40°			Entropy from −40°		Temp.
°F	Abs. lb/in²	Gage lb/in³	Liquid ft³/lb	Vapor ft³/lb	Liquid lb/ft³	Vapor lb/ft³	Liquid Btu/lb	Latent Btu/lb	Vapor Btu/lb	Liquid Btu/lb/°F	Vapor Btu/lb/°F	°F
t	p	p_d	v_f	v_g	$1/v_f$	$1/v_g$	h_f	h	h_g	s_f	s_g	t
−40	9.32	10.92*	0.0106	3.911	94.58	0.2557	0.00	73.50	73.50	0.00000	0.17517	−40
−38	9.82	9.91*	0.0106	3.727	94.39	0.2683	0.40	73.34	73.74	0.00094	0.17490	−38
−36	10.34	8.87*	0.0106	3.553	94.20	0.2815	0.81	73.17	73.98	0.00188	0.17463	−36
−34	10.87	7.80*	0.0106	3.389	93.99	0.2951	1.21	73.01	74.22	0.00282	0.17438	−34
−32	11.43	6.66*	0.0107	3.234	93.79	0.3092	1.62	72.84	74.46	0.00376	0.17412	−32
−30	12.02	5.45*	0.0107	3.088	93.59	0.3238	2.03	72.67	74.70	0.00471	0.17387	−30
−28	12.62	4.23*	0.0107	2.950	93.39	0.3390	2.44	72.50	74.94	0.00565	0.17364	−28
−26	13.26	2.93*	0.0107	2.820	93.18	0.3546	2.85	72.33	75.18	0.00659	0.17340	−26
−24	13.90	1.63*	0.0108	2.698	92.98	0.3706	3.25	72.16	75.41	0.00753	0.17317	−24
−22	14.58	0.24*	0.0108	2.583	92.78	0.3871	3.66	71.98	75.64	0.00846	0.17296	−22
−20	15.28	0.58	0.0108	2.474	92.58	0.4042	4.07	71.80	75.87	0.00940	0.17275	−20
−18	16.01	1.31	0.0108	2.370	92.38	0.4219	4.48	71.63	76.11	0.01033	0.17253	−18
−16	16.77	2.07	0.0108	2.271	92.18	0.4403	4.89	71.45	76.34	0.01126	0.17232	−16
−14	17.55	2.85	0.0109	2.177	91.97	0.4593	5.30	71.27	76.57	0.01218	0.17212	−14
−12	18.37	3.67	0.0109	2.088	91.77	0.4789	5.72	71.09	76.81	0.01310	0.17194	−12
−10	19.20	4.50	0.0109	2.003	91.57	0.4993	6.14	70.91	77.05	0.01403	0.17175	−10
−8	20.08	5.38	0.0109	1.922	91.35	0.5203	6.57	70.72	77.29	0.01496	0.17158	−8
−6	20.98	6.28	0.0110	1.845	91.14	0.5420	6.99	70.53	77.52	0.01589	0.17140	−6
−4	21.91	7.21	0.0110	1.772	90.93	0.5644	7.41	70.34	77.75	0.01682	0.17123	−4
−2	22.87	8.17	0.0110	1.703	90.72	0.5872	7.83	70.15	77.98	0.01775	0.17107	−2
0	23.87	9.17	0.0110	1.637	90.52	0.6109	8.25	69.96	78.21	0.01869	0.17091	0
2	24.89	10.19	0.0110	1.574	90.31	0.6352	8.67	69.77	78.44	0.01961	0.17075	2
4	25.96	11.26	0.0111	1.514	90.11	0.6606	9.10	69.57	78.67	0.02052	0.17060	4
5†	26.51	11.81	0.0111	1.485	90.00	0.6735	9.32	69.47	78.79	0.02097	0.17052	5†

*Inches of mercury below one atmosphere.
†Standard ton temperatures.

Table 3-1. Freon-12 Properties of Saturated Vapors (Cont'd)

Temp. °F t	Pressure Abs. lb/in² p	Pressure Gage lb/in³ p_d	Volume Liquid ft³/lb v_f	Volume Vapor ft³/lb v_g	Density Liquid lb/ft³ $1/v_f$	Density Vapor lb/ft³ $1/v_g$	Heat Content from −40° Liquid Btu/lb h_f	Heat Content from −40° Latent Btu/lb h	Heat Content from −40° Vapor Btu/lb h_g	Entropy from −40° Liquid Btu/lb/°F s_f	Entropy from −40° Vapor Btu/lb/°F s_g	Temp. °F t
6	27.05	12.35	0.0111	1.457	89.88	0.6864	9.53	69.37	78.90	0.02143	0.17045	6
8	28.18	13.48	0.0111	1.403	89.68	0.7129	9.96	69.17	79.13	0.02235	0.17030	8
10	29.35	14.65	0.0112	1.351	89.45	0.7402	10.39	68.97	79.36	0.02328	0.17015	10
12	30.56	15.86	0.0112	1.301	89.24	0.7687	10.82	68.77	79.59	0.02419	0.17001	12
14	31.80	17.10	0.0112	1.253	89.03	0.7981	11.26	68.56	79.82	0.02510	0.16987	14
16	33.08	18.38	0.0112	1.207	88.81	0.8288	11.70	68.35	80.05	0.02601	0.16974	16
18	34.40	19.70	0.0113	1.163	88.58	0.8598	12.12	68.15	80.27	0.02692	0.16961	18
20	35.75	21.05	0.0113	1.121	88.37	0.8921	12.55	67.94	80.49	0.02783	0.16949	20
22	37.15	22.45	0.0113	1.081	88.13	0.9251	13.00	67.72	80.72	0.02873	0.16938	22
24	38.58	23.88	0.0113	1.043	87.91	0.9588	13.44	67.51	80.95	0.02963	0.16926	24
26	40.07	25.37	0.0114	1.007	87.68	0.9930	13.88	67.29	81.17	0.03053	0.16913	26
28	41.59	26.89	0.0114	0.973	87.47	1.028	14.32	67.07	81.39	0.03143	0.16900	28
30	43.16	28.46	0.0115	0.939	87.24	1.065	14.76	66.85	81.61	0.03233	0.16887	30
32	44.77	30.07	0.0115	0.908	87.02	1.102	15.21	66.62	81.83	0.03323	0.16876	32
34	46.42	31.72	0.0115	0.877	86.78	1.140	15.65	66.40	82.05	0.03413	0.16865	34
36	48.13	33.43	0.0116	0.848	86.55	1.180	16.10	66.17	82.27	0.03502	0.16854	36
38	49.88	35.18	0.0116	0.819	86.33	1.221	16.55	65.94	82.49	0.03591	0.16843	38
40	51.68	36.98	0.0116	0.792	86.10	1.263	17.00	65.71	82.71	0.03680	0.16833	40
42	53.51	38.81	0.0116	0.767	85.88	1.304	17.46	65.47	82.93	0.03770	0.16823	42
44	55.40	40.70	0.0117	0.742	85.66	1.349	17.91	65.24	83.15	0.03859	0.16813	44
46	57.35	42.65	0.0117	0.718	85.43	1.393	18.36	65.00	83.36	0.03948	0.16803	46
48	59.35	44.65	0.0117	0.695	85.19	1.438	18.82	64.74	83.57	0.04037	0.16794	48
50	61.39	46.69	0.0118	0.673	84.94	1.485	19.27	64.51	83.78	0.04126	0.16785	50
52	63.49	48.79	0.0118	0.652	84.71	1.534	19.72	64.27	83.99	0.04215	0.16776	52

Table 3-1. Freon-12 Properties of Saturated Vapors (Cont'd)

Temp.	Pressure		Volume		Density		Heat Content from −40°			Entropy from −40°		Temp.
°F t	Abs. lb/in² p	Gage lb/in² p_d	Liquid ft³/lb v_f	Vapor ft³/lb v_g	Liquid lb/ft³ $1/v_f$	Vapor lb/ft³ $1/v_g$	Liquid Btu/lb h_f	Latent Btu/lb h	Vapor Btu/lb h_g	Liquid Btu/lb/°F s_f	Vapor Btu/lb/°F s_g	°F t
54	65.63	50.93	0.0118	0.632	84.50	1.583	20.18	64.02	84.20	0.04304	0.16767	54
56	67.84	53.14	0.0119	0.612	84.28	1.633	20.64	63.77	84.41	0.04392	0.16758	56
58	70.10	55.40	0.0119	0.593	84.04	1.686	21.11	63.51	84.62	0.04480	0.16749	58
60	72.41	57.71	0.0119	0.575	83.78	1.740	21.57	63.25	84.82	0.04568	0.16741	60
62	74.77	60.07	0.0120	0.557	83.57	1.795	22.03	62.99	85.02	0.04657	0.16733	62
64	77.20	62.50	0.0120	0.540	83.34	1.851	22.49	62.73	85.22	0.04745	0.16725	64
66	79.67	64.97	0.0120	0.524	83.10	1.909	22.95	62.47	85.42	0.04833	0.16717	66
68	82.24	67.54	0.0121	0.508	82.86	1.968	23.42	62.20	85.62	0.04921	0.16709	68
70	84.82	70.12	0.0121	0.493	82.60	2.028	23.90	61.92	85.82	0.05009	0.16701	70
72	87.50	72.80	0.0121	0.479	82.37	2.090	24.37	61.65	86.02	0.05097	0.16693	72
74	90.20	75.50	0.0122	0.464	82.12	2.153	24.84	61.38	86.22	0.05185	0.16685	74
76	93.00	78.30	0.0122	0.451	81.87	2.218	25.32	61.10	86.42	0.05272	0.16677	76
78	95.85	81.15	0.0123	0.438	81.62	2.284	25.80	60.81	86.61	0.05359	0.16669	78
80	98.76	84.06	0.0123	0.425	81.39	2.353	26.28	60.52	86.80	0.05446	0.16662	80
82	101.70	87.00	0.0123	0.413	81.12	2.423	26.76	60.23	86.99	0.05534	0.16655	82
84	104.80	90.10	0.0124	0.401	80.87	2.495	27.24	59.94	87.18	0.05621	0.16648	84
86†	107.90	93.20	0.0124	0.389	80.63	2.569	27.72	59.65	87.37	0.05708	0.16640	86
88	111.10	96.40	0.0124	0.378	80.37	2.645	28.21	59.35	87.56	0.05795	0.16632	88†
90	114.30	99.60	0.0125	0.368	80.11	2.721	28.70	59.04	87.74	0.05882	0.16624	90
92	117.70	103.00	0.0125	0.357	79.86	2.799	29.19	58.73	87.92	0.05969	0.16616	92
94	121.00	106.30	0.0126	0.347	79.60	2.880	29.68	58.42	88.10	0.06056	0.16608	94
96	124.50	109.80	0.0126	0.338	79.32	2.963	30.18	58.10	88.28	0.06143	0.16600	96
98	128.00	113.30	0.0126	0.328	79.06	3.048	30.67	57.78	88.45	0.06230	0.16592	98

†Standard ton temperatures.

Table 3-1. Freon-12 Properties of Saturated Vapors (Cont'd)

Temp.	Pressure		Volume		Density		Heat Content from -40°			Entropy from -40°		Temp.
°F t	Abs. lb/in² p	Gage lb/in³ p_d	Liquid ft³/lb v_f	Vapor ft³/lb v_g	Liquid lb/ft³ $1/v_f$	Vapor lb/ft³ $1/v_g$	Liquid Btu/lb h_f	Latent Btu/lb h	Vapor Btu/lb h_g	Liquid Btu/lb/°F s_f	Vapor Btu/lb/°F s_g	°F t
100	131.6	116.9	0.0127	0.319	78.80	3.135	31.16	57.46	88.62	0.06316	0.16584	100
102	135.3	120.6	0.0127	0.310	78.54	3.224	31.65	57.14	88.79	0.06403	0.16576	102
104	139.0	124.3	0.0128	0.302	78.27	3.316	32.15	56.80	88.95	0.06490	0.16568	104
106	142.8	128.1	0.0128	0.293	78.00	3.411	32.65	56.46	89.11	0.06577	0.16560	106
108	146.8	132.1	0.0129	0.285	77.73	3.509	33.15	56.12	89.27	0.06663	0.16551	108
110	150.7	136.0	0.0129	0.277	77.46	3.610	33.65	55.78	89.43	0.06749	0.16542	110
112	154.8	140.1	0.0130	0.269	77.18	3.714	34.15	55.43	89.58	0.06836	0.16533	112
114	158.9	144.2	0.0130	0.262	76.89	3.823	34.65	55.08	89.73	0.06922	0.16524	114
116	163.1	148.4	0.0131	0.254	76.60	3.934	35.15	54.72	89.87	0.07008	0.16515	116
118	167.4	152.7	0.0131	0.247	76.32	4.049	35.65	54.36	90.01	0.07094	0.16505	118
120	171.8	157.1	0.0132	0.240	76.02	4.167	36.16	53.99	90.15	0.07180	0.16495	120
122	176.2	161.5	0.0132	0.233	75.72	4.288	36.66	53.62	90.28	0.07266	0.16484	122
124	180.8	166.1	0.0133	0.227	75.40	4.413	37.16	53.24	90.40	0.07352	0.16473	124
126	185.4	170.7	0.0133	0.220	75.10	4.541	37.67	52.85	90.52	0.07437	0.16462	126
128	190.1	175.4	0.0134	0.214	74.78	4.673	38.18	52.46	90.64	0.07522	0.16450	128
130	194.9	180.2	0.0134	0.208	74.46	4.808	38.69	52.07	90.76	0.07607	0.16438	130
132	199.8	185.1	0.0135	0.202	74.14	4.948	39.19	51.67	90.86	0.07691	0.16425	132
134	204.8	190.1	0.0135	0.196	73.81	5.094	39.70	51.26	90.96	0.07775	0.16411	134
136	209.9	195.2	0.0136	0.191	73.46	5.247	40.21	50.85	91.06	0.07858	0.16396	136
138	215.0	200.3	0.0137	0.185	73.10	5.405	40.72	50.43	91.15	0.07941	0.16380	138
140	220.2	205.5	0.0138	0.180	72.73	5.571	41.24	50.00	91.24	0.08024	0.16363	140

Example—Determine the net refrigerating effect for Freon-12 refrigerant under standard ton conditions, 5°F evaporator, 86°F liquid in receiver.

Solution—With reference to Table 3-1 we obtain:

$$\text{Column } h_f = \text{latent heat of evaporation}$$
$$\text{at } 5°F$$
$$= 69.47 \text{ Btu/lb}$$

Heat of liquid at 86°F, column h_1 = 27.72 Btu/lb
Heat of liquid at 5°F, column h_1 = 9.32 Btu/lb

Therefore, it requires 27.72 – 9.32 = 18.40 Btu/lb to cool 86°F liquid to 5°F liquid. Note that 9.32 Btu of heat of the liquid at 5°F remain, as the liquid is not cooled below this point. The net refrigerating effect is, therefore, 69.47 – 18.40 = 51.07 Btu/lb (approximately).

SAFETY PRECAUTIONS

The compressor is largely air-cooled. It must not be used as an air compressor because the temperatures of compression of air are extremely high and proper cooling cannot be obtained. A first ruler of operation is that the compressor is designed as a gas pump. It must never be permitted to pump liquid refrigerant or oil. Oil lifting and oil pumping will occur if the compressor is operated with suction valve B closed (Fig. 3-12). Liquid refrigerant getting back to the compressor causes hammering and damage to the valves. It reduces compressor efficiency and will cause oil to pump out of the crankcase. Liquid pumping is usually caused by a defective or improperly regulated thermal expansion valve. It can come about through a leaky or improperly seating electric valve, which permits refrigerant flow during a shutdown period.

If the compressor pounds due to lifting oil or liquid coming back, try to correct the condition rather than permit the pounding to continue. Liquid coming back to the compressor can be checked by momentarily throttling the compressor suction stop valve B. The suction gas should be superheated and free of liquid.

A heat exchanger may be needed in the suction line. **Never use a torch or attempt a repair on a line containing refrigerant. Never tamper with or attempt readjustment of a factory-set safety-relief valve. Never operate without the oil showing at least half full or higher in the compressor-gage glass. Never bridge an overload or any protective device because it kicks out in operation.** Find the trouble and make the proper repair. Never permit the brine in an ice-making tank or the water in a water cooler to freeze. If operating at pressure below the freezing point of the brine or water in this tank, shut down for a while until there is a heavier load. If, at any time, the plant is not in service and temperatures are below the freezing point, be sure to drain the water from the condensers, pump, compressor-head jacketed control plate, seal, and connecting piping.

Halocarbon Refrigerant

Since halocarbons are practically odorless and nontoxic, it will not be necessary to wear a gas mask when servicing the equipment unless there is a high concentration of refrigerant or an open flame or spark. *However, it is very essential that proper protection be afforded the eyes by the use of goggles or large-lensed spectacles to eliminate the possibility of liquid refrigerant coming in contact with the eyes and causing possible injury by freezing the moisture in them.* This protection is necessary and should be taken whenever loosening a connection in which refrigerant is confined.

SUMMARY

Compressors can be classified as single-cylinder or multicylinder. They may also be called reciprocating, rotary or centrifugal. They may use either a direct drive or belt drive for their power.

The passage of the refrigerant to and from the compressor is controlled by means of a discharge and suction valve located on a specially designed valve plate. The flapping action of the valve permits the flow of refrigerant out only through the discharge valve port and flow in only through the suction valve port.

Reciprocating compressors derive their name from the reciprocating or back and forth action of the piston. Reciprocating compressors usually use a force-feed lubrication system. The function of the shaft seal in a refrigeration compressor is to prevent gas from escaping from the compressor at the point where the shaft leaves the compressor housing.

Pistons are usually made of cast iron and polished to fit precisely the cylinder in which they are placed. Piston rings are used to make sure the gas does not escape during compression. Compressor connecting rods are similar to those found in an automobile engine. Some compressors are equipped with what is known as an eccentric shaft which is a different application of the crankshaft principle.

Valves in a compressor direct the flow of refrigerants through the unit.

The rotary compressor compresses gas by the movement of the blades in relation to a pump chamber.

Hermetic compressors are sealed and directly connected to an electric motor. The motor and compressor operate on the same shaft and are enclosed in a common casing.

Centrifugal compressors are relatively high-speed machines that have a continuous stream of gas compressed by centrifugal force. They are large volume machines that handle large requirements for refrigeration. They are available in single or multistage arrangements.

Compressor maintenance requires both sight and sound observations. Oil level, oil pressure and other indicators of proper or improper operation should be checked frequently.

One of the biggest problems with compressors is the leakage of refrigerant since the shaft seal does not always operate properly over long periods of time.

Compressor efficiency can be found by taking the output and dividing it by the input.

A compressor can stick because of moisture in the system or lack of lubrication.

A knock in the compressor may be caused by a loose connecting rod, eccentric strap and rod, eccentric disk, piston pin, crankshaft, or too much oil in the system.

Conventional reciprocating compressors are lubricated by using the splash method. Special dippers or slingers are fastened to the crankshaft to distribute the crankcase oil to the pistons, pin bearings, cylinder walls, crankshaft bearings and seal.

The importance of proper belt alignment in open units must be kept constantly in mind when service necessitates changing a compressor, motor or motor pulley.

The compression ratio is figured by dividing the absolute head pressure by the absolute suction pressure.

Speed, clearance and temperature conditions all affect the type of lubrication needed for compressors.

Never use a torch or attempt a repair on a line containing refrigerant. Always use proper protective gear when servicing a unit.

REVIEW QUESTIONS

1. What is the function of a compressor in the mechanical refrigeration system?
2. Name various types of compressors.
3. Name the component parts of a reciprocating compressor.
4. What is meant by forced-feed lubrication?
5. What is the function of the shaft seal?
6. What is a centrifugal compressor and how does it operate?
7. What is a rotary compressor and how does it work?
8. State the difference between a rotary and hermetic compressor.
9. Define maintenance and care of compressors.
10. Define compressor mechanical efficiency.
11. What is meant by compressor ratio?
12. Define volumetric efficiency, compressor capacity, and performance factor.
13. Calculate the theoretical capacity of a compressor equipped with two cylinders, each having a 3-inch bore and a 4-inch stroke, and operating at 600 rpm.
14. What is the purpose of piston rings in a compressor?
15. What is a connecting rod?
16. What is an eccentric shaft? Where is it used?
17. What is a poppet valve used for?
18. What separates the suction port from the discharge port in a rotary compressor?
19. Where are centrifugal compressors used?
20. What is meant by multistage compressors?

21. What is the purpose of a suction strainer?
22. What is the formula used for figuring compressor efficiency?
23. What does a knock in a compressor signify?
24. What is the difference between a dipper and a slinger?
25. How are compressors usually lubricated?

CHAPTER 4

Domestic Refrigeration

Domestic and commercial refrigeration principles are very similar, except that in domestic refrigeration we are dealing with smaller refrigeration units. Most of them use hermetically or semi-hermetically sealed compressors. A few have service valves, but the majority of domestic refrigerators are completely sealed, without any service valves.

The process of refrigeration is most commonly accomplished by the evaporation of a liquid refrigerant, thereby extracting heat from the medium to be cooled. The refrigeration cycle is then composed chiefly of four further steps, which remove this heat from the evaporating refrigerant by again putting it in the liquid state in order to be used repeatedly in a continuous process.

The types of compressors generally used in home refrigerators may be grouped according to their construction as *reciprocating* and *rotary*. The principal components of a compression refrigeration system are:

1. Compressor
2. Condenser
3. Drier-strainer
4. Capillary tube
5. Evaporator (freezer unit)
6. Accumulator

In modern systems, the capillary-tube method is employed to control the flow of refrigerant to the evaporator, whereas older systems are generally provided with an expansion valve for refrigerant control or metering purposes. This applies only to domestic refrigeration, not to commercial or industrial refrigeration.

THE REFRIGERATION CYCLE

A thorough understanding of the cycle of operation of a refrigerator is necessary before a correct diagnosis of any service problems can be made. Thus, only by a thorough study of the fundamentals will one be able to master the field of refrigeration.

A *cycle*, by definition, is an interval or period of time occupied by one round or course of events in the same order or series. The word cycle, as applied here, means a series of operations in which heat is first absorbed by the refrigerant, changing it from liquid to a gas, and then the gas is compressed and forced into the condenser where the heat is absorbed by the circulating air, thus bringing the refrigerant back to its original or liquid state. With reference to Fig. 4-1, the cycle of operation consists of the following steps:

1. The compressor pumps refrigerant through the entire system. It draws cool refrigerant gas in through the suction line from the evaporator freezer coils. At the same time, it compresses the gas and pumps it into the discharge line. The compressed gas sharply rises in temperature and enters the condenser.
2. The condenser performs a function similar to that of the radiator in an automobile; that is, the condenser is the cooling coil for the hot refrigerant gas. In the condenser, the heat is expelled into the room air outside the cabinet.

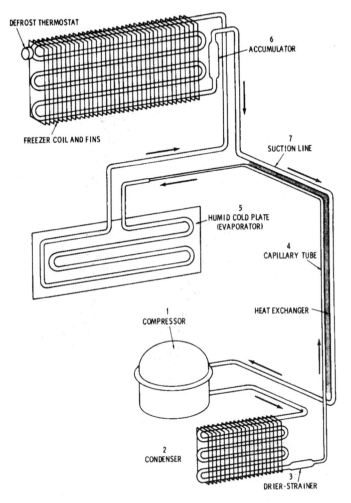

DEFROST THERMOSTAT

6
ACCUMULATOR

FREEZER COIL AND FINS

7
SUCTION LINE

5
HUMID COLD PLATE
(EVAPORATOR)

4
CAPILLARY TUBE

1
COMPRESSOR

HEAT EXCHANGER

2
CONDENSER

3
DRIER-STRAINER

Fig. 4-1. Typical refrigerant flow diagram.

During this process, the refrigerant gas gives up the heat it removed from inside the cabinet and changes into a liquid state.

3. As the hot refrigerant liquid leaves the condenser to enter the capillary tube, a drier-strainer removes any moisture or impurities.

137

4. The capillary tube is carefully calibrated in length and inside diameter to meter the exact amount of liquid refrigerant flow required for each unit. A predetermined length of the capillary tube is usually soldered along the exterior of the suction line, forming a heat exchanger, which helps to cool the hot liquid refrigerant in the capillary tube. The capillary tube then connects to the larger diameter tubing of the evaporator.

5. As the refrigerant leaves the capillary tube and enters the larger tubing of the humid plate and evaporator, the sudden increase in tubing diameter forms a low-pressure area and the temperature of the refrigerant drops rapidly as it changes to a mixture of liquid and gas. In the process of passing through the evaporator, the refrigerant absorbs heat from the storage area and is gradually changed from a liquid and gas mixture to a gas.

6. The low-pressure refrigerant gas leaving the evaporator coil now enters the accumulator, which is a large cylinder designed to trap any refrigerant liquid that may not have changed to gas in the evaporator. Since it is impossible to compress a liquid, the accumulator prevents any liquid from returning to the compressor.

7. As the refrigerant gas leaves the accumulator, it returns to the compressor through the suction line, which is part of the heat exchanger, thus completing the cycle.

High-Pressure Side

The high-pressure side of the system is the side containing the high-pressure refrigerant. It consists of the condenser, capillary tube, and compressor.

Low-Pressure Side

The low-pressure side is that part of the system where the refrigerant is in a gaseous state at low temperature and pressure. It consists of the evaporator and suction line. The low- and high-pressure sides of a refrigeration system are shown in Fig. 4-2.

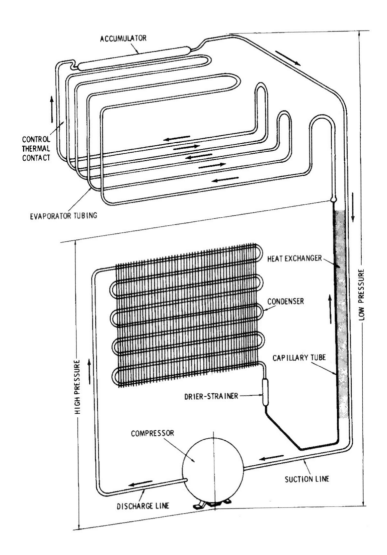

Fig. 4-2. Low- and high-pressure sides of a refrigeration system.

COMPRESSORS

Compressors are usually of the hermetic, or sealed, type, although numerous refrigerators of the older models are of the so-called open type. The open-type compressor is one in which the motor is connected to the compressor by means of a belt, whereas in the hermetic type the motor and compressor are connected directly to the same shaft and sealed into the same compartment to furnish an airtight and dustproof assembly. A hermetic or sealed compressor is shown in Fig. 4-3. The hermetically-sealed unit provides a compact assembly and, as such, will require less space, with the additional advantage of the elimination of a shaft seal and stuffing boxes. Also, the elimination of the belt and pulley will result in reduced noise and maintenance.

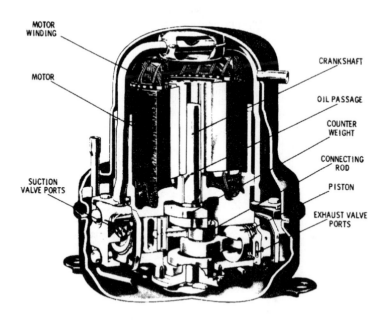

Tecumseh Products Company

Fig. 4-3. Cutaway view of a typical twin-cylinder sealed compressor used in household refrigerators.

140

CONDENSERS

Since the refrigerant leaves the compressor in the form of high-pressure vapor, some method must be found to bring the vapor back into liquid form. It is the function of the condensing unit to condense the vapor back to a liquid so that it can be reused in the refrigeration cycle. Condensers are of two main types—air-cooled by means of a forced fan draft, and air-cooled by natural air circulation.

The most common condenser construction is shown in Fig. 4-4. It consists of copper tubing or coils upon which fixed fins have been inserted to assist in removing the rapid accumulation of heat. This cooling process is further provided by means of a fan

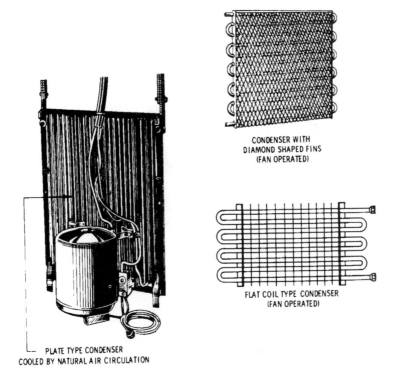

CONDENSER WITH
DIAMOND SHAPED FINS
(FAN OPERATED)

FLAT COIL TYPE CONDENSER
(FAN OPERATED)

PLATE TYPE CONDENSER
COOLED BY NATURAL AIR CIRCULATION

Fig. 4-4. Various types of condensers showing construction principles.

(driven either by the compressor motor or by an independent motor), which forces air through the condenser.

Other condensers are made up of corrugated plates welded together, the corrugations forming tubes in which the gas is condensed. In this type of condenser, air passes over the outside surface of the plates and removes the heat by condensation from the refrigerant gas.

DRIER-STRAINER

As previously noted, the function of the drier-strainer is to remove moisture and impurities from the refrigeration system. A typical drier-strainer (Fig. 4-5) consists essentially of a tubular metal container or housing arranged for connection into the refrigerant circuit. The drying and purifying agent is usually silica gel, which, in addition to a cup-shaped inlet screen, provides the filtering and drying action.

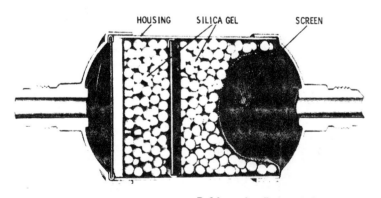

Refrigeration Research Company, Inc.

Fig. 4-5. Typical drier-strainer.

CAPILLARY TUBES

The capillary tube is essentially an expansion device used as a part of the refrigerant circuit. It consists normally of a miniature tube, the length of which depends upon the size of the condensing unit and the kind of refrigerant used. The bore or inner diameter is very small, and the length varies greatly from a few inches up to several feet. Because the capillary tube offers a restricted passage, the resistance to the refrigerant flow is sufficient to build up a high enough head pressure to produce condensation of the gas. The operating balance is obtained by properly proportioning the size and length of the tube to the particular unit on which it is to be used.

The refrigerant employed has a direct influence on both the length and diameter of the capillary tube. In any event, the size of the inner tube diameter must be such as to keep the tube full of liquid under normal operating conditions. Because of the minute size of the tube bore, it is important that the refrigerant circuit be kept free from dirt, grease, and any foreign matter since these obstructions may close up the tube and thus make the unit inoperative. To prevent obstructions to the free flow of refrigerants, a drier-strainer is provided, as noted in the refrigerant circuit diagram (Fig. 4-6). If the refrigerant tube becomes plugged, the evaporator will defrost and the unit may run continuously, or the thermal relay cuts out, in which case both the suction and discharge pressures will be abnormally high.

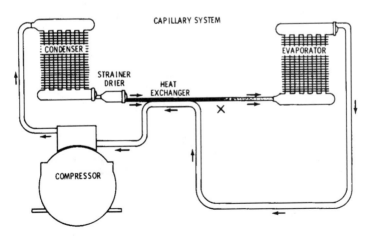

Fig. 4-6. Installation of a capillary system in a typical refrigerating circuit.

EVAPORATORS

The function of the evaporator is to absorb heat from the refrigerator cabinet, the heat being introduced by food placed in the refrigerator, insulation loss, and door openings. Evaporators used in present-day designs are of the direct-expansion type because of its simple construction, low cost, and compactness and also because it provides a more uniform temperature and rapid cooling. The evaporator consists simply of metal tubing, which is formed around the freezer compartment to produce a cooling area for freezing ice cubes and provide the desired cooling effect for the food-storage compartments. Figure 4-7 shows various types of frozen-food compartments.

In operation, when the refrigerant leaves the capillary tube and enters the larger tubing of the freezer shelves, the sudden increase in tubing diameter forms a low-pressure area and the temperature of the refrigerant drops rapidly as it changes to a mixture of liquid and gas. This cold mixture passes through the top shelf (or top freezer plate in some models) and then to the bottom shelf, descending through each additional shelf until it reaches the accumulator. In the process of passing through the shelf tubing, the refrigerant will absorb heat from the food-storage area and will gradually change from a liquid and gas mixture to a gas.

ACCUMULATORS

The accumulator is a large cylindrical vessel designed to trap any refrigerant liquid that may not have changed to gas in the evaporator. In this manner any liquid refrigerant remaining in the low side of the system is prevented from entering the suction line to the compressor.

Diagrams of current changes in late-model refrigerators in line with current energy conservation programs are shown in Figs. 4-8 to 4-12.

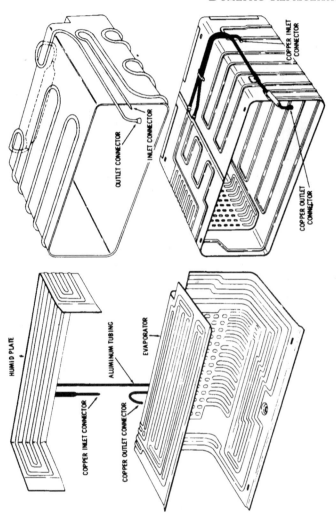

Fig. 4-7. Typical evaporator arrangement in household refrigerators.

Post-Loop-Condenser Design Change

The energy-saving post-loop condenser already employed in many of our current model refrigerators has been expanded in its use so that a portion of it is incorporated in the center vertical

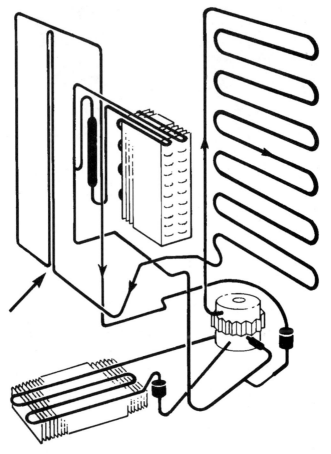

Frigidaire Corporation Div. of General Motors

Fig. 4-8. Model FGI-16V refrigeration system.

mullion. This eliminates the need for a center vertical drier, thus conserving more electrical energy. This improvement became effective on all side-by-side model refrigerators starting with 1976 production.

The post-loop routing for the FCI-16V model is shown in Fig. 4-8. From the condenser outlet, the post-loop tube is routed to the bottom of the food compartment and then to the center mullion. From here it goes to the top of the center mullion, makes

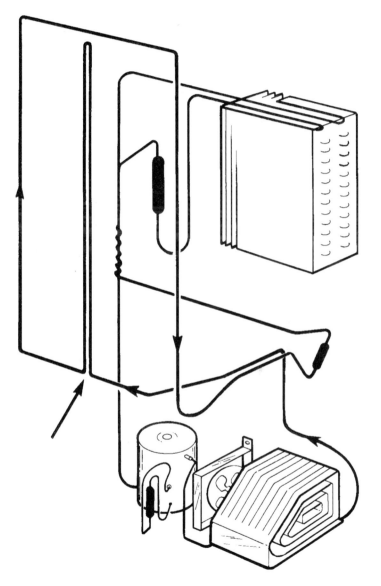

Frigidaire Corporation Div. of General Motors

Fig. 4-9. Model FGI-20V, 2OV3 refrigeration system.

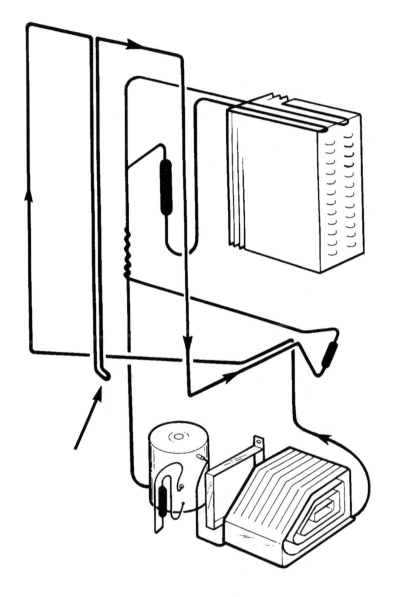

Frigidaire Corporation Div. of General Motors

Fig. 4-10. Model FGI-22V refrigeration system.

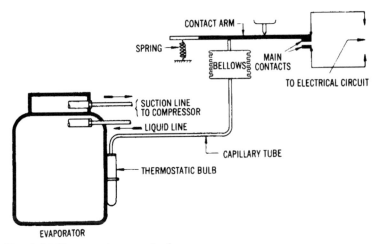

Fig. 4-11. Temperature control.

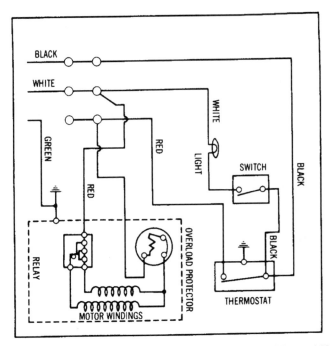

Frigidaire Corporation Div. of General Motors

Fig. 4-12. Schematic diagram for models D-43 and I-43.

149

a loop, and back down again across the bottom of the freezer compartment. From here it goes around the perimeter of the cabinet shell at the front as on former models.

For the FCI-20V and FCI-20V3 models, the routing is approximately the same as it is on the FCI-16V model (Fig. 4-9). The difference is the FCI-16V utilizes a static condenser system while the FCI-20V models have a forced-air system.

The basic difference between the 20V and 22V is that the center mullion post-loop section enters at the top on the 22V models (Fig. 4-10).

CONTROL METHODS

The most common method of temperature control presently employed in household refrigeration units is the *thermostatic control*. The apparatus consists principally of a thermostatic bulb (Fig. 4-11), which is fastened to the evaporator (chilling unit) and connected to the bellows by means of a capillary tube.

The bulb and tube are charged with a highly volatile fluid. As the temperature of the bulb increases, gas pressure in the bulb-bellows assembly increases and the bellows pushes the operating shaft upward against the two spring pressures. The shaft operates the toggle, or snap, mechanism. Consequently the upward travel of the shaft finally pushes the toggle mechanism off center, and the switch snaps closed, starting the motor. As the motor runs, the control bulb is cooled, gradually reducing the pressure in the bulb-bellows system. This reduction of bellows pressure allows the spring to push the shaft slowly downward until it has finally traveled far enough to push the toggle mechanism off center in the opposite direction, snapping the switch open and stopping the motor. The control bulb then slowly warms up until the motor again starts and the cycle repeats itself.

A typical schematic wiring diagram is shown in Fig. 4-12. Schematic wiring diagrams will be most helpful to you once you are able to read them. The wiring is not laid out in the manner that it is installed in the refrigerator, but you may see at a glance how each part is connected in the electrical system, and from there tracing out the wiring in the refrigerator becomes simple.

SUMMARY

The types of compressors generally used in home refrigerators may be grouped according to their construction as reciprocating and rotary. The refrigeration cycle is an interval, or period of time, occupied by one round or course of events in the same order or series. The cycle in refrigeration consists of taking a liquid, making it a gas by way of heating it, and then removing the heat and moving the liquid back to the heat source to once again remove the heat by causing the liquid to become gaseous.

The high-pressure side of the system is the side that contains the high-pressure refrigerant; it consists of the condenser, capillary tube, and the compressor.

The low-pressure side of the system is that part where the refrigerant is in a gaseous state at low temperature and pressure; it consists of the evaporator and suction line. The low- and high-pressure sides of a refrigeration system go together to make up the complete cooling cycle.

The function of the evaporator is to absorb heat from the refrigerator cabinet. The compressor causes the gas to be converted to liquid as it is cooled by the condenser and put under pressure again by the compressor.

REVIEW QUESTIONS

1. List the principal components of a compression refrigeration system.
2. What is a refrigeration cycle?
3. Describe the flow of refrigerant from the compressor through the entire system.
4. What components make up the high-pressure side of the refrigeration system?
5. What components make up the low-pressure side of the refrigeration system?
6. What is the purpose of a compressor?
7. What is the purpose of a condenser?
8. What purpose does the evaporator serve in a refrigeration system?
9. What is a drier-strainer used for?
10. What is an accumulator?

CHAPTER 5

Absorption System for Domestic Refrigeration

Absorption-type refrigeration uses heat and a refrigerant that may be absorbed into water. Domestic systems use ammonia, and industrial systems may use ammonia or lithium bromide. Commercial absorption refrigeration is discussed in detail in Chapter 17.

The absorption system of refrigeration differs from that of the conventional compression type mainly in that it uses heat energy instead of mechanical energy to change the condition required in the refrigeration cycle. The heat energy required may be obtained from a gas flame, an electric heater, or a kerosene flame. The principal advantages in the absorption system of refrigeration lies in the fact that since no moving parts are involved, the repairs and maintenance cost will be minimal.

The comparison between the functioning of the *absorption* and *compression* systems of refrigeration is principally as follows: *The absorption system uses heat (usually a gas flame) to circulate the refrigerant, while the compression system employs a compressor to circulate the refrigerant.* Some absorption systems may be operated by gas or electricity. You will find many small refrigerators in travel trailers, in which the boiler or generator may be operated by either propane gas or electricity at the will of the owner.

Three components of the absorption system compare with the four components of the compression system, as follows:

1. The action of the boiler or generator compares with the stroke of the compressor.
2. The condenser and evaporator serve the same purpose in an absorption system as in the compression system.
3. The absorber compares to the low, or suction, side of the pump.

PRINCIPLE OF OPERATION

After the proper heat has been applied to the generator, the ammonia will vaporize from the water. The vapor bubbles, in trying to escape, will carry water up the percolator tube. The vapor and water are allowed to separate so that the vapor is free to continue upward into the condenser. Here, with proper air circulation, the ammonia vapor will be condensed to a liquid, which then flows through a liquid trap into the evaporator. When the evaporator shelf is level, the proper slope is established in all coils to induce a gravity flow downward.

At the time the unit is charged, a small amount of hydrogen is introduced. At this point of the cycle of the unit, hydrogen flows upward into the evaporator and tends to mix with the ammonia vapor, encouraging more evaporation. It is this evaporation process which produces refrigeration. Since the mixture of hydrogen and ammonia vapor is considerably heavier than hydrogen alone, the normal tendency for this mixture is to flow downward. It is encouraged to do this, and in so doing it is forced to pass upward through the absorber.

Water that has been separated from the ammonia by heat flows downward through the absorber. The water temperature has been reduced so that it will again absorb the ammonia quite readily, and the water and ammonia solution flows back to the generator for recirculation. Since the hydrogen has been washed free of ammonia (and lightened), it flows upward again through the evaporator. When the absorption system is working normally, all these actions are continuous. A thermostatically controlled gas valve, with the sensor attached to the evaporator coil, will vary the heat input and consequently the amount of refrigeration that the load of the refrigerator requires.

ABSORPTION-REFRIGERATION CYCLE

The freezing system of the gas refrigerator is made up of a number of steel vessels and pipes welded together to form a hermetically-sealed system. All the spaces of the system are in open and unrestricted communication so that all parts are at the same total pressure. The charge includes an aqua-ammonia solution of a strength of about 30% concentration (ammonia by weight) and hydrogen.

The elements of the system include a *generator* (sometimes called *boiler*, or *still*), a *condenser*, an *evaporator*, and an *absorber*. There are three distinct fluid circuits in the system: an *ammonia circuit*, including the generator, condenser, evaporator, and absorber; a *hydrogen circuit*, including the evaporator and absorber; and a *solution circuit*, including the generator and absorber. A diagram of this circuit is shown in Fig. 5-1.

Heat is applied by a gas burner or other source of heat to expel the ammonia from solution. The ammonia vapor thus generated flows to the condenser. In the path of flow of ammonia from the generator to the condenser are interposed an *analyzer* and a *rectifier*. Some water vapor will be carried along with the ammonia vapor from the generator. The analyzer and rectifier serve to remove this water vapor from the ammonia vapor. In the analyzer, the ammonia passes through a strong solution that is on its way from the absorber to the generator. This reduces the temperature of the generated vapor somewhat to condense water

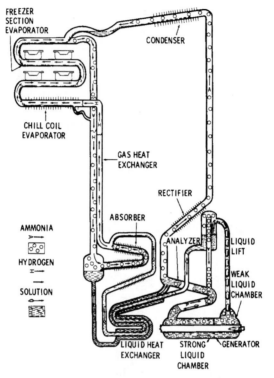

FREEZER
SECTION
EVAPORATOR

CONDENSER

CHILL COIL
EVAPORATOR

GAS HEAT
EXCHANGER

RECTIFIER

ABSORBER

AMMONIA

ANALYZER   LIQUID
            LIFT

HYDROGEN

WEAK
LIQUID
CHAMBER

SOLUTION

LIQUID HEAT   STRONG   GENERATOR
EXCHANGER   LIQUID
               CHAMBER

Whirlpool Corporation

Fig. 5-1. Operating cycle of a gas-absorption refrigerating system.

vapor, and the resulting heating of the strong solution expels some ammonia vapor without additional heat input. The ammonia vapor then passes through the rectifier, where the residual small amount of water vapor is condensed by atmospheric cooling and drains to the generator by way of the analyzer.

The ammonia vapor, which is still warm, passed on to the section of the condenser, where it is liquefied by air cooling. The condenser is provided with fins for this purpose. The ammonia thus liquefied flows into the evaporator. A liquid trap is interposed between the condenser section and evaporator to prevent hydrogen from entering the condenser. Hydrogen gas enters the lower evaporator section and flows upward, in counterflow to the downward-flowing liquid ammonia. The effect of placing a hydrogen atmosphere above the liquid ammonia in the evaporator is to reduce the partial pressure of the ammonia vapor in accordance with *Dalton's law of partial pressures.*

Under Dalton's law, the total pressure of a gas mixture is equal to the sum of the partial pressures of the hydrogen. Consequently, in the evaporator, the partial ammonia-vapor pressure is less than the total pressure by the value of partial pressure of the hydrogen. The lesser ammonia-vapor pressure results in evaporation of the ammonia, with consequent absorption and cooling of the surroundings, which are in a well-insulated enclosure. The cool, heavy gas mixture of hydrogen and ammonia vapor formed in the evaporator leaves the top of the evaporator and passes downward through the gas heat exchanger to the absorber.

Since the weight of a gas is proportional to its molecular weight, and since the molecular weight of ammonia is 17 and the molecular weight of hydrogen is 2, the specific weight of the strong gas is greater than that of the weak gas. This difference in specific weights is sufficient to initiate and maintain circulation between the evaporator and the absorber. The gas heat exchanger transfers heat from one gas to the other. This saves some cooling in the evaporator by precooling the entering gas. A liquid drain at the bottom of the evaporator is connected to the downflow space of the gas heat exchanger. In the absorber, a flow of weak solution (water weak in ammonia) comes in direct contact with the strong gas. Ammonia is absorbed by water, and the hydrogen (which is practically insoluble) passes upward from the top of the absorber through another chamber of the gas heat exchanger into the evaporator. The liquid and gas flow in opposite directions.

From the absorber, the strong aqua-ammonia solution flows through the liquid heat exchanger to the analyzer and then to the strong liquid chamber of the generator. The liquid heat exchanger precools the liquid entering the absorber and preheats the liquid entering the generator. Further precooling of the weak solution is obtained in the finned air-cooled loop between the liquid heat exchanger and absorber. Actual unit construction includes additional refinements that increase efficiency, but the cycle described previously is typical of the cycle in any absorption refrigerator. Refrigeration is accomplished by a continuation of the cycle described herein.

THE FLUE SYSTEM

The flue system is composed of a combustion chamber where the gas-air mixture is burned. The flue extension is to provide a draft for an adequate flow of secondary air and a dilution flue to cool the products of combustion and carry them to the top of the refrigerator.

The combustion chamber is properly known as the *generator flue* because it is attached to the generator. A *flue baffle* is normally located at the back or top end of the generator. The purpose of the flue baffle is to distribute the heat, and it should always be located in its correct position, otherwise the efficiency of the unit will be affected.

The flue extension is made of nonmetallic materials and is attached to the generator flue. On horizontal generator-flue systems, a cleanout cover is located at the rear of the unit. The flue baffle may be removed through the opening when the cleanout cover is removed. When the unit is in operation, the cleanout cover must be in place for proper draft.

The dilution flue is a nonmetallic pipe, which extends from the top of the generator flue extension to the top of-the cabinet. Air for dilution is induced at the bottom of the loose-fitting dilution flue and mixes with products of combustion as it travels to the top of the cabinet.

GAS-CONTROL DEVICES

An absorption-system refrigerator requires various gas control devices for its operation, which are shown in Fig. 5-2.

1. Shutoff valve
2. Pressure regulator (for piped gas only)
3. Input control valves
4. Gas burner
5. Thermostat

The gas components may be removed easily as an assembly for service when required, provided that the gas shutoff valve is closed and the electric supply is disconnected.

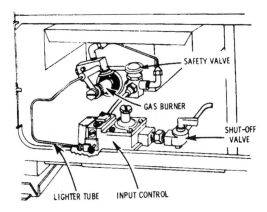

Fig. 5-2. Gas components in typical absorption in refrigeration system.

Gas Shutoff Valve

The gas shutoff valve, as the name implies, is a simple *on* and *off* valve to enable the serviceman to cut off the gas supply when necessary for repairs and service.

Pressure Regulators

Pressure regulators are normally provided on all piped-gas refrigerating units, except those using straight liquid petroleum (LP) gas since LP gas containers are equipped with individual pressure regulators. The gas pressure regulator (Fig. 5-3) functions to provide a constant, even gas pressure to the gas burner during all main-line pressure fluctuations.

The action of the gas pressure regulator is very simple. A valve is suspended on a flexible diaphragm in a way that will cause the valve to come in contact with the valve seat when the diaphragm opens, allowing the gas to pass. The pressure of the gas will move the diaphragm up or down, controlling the amount of gas. Pressure applied from above the diaphragm, either by spring pressure or by weights (depending upon the type of regulator), will tend to open the valve. Therefore, the outlet pressure of the regulator can be adjusted by turning a screw to change the spring tension or (on dead-weight regulators) by adding or removing weight.

159

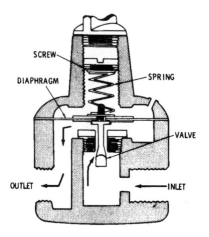

SCREW

DIAPHRAGM

SPRING

VALVE

OUTLET

INLET

Fig. 5-3. Gas-pressure regulator.

Input Controls

The input control consists essentially of two automatically operated gas valves that are electrically controlled by means of the refrigerator thermostat. The most common types of controls are manufactured by General Controls or White Rodgers, as shown in Fig. 5-4. These controls function in a similar manner, although they differ in operation. The input control consists of:

1. A felt filter to remove gum and foreign materials from the gas
2. A diaphragm-type pressure regulator to adjust maximum flame gas pressure when the control is used on piped gases
3. Two identical electrically-operated gas valves, which alternate opening and closing as the refrigerator thermostat cycles (one valve supplies gas for a maximum burner flame— the other releases gas for an intermediate flame)
4. A bypass to supply gas for a burner pilot flame when both valves are closed due to no electrical power
5. A plate containing two orifices to reduce the gas flow for the intermediate and pilot flames
6. A pushbutton-valve gas release for burner lighting

When the refrigerator is to be operated on LP gas, the pressure-regulator capscrew is replaced with one having a long narrow pin. The pin locks the pressure regulator in an open position. The gas is regulated at the supply tank.

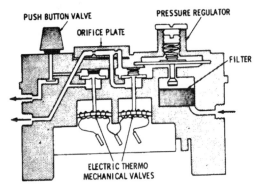

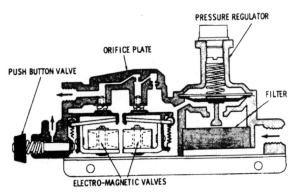

General Controls and White Rodgers

Fig. 5-4. Components of a gas-input-control assembly.

Gas Burners

The gas-burner assembly (Fig. 5-5) consists essentially of a *turbulator, orifice spud*, and *thermal safety valve*. The function of the turbulator, as its name implies, is to create turbulence in the gas stream as it enters the mixing tube. This turbulence will have two effects: the one that is apparent and readily understood is that

161

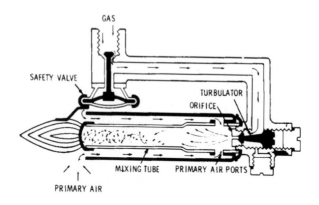

Fig. 5-5. Typical gas-burner operation.

it aids in mixing primary air and gas; the second effect is that the air-gas ratio is automatically adjusted so that with proper air-shutter adjustment, the flame will be all blue on maximum flame and have stability on pilot flame. All piped gas-burner turbulators have two grooves in the conical end, and the head of the turbulator must always fit tightly into the rear side of the orifice. All LP gas-burner turbulators have a single groove in the conical end and are spring-loaded for adjustment to the gas pressure.

The burner orifice required on gas refrigerators is so small that a slight change in its dimensions changes its capacity. For this reason, each orifice spud is individually flow tested and rated to capacity. All orifice spuds having the same rated capacity have a comparative flow accuracy of plus or minus 3 percent for orifice spuds used with piped gas. The number marked on the orifice spud is not the hole diameter or drill size. In fact, the hole diameter may vary somewhat to compensate for other factors, such as surface resistance, which would affect the capacity.

The safety valve is a protective device designed to stop the gas flow to the burner in case the flame should leave the generator flue. This valve is similar in appearance to the thermal valve on the burner. The valve consists of a bimetallic disk to which a valve stem is attached. The valve is open at room temperature. However, by overheating the valve body or the heat conductor, the disk will snap the valve closed. The safety valve will reset auto-matically as it cools down in temperature.

Thermostats

The thermostat is an automatic gas valve in the burner supply line that controls the flow of gas to the burner to supply either a maximum or minimum flame, depending upon a control-dial setting and thermostat-bulb temperature. Figure 5-6 identifies the different parts of the thermostat and their relative position to each other.

The working parts of the thermostat are enclosed in an outer case. In the lower part of the case is a gas passage, in which a gas valve is imposed. Under the valve is a spring that pushes the valve toward a closed position. Above the gas passage is a power element consisting of a bulb connected to a flexible diaphragm by a capillary tube.

In this system is a small quantity of gaseous hydrocarbon. Motion of the power element is transmitted to the gas valve by the valve pin. When the bulb is warm, the gas expands and exerts a pressure in the power element, which pushes the valve open. When the bulb is cold, the gas contracts, the pressure inside the

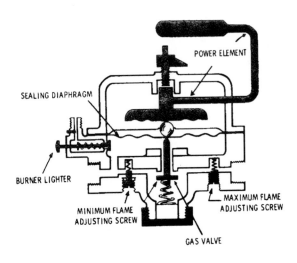

Fig. 5-6. Absorption-refrigerator thermostat valve.

power element is reduced, and the valve spring pushes the valve closed. From the foregoing description, it is evident that there is a throttling action and that the valve may be in any position between wide open and fully closed, depending upon the thermostat dial setting and the thermostat-bulb temperature.

To prevent the gas supply to the burner from being completely closed off when the valve is closed, a bypass or minimum flame passage is provided. Adjustment screws on the thermostat control the volume of gas flow through the passages that supply gas for maximum and minimum flame. There is also an adjustment screw on the burner lighter valve. This valve is used only when lighting the burner. Control of temperatures inside the refrigerator is accomplished by turning the control dial. The highest number on the dial is the coldest position. A defrost position is provided for manual defrosting.

ELECTRICAL ACCESSORIES

Gas refrigerating units are capable of operating without any electrical connections, which is usually the case where the refrigerator is operated on LP gas and electricity is not available. Electrically equipped absorption units, however, will add certain desirable features, such as:

1. Cabinet lights
2. Automatic defrosters
3. Air-circulator fans
4. Air-circulator thermostat

Cabinet Lights

Cabinet lights are wired in the conventional manner, the interior light bulb being energized by the light switch when the cabinet door is opened.

Defrost Timers

Defrost timers are manufactured to operate the electrical defrost heater at certain time intervals, usually once every 12 or

24 hours. Figure 5-7 shows a typical self-starting electrical defrost timer operating upon the well-known electric-clock principle. The dial on the timer should be set to the correct time of day when the refrigerator is installed and reset to the correct time any time the electric service is interrupted.

The 24-hour defrost timer contains an electric switch, which makes contact for a period of about 20 minutes, beginning at the 2:00 A.M. position on the timer. This means that if the dial is set to the correct time, defrosting will occur automatically at 2:00 A.M. each day. The defrost water will drain from the chill coil onto the lower surface of the chill cover and then into a drain elbow, which directs the water to an evaporation pan.

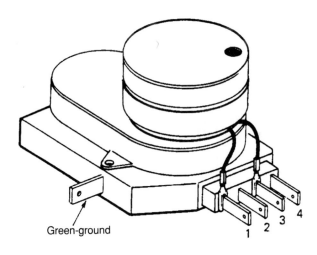

Fig. 5-7. Typical defrost timer.

Air-Circulation Fans and Thermostats

The air-circulation fan, as the name implies, is generally designed to become operative when the ambient temperature becomes unduly high. The additional air movement provided by the air circulator will provide more cool air for cooling the condenser. At the same time, the movement of cooler air at the rear of the cabinet will reduce the amount of heat transfer to the inside of the cabinet, as shown in Fig. 5-8.

The thermostat that controls the air-circulator fan is set to "cut in" at approximately 100°F and to "cut out" at approximately 80°F. This means that the fan will not operate until the temperature of the thermostat reaches 100°F and once it has started to operate, the fan will remain in operation until the thermostat temperature has dropped to 80°F. The electrical components are shown in Fig. 5-9.

INSTALLATION AND SERVICE

Before installing the refrigerator, turn off all other gas appliances, such as gas ranges, gas furnaces, gas water heaters, and gas spaceheaters. Next, shut off the gas supply at the main inlet or outlet at the gas meter.

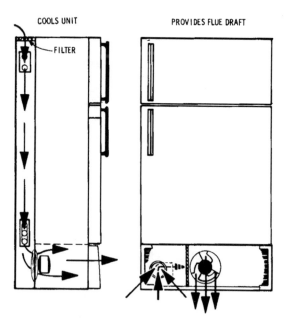

Whirlpool Corporation

Fig. 5-8. Air-circulation-fan arrangement in a typical refrigerator.

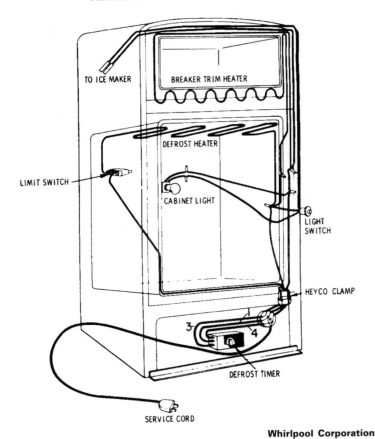

TO ICE MAKER

BREAKER TRIM HEATER

DEFROST HEATER

LIMIT SWITCH

CABINET LIGHT

LIGHT SWITCH

HEYCO CLAMP

DEFROST TIMER

SERVICE CORD

Whirlpool Corporation

Fig. 5-9. Arrangement of electrical components used in an automatic-defrost refrigerator.

Air Circulation

All refrigerators must have proper air circulation for proper operation. While the refrigerator is in operation, air enters from the bottom and travels upward through the rear section of the cabinet and out at the top, as shown in Fig. 5-10. This air circulation takes place naturally unless prevented by insufficient clearance or blocked air passages. Proper air circulation will usually result when the refrigerator is installed indoors directly in front of a wall, allowing a minimum distance of 2 inches of

167

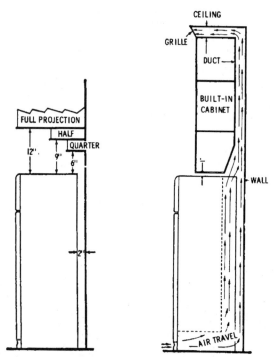

Fig. 5-10. Correct clearance arrangement for proper air circulation.

clearance from the back of the refrigerator to the wall and at least a 12-inch clearance above the top of the refrigerator.

When the recommended top clearances cannot be obtained, an air duct should be installed. The air duct should have about the same width as the refrigerator and be approximately 7 inches deep. This duct may return the air to the room at ceiling height or exhaust it through a suitable vent in the roof.

Leveling

The equal distribution of the liquid within the freezing compartment requires the unit to be installed and maintained in a level position, both front, back, and side to side. The leveling of the refrigerator can be accomplished conveniently by shimming with small wooden strips or other available material.

Gas-Line Connection

When connecting the gas line, use tubing and fittings as prescribed by local codes. Install the gas line so that the refrigerator can be disconnected at the inlet valve without damage to the controls. On liquid-petroleum (LP) gas installations, a gas pressure regulator in the gas line is not needed since the regulator at the gas supply tanks should maintain constant gas pressure at the burner. Figure 5-11 shows typical gas-connection methods.

The installation shown in Fig. 5-11 will not permit the refrigerator to be disconnected between the gas cock and thermostat. Therefore, to disconnect the refrigerator, it is necessary to shut off the gas supply from the meter. By using a tubing connection between the gas cock and gas filter, or between the gas filter and

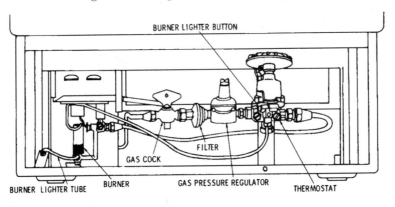

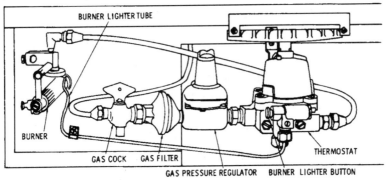

Fig. 5-11. Gas connections for vertical and horizontal installations.

169

thermostat, the controls can be installed in a way that will permit the refrigerator to be easily disconnected for service.

Piping Material—Standard-weight wrought-iron pipe coupling with the American Standard wrought-steel and wrought-iron pipe shall be used in installation of appliances supplied with utility gases. Threaded copper or brass pipe in iron-pipe sizes may be used with gases not corrosive to such material. Gas pipe shall not be bent. Fittings shall be used when making turns in gas pipe.

The connection of steel or wrought-iron pipe by welding is permissible. Threaded pipe fittings (except stopcocks or valves) shall be of malleable iron or steel when used with steel or wrought-iron pipe and shall be copper or brass when used with copper or brass pipe. When approved by the authority having jurisdiction, special fittings may be used to connect either steel or wrought-iron pipe.

Piping Material for LP Gases—Gas piping for use with undiluted LP gases shall be of steel or wrought-iron pipe, complying with the American Standard for wrought-steel and wrought-iron pipe, and brass or copper pipe, or seamless copper, brass, steel, or aluminum tubing. All pipe or tubing shall be suitable for working pressure of not less than 125 lb/in.² Copper tubing may be of the standard K or L grade, or the equivalent, having a minimum wall thickness of 0.032 inch. Aluminum tubing shall not be used in exterior locations or where it is in contact with masonry, plaster walls, or insulation.

Defective Material—Gas pipe or tubing and fittings shall be clear and free from burrs and defects in structure or threading and shall be thoroughly brushed, chipped, and scale-blown. Defects in pipe, tubing, or fittings shall not be repaired. When defective pipe, tubing, or fittings are located in a system, the defective material shall be replaced.

Gas pipe, tubing, fittings, and valves removed from any existing installation shall not be used again until they have been thoroughly cleaned, inspected, and ascertained to be equivalent to new material. Joint compounds shall be applied sparingly and only to the male threads of pipe joints. Such compounds shall be resistant to the action of LP gases.

BURNER-ORIFICE CLEANING

The burner orifice must be cleaned when the refrigerator is installed and each time a new orifice is installed. This is necessary because the orifice is so small that a particle of dirt can materially change its operating characteristics. Dirt in the orifice can also cause an unstable flame and burner outage.

Clean the orifice after it has been installed on the burner body. The air-shutter barrel, mixing tube, seal screw, and turbulator must be removed. Use a round toothpick, as shown in Fig. 5-12, sharpened on an emery board, to clean the orifice. Insert in the orifice and rotate with light pressure. Blow away any dust, but do

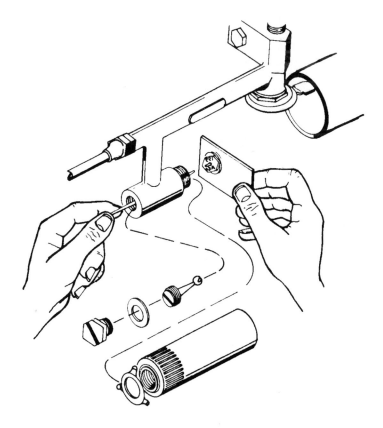

Fig. 5-12. Typical method for orifice cleaning.

not rub the orifice with the fingers. If the burner is installed on a refrigerator, use a mirror for observation.

BURNER-AIR ADJUSTMENT

The entire flame must be free of all yellow after the adjustments have been completed. Do not make this adjustment if there is an abnormal flue draft. On models having a filter above the condenser, check to be sure the filter is clean and in place. Tighten the air-shutter barrel securely on the burner body. This position should result in a yellow flame. Rotate the air-shutter barrel in the opposite direction, increasing the primary air until the yellow just disappears from the entire flame. From this point, continue to turn the air-shutter barrel in the same direction until the flame becomes noisy or unstable, or until the stop is reached. The desired setting for proper combustion is approximately midway between these two points.

THERMOSTAT-VALVE CLEANING

To clean the thermostat valve, turn off the gas supply to the refrigerator. If the cleanout cap is inaccessible, disconnect the flare nuts at the inlet and outlet of the thermostat. Remove the screws attaching the thermostat to the cabinet, and turn the thermostat until the cleanout cap is accessible. Be careful not to pinch or break the capillary. Remove the cleanout cap and gasket, spacer, valve spring, and valve. Clean the valve seat in the thermostat, using the eraser end of a pencil covered with a clean cloth. Blow into the valve seat to make certain all dirt has been removed. Clean the flat face of the valve with a clean cloth.

GAS-PRESSURE ADJUSTMENT

The maximum-flame gas pressure must be determined and adjusted for each refrigerator when installed or reinstated in a different location. The pressure was determined at the time the burner orifice spud was selected. The maximum flame provides the correct heat input for the greatest capacity of the refrigerator unit. The unit cooling capacity will be decreased if the heat input is more or less than the specified rating.

Adjustment to the intermediate and pilot-flame gas pressure is not necessary, and provision for such has not been made. The orifice plate on the input control meters the gas flow for the lower flame settings. To adjust the maximum-flame gas pressure, proceed as follows:

1. Make sure the appliance cord is connected to a power source and the thermostat is turned to a numbered position. Maximum-flame valve must be open in the input control.
2. Remove the seal screw from the side of the burner, and connect a water-filled manometer, as shown in Fig. 5-13.
3. Light the burner, and allow the generator flue to warm.

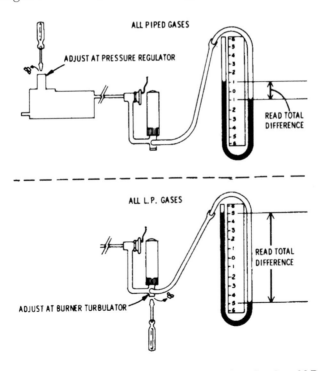

Fig. 5-13. Manometer method of gas pressure for piped and LP gas.

4. On *piped gas*: Remove the seal screw, and turn the adjusting screw in the pressure regulator on the input control counterclockwise (to decrease the pressure) or clockwise (to increase the pressure).

FLAME STABILITY

After completing the control adjustments, always check the flame for stability. Disconnect the appliance cord from the power source, thus reducing the flame to a pilot. If the pilot flame appears jumpy or wobbly, rap the bottom cross rail of the cabinet several times lightly, using a wrench. If the flame is unstable, the rapping will cause the flame to go out. If the flame Is stable, the rapping will not affect it. An unstable flame is usually caused by incorrect primary air adjustment, loose turbulator, and/or dirt in the orifice. When the test for flame stability has been completed, reconnect the appliance cord.

REFRIGERATOR-UNIT
REMOVAL AND REPLACEMENT

To remove and replace the refrigerator unit, proceed as follows:

1. Remove all trays, baffles, and other parts from the freezing compartment to permit its removal through the freezing-compartment opening at the rear of the cabinet. If the unit has a separate food-compartment cooling coil, remove the grille and thermostat bulb.
2. On models with a permanently insulated section between the freezer compartment and food compartment, remove the trim around the freezer compartment, remove the rubber plug from the drain hole, and remove the screws on the sides of the freezer.
3. Move the refrigerator away from the wall, and remove the rear panel, dilution flue, and top louver.
4. On models having pushbutton defrost, disconnect the electric cord from the wall receptacle. Carefully pry out the

complete pushbutton assembly from the trim. Remove the side- and top-door opening trim. Remove the defrost-thermostat feeler from the clamp on the freezer by loosening the screws from the inside of the freezer. Remove the thermostat support screws from the liner. Pull the assembly downward out of the cabinet. Now remove all the wires from the thermostat terminals. The defrost heaters, when used, may be removed from the freezer coils after the unit is out of the cabinet by prying off the retaining clamps. Install the heater on the replacement unit while the unit is out of the cabinet by reversing this procedure.

5. Remove the unit retaining bolts, and, on models with a gas heat exchanger (or tubes) embedded in the cabinet insulation, remove the screws from the cover plate. Carefully lift the unit out of the cabinet.

6. Remove and install on the replacement unit the styrofoam insulation and freezer drain pan, if present. When replacing the unit, it is necessary to disconnect the front chill coil supports in order to remove the plastic drain chute from beneath the freezer section. After the drain chute has been removed from the inoperative unit, reconnect and securely tighten the supports on both units.

7. Install the replacement unit by reversing the removal procedure. Make sure that the gasket and/or sealing compound around the freezing compartment opening is properly placed in position to ensure a satisfactory seal. On models with embedded gas heat exchangers, place sealing compound around the edge of the cover plate and attach the plate securely to the cabinet. The tubes protruding through this plate should also be carefully sealed.

8. On models equipped for automatic defrosting, the drain tube should be properly sealed around the tube and cover plate on the cabinet side.

REFRIGERATOR MAINTENANCE

The following pointers should be observed for proper adjustments and maintenance of gas refrigerators.

Lighting the Burner

Lighting the burner is a rather simple operation provided the manufacturer's instructions accompanying the unit are followed. The lighting procedure is generally as follows:

1. Be sure the gas valve is turned on.
2. Push the lighter button.
3. Light the gas at the end of the burner tube.
4. Continue to push the lighter button until the burner valve clicks open and the burner flame ignites. Should the burner flame go out, wait 5 minutes before relighting.

Do not leave the burner lighted unless all final adjustments have been made. The burner flame should burn a blue color and must enter the flue opening.

Cold Control

The position of the thermostat dial should depend upon the refrigeration load. When the food load is heavy, turn the dial toward a colder position; when the food load is light, turn the dial toward a less cold position. A colder setting will be required in summer than in winter. Experience and observing the effect of the various thermostat settings will usually provide the operator with the required knowledge after a short period of time.

Stopping Refrigeration

When necessary to discontinue refrigeration for any length of time, turn off the gas supply. Remove the ice-cube trays, and empty and dry them. Dry the interior of the refrigerator. Leave the door partly open to ventilate the cabinet interior and keep it fresh.

Refrigerator Too Cold

Low refrigerator temperature may be caused by any combination of the following:

1. If the minimum flame is set too large, the thermostat will not be able to control the temperature in the food-storage

compartment because some refrigeration will be taking place even when the thermostat valve is completely closed.

2. Dirt on the valve seat will keep the valve open slightly. Gas passing through the valve will cause the minimum flame to be too large, giving the same effect as just described.

3. The refrigerator is designed for operation at normal room temperatures. Location in unheated rooms will cause foods to freeze.

4. The highest number on the control dial is the coldest setting. If the food-storage-compartment temperature is too cold, turn the dial to a warmer setting.

Refrigerator Not Cold Enough

If any of the basic adjustments are incorrect, the efficiency of the unit may be impaired. Therefore, when a refrigerator is not cold enough, check the basic adjustments for proper air circulation, level, and right burner flame. In addition, the following items may cause a complaint of poor refrigeration:

1. Check the door gasket seal and repair if necessary.

2. If a built-in refrigerator is improperly vented, there may not be enough air circulation over the condenser to cool the unit efficiently. Check for proper air circulation.

3. Check for leaky cabinet seals when the service history shows that the unit has been changed or if there is evidence pointing to wet insulation.

4. On models that depend on circulation of air around the evaporator for cabinet cooling, the efficiency of the unit will decrease as ice builds up on the evaporator. Some models located in extremely humid climates may need defrosting several times a week for peak efficiency.

5. Dirty condenser fins will prevent the air from circulating through the fins for proper cooling, and, as a result, the efficiency of the unit will decrease in proportion to the amount of dust and lint on the fins.

6. The flue must be clean and have proper draft. A visual inspection of the flue will show whether the flame is entering the flue.

7. The highest number on the control dial is the coldest position. If the complaint is that the refrigerator is not cold enough, the control dial may not be set to a cold enough position.
8. If the maximum-flame pressure setting is in excess of the recommended pressures, the unit may be overloaded and will not perform as efficiently as it should.
9. Dirt in the orifice spud may partially close the opening and limit the amount of gas which can be burned.

No Refrigeration

A complaint of no refrigeration can be caused by the following:

1. Check for burner flame outage.
2. If the evaporator is warm and the control dial is set on the coldest position, it should not be difficult to adjust the maximum-flame screw to obtain the correct pressure at the burner; if this is impossible, change the thermostat.

Unit failure will seldom be experienced. Before condemning a unit, be sure that the following requirements have been fulfilled:

1. Clean the condenser and be sure that there is enough clearance above the refrigerator for proper air circulation.
2. Level the unit front to back and side to side, checking from the bottom shelf of the freezer section.
3. Adjust burner.

Burner Flame Goes Out

The burner will react to adverse conditions more noticeably when it is operating on minimum flame. Any difficulty with flame outage can usually be discovered by looking at the minimum flame and by checking minimum-flame pressures. Burner outage is usually caused by the following:

1. If the burner is spaced too close to the generator, the minimum flame will be pulled away from the heat conductor, causing the heat conductor to cool off and close the automatic shutoff valve. The burner spacing arrangements are shown in Fig. 5-14.

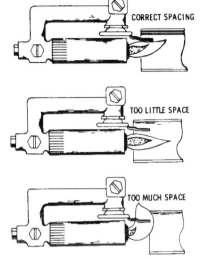

Fig. 5-14. Correct and incorrect burner-to-generator spacing arrangement for proper flame.

2. Check to be sure that the correct orifice is in the burner, and use a manometer to set the recommended pressure.
3. Dirt in the orifice spud may partially close the opening. A visual inspection of the orifice and cleaning are necessary to correct this condition.
4. Always check for flame stability after working on a burner or when there is a complaint of burner outage. To test flame stability, turn the control dial within its operating range until the flame is maximum size; then gradually turn the dial in the opposite direction until the flame begins to reduce in size. Observe the flame as it reduces from maximum flame to minimum flame. If the flame is jumpy or distorted as it passes from maximum to minimum size, it is probably unstable. A further check for stability can be made by snapping the fingers or tapping two wrenches together. An unstable flame will react to the sound waves.

Flame Burns Outside of Generator Flue

The flame must burn inside the generator flue at all times for proper combustion and proper refrigeration. If the flame is allowed to burn outside the generator flue for even 5 minutes, the cabinet seal may be damaged and the insulated section heated by the flame may produce odors due to the insulation being scorched. To correct this complaint, replace the insulation and reseal the cabinet at the points affected by the flame. The following items may keep the flame from entering the flue:

1. The flue may be restricted by objects dropped into it or by deposits that are the result of combustion (sulfur deposits, carbon, etc).
2. Clean out covers, flue extensions, and flue elbows that are corroded, loose, or missing.
3. A natural draft will not be established in the flue until it has become warm. Turning the thermostat to the coldest position and setting the air adjustment for a hard flame will help the flame to enter the flue. Blowing through a short length of tube into the generator flue opening will also cause the flame to enter unless the flue is stopped up. Final air adjustment should be made after the flame enters the flue.
4. A burner spaced too far away from the generator flue will cause the flame to back out when it is controlling at minimum flame.
5. If the minimum flame is too small, the generator flue may cool off, stopping the natural draft; this can also occur if the refrigerator is located in an extremely cold room even with the correct minimum flame.
6. The refrigerator should not be vented to another room, the attic, or the outside. Downdrafts caused by improper venting may force the flame out of the flue.

Manual Defrosting

Most complaints of "incomplete defrosting" or "defrosting takes too long" are usually caused by too large a minimum flame. To correct this condition, check to be sure that the burner has the correct orifice spud and that the minimum-flame pressure is correct. Dirt on the thermostat valve seat may cause the minimum flame pressure to be incorrect. If it is impossible to obtain the correct minimum-flame pressure by turning the minimum-flame adjustment screw, clean the thermostat valve and valve seat.

Automatic Defrosting

The most common causes of defrost problems on 24-hour automatic and pushbutton models are combined in this section because of the related nature of the two types of automatic defrosting.

1. Connect an ohmmeter or test cord to the heater, and check for continuity or for a short; if the heater is found to be defective, replace it.
2. The defrost heater must have good contact with the section of the evaporator which is to be defrosted. Check to be sure that the heater clamps are holding the heater in contact with the section of the evaporator to be defrosted.
3. Failure of the electrical power supply to the heater could result in complaints of no defrosting. This could be caused by an electrical short, an unplugged service cord, or a blown fuse.
4. Place a long screwdriver against the timer case, and press an ear to the handle. If the power cord is plugged in and the clock is running, the gear train sound should be audible. To test the timer electrically, disconnect the power cord and remove the timer. Check for continuity between terminals 1 and 4. This is the clock circuit, and it should have continuity at all times. If no continuity exists, replace the timer. Check for continuity between terminals 3 and 4. Continuity should exist when the dial is turned to the 2:00 A.M. position. If no continuity is found, replace the timer.
5. The pushbutton defrost control is a "fail-safe" mechanism and will not start a defrost cycle if the power element has lost its charge. If the refrigerator is located at an altitude greater than 5000 feet above sea level, the defrost mechanism must be recalibrated to accomplish a complete defrosting cycle.

Odors Inside the Refrigerator

Odors inside the refrigerator may be the result of storing odorous unwrapped foods or infrequent cleaning. Odors will also develop if a refrigerator has been in use and is stored for any length of time with the door closed. If, by any chance, the flame has burned outside of the generator flue, it is possible that a phenolic odor due to burned insulation may develop in the cabinet. To remove this odor, the section of the burned insulation must be replaced.

To remove any of these odors, first remove the source, and then wash with a solution of baking soda and water (approximately 1 tablespoon to 1 quart of water). In extreme cases, use activated charcoal to remove the odor.

Odors Outside the Refrigerator

Odors outside the refrigerator are usually the result of poor combustion caused by the flame burning outside the flue, touching the generator flue, or wrong primary air adjustment. Gas leaks can also cause odors outside the refrigerator. Use a soap-bubble solution to find gas leaks.

Frost Forms Rapidly

Rapid formation of frost can be caused by a bad door seal, uncovered foods or liquids, high humidity, excessive usage, or bad cabinet seal. Keep in mind that the moisture that forms the frost must enter the cabinet in one of the ways mentioned previously. To reduce the amount of frost, the user can be cautioned against excessive door openings on hot humid days. All foods and liquids should be covered before being placed in the refrigerator.

Loose Gas Connection

All gas connections should be leak-tested with a soap-bubble solution before putting the refrigerator into operation. Gas leaks, no matter how small, should be corrected. Be sure that all seal washers and seal screws are tightened on burners, thermostats, etc.

Inoperative Light Switches

A visible check of the light switch may be made when the power cord is plugged in by manually pushing the button in and out to see if the switch is sticking and if the light is going off and on.

ABSORPTION-REFRIGERATION TROUBLESHOOTING GUIDE

On the following pages are listed the most common troubles encountered in the repairs of absorption-type household refrigerators. Each trouble and possible cause are given with the method used for remedying the defect.

Symptoms and Possible Cause *Possible Remedy*

No Refrigeration

(a) Burner	(a) Check to see that the flame provides proper heat input.
(b) Electrical power (when available)	(b) Check for 115 V AC at front terminal board.

Partial Refrigeration

(a) Unlevel unit	(a) Check unit for levelness; adjust levelers at bottom corners, if necessary.
(b) Air filter	(b) Check air filter above condenser. If dirty, wash in detergent and instruct user. If missing, replace and check for dirty condenser.
(c) Air circulation	(c) Check to see that cabinet fans are operative.
(d) Thermostat	(d) Check to see that thermostat is on numbered position calling for refrigeration.

Symptoms and Probable Cause   Possible Remedy

Partial Refrigeration (cont.)

(e) Usage

(e) Instruct user regarding excessive door openings, excessive ice usage, improper loading and storage, and commercial usage.

(f) Improper defrosting

(f) Check to see that evaporator air chambers are defrosted (airflow not restricted). Check to see that defrost heater circuit is not energized.

(g) Flue baffle

(g) Check for omission.

(h) Vapor seal

(h) Check door gasket for good seal. If interior liner sweats or frosts, check and reseal all cabinet openings for good vapor seal.

(i) Cabinet lights

(i) Check to see that lights do not burn when door is closed.

(j) Gas pressure

(j) Check to see that maximum-flame gas pressure is correct. Be sure maximum flame valve is open on input control.

(k) Burner orifice

(k) Disassemble burner and clean orifice. Dirty orifice will restrict gas flow and reduce heat input.

(l) Inefficient unit

(l) Performance-test unit before replacing.

Symptoms and Possible Cause   Possible Remedy

Partial Refrigeration (cont.)

(m) Burner thermal valve

(n) Hi-Temp safety valve

(m) Check to see that valve opens during burner lighting.

(n) Check to see that valve has not failed in closed position.

Burner Outage*

(a) Burner-valve heat conductor

(a) Check to see that heat conductor impinges on intermediate flame. If necessary, turn thermostat dial to "off" position to obtain intermediate flame.

(b) Burner orifice

(b) Disconnect appliance cord from power supply. Check pilot flame for stability.

(c) Air circulation

(c) Check to see that absorber fan is operative. Check to see that air filter above condenser is not dirty or restricted. *Note:* A restricted filter will increase generator draft, pulling flame from burner; this will give flame characteristic of being noisy.

(d) Input control

(d) Check to see that both valves are operative and there are no internal leaks. Check to see that orifice plate is for type of gas used and orifices are not restricted or oversized.

*Customer complaint may be "no refrigeration"

Symptoms and Possible Cause   *Possible Remedy*

Improper Defrosting*

(a) Defrost timer

(b) Defrost heater

(c) Defrost switch

(d) Safety switch

(a) Check to see that motor runs and defrost heater circuit is energized at 2:00 A.M. position.

(b) Check for resistance and proper thermal contact with unit defrost liquid trap. *Note:* Heater will glow if contact is inadequate.

(c) Check for thermal contact with freezer evaporator. Replace switch if inoperative. If stuck closed, heater will be energized too long during defrosting and will cycle on the safety switch. If stuck open, freezer fan will not operate at any time.

(d) Check for thermal contact with heater head conductor. Check for continuity.

Operates Too Cold

(a) Thermostat

(a) Check numbered position. Set on lower number if possible. Instruct user. Check that capillary tube is inserted full length of well evaporator. Check calibration.

*Improper defrosting will probably not be noticed by user. The complaint would likely be "not enough or no refrigeration."

186

Symptoms and Possible Cause	Possible Remedy

Exterior Condensation

(a) Location

(b) Insulation

(a) Instruct user regarding very damp location (e.g., basement).

(b) Check adjacent insulation area for void. Fill void with fiberglass or equivalent.

Interior Condensation

(a) Heaters

(b) Usage

(c) Vapor seal

(a) Check adjacent antisweat heaters for resistance. Replace if inoperative.

(b) Instruct user regarding uncovered foods and liquids.

(c) Check and adjust door gasket for good seal. Check and reseal any cabinet openings (unit, evaporator, cabinet, fans, etc.).

Odor

(a) New appliance

(a) Inside storage compartment will have "new" odor for a few minutes when door is first opened. Instruct user this is normal. When burner is lighted and heated for first time, odor will occur; this is normal. Advise user.

Inoperative Freezer Fan

(a) Defrost switch

(a) Check to see that thermal switch in series with fan

Symptoms and Probable Cause   Possible Remedy

Inoperative Freezer Fan (cont.)

	motor has cooled to "closing" temperature. Advise user it is normal for fan to be temporarily inoperative because of switch following installation, defrosting, and restarting when warm. Check for thermal contact and continuity when cold.
(b) Fan motors	(b) Check to see that motor is operative.

Noisy Burner

(a) Dirty air filter	(a) Wash filter. Do not adjust burner primary air until filter has been checked.
(b) Burner primary air	(b) Rotate burner air-shutter barrel to reduce primary air. Flame must not be yellow when adjustment is completed.
(d) Excessive generator draft	(c) Check for obstruction in area of condenser air intake.

SUMMARY

Absorption-type refrigeration uses heat and a refrigerant that may be absorbed into water. Domestic systems use ammonia, and industrial systems use ammonia or lithium bromide.

Absorption-type refrigeration differs from conventional compression-type refrigeration mainly in that it uses heat energy instead of mechanical energy to change the condition required in the refrigeration cycle.

The freezing system of the gas refrigerator is made up of a number of steel vessels and pipes welded together to form a hermetically-sealed system. The system uses a generator, condenser, evaporator, and absorber. Heat is applied by a gas burner or other source to expel the ammonia from the solution. The ammonia vapor thus generated flows to the condenser. In the path of flow of ammonia from the generator to the condenser are interposed an analyzer and a rectifier.

Dalton's law of partial pressures is important in understanding how the absorption-type refrigeration system operates. An absorption system requires various gas control devices for its operation. There are pressure regulators, shutoff valves for the gas, and input controls. The gas burner consists of a turbulator, orifice spud, and thermal safety valve. A thermostat is used in the form of an automatic gas valve in the burner supply line. It controls the flow of gas to the burner to supply either a maximum or minimum flow.

Defrosting is ordinarily done by electrical means, using a timer and electric heater elements to control the frost buildup. Air is circulated by electric fans.

REVIEW QUESTIONS

1. What refrigerant is used in the absorption-type refrigeration system?
2. Where does the absorption-type refrigerator obtain its heat?
3. Explain how heat makes refrigeration.
4. Explain the absorption cycle.
5. What is a condenser?
6. What is an absorber?
7. What is a rectifier?
8. What does the liquid heat exchanger do?
9. Why is a strong liquid chamber needed?
10. What is the function of the generator?
11. What is the difference between the chill-coil evaporator and the freezer-section evaporator?

12. What does Dalton's law of partial pressures have to do with absorption-type refrigeration?

13. Why do you need a flue system with an absorption-type refrigerator?

14. What are some of the gas control devices used in absorption-type refrigeration?

15. How does the absorption-refrigerator thermostat valve work?

CHAPTER 6

Thermoelectric
Cooling

The type of refrigeration known as *thermoelectric cooling* may
also be used as a heat pump.[*]

PERFORMANCE OF
THERMOELECTRIC COUPLES

A complete circuit consisting of two dissimilar thermoelectric
materials is referred to as a *couple*. A typical couple is shown in
Fig. 6-1. The two thermoelectric materials are represented by n
and p. The n type has a negative Seebeck coefficient and an excess
of electrons. The p type has a positive coefficient and a deficiency
of electrons. The current is shown in the conventional direction;
electrons actually flow in the opposite direction. While there are

---

[*]Some of the information in this chapter is taken from *ASHRAE Handbook of Fundamentals*, Washington, D.C. Reprinted by permission.

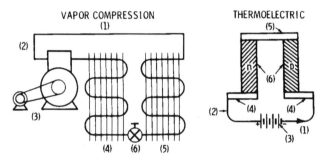

Fig. 6-1. Comparison of vapor compression cooling to thermoelectric cooling.

a total of four connections between the thermoelectric materials and the copper straps, there are only two thermoelectric junctions: the upper, or cold, one, at which heat is absorbed; and the lower, or hot, one, from which heat is evolved. If the direction of the current reverses, the upper junction would evolve heat and the lower would absorb it.

For many years the practical application of the thermoelectric effect was restricted almost exclusively to thermocouples for the measurement of temperature because the metals exhibited a comparatively small Seebeck effect.

The *Seebeck effect* is named for Thomas J. Seebeck who, in 1822, found that when two different metals are used in a circuit and one junction point is hotter than the other, there is an electric current in the circuit. A device consisting of wires or strips of two

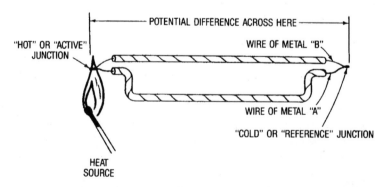

Fig. 6-2. Hot and cold junction points illustrate the Seebeck effect.

different metals, which are in contact at two junction points, as shown in Fig. 6-2, is called a thermocouple. When there is a temperature difference between the two junctions, a difference of potential is generated in proportion to the temperature difference, which produces a current in the thermocouple. The voltages and the currents produced by thermocouples are quite small. However, the Seebeck effect in semiconductors can be considerably greater. The advent of the transistor and other semiconductor devices has stimulated research pertaining to properties of semiconductors in general, and from this have come materials in which the thermoelectric effects are of sufficient magnitude so that the fabrication of useful devices has become a reality (Fig. 6-3).

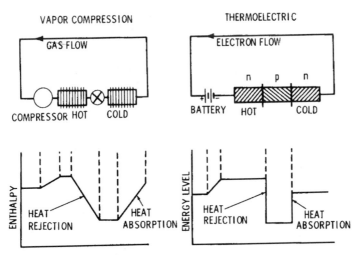

Fig. 6-3. Energy-level analogies.

ADVANTAGES AND APPLICATIONS
OF THERMOELECTRIC COOLING

The heating and cooling functions of the condenser and evaporator in a mechanical system can be interchanged by reversing the direction of refrigerant flow. This cannot be achieved without considerable difficulty and expense. Since the motor and compressor involve rotary and reciprocating motion, worn parts and noise may also present a problem. The containing of the refrigerant requires a hermetic system that must be leakproof. A further inherent limitation of this system is that it cannot be readily made in miniature to economically provide a small amount of refrigeration.

The heating and cooling functions of a thermoelectric system can be easily interchanged by reversing the polarity of the direct current applied to it. Since there are no moving parts, there is nothing to wear out and nothing to generate noise. There is no refrigerant to contain, and the pressure-tight tubing has been replaced by electrical wiring. Since the cooling capacity of a single thermoelectric couple is very small, it is practical to make a very low refrigeration capacity. The thermoelectric system can also be made in large capacities by using many couples.

Capacity control in a thermoelectric system can be achieved by varying the voltage applied to the couples. The current flow is thus changed and the capacity changes. Another advantage is that a thermoelectric system will operate under zero gravity or at many times the force of earth gravity, and it will operate in any position. This is important in considering cooling applications for use in space travel. A thermoelectric system also operates at temperatures up to 220°F.

Along with the advantages, there are some disadvantages. One is that, at this time, a thermoelectric system requires more power than a compressor system to produce a given cooling capacity. Also, since a thermoelectric system has no moving parts and nothing to wear out, it might be expected to have an infinite life. Individual couples and some devices have operated without a noticeable change for some years. However, there are some things that, if not controlled, may cause degradation of performance. It is known that copper will diffuse into a thermoelectric material and adversely affect its thermoelectric properties. The solders for joining thermoelectric material to copper straps must be carefully selected so that they do not diffuse into the thermoelectric material and eventually result in a joint failure. These effects can be minimized, and probably eliminated, by nickel-plating the ends of the couple legs prior to soldering. The solders must also be selected with regard to the temperature range of operation of the system.

Thermal expansion and contraction of the materials on the opposite faces of a system will induce stress. If the function of a particular system is to alternately cool and heat, there will be a reversal of these stresses. Materials must be selected to withstand this.

For many applications, the advantages of thermoelectric cooling will outweigh its chief disadvantage of a relatively low coefficient of performance. For some cooling requirements, a thermoelectric system is the only practical means.

Some of the applications of thermoelectric cooling are: air conditioning and refrigeration systems for use on submarines, temperature control of electronic components, cooling of scientific instrument components, and small specialty appliances (Fig. 6-4).

The U.S. government has sponsored the development of many thermoelectric systems. Among these are: an 8500-Btu unit for possible use as a refrigerating method for frozen food storage rooms on nuclear submarines, a 9-ton air conditioner for a small submarine, temperature control of crystal oscillators in frequency standards to improve their life and sensitivity, a 27,300-Btu direct-transfer air/water system, and a 24,000-Btu direct-transfer air/water system that has undergone trials on a U.S. Navy destroyer.

Thermoelectric applications in the scientific instrument field include: immersion cooler for cooling liquids in a beaker, thermocouple reference junction at 32°F, an instrument for measuring dew point, and a microscope stage cooler (Fig. 6-5).

There are also applications in commercial and appliance refrigeration. Some of these are: a drinking-water cooler for use on diesel locomotives, a small bottle-type cooler for offices, and a small ice maker for use in hotel rooms (Fig. 6-6).

The latter unit was capable of freezing one tray of water (approximately 1 lb) in 6 hours. Small portable refrigerators with less than 1 cu ft internal volume have been marketed by two manufacturers. Both of these could be operated from a 120V AC source or from a 12V DC automobile electrical system. A unit with a 2 cu ft internal volume, which can be used as either a refrigerator or a warming oven, has been produced. The cabinet air temperature is 35°F on the cooling cycle and 160°F on the warming cycle (Fig. 6-7).

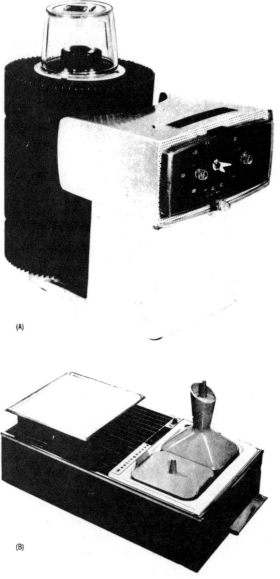

(A)

(B)

Fig. 6-4. (A) Thermoelectric baby bottle cooler and warmer; (B) thermoelectric buffet server featuring cold and hot surfaces—feasibility model.

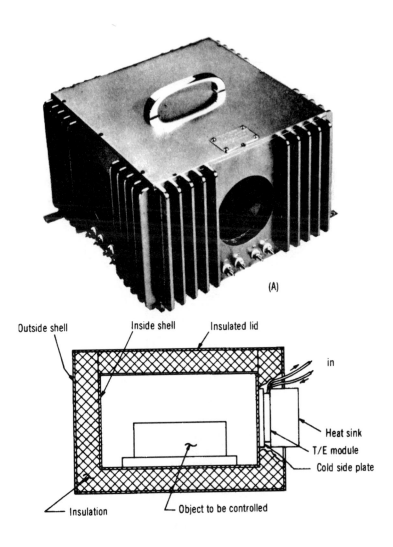

(A)

Outside shell   Inside shell   Insulated lid

in

Heat sink

T/E module

Cold side plate

Insulation

Object to be controlled

Westinghouse

Fig. 6-5. (A) Controlled-temperature cover for reference elements; (B) cooling-chamber details.

197

Fig. 6-6. Thermoelectric dehumidifier—feasibility model.

Westinghouse

Fig. 6-7. Prototype 9 cu ft thermoelectric refrigerator.

SUMMARY

The semiconductor theory and practice used for transistors and diodes is being used for thermocouples that can cool and warm. The direction of current flow through the thermocouple determines whether it warms or cools. This Seebeck effect was discovered many years ago. Many practical applications for the thermoelectric effect are very small in size. However, as more advances are being made in semiconductor products, the likelihood of advancements in thermoelectric cooling will be realized.

The thermoelectric type of cooling means there are no moving parts, which increases efficiency. It also allows for refrigeration where there would be limitations on the conventional type of refrigeration. No high-pressure fluids or gases are needed. Just a small electric current at low voltage will do the job. However, more power is required to produce the same amount of cooling by thermoelectric means than by running a compressor-type refrigeration system. For some types of cooling the thermoelectric method offers the only available source of temperature control.

Commercial and laboratory uses for thermoelectric devices are just being developed. However, they use the same principle of operation as semiconductor materials.

REVIEW QUESTIONS

1. What is the Seebeck effect?
2. How does the thermoelectric method of cooling work?
3. How is the heating effect brought about in a thermoelectric device?
4. What are the advantages of thermoelectric heating and cooling over conventional heating and cooling systems?
5. How efficient is the thermoelectric cooling method?
6. What are some of the disadvantages of the thermoelectric cooling system?
7. List some applications for the thermoelectric method of cooling.
8. What is the difference between p type and n type materials?
9. Why is the Seebeck effect useful in both heating and refrigeration equipment?
10. What is the upper limit in temperature at which the thermoelectric system works?
11. Why is this type of refrigeration system so useful in outer space and in submarines?
12. How long does it take the thermoelectric system to freeze water?

CHAPTER 7

Refrigeration Service Equipment and Tools

This chapter covers much of the equipment and tools needed to service both domestic and commercial refrigeration systems. Thus they will be applicable in much of your servicing of refrigeration work.

Domestic hermetically-sealed units have been provided with several types of methods of adding or drying out refrigerants that have accumulated moisture. Some require a special adapter for connection to the hermetically-sealed compressor, some have a service valve, and some require a special adapter to connect the suction line to the compressor with a valve incorporated with it. Domestic refrigeration will be explained in Chapter 8. It is the intent of this chapter to make you aware of the facilities that are available to you for refrigeration servicing.

Domestic refrigerators have a certain number of ounces of refrigerants that the unit is to be charged with. Practically 100 percent of commercial refrigerators have service valves on both the discharge and suction sides of the compressors. An exception to this may be self-contained water coolers, which may or may not have service valves.

SERVICE VALVES

Service valves referred to in service operations are named according to their function:

1. Compressor suction
2. Compressor discharge $\Big\}$ Service shutoff valves
3. Receiver service
4. Evaporator liquid-line service
5. Evaporator suction-line service

The compressor suction and discharge valves (Fig. 7-1) are located at the compressor inlet and outlet connection, respectively. These are usually dual valves, commonly called *two-way valves*, which open or close with the valve stem screwed all the way out or all the way in. In this manner, turning the valve stem will control two outlets from one valve. The side-port opening may be used for charging, dehydrating, testing, etc., after the proper connections have been made. The side port is closed by turning the valve stem, which is covered by a brass cap. The port opening is sealed with a small plug, which should never be removed until the valve is checked. The main opening of the valve is closed by turning the stem all the way in (to the right).

After making gage, charging, or dehydrating connections, the valve should be opened to the side port by turning the valve stem one and one-half turns. Do not turn the stem all the way because this would close the main opening and stop the flow of refrigerant. Fig. 7-2 shows the various positions of service shutoff valves. The valve stem is provided with a packing gland, which may be tightened in case of a leak around the stem. The port plug and valve cap should always be replaced after adjustments are made to prevent any gas from escaping. The receiver service valve is usually located at the receiver on the liquid line to the evaporator and is generally a single-seated-type valve, having one inlet, one outlet, and no opening for a gage.

The evaporator liquid- and suction-line service valves are used only on evaporators having a float-valve header. Frequently, the valves are equipped with a built-in strainer to remove foreign particles before entering the evaporator, which is located at the evaporator header in the liquid-line connection between the receiver and the evaporator. The evaporator suction-line service valve is also located at the evaporator header in the suction line and is connected between the compressor and the evaporator.

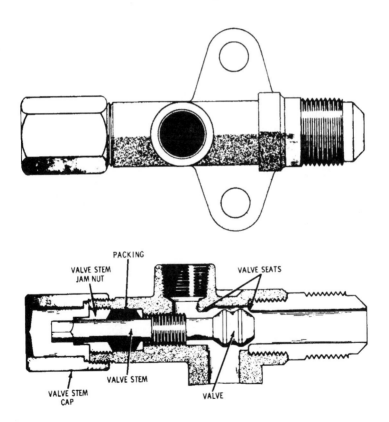

Fig. 7-1. Exterior and interior view of a typical compressor suction or discharge valve.

To Crack a Valve

The term *crack*, as used in servicing refrigeration units, means to open slightly and close again. Thus, to *crack* the liquid-line shutoff valve, for example, open it slightly to allow a tiny spurt of liquid to flow through, and then close it. The same procedure applies to cracking any valve or valve port in the system. This is important when purging the gage on the charging line of air and to avoid releasing an excessive amount of refrigerant.

COMBINATION GAGE SET

A test gage set, or testing manifold (Fig. 7-3), consists of two tee valves built into one valve body, with the valve stems extending

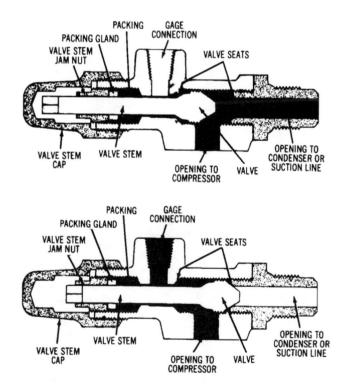

Fig. 7-2. Service shutoff valve in open and closed positions.

out on each end for wrench or hand-wheel operations. The tee valves are constructed so that the valve works only on the leg that is attached to the tee. The opening or closing of the stem only affects the one opening; the other two openings on the valve will remain open.

Mueller Brass Company

Fig. 7-3. Typical combination gage set.

Importance of Gages and Combination-Gage Set

Since most household refrigeration units are not equipped with gages, a serviceman must insert his own gage for the various testing and service operations that must be dealt with in solving day-to-day problems. Numerous important operations and tests are performed by the test gage set, the most important being the following:

1. Observation of operating pressure
2. Charge refrigerant through compressor, etc.

3. Purge receiver
4. Charge liquid into high side
5. Build up pressure in low side for control setting or to test for leaks
6. Charge oil through the compressor, etc.

Types of Gages Used

The gages used in refrigeration servicing are the *low-pressure (compound) gage* and the *high-pressure gage*. The standard type of low-pressure gage is often called a *compound gage* because its construction permits a reading of both pressure in pounds per square inch and vacuum in inches of mercury, as shown in Fig. 7-4. A standard compound-gage dial is graduated to record a pressure range of from 30 inches of vacuum to 60 psi. The high-pressure gage is equipped with a dial for measuring pressures of from 0 to 300 psi. Gages used in refrigeration service should read approximately double the actual working pressure.

SEALED SYSTEMS

The first step when charging a sealed system with refrigerants is to have the right charging and evacuating equipment, which will enable the serviceman to charge the unit with the exact amount of refrigerant required. Although several charging methods are used, the sight-glass charging cylinder is recommended. This method of refrigerant charge will be accurate regardless of ambient temperature.

Evacuating and Recharging

Generally, every refrigerator unit is properly charged with refrigerant upon delivery; the main service due to refrigerant troubles usually consists of evacuation and recharging. A step-by-step procedure after the charging equipment has been installed, as shown in Fig. 7-5, will be as follows:

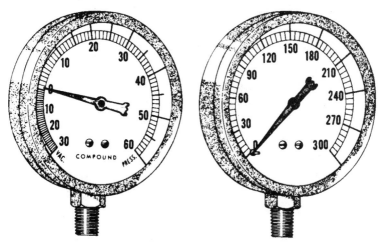

Fig. 7-4. Typical compound- and plain-pressure gage.

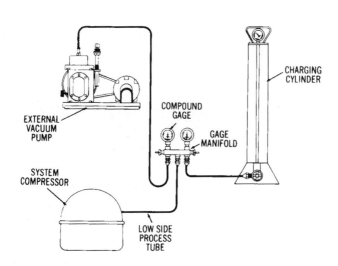

Fig. 7-5. Typical charging and evacuation arrangement when installed.

207

1. Discharge the system and adapt the proper replacement components.
2. Evacuate the system thoroughly.
3. Charge accurately through the low side with the proper refrigerant.
4. Leak-test carefully.

Figs. 7-6 to 7-10 are schematic illustrations of typical external vacuum-pump and charging-cylinder connections through a gage manifold and the compressor process tube.

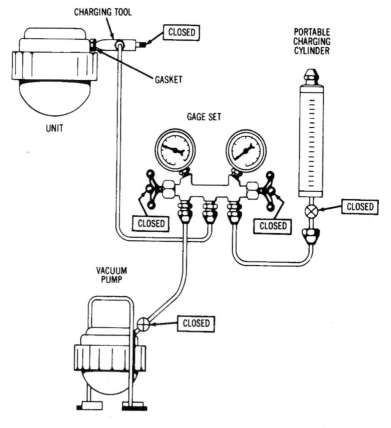

Fig. 7-6. Arrangement of charging equipment when installed and ready for use with all valves in closed position.

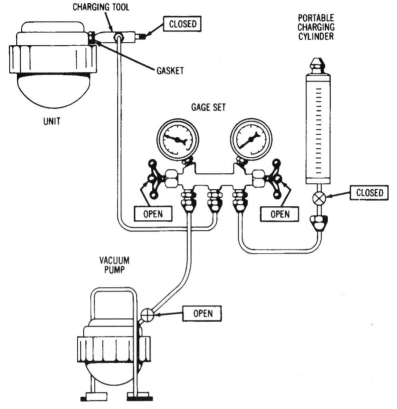

Fig. 7-7. To evacuate lines, open gage set and gage manifold. Close the vacuum-pump valve and check to see that the vacuum holds.

Evacuation of Sealed System

It is a good rule, whenever a sealed system is opened and the refrigerant charge removed, to evacuate the system and install a new service drier before recharging the system. To evacuate the refrigerator, proceed as follows:

1. Use a good external vacuum pump and change the oil often for efficient operation.
2. Install a service valve on the low-side process tube of the compressor.

3. Connect the vacuum-pump hose to the service valve. Leave the valve in a closed position.

4. Start the vacuum pump, open its discharge valve, and slowly open the service valve on the system compressor.

5. If an extremely efficient vacuum pump is used, crack the suction valve on the pump for the first minute, and then slowly open it. This procedure will prevent the oil in the system from foaming and being sucked into the vacuum pump in large quantities.

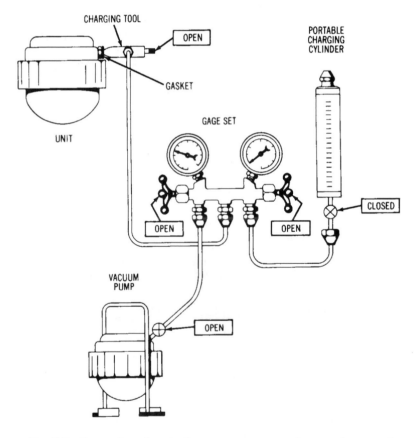

Fig. 7-8. Open charging tool, vacuum-pump valves and evacuate assembly for about 20 min. Close the vacuum-pump valve and check to see that the vacuum holds for about 5 minutes.

6. Next, pull a vacuum for about 20 minutes, which should give a reading of 500μ, or 29.6 in. Hg, on the compound gage. At the end of the 20-minute vacuum time, the vacuum pump should be valved off and the micron gage left in the system. *It is important to observe the gage.* If the gage reading rises, there could be a leak in the system.

7. Close the service valve on the system compressor; then stop the vacuum pump.

8. Connect a sight-glass charging cylinder hose to the service valve (Fig. 7-5) and purge the charging hose. Induce a

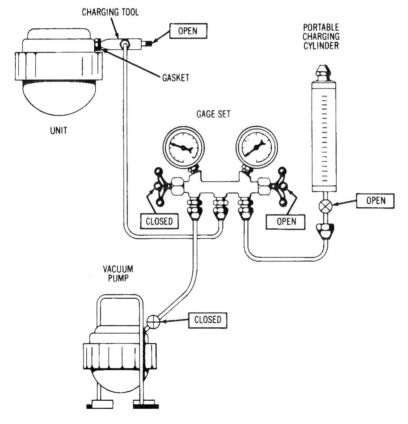

Fig. 7-9. To charge the unit, close the vacuum-pump line and open the charging cylinder valve; allow the refrigerant to be drawn into the compressor.

charge of refrigerant into the system until the low-pressure side reads 30 to 40 psig; then leak-test the low side. After the low-side leak tests are completed, run the system compressor for a few minutes.

9. Leak-test the high side of the refrigerant system.

10. After completion of the leak tests, purge this temporary refrigerant charge out of the low side. This refrigerant charge will assist in removing moisture within the system.

11. Repeat the vacuum pump procedure for another 30 min-

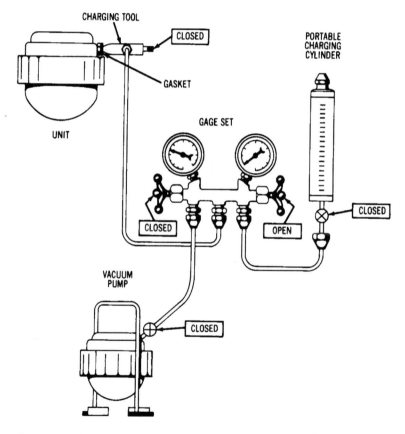

Fig. 7-10. After the compressor has been fully charged, close the charging valve on the compressor and remove the charging tool. Check the charging screw for tightness.

utes, which should give a reading of 500μ or 29.6 in. Hg. If the discharge tube of the vacuum pump is in a container of refrigeration oil, there should be no bubbles in the oil after completion of the foregoing procedure, thus indicating that the system has been evacuated properly.

You have noticed that the figure 29.6 in. Hg is given in many instances. Remember that this is at sea level. In the early portion of this book, we gave the inches of mercury values to be expected at other altitudes. This value is approximately 1 in. Hg less per 1000 feet of altitude. Thus, at 5000 feet, you would observe 24.6 in. Hg.

Charging of Sealed System

The equipment for charging refrigerant is the same as that used for evacuation. The charging procedure is normally as follows:

1. The first step in charging a system is to have the proper equipment so that the refrigerant can be accurately determined within ± ¼ ounce. Although there are several methods of charging, the use of a sight-glass charging cylinder is preferred since this method is accurate regardless of ambient temperature.
2. Always charge through the low side of the system either through the process tube or the suction line. Charge the system with the compressor off since most of the refrigerant will enter in liquid state. Allow about 5 minutes after the charge has entered the system before starting the compressor. The efficiency of charging can be increased by elevating the charging cylinder; however, refrigerant should always be introduced slowly.
3. Connect the charging hose of the charger to the service valve on the compressor.
4. Open the charger valve and purge the hose. As soon as the refrigerant has stabilized in the charging cylinder, check the pressure reading of the cylinder and rotate the sleeve to correct the setting for pressure and the type of refrigerant being used. Control the flow of refrigerant with the service valve, not the charger valve.

5. When charging the refrigeration system, some bubbles may appear in the charging cylinder. These can be eliminated by closing the service valve and tipping the charging cylinder upside down momentarily. Then continue with charging until the correct charge has entered the system.

6. If, at any time, it is required to raise the internal pressure of a charging cylinder, use a bucket of warm water (not over 125°F) and place the cylinder in the bucket. *Caution: Never place a flame on any refrigerant cylinder.* This can cause hydrostatic pressure to build up to a dangerous level, and, in extreme cases, it can cause the cylinder to rupture like a bomb. Extreme care should always be taken when handling refrigerant cylinders.

7. When it has been ascertained that the system has the correct refrigerant charge, use the pinchoff tool and make the final joint. If possible, leave enough round tubing between the compressor and the pinchoff crimp in order that the process tube may be used at a later date.

EXPANSION-VALVE UNITS

Because numerous serviceable refrigerators are equipped with expansion valves for refrigerant control, the more common service operations on these units are described in the following paragraphs:

Use of Combination-Gage Set to Add Oil

With reference to Fig. 7-11, oil may be added to the compressor as follows:

Connect a piece of ¼-in. copper tubing between the oil supply and valve E. Pour the refrigeration oil into a clean dry bottle. Put the compound gage on the suction-line valve and close the liquid line at the receiver. When a sufficient amount of oil has been added, close the valve at point A and open valve D. The compressor may now be put into normal operation. *Caution: Make sure that the end of the oil-supply tubing is kept*

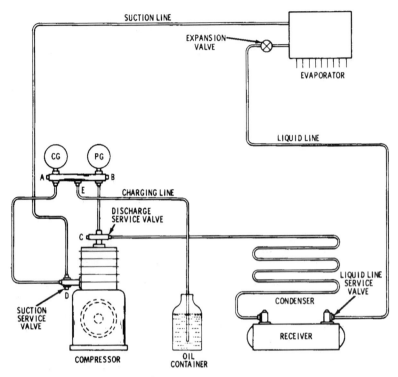

Fig. 7-11. Schematic diagram showing connections for adding oil to the compressor.

submerged below the oil level in the container so that air will not be drawn into the system during the operation.

To Charge Refrigerant through Low Side

With reference to Fig. 7-12, refrigerant may be added as follows:

Connect the suction-line valve port and the refrigerant tank with a piece of ¼-in. copper tubing with gages connected as shown. The cylinder should be placed on a suitable scale in an upright position, with the scale reading noted. The ¼-in. tube between the manifold and the charging drum must be long enough so as not to affect the scale reading. Purge the air out

215

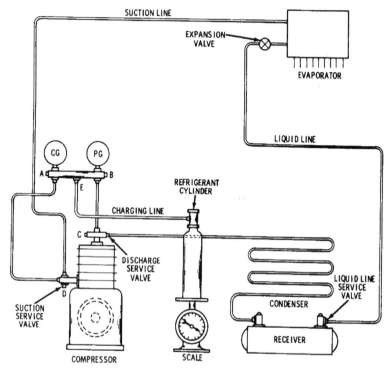

Fig. 7-12. Schematic diagram showing connections for adding refrigerant through the low side.

of the charging line by turning on the gas at the tank valve and cracking the valve unit at valve B. When the line has been purged, open the suction-line valve and start the compressor. A back pressure of about 5 lb is maintained in charging with sulfur dioxide (SO_2) and about 10 lb for Freon (F-12).

If the cylinder is cold, apply heat to the gas cylinder by means of a hot cloth or a pail of hot water, which will assist in the transfer of refrigerant to the unit. When the required amount of refrigerant has been added, close the valve on the gas drum and allow the pressure to fall to zero on the gage; then closer the valve at the suction line. Disconnect the lines and replace the plugs. Put the unit into normal operation.

To Charge Refrigerant through High Side

In some cases, where a large amount of refrigerant is to be added, it is advantageous and time saving to charge the refrigerant as a liquid into the high side of the system instead of pumping it into the low side. In such cases, it is recommended that only a known required quantity of refrigerant be contained in the charging cylinder. To accomplish a charge, a reference to Fig. 7-13 will be of assistance. Proceed as follows:

Connect the refrigerant cylinder by means of ¼-in. copper tubing to the manifold, as shown. The compressor should be stopped, and the condenser allowed to cool to room temperature. The charging line should be purged in the manner previously described. The refrigerant cylinder should be heated,

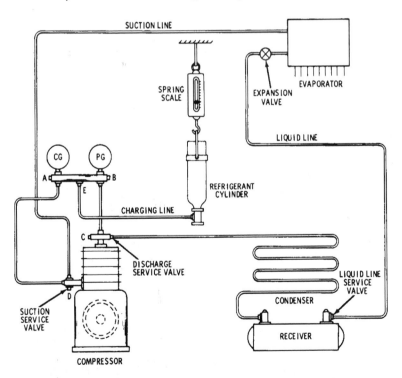

Fig. 7-13. Schematic diagram showing connections for adding refrigerant through the high side.

inverted, and securely fastened, after which the cylinder valve is opened. While the cylinder is being discharged, it may be necessary to supply additional heat to maintain the flow of refrigerant to the unit. When the required amount of gas has been added, close both the valve on the gas container and valve A. Valve B should be opened to bypass the pressure from the charging line to the low side. Finally, after the charging line records zero, valve B should be closed, after which the manifold connection to the service drum is capped and the unit put into operation.

To Evacuate Air from Entire System

During normal operations, both the back and head pressures are above atmospheric pressure, and the only way in which air can enter the system is during service operations or due to a leak in the system. When the high side of the system has a leak, it will expel gas until atmospheric pressure is reached, and the back pressure will drop below atmospheric pressure. When the unit is operated below atmospheric pressure, air will be drawn into the system. Air in the system will cause excessive high-side pressure and will nearly always stop the unit by overloading the motor, which will trip the overload cutout.

When evacuating air from a system, it is important to first ascertain that all leaks have been stopped; otherwise air will reenter the line. To purge air from the system, refer to Fig. 7-14 (which shows a diagram of the connections) and proceed as follows:

Connect the suction side of the compressor to the compound gage. The compressor discharge service valve should be connected to a glass jar filled with water, by means of ¼-in. copper tubing. To prevent any loss of oil, which may escape during the pumping operation, the oil should be carefully measured and returned to the compressor after the purging operation. Check the compressor discharge-valve port to make sure it is closed, after which the compressor suction-valve ports should be opened to both the gage set and the evaporator. The receiver liquid line should be checked to make sure it is opened, and the compressor should be put under intermittent

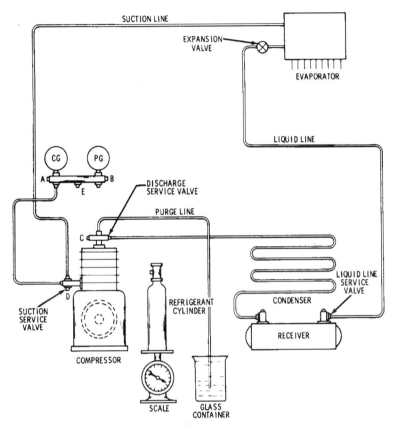

Fig. 7-14. Schematic diagram showing connections for purging air from the entire system.

operation until a maximum amount of vacuum is obtained, as registered by the compound gage.

If, after the unit has been stopped for a period of 4 or 5 minutes without any change in the gage reading, the unit is free from air, the glass jar end of the copper tubing should be connected to the middle leg of the manifold. At this time, a charged refrigerant gas drum should be connected to the manifold at valve B and the line connection from the gas drum purged. When the gas-drum valve and manifold valve are opened, the gas charge enters the system and the pressure is built up to a positive value of about 10 lb. When the required amount of refrigerant has been added, close the valve on the gas drum and add any oil which may have escaped during the purging operation. After this operation, a final check for leaks should be made and the unit put back into normal operation.

Purging Air from Condenser

Air in the condenser will cause excessive high-side pressure, resulting in long operating periods with accompanying waste of power. The unit should be shut down during this test and during the purging operation. Allow the condenser to cool down to room temperature. With reference to the diagram, proceed as follows:

Attach a purging line ($\frac{1}{4}$-in. copper tubing) to the service connection valve E on the manifold, as shown in Fig. 7-15. If the refrigerant is sulfur dioxide (SO_2), place the end of the tubing into the bottom of a container in which $1\frac{1}{2}$ lb of concentrated lye (NaOH) has been dissolved in 1 gallon of water. Then crack valve B (loosen and immediately tighten), and purge slowly for several minutes. Observe the pressure drop registered on the pressure gage; when it has returned to normal, the purging should be discontinued, valve B should be tightened, the purging line removed, and the unit returned to normal operation.

Evacuating Entire Refrigerant Charge

When it becomes necessary to remove the entire refrigerant charge from the unit (as in the case when an exchange of the compressor or any other component of the condensing unit is necessary), it is often desirable to salvage the refrigerant by evacuating (pumping) it into an empty service drum. With reference to Fig. 7-16, the procedure is as follows:

To pump the gas into an empty cylinder or drum, put the cylinder in a bucket of cold water (preferably ice water), as shown. Connect the empty cylinder to the middle leg of the manifold by means of $\frac{1}{4}$-in. copper tubing. The compressor discharge service valve should be closed, and valve B should be

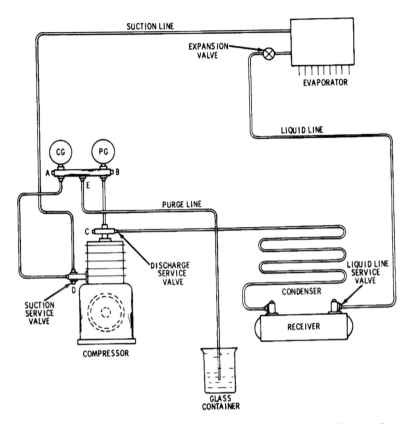

Fig. 7-15. Schematic diagram showing the connections for purging air from the condenser.

opened; the valve on the cylinder should also be open. After the compressor suction-line service valve and the receiver liquid-line valve are opened, the compressor is put into operation. Keep the cylinder as cold as possible until the operation is complete. The compound gage attached to the suction shutoff valve will show when the charge has been completely removed. If, during this pumping operation, the pressure gage indicates an abnormally high pressure, the unit should be stopped, allowing the compressor to cool down. An indication of a complete evacuation of the system will be obtained by closing the cylinder valve. If an observation of the gage indicates a

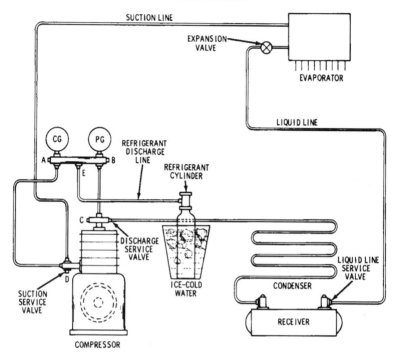

Fig. 7-16. Schematic diagram showing connections for evacuating the entire refrigerant charge.

constant high vacuum for any length of time, the evacuation is complete and the service attachments, including the gas cylinder, may be removed from the unit.

LEAK TESTING

The old soap-bubble and oil method of detecting leaks in refrigerators using modern refrigerants is, at best, only makeshift and should be used only in the absence of proper detectors. There are several types of leak detectors available, such as the *halide torches*, or the *electronic* type presently manufactured by the General Electric Company (Fig. 7-17). The halide type, however, is probably the most widely employed and is sufficiently sensitive for household refrigeration systems.

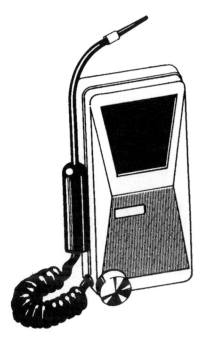

Fig. 7-17 Electronic halogen leak detector.

Leaks in a refrigeration system will usually result in an undercharge of refrigerant. To add refrigerant without first locating and repairing the leak would only be a temporary solution since it will not correct the difficulty. The leak must be located and repaired, if possible, after which the entire system must be evacuated and recharged with the proper amount of refrigerant. Whenever a new charge of refrigerant is added, it is necessary to install a new drier. Any leak, regardless of its size, must be located before a determination can be made of the operative status of the system components. Do not replace a component because the system is short of gas unless a nonrepairable leak is found.

If the analysis indicates a leak, find it before opening the system. It is better to locate the leak before discharging than if the surrounding air is contaminated with refrigerant from a newly opened system. The presence of oil around a tubing joint usually indicates a leak, but do not let this be the determining factor. Always check the area with a leak detector to make sure.

Soap-Bubble Method

Soap bubbles can be used to detect small leaks in the following manner: Brush liquid detergent over the suspected area and watch for the formation of bubbles as the gas escapes. Sometimes, if the leak is of a slight intensity, several minutes must elapse for a bubble to appear. *Caution: Use the bubble method only when it is certain that the system has positive pressure.* Using it where a vacuum is present could pull moisture or soap bubbles into the system. A joint that is suspected of leaking can be enclosed in an envelope of cellophane film. Tightly tape both ends and any openings to make it gas-tight. After about 1 hour, pierce one end of the film for the probe and pierce the other end for air to enter. If a response is obtained, the joint should be repaired.

Halide Leak Detector

When testing with the halide torch, make sure the room is free from refrigerant vapors. Watch the flame for the slightest change in color. A very faint *green* color indicates a small leak. The flame will be unmistakably *green* to *purple* when large leaks are encountered. To simplify the leak detection, keep the system pressurized to a minimum of 75 psi. This is easily accomplished for the high side by merely running the compressor. To pressurize the low side, allow the entire system to warm up to room temperature.

The halide-leak-detector method has been used for many years by air conditioning and refrigeration servicemen and consists of a combustible gas supply with a small burner on top. A tube from the bottom of this burner forms the probe. When this probe is moved near a large leak, the refrigerant going into the tube combines with the flame to cause perceptible change in the flame color. Disadvantages of using the halide torch include the following:

1. It is slow and relatively insensitive.
2. In a brightly lighted area, it is almost impossible to locate leaks smaller than about 6 ounces per year.
3. If the area is darkened, a smaller leak can be located; however, the change in the color of the flame is difficult to

interpret because other gases and dust cause the flame to change color.

Electronic-Type Leak Detector

Although leak testing with a halide torch is considered satisfactory in most instances, a more reliable test is made by means of the electronic-type tester shown in Fig. 7-17. This easy-to-use instrument reduces the guesswork from leak testing because it is more sensitive, faster responding, and capable of detecting a leak even though the surrounding air may be contaminated.

A number of halogen leak detectors are available. However, some of the newer electronic detectors are very sensitive. Figure 7-17 shows an automatic type that has a permanently-sealed, miniature battery-operated pump that produces a computer-like beeping signal that changes in both speed and frequency as the leak source (halogen or vapor) is approached. It can detect leaks as small as 0.5 ounce per year. An additional feature is that this instrument is capable of calibrating itself automatically while in use. It operates on two C cells. The detector sensor is not ruined by large doses of refrigerant.

To make a leak test by means of the electronic detector, proceed as follows:

1. Turn on the test detector.
2. Probe for leaks.
3. Recheck the suspected leaks for confirmation.

After replacing a component in the refrigerating system, always leak-test all joints before recharging. The extra time it takes is negligible compared to the loss of charge due to a faulty connection. Be sure to clean the excess soldering flux (if used) from the new joints before leak testing since it could seal off pinhole leaks that would show up later.

Service instruction charts have been worked out, which will be of help in locating and repairing faults likely to be encountered in day-to-day service work.

SERVICE TOOLS

Some of the tools illustrated will be used mostly with commercial refrigeration work, while some application will also be found in home refrigeration servicing.

A careless repair job is dangerous and has no place in refrigeration service work. Properly chosen tools will not only save time but will, in addition, insure a neat, dependable, and satisfactory job which will result in mutual benefits to the serviceman and customer alike.

There are many types of tools on the market. Some of them are so poorly constructed that they are expensive at any price. A good tool warrants a fair price, because it is properly designed and will stand up under service. A list of special tools is given in this section.

Test-Gage Set

One of the handiest aids in servicing small machines is the test gage. Most small units are not equipped with gages; the serviceman must apply his own so that the pressures existing in the high and low sides can be determined. A test-gage set consists essentially of two *tee valves* and a special *tee*, as shown in Fig. 7-18.

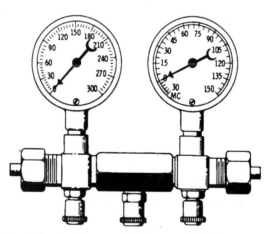

Fig. 7-18. Combination gage set. High-pressure side gages are usually scaled for pressures up to 300 psi. The low-pressure gage is of the compound type and is used to measure inches of vacuum and pressure in pounds per square inch.

The tee valves are constructed so that the valve stem works only on the leg attached to the tee. Therefore, opening or closing the stem affects only this one opening; the other two openings of the valve remain open at all times. This fact is plainly shown in the sectional view of the valve in Fig. 7-19. This unit can be used to obtain pressure readings on both high and low sides when adding oil, adding or recharging refrigerant, purging off excess refrigerant or air, or pumping refrigerant from one side to the other.

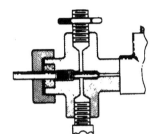

Fig. 7-19. Sectional view of a refrigerator tee valve.

Inspection Mirrors

Magnifying mirrors are used by many servicemen for observing leaks in refrigerating machines where soap and water or oil is used as a leak detector. The long handles permit these mirrors to be used in difficult or cramped quarters. Some compressor assemblies are designed so that these mirrors offer the only means of seeing certain bolts, inserting other pins, or aligning parts.

Pipe Cutters

Pipe can be cut in two ways: by hacksaw or by pipe cutter. The former method is often slow and laborious. Pipe cut by a hacksaw is not beaded or provided with a thickened lip, as is the case when pipe cutters are used. If a three-wheel cutter is used, as shown in Fig. 7-20, the outer edge of the pipe must be filed down before a die can successfully be applied to the pipe.

When a single cutter wheel is used in conjunction with two rollers, the outside of the pipe stays close to its original diameter, but the inside of the pipe is given a lip, or burr, that must be removed with a pipe reamer. If this burr is allowed to remain, it will decrease the free internal diameter and offer a resistance to the passage of the fluid that is to flow through the pipe. Such burrs collect dirt and often result in subsequent binding or clogging at such obstructions.

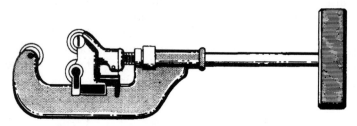

Fig. 7-20. Adjustable three-wheel pipe cutter for use on 1/8-inch to 1/4-inch pipe or tubing.

Pipes severed with a hacksaw do not require reaming, and the die can be applied without the necessity of filing the outer edge. In effecting a cut with a hacksaw, it is important that a square cut be made. If an angular cut is made, the die will not cut straight threads. When such a threaded end is screwed into a fitting, the pipe will point out at an angle and it will be difficult to assemble a run of pipe with such threads, especially close or short lengths. Pipe cutters offer the quickest severance and, if properly applied, result in a square cut, so that the die will follow the pipe without trouble.

Tube Cutters

The larger installations maker use of steel or iron pipe for connections, but the smaller machines utilize copper tubing. Copper tubing can be easily installed because it can be bent around obstructions and eliminates many elbows. Since tubing is made in fairly long lengths, most small installations can be completed by the use of a single length of tubing. For instance, the evaporator can be connected to the receiver by means of a ¼-in. tube in one single piece even when the two units are separated by a distance of 100 feet. Therefore, sharp bends and all chances of clogging at constrictions or bends are eliminated, while the friction factor is greatly reduced.

Copper tubing can be cut by means of a saw and flaring block, as shown in Fig. 7-21. The tube is held in the flaring block for the purpose of securing a square end. It is important to secure a square end; otherwise the operator will be obliged to make careful user of a file. A much easier and quicker method is to use a tubing cutter, as shown in Fig. 7-22. This tool is a small version of the pipe cutter shown previously.

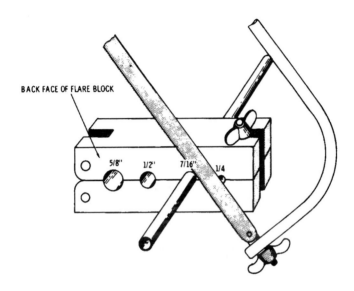

BACK FACE OF FLARE BLOCK

5/8" 1/2" 7/16" 1/4"

Fig. 7-21. Method of cutting tubing with a hacksaw in a flaring block.

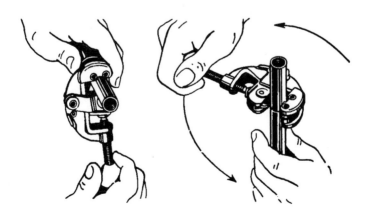

Fig. 7-22. Using a tubing cutter to cut tubing.

Bending Tools

Where accurate bends of a specified radius must be made, a bending tool should be used. This bending tool, shown in Fig. 7-23, consists of a stationary arm on which is located the radius guide. The radius guide is usually provided with a pin, so that the bending arm can be attached and used to make the bend. The bending shoe is also grooved and is furnished with a shoe block or a grooved wheel. The tool makes a perfect bend without crimping or flattening the tube.

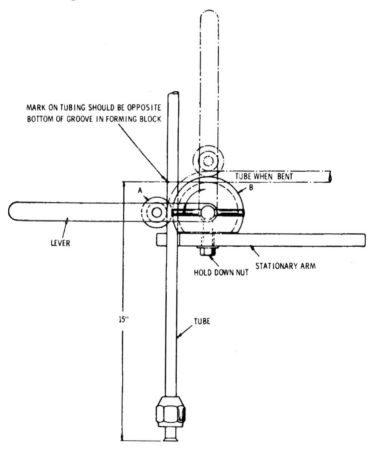

MARK ON TUBING SHOULD BE OPPOSITE
BOTTOM OF GROOVE IN FORMING BLOCK

TUBE WHEN BENT

A

B

LEVER

HOLD DOWN NUT   STATIONARY ARM

15"   TUBE

Fig. 7-23. Bending tubing by means of a specially designed bending tool.

Pinchoff Tools

Most of the larger apparatus are equipped with sufficient valves to enable any portion or section to be closed off and worked on without loss of the refrigerant charge. Some of the smaller and simpler systems have only a few valves. This is an economy measure so that the cost of the apparatus can be kept to a minimum. While the elimination of valves makes a simpler, lower-priced piece of equipment, it makes servicing somewhat harder. Where there are only a few or no valves in a system, the pinchoff tool offers a means of closing the copper lines so that the defective or worn part can be removed and replaced.

A simple pinchoff tool is shown in Fig. 7-24. Notice the jaws for the purpose of pinching or squeezing a tube shut. These jaws do not come together when the two blocks are flush; if they did, the tubing would be severed. The pinching jaws are so designed that the tubing is pinched shut but not mashed. The round opening is used to round up the pinched section when the repair or replacement has been completed.

Valve Keys

It will be found that some refrigeration apparatus is equipped with valves that have no handles. The stems are provided with a square end, so that a handle can be applied or a key fitted to the valve stem, as shown in Fig. 7-25. When opening or closing such a valve, a *valve key* or *tee wrench* is sometimes used. Common sizes have ³⁄₁₆-, ⁷⁄₃₂-, ¼-, and ⁵⁄₁₆-in. openings. These sizes can be obtained in the *tee* form or with a square end so the tool can be used with a ratchet wrench.

Packing Keys

Some valves are packed with a resilient material that serves as a seal to prevent the refrigerant from escaping around the stem. A packing nut can be tightened by means of a wrench. When tightening external packing nuts, use a 12-point wrench if one is available, or make use of the proper open-end wrench.

The packing nut is generally the internal type and is provided with a slot across its face for tightening. Unfortunately, the stem of the valve projects through this nut and makes it necessary to use

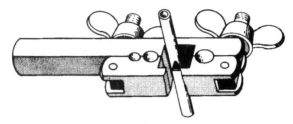

Fig. 7-24. Typical pinchoff tool. Block is cut away to show working principles.

Fig. 7-25. Various valve keys.

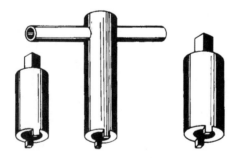

Fig. 7-26. Typical packing keys.

a special gland or packing-nut key, such as shown in Fig. 7-26. This key is hollow in the center, permitting the valve stem to extend through the tool, and is provided with two teeth that engage the slot in the packing nut.

Flaring Tools

A typical flaring tool is shown in Fig. 7-27. It consists of two bars held together by a wing nut and bolt; the bars are provided with holes for the various sizes of tubing. A yoke containing the forming die is slipped on the bars, and the handle is rotated to produce a flare. This tool has been very widely used because of its simplicity and ease of performing the flaring operation, and because the flares produced by it are uniformly excellent in producing a tight seal.

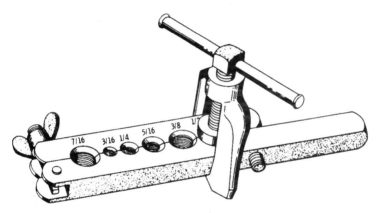

Fig. 7-27. Flaring tool for use on various tube diameters.

Before applying pressure with the forming tool, place a drop of oil on the end of the tube. Be sure to use only refrigeration oil. Any other type of oil may enter the tube and give trouble by reacting with the refrigerant at a later date. The proper use of the flaring tool is shown in Fig. 7-28.

Gasket-Tacking Machine

The serviceman should notice the condition of the packing around ducts and cabinet doors. If it is worn or collapsed so that air is permitted to pass, the packing strips should be renewed or replaced. One of the best labor-saving devices is the use of a tacking or stapling machine when a wooden door or strip is provided for the attachment of the gasket material.

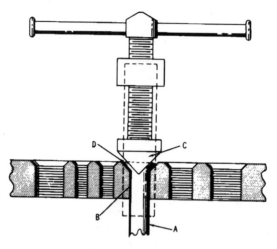

Fig. 7-28. Proper use of flaring tool: (A) Tubing in place; (B) teeth in the die, holding tubing securely; (C) compressor making a perfect flare; (D) black section shows the flare formed at the top of the tubing.

MAKING A HAND BEND

Occasionally a serviceman will find it necessary to remove a battered, leaking, plugged, or defective piece of tubing and replace it with a new piece or section. If no mechanical bending tool or spring is available, the bend must be made entirely by hand. Five different procedures can be used in bending copper tubing, which will be given in the following sections:

Procedure No. 1

1. Make sure the piece of tubing is straight.
2. Mark the tubing where the bend is to be made.
3. If possible, use the old piece of tubing as a guide, or use a piece of wire that has been bent to the desired form and cut to the proper length.
4. Grip the tubing with both hands, one on each side of the mark; place the thumbs directly underneath the point at which the tubing is to be bent.

234

5. Draw the tubing with both hands, bending it over the thumbs.
6. Bend slowly and watch for indications of buckling or kinking.
7. Keep moving thumbs and bend slowly so that the curve takes shape evenly.
8. Never attempt to make a very close bend with the hands.

Figure 7-29 shows three types of bends used in refrigeration.

Precautions—If the bend to be replaced is a close one, use iron or any stiff wire to measure off a wide swing for the replacement; a close bend cannot be made by hand and should not be tempted.

Inside and Outside Springs—An inside spring can be used for making a bend in the tubing. The inside spring can only be used where the bend is to be or near the end of the tubing. If a bend is to be made in the middle or at any distance from the end, an outside spring or bending tool must be employed.

A No. 4 screen-door spring with the paint removed can be used for an inside bending tool. The spring is prone to gather dust, dirt, and grime between its coils, and this can result in dirt being

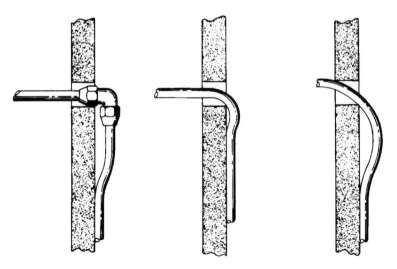

Fig. 7-29. Three types of bends as required in refrigeration service.

introduced into a system. Make sure the spring is clean before using it. One end of the spring should be enlarged or a ring of heavy wire attached to one end so that it can be removed from the tubing. An inside spring will give a closer bend than an outside spring. An inside spring can only be used successfully on ½-in. or larger sizes of tubing because an inside spring is much harder to remove when the diameter is less than ½-in. in size. A tool kit can be purchased that will include not only the proper tube cutter and reamer, but also the tube bending springs.

Procedure No. 2

1. Straighten the piece of tubing. Ream the end.
2. Mark the tubing at the point where the bend is to be made.
3. Insert a clean, oiled spring, making sure that the spring extends well beyond the bend point.
4. Grasp the tubing with both hands, and bend over the thumbs. Move the thumbs as the bend progresses so that a bend of the desired degree is obtained.
5. Bend a little beyond the degree desired, and then bring it back to the desired radius. This will loosen the spring so it can be removed.

Precautions—When a very close bend is required at the end of a piece of tubing and no mechanical bending tool is available, use both inside and outside springs at the same time. Be sure that the spring is clean. Do not drop it on the floor or place the spring where it can gather dirt between its convolutions. This dirt will drop out in the copper tubing and be introduced into the refrigerant system, where it will cause operating defects. Do not attempt too close a bend. Move the thumbs after each slight bend and examine the bend.

When a bend collapses or flattens and the spring cannot be removed easily, tap the tube round with a hammer. This will release the spring. If done carefully, the tube bend will not be badly battered and can be placed in use if it is not subjected to machine vibration. Use a flat piece of iron or wood between the hammer and tube so that it will not be dented or battered. Oil the spring with refrigeration oil. Do not use any other type of oil. If no refrigeration oil is on hand, use the spring dry. Oil makes the spring easier to remove and the bend somewhat less strenuous to form.

Outside bending springs are made in all sizes for use on standard tubing. Servicemen have more or less adopted the outside tube spring. A miniature pipe bender is used on production jobs where a great number of the same-sized bends are required, but this tool is not used to any great extent in the field. Outside springs should be used to acquire practice in the forming of bends, as shown in Fig. 7-30.

Procedure No. 3

1. Mark the tubing where the bend is desired.
2. Make another mark on the tubing one-half the length of the spring.
3. Slide the outside spring up to the second mark. The first mark (where the bend is desired) will be covered, but it is known to be at the center of the bending spring.

Precautions—Determine the size of the bend required, and make the bend in one smooth motion. If bent too much or too little at a time, the repeated working will harden the tubing and make bending difficult. Such a bend can crystallize and break the

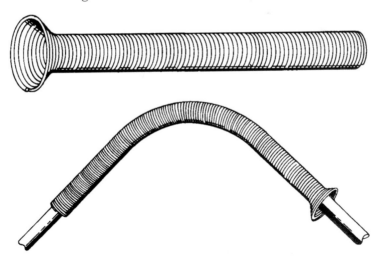

Fig. 7-30. Typical outside spring and method of application to tubing.

tubing if subjected to vibration in use. Some springs have plain ends, while others are provided with a flare or funnel-shaped end so that the spring can be easily slipped on and off the tube. When pulling the spring off the tube, grasp the spring and pull on the flared end. Springs will slip on and off much easier if they are slightly oiled. Wipe each tube after bending to remove any oil squeezed from between the convolutions of the spring. Use the springs with care or they will become sprung or kinked. A spring should last for years with proper care.

MAKING A RELIEVED BEND

This procedure is for making a bend that will hug the corners of a wall when rounding a corner, as shown in Fig. 7-31. Notice that any ordinary 90° bend will not touch the walls and is likely to be dislodged by any moving object coming into contact with it. The relieved bend will hug the wall tightly and permit proper attachment. Use this type of bend for a neat, satisfactory jog.

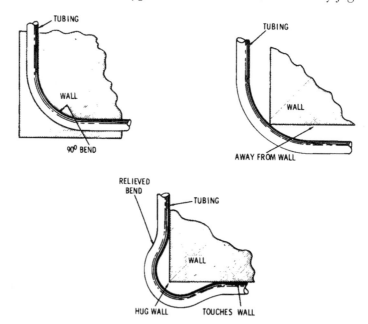

Fig. 7-31. Use of relief bend when close application to a wall is desirable.

Procedure No. 4

1. Make a 90° bend as outlined in the previous procedure.
2. Move the spring to one leg and make a relieved bend.
3. Do the same with the other leg.
4. With both legs relieved, try the bend on a corner and determine whether it hugs both walls properly.

USE OF RESEATING TOOLS

The reseating tool is used for resurfacing the ends of flared fittings, which, due to excessive use or abuse, would otherwise have to be replaced with new fittings. Some fittings are specially designed. Through continued use, or more often through abuse, the flared end or *nose* becomes rounded, scored, or grooved to the extent that it cannot be made gas tight. Through the use of a reseating tool, these special (as well as ordinary) fittings can be reconditioned and used again with the assurance that they are as good as new insofar as the flare nose is concerned.

There are several types of reseating tools on the market. One type is equipped with a cutter guide, removable bushings, cutter feed, and automatic tension and is an excellent tool for use in the shop where considerable resurfacing is done, such as in the case of general overhauling work.

Procedure No. 5

1. Insert the damaged fitting in an adapter of proper size. Screw the fitting into the adapter from the rear and allow it to run in until the fitting seats on the shoulder. Use a 12-point wrench where a hex shoulder is available. Do not damage the threads on other parts of the fittings.
2. Place a drop or two of refrigeration oil on the reamer body and insert the reamer in the proper adapter.
3. With a firm, light pressure, rotate the reamer clockwise so that the end of the fitting is smooth.
4. If the fitting nose is badly scored or ridged, it may take several minutes to cut a clean, smooth face.

5. Remove the reamer and inspect the face at regular intervals. When a perfect seat is obtained, do not cut away any metal.
6. When the face is smooth, remove the reamer and take the fitting out of the adapter.
7. If the end is clean and satisfactory, proceed with the operation; if not, make further use of the reseating tool.
8. With the flare nose clean and smooth, run a strip of cloth through the fitting to remove metal shavings. Do not blow through the fitting to clear it of shavings. Moisture in the breath can condense in the fitting.

Precautions—When using a reamer, a light, steady pressure must be applied. If too heavy a cut is attempted, or if the tool is applied too lightly, the reamer may skip or chatter and produce a wavy surface almost impossible to remove by subsequent treatment with hand tools. A lathe can be employed to obtain a true, smooth surface, but this course is open only to the man in the repair shop, not to the serviceman. Bear in mind that an SAE flared fitting must have a projecting nose. If the operation continues to cut away an excessive amount of metal, the connection will not be airtight and a new flare must be made.

COPPER TUBING

It is necessary, especially in smaller systems, to use dehydrated tubing that is factory sealed, which prevents moisture from entering the system at the time of installation. Never permit unused tubing to remain unsealed in stock. A safe procedure is to seal the ends immediately after cutting or flatten the tube end while the stock is not in use. The sizes and dimensions of various copper tubing used in the refrigeration industry are given in Table 7-1.

The most popular type of fittings used to make the connections on copper-tubing installations is the flared type because of the ease of assembly and maintenance of tight joints, and because such a fitting can be taken apart and used several times without requiring replacement. Valves are made in a variety of forms. For the small unit, packed and bellows-sealed valves are employed; for larger units, the packed variety is used almost exclusively because of the larger sizes employed.

Table 7-1. Copper Tubing Sizes and Dimensions

Sizes (O.D.)	Stubs (Gage)	Wall Thickness (in.)	Weight (lb/1000 ft.)	Approx. No. ft/lb
1/8	20	0.035	38	25.4
3/16	20	0.035	65	15.4
1/4	20	0.035	92	11.0
5/16	20	0.035	119	8.5
3/8	20	0.035	145	7.0
7/16	20	0.035	171	6.0
1/2	20	0.035	198	5.0
5/8	20	0.035	251	4.0
3/4	19	0.042	362	2.8

A valve formed of forged brass is shown in Fig. 7-32 and shows the manner of construction employed in small-valve design. A comparison of the two types of valves is illustrated by examining the two angle valves shown in Fig. 7-33. One employs the usual

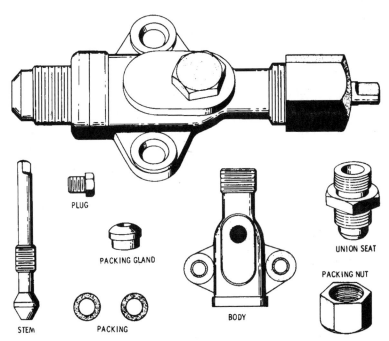

PLUG

PACKING GLAND

UNION SEAT

PACKING NUT

BODY

STEM

PACKING

Fig. 7-32. Forged brass valve showing assembly and parts.

241

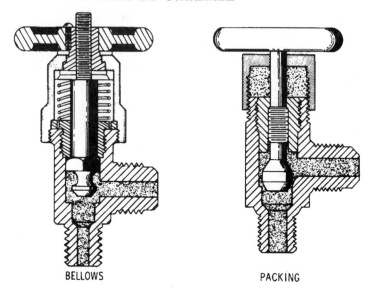

BELLOWS   PACKING

Fig. 7-33. Sectional views of two types of valves.

packing, and the other makes use of the corrugated sheet-metal seal or syphon bellows. Both valves are backsealed, so that when fully opened, the valve itself is sealed against any possible leakage. New safety-code specifications, which are being adopted in almost every city, require that fittings, fastenings, valves, and safety devices conform to certain standards. One of the requirements in flare-tube fittings is that intermediate bushings are used.

Two types of safety devices are shown in Fig. 7-34. One is rupturable; and the other depends on heat to effect a pressure release. The rupture devices usually utilize a thin silver disk of a certain diameter and thickness, which, in the event of high pressure being generated due to any cause in the system, will rupture and release the refrigerant, leaving the unit without any chance of its being destroyed or strained in any way. The fusible type of safety device is usually made in the form of a plug intended to be screwed into a boss or fitting and is operative only by heat. Regulations require the use of such plugs on refrigerant storage or charging drums, receivers, and isolated portions of apparatus that are filled with refrigerant. In the event of fire, the gas will be released without danger of explosion.

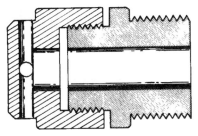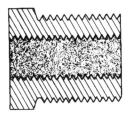

Fig. 7-34. Rupturable safety devices to relieve excessive pressure; the plugs melt at excessive temperatures.

One type of connector developed and adopted by the trade is the capillary-soldered type. It is lighter, simpler, stronger, and more economical than the threaded types. These special fittings are made in the shapes and sizes familiar to the trade and differ only in the method of joining the tubing. The tubing is inserted into the fitting until it rests against a shoulder. Then wire solder is applied through a small feed hole in the fitting under heat from a blowtorch. The solder is drawn into the fitting and evenly distributed by capillary attraction, resulting in a firm joint. This method of effecting a joint is shown in Fig. 7-35, where the wire solder is fed into the hole in the side of the fitting (a *tee*, in this instance) while a blowtorch heats the fitting. A sectional view of the completed joint indicates the capillary channel holes and shoulder; it also shows the larger surface between the fitting and tube joined by the solder.

SILVER BRAZING

The use of high-melting-point alloys has increased considerably in recent years, particularly in supermarkets, locker plants, air conditioning, or wherever a large number of joints is used. The inherent mechanical strength of the joint and the reduction of leak possibilities have made this a desirable type. In addition, code requirements in some areas have forced the use of these alloys.

In order for brazing alloys to melt and flow properly, red heat (1150-1450°F) is required. Copper will react with the oxygen in the air at these temperatures to form a scale of copper oxide on the inner walls of the tubing and fittings. The scale is broken off into flakes by the turbulence of flowing liquid refrigerant. The flakes quickly break up into a fine powder, which plugs driers, strainers, and capillary tubes.

Copper oxide is so finely divided that a filter bed of other materials is usually required in driers before the filtration system in the drier can remove it, thus accounting for the approximately 48 hours needed to remove this material. Driers are sometimes returned as defective because they have shown an undesirable pressure drop on a new system. In conjunction with other dirt, a substantial accumulation of copper oxide is often the explanation. Chemical analysis readily confirms the condition. The drier is not defective. It has done exactly what it was designed to do—take out the dirt.

If the air in the line being brazed is replaced with an inert gas, such as dry nitrogen, the formation of copper oxide can be eliminated. The line should be purged thoroughly and a slow

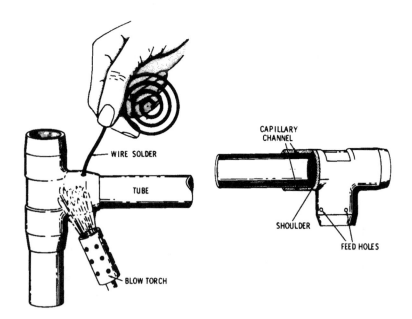

Fig. 7-35. Soldering a fitting in place.

steady flow of nitrogen maintained by means of a pressure-reducing valve. High-pressure gases (nitrogen or carbon dioxide) should never be connected directly to a system but must be metered through a pressure-reducing valve for the protection of the user. Too high a rate of flow is undesirable because of the cooling effect of the gas. The inert-gas method will keep the inside of the copper lines and fittings clean and bright. When it is not completely successful, the trouble is usually attributed to applying heat before all of the air has been displaced.

On large copper pipe, the end of the tubing should be covered with some type of elastic material, such as a rubber balloon, and a small slit made in the material to provide an escape valve. In this manner, the line can be kept full of nitrogen without use of excessive quantities. The inert-gas method, wherever practical, is recommended for repair work. The nitrogen clears the line of refrigerant vapor, preventing acid formation due to refrigerant breakdown as well as copper oxide formation.

Many servicemen find it impractical to transport heavy nitrogen cylinders in and out of buildings and have solved the problem by permanently mounting the cylinder on their service trucks. The cylinder is equipped with a pressure-reduction valve and flexible lines long enough to reach most of their jobs. Purge hose or a similar type of material is satisfactory for the flexible line. The line pressure is held to an absolute minimum, with the gas flow just high enough to maintain the required slow, steady rate.

SUMMARY

Domestic hermetically-sealed units have been provided with several types of methods of adding or drying out refrigerants that have accumulated moisture. Some require a special adapter for connection to the hermetically-sealed compressor, some have a service valve, and some require a special adapter connecting the suction line to the compressor with a valve incorporated with it. Domestic refrigerators have a certain number of ounces of refrigerants that the unit is to be charged with. Practically 100 percent of commercial refrigerators have service valves on both the discharge and suction sides of the compressors. An exception to this may be self-contained water coolers, which may or may not have service valves.

Since most household refrigeration units are not equipped with gages, a serviceman must insert his own for the various testing and service operations that must be dealt with in solving day-today problems. A number of operations and tests may be performed by the use of a test-gage set.

The first step in charging a sealed system with refrigerants is to have the right charging and evacuating equipment. This will enable the serviceman to charge the unit with the exact amount of refrigerant. It is a good rule, whenever a sealed system is opened and the refrigerant charge removed, to evacuate the system and install a new service drier before recharging the system.

Refrigerant can be added on the low or high side. When it becomes necessary to remove the whole charge from a system, it is best to use a pump and evacuate the refrigerant into a drum so it can be used again.

Leaks can be detected by the soap-bubble method or by using an electronic device.

REVIEW QUESTIONS

1. Where are the compressor suction and discharge valves located?
2. What is another name for a test-gage set?
3. What are the two types of gages used in refrigeration servicing?
4. Why is it a good rule to evacuate a system and remove the charge whenever it is opened?
5. What are the steps in charging a closed system?
6. What are the steps used to evacuate a closed system?
7. Describe charging a system through the low side.
8. Describe charging a system through the high side.
9. How do you evacuate air from the entire system?
10. How do you test for leaks in a closed system?
11. What is the difference between the halide leak detector and the soap-bubble method?

12. What is meant by a halogen gas?
13. Where are magnifying mirrors used by a refrigeration serviceman?
14. What is a pinchoff tool?
15. What are valve keys used for?
16. What are packing keys used for?
17. Why do you need to use a flaring tool?
18. What is a relieved bend?
19. What is a reseating tool?
20. How is silver brazing different from soldering?

CHAPTER 8

Domestic Refrigeration Operation and Service

In order to enable the serviceman to render intelligent and efficient service, it is necessary that he understand the operation of the different parts that make up a complete refrigeration system. Although some service problems can be detected and eliminated in a few minutes, others may require considerable work before being properly diagnosed and corrected. The importance of a thorough diagnosis of the faulty unit cannot be too strongly emphasized. Do not simply guess at the remedy to be applied until the actual trouble has been determined.

SEALED-SYSTEM OPERATION

The correct operation of the sealed system is dependent upon the proper functioning of each part that makes up the system. If the system does not operate correctly due to abnormally long running periods or warmer than normal interior temperatures, the trouble may be caused by one of the following conditions:

1. Restricted capillary tube
2. Incorrect refrigerant charge
3. Partial restriction in evaporator
4. Defective compressor

Restricted Capillary Tube

A restriction in the capillary tube may be caused by moisture in the system, kinked tubing, or foreign particles lodged in the lines. Each of these conditions will cause symptoms to occur. The evaporator will have little or no frost formation, and the compressor will run for an extended period of time and eventually cycle on the overload protector. Moisture in the system will usually freeze at the outlet of the capillary tube where it joins the evaporator tubing. The tubing in the immediate vicinity of the freezeup may be heavily frosted, but the remaining evaporator tubing will be free from frost.

If, when the refrigerator is inspected, the compressor is running but the evaporator is not frosted, stop the compressor and remove the frame. Apply heat to the end of the capillary tubing; usually a match held under the tubing at this point will be sufficient. If an ice block is present, the application of heat will melt the ice and a gurgling sound can be heard as the refrigerant surges through the tubing. If, after applying heat to the capillary tube, the restriction has not been eliminated, check for kinks in the tubing or an undercharged system. In the event the application of heat melts the ice block, the drier should absorb and retain enough moisture to eliminate future freezeups.

If the freezing reoccurs and the unit is under warranty, the entire unit should be replaced. If it is not under warranty, there are several procedures that you may take to remedy the problem.

Usually small domestic units have no service valves. A procedure that works in most cases involves attaching a valve to the suction line. This valve clamps over the suction line and has neoprene washers that press against the suction line, forming a non-leaking gasket. You then screw down the valve, which has a needle point that punctures the suction line. The refrigerant gas in the unit is then released through this valve. Connect a vacuum pump to the valve, and a line from the discharge side of the pump should be inserted in a container of refrigeration oil. When the vacuum pump is running, bubbles will pass through the oil as long as any gas or water vapor is being pumped out. A vacuum gage should be installed so that you may check the actual vacuum being developed.

The door of the refrigerator should be left open, and heat should be applied to the tubing and the evaporator. When the bubbles stop showing up in the oil, the vacuum pump should be left operating for an additional ½ hour. The purpose of the application of heat and the long pumping time is to be certain that any moisture in the system will be removed.

Then close the valve that you have installed on the suction line, or if the unit happens to have service valves, close the suction line completely. Disconnect the vacuum pump, and connect a tank of refrigerant to the suction valve. Insert a compound-gage assembly between the tank of refrigerant and the suction valve. Crack the valve on the refrigerant tank slightly, and loosen the flexible connection hose that you should be using from the compound-gage assembly to the suction valve on the compressor. This will allow some refrigerant to escape into the room, thus purging air and moisture from the compound gage and the refrigerant lines to the compressor. Open the suction valve at the compressor and allow enough gas to enter so that the gage shows 1 to 2 lb pressure, then close all valves.

Install a new drier-filter, discarding the old one. This can be accomplished by cutting the lines close to the old filter and using flare nuts and nut fittings, or by connecting the lines by slipping one over the other and using a Phasco fluxless brazing rod. The lines only need be heated cherry red to apply the Phasco rod. Silver soldering may be done instead of using a Phasco rod, but is more complicated to use. When cutting the old drier out, remember that you have 1 lb or so of pressure within the unit.

After the drier-filter has been changed, reconnect the vacuum pump, as outlined previously, and run it for ½ hour after bubbles stop appearing in the oil bath. Close the valve that you have put on the suction line or the service valve on the compressor (if it has one) before stopping the vacuum pump.

You are now ready to recharge the system with refrigerant. Use a recharging cylinder with a glass front, marked in ounces for the most commonly used refrigerants. The nameplate on the refrigeration unit will tell you how many ounces of what type refrigerant to use. With the same compound-gage arrangement that you used before, connect to the recharging cylinder instead of the compressor. Close the valve at the bottom of the cylinder, crack the valve on the refrigerant tank, turn the tank upside down, and crack the connection at the recharging cylinder to purge air and moisture out of the lines. Then tighten the connection at the recharging cylinder, open the valve at the bottom of the cylinder slightly, and you can see the liquid refrigerant entering the cylinder by looking through the glass front.

There is a purging valve at the top of the recharging cylinder to remove air and moisture from the cylinder. Add a few ounces more than is required to charge the system. Close the refrigerant tank valve and recharging cylinder valve. Disconnect the charging hose from the compound gage, being careful, as liquid refrigerant will come out and you may freeze your hands or get it in your eyes.

Connect the end of the charging hose to the suction valve on the compressor, leaving it slightly loose. Open the valve at the bottom of the recharging cylinder slightly to purge the recharging line; then tighten the connection of the charging line at the compressor. You are now ready to recharge the system.

Note how many ounces of refrigerant you have in the cylinder, and subtract the number of ounces required to properly charge the system. This will give you a stopping point on the recharging cylinder. Start the refrigerator, and open very slightly the valve on the recharging cylinder. Remember that you are adding liquid refrigerant to the compressor, and this must be done slowly to prevent oil from leaving the crankcase with the refrigerant and going into the system. In other words, *add the new refrigerant slowly*. When you have added the correct amount as shown on the glass front, shut the suction valve at the compressor and then the valve at the recharging cylinder. Leave all equipment in place, and observe the operation of the refrigerator.

If all is well, you should see frost start to form at the inlet to the evaporator and slowly expand, eventually covering the entire evaporator. You will also notice a hissing sound as the liquid enters the evaporator. If the refrigerator frosts back to the compressor, you have overcharged it and should slowly purge a little of the gas out to correct this situation.

The chances of having repaired the cause of the restricted capillary tube are quite good. If the capillary tube is still plugged, the following information will be useful. The vacuum pumping and recharging of the system will not be repeated. A schematic diagram of refrigeration tubing is shown in Fig. 8-1.

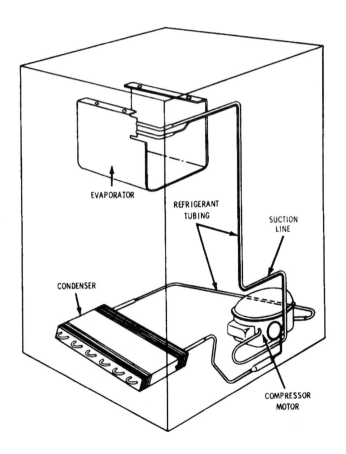

Fig. 8-1. Schematic diagram showing arrangement of refrigerant tubing in a typical household refrigerator.

A kinked capillary tube will stop the flow of refrigerant, and the evaporator will be free from frost. The compressor will run continuously and eventually cycle on the overload protector. Check the capillary tube along its entire length and, if possible, straighten the tubing enough to alleviate the difficulty. Foreign particles lodged in the capillary tube will also stop the flow of refrigerant, and the system will exhibit the same symptoms as with a kinked tube. If checks have been made to eliminate the possibility of a moisture freezeup or a kinked capillary tube, it may be assumed that a foreign particle is lodged at the entrance to the capillary tube.

A restriction usually occurs in the capillary tube because of its small diameter. The symptoms of a restricted capillary are:

1. Lack of frost on the evaporator. The compressor may operate for a short period of time and then cycle on overload.
2. Moisture freezeup may cause a restriction in the line, and it usually occurs at the outlet end of the capillary tube. Normally, a frost buildup can be detected in this area, but insulation wrapped around the tubing may conceal or limit the amount of frost accumulation. Expose the discharge end of the capillary tube, and apply heat at this point. If there is enough head pressure, and if the restriction is caused by moisture freezeup, a gurgling noise will be heard as the heat releases the refrigerant through the tubing. It is possible that this moisture will be absorbed by the drier and remedy the problem. However, if the freezeup recurs, the drier must be replaced. If this does not remedy the problem, replace the heat exchanger.
3. A kink in the capillary tube will reveal about the same symptom as a moisture freezeup except for the accumulation of frost. Check the entire length of the capillary tube and, if possible, straighten the kink to relieve the restriction. Check the unit operation, and, if the trouble persists, replace the defective part.
4. If there is no freezeup or kink in the capillary tube, it can be assumed that a foreign particle is causing the restriction.

If you find that the capillary tube is plugged and the unit is under guarantee, replace the entire unit. If it is out of guarantee, there is a high-pressure hand pump available for blowing out capillary tubes. Cut the inlet line, allowing the refrigerant to escape, and then apply pressure with the hand pump, blowing out the restriction. This operation is mentioned assuming that the lines are copper and not aluminum lines. If this does not work, replace the heat exchanger, making sure that the capillary tube attached is of the proper dimensions for the refrigerator.

It is a good practice to replace the drier-filter when repairing the capillary tube. When repairs are completed, attach a valve as discussed previously. Use the vacuum pump as before, and recharge the unit.

Incorrect Refrigerant Charge

A refrigerator system that is *undercharged* with refrigerants will produce various conditions, depending upon the degree of undercharge. During normal operation, a system that is fully charged will have frost covering the entire evaporator and accumulator. Any degree of undercharge or a gradual leak of refrigerant will be noticed first by the absence of frost on the accumulator. The compressor will run for long periods of time, and the food compartment temperature may be colder than normal. As the leakage of refrigerant progresses, the last few passages in the evaporator will be free of frost. The compressor will run continually since the temperature of the evaporator at the control contact location fails to descend to the control cutout point.

A system that is *overcharged* will have a frostback condition on the suction line under the cabinet during the *on* period. The frost will then melt and drip on the floor during the *off* period. This situation in a slightly overcharged system may be remedied by wrapping the suction line with insulation tape or its equivalent. Should moisture continue to drip on the floor, replace the sealed system.

Overcharge—The symptoms of an overcharged refrigerator are usually as follows:

1. A frostback condition on the suction line to the extent that it can be noticed on the line back to the compressor unit.
2. If the overcharge is great enough, the control contact may not get cold enough to cut out the compressor, resulting in

long periods of operation.
3. Suction and head pressure will be high both at cutout and when the system has stabilized.
4. An overcharged system must be evacuated and recharged with the proper amount of refrigerant.

Undercharge—When too little refrigerant is present in the refrigerator, the system is undercharged. This defect can be diagnosed as follows:

1. In an undercharged system the refrigerant system will be colder than normal. The freezer compartment may or may not be cold, depending upon the amount of undercharge.
2. The compressor may run continuously because the cold-control contact fails to descend to the control cutout point.
3. Suction charge will be low (possibly a vacuum), depending upon the degree of undercharge.
4. Head pressure will be lower than normal at cutout or when the system has stabilized.
5. An undercharged system must be evacuated and recharged with the proper amount of refrigerant.

Partial Restriction in Evaporator

An accumulation of moisture or foreign particles may freeze or lodge in the evaporator tubing and cause a partial restriction at that point. This condition is usually indicated by the evaporator being heavily frosted for a few passes ahead of the restriction and bare of frost on the first few passes behind the restriction. A restriction of this nature tends to act as a second capillary tube. Increased pressure in the evaporator line toward the high side will cause warmer temperatures; decreasing the pressure as the refrigerant passes toward the low side will cause colder temperatures. The evaporator tubing on the high side of the restriction will be free of frost, and the tubing on the low side will be heavily frosted.

If the restriction occurs on the low side of the control feeler-tube contact point, the compressor will run continuously since the control cutout point is never reached. Should the restriction occur on the high side of the control feeler-tube contact point, the compressor will cycle frequently, but running time will be short and the food compartment temperatures will be warmer than usual. A partial restriction in the evaporator tubing will require replacement of the sealed system.

Defective Compressor

A compressor that is not pumping adequately will produce very little cooling effect. The evaporator may be covered with a thin film of frost, but the evaporator temperature will not descend to the cutout temperature of the control, even with continuous running of the compressor. Place your hand on the evaporator surface for 2 or 3 seconds, then examine the surface. If all of the frost has melted where the evaporator was touched, suspicions of a defective compressor will be confirmed and the sealed system must be replaced.

An inefficient or defective compressor will affect the operation as follows:

1. A compressor that is not pumping correctly will produce very little cooling effect. All cooling surfaces may be covered with a thin film of frost.
2. A popular checking method is to place a hand on the accumulator for 2 or 3 seconds and then examine the surface. If all of the frost is melted at the contact surface, operating pressures should be checked.
3. If high-side pressures are lower than normal and low-side pressures higher than normal, suspicions of a defective compressor are confirmed and the compressor unit must be replaced.

ELECTRICAL TESTING

In normal operation the compressor is actuated by the thermostat. When the thermostat contacts close, the motor running winding and relay coil are energized. The heavy surge of current that will be drawn by the motor attempting to start will create an electromagnetic force strong enough to pick up the relay arma-ture and close the starting contacts of the relay, which will energize the running windings. A fraction of a second later, as the motor comes up to full speed, the running current decreases to normal value and allows the relay armature to drop and open the starting contacts.

The starting winding, which was needed to provide the extra torque to overcome the inertia of the compressor, is now out of the circuit. The motor continues to operate on only the running winding. The overload protector has a set of bimetal contacts that are normally in the closed position and are in series with a resistance heater. If there is difficulty in starting the motor or if the compressor becomes too hot for any reason, the excess heat will cause the contacts to open and interrupt all electrical power to the compressor motor.

Testing with Volt-Ohmmeters

Test lamps, as shown in Fig. 8-2, were formerly used to a great degree for electrical troubleshooting. They can be dangerous because of bulbs breaking and electrical shocks. Now practically all servicemen use volt-ohmmeters. The continuity tests are done on the ohmmeter scale. These instruments have self-contained batteries of low voltage.

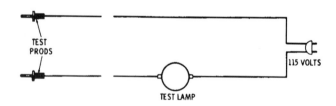

Fig. 8-2. Typical test-lamp arrangement.

Starting Winding—To test for continuity in the motor start-ing winding, set the ohmmeter on the low-ohms scale and place the test leads on the terminal S (starting) and C (common), as shown in Fig. 8-3. If the ohmmeter stays on 0 ohms, the starting winding is open.

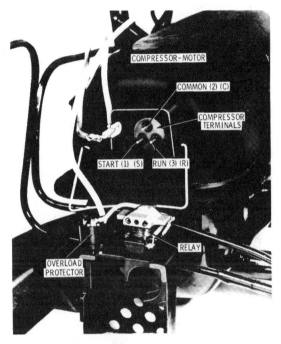

Hotpoint Company

Fig. 8-3. Compressor-motor terminal arrangement in a typical household refrigerator.

Running Winding—To test the running winding, place the ohmmeter-test probes on compressor terminals R (running) and C (common). If the ohmmeter reads 0 ohms, the running winding is open. If either the starting or running windings are open, the entire unit must be replaced.

Overload Protector—Before an accurate continuity check can be made on the overload protector, it must be determined that the unit has not been on overload during the past hour. If the unit is on overload, the protector contacts will be open and there will be no circuit continuity. If, however, the unit has not cycled on the overload, the continuity check will determine whether or not the overload protector is inoperative. To test the protector, place the ohmmeter probes on protector terminals C (common) and 3 on line, depending upon the wiring arrangement. If the ohmmeter registers 0 ohms, either the overload protector or the resistance heater is open. The overload protector is actuated with every start, and therefore is a very likely source of trouble.

Units with Capacitors—Capacitors are used on the larger refrigerators and are in series with the starting winding to give more starting torque. They occasionally go bad. The best test for capacitance is to use a good capacitor-checker or temporarily replace the capacitor with a good one. Sometimes it is well to replace the overload protector when replacing a capacitor, for preventive maintenance.

COMPRESSOR TESTING WITH DIRECT POWER

Before proceeding with test, be sure the compressor is level. To test the compressor without any cabinet wiring in the circuit, disconnect the compressor lead cord from the junction block and connect it directly to the electrical wall outlet. If necessary, use a short extension cord to reach the outlet. If the compressor starts, the trouble is not in the compressor, relay, or overload protector. Plug the compressor lead cord back into the junction block, and check the temperature control and all cabinet wiring for an open circuit or poor connections. If the compressor does not start, check the overload protector.

The overload protector is mounted on the compressor under the terminal cover. The protector trips open when the compressor is overheated and/or when excessive current is drawn. Cycling on the overload may be the result of poor air circulation around the compressor and condenser. Do not attempt to start the compressor until the refrigeration system has become equalized, which takes about 3 to 5 minutes after the previous run. At least 100V are required at the compressor terminals during the starting interval. A typical compressor motor circuit is shown in Fig. 8-4. Quite often the overload device is not a separate piece, as shown in Fig. 8-4, but is included in the starting relay.

If the compressor repeatedly starts and runs for 5 or 6 seconds

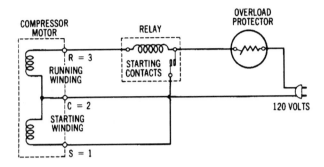

Fig. 8-4. Schematic wiring diagram of a compressor motor.

and then cycles on the overload protector, the starting relay contacts may be stuck closed and the excess current is tripping the overload. Disconnect the wire from compressor terminal S, and make a momentary short between wire terminals R and S (Figs. 8-3 and 8-4). If the compressor starts and runs without cycling on the overload protector, replace with a new complete starting relay.

To check for an open overload protector, short across the terminals (Figs. 8-3 and 8-4). If the compressor starts after shorting the protector terminals, replace the overload with one having the same part number. If the compressor does not start, look for other trouble such as a low line voltage, defective starting relay, and defective operating protective relay or compressor. Other possibilities are a blown fuse, broken service cord, inoperative thermostat, or faulty capacitor.

COMPRESSOR MOTOR RELAYS

It is important to locate the terminals you are to work with in a compressor before you start any testing procedures. There are two types of relays used on compressors—the potential and the current.

The potential relay is generally used with large commercial and air conditioning compressors. The current relay is generally used with small refrigeration compressors up to ¾ horsepower. The potential relay is shown in Fig. 8-5.

261

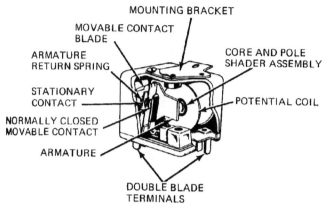

MOUNTING BRACKET

MOVABLE CONTACT
BLADE

ARMATURE
RETURN SPRING

CORE AND POLE
SHADER ASSEMBLY

STATIONARY
CONTACT

POTENTIAL COIL

NORMALLY CLOSED
MOVABLE CONTACT

ARMATURE

DOUBLE BLADE
TERMINALS

Tecumseh Products Company

Fig. 8-5. Potential relay. This is generally used with large commercial and air conditioning compressors with capacitor-start, capacitor-run motors up to 5 horsepower.

Potential Relay

The potential relay may be used with motors that are of the capacitor-start, capacitor-run types up to 5 horsepower. Relay contacts are normally closed. The relay coil is wired across the start winding. It senses voltage change. Start winding voltage increases with motor speed: as the voltage increases to the specific pick-up value, the armature pulls up, opening the relay contacts and de-energizing the start winding. After switching, there is still sufficient voltage induced in the start winding to keep the relay coil energized and the relay starting contacts open. When power is shut off to the motor, the voltage drops to zero. The coil is de-energized and the start contacts reset (Fig. 8-5).

Many of these relays are extremely position sensitive. When changing a compressor relay, care should be taken to install the replacement in the same position as the original. Never select a replacement relay solely by horsepower or other generalized rating. Select the correct relay from the parts guide book furnished by the manufacturer.

Current Relay

When power is applied to the compressor motor, the relay solenoid coil attracts the relay armature upward. This causes the

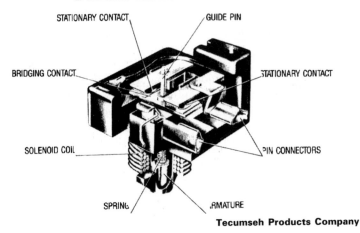

STATIONARY CONTACT · GUIDE PIN · BRIDGING CONTACT · STATIONARY CONTACT · SOLENOID COIL · PIN CONNECTORS · SPRING · ARMATURE

Tecumseh Products Company

Fig. 8-6. Current relay. This is generally used with small refrigeration compressors up to 3/4 horsepower.

bridging contact and stationary contact to engage (Fig. 8-6). This energizes the motor start winding. When the compressor motor attains running speed, the motor main winding current is such that the relay solenoid coil de-energizes. This allows the relay contacts to drop open. This, in turn, disconnects the motor start winding.

The relay must be mounted in true vertical position so that the armature and bridging contact will drop free when the relay solenoid is de-energized.

COMPRESSOR TERMINALS

Terminals are mounted on the outside of the hermetically-enclosed compressor motor. Therefore it is possible to make connections to the compressor motor inside the shell by using these terminals. There are several different types of terminals used on the various models of Tecumseh compressors. Other manufacturers may vary also. Make sure the proper terminals are located because guessing can cause much damage.

Tecumseh terminals are *always* thought of in the order of common, start, run. Read the terminals in the same way you would read the sentences on a book's page. Start at the top left-hand corner and read across the first line from left to right. Then, read the second line from the left to right. In some cases three lines must be read to complete the identification process. Figure 8-7 shows the different arrangement of terminals. All Tecumseh compressors, except one model, follow one of these patterns. The exception is the old twin-cylinder, internal-mount compressor built at Marion. This was a 90° piston model designated with an "H" at the beginning of the model number (that is, HA100). The terminals were reversed on the H models and read run, start, and common.

Figure 8-8 shows the H model terminal sequence. These compressors were replaced by the J-model series in 1955. All J models follow the usual pattern for common, start, and run.

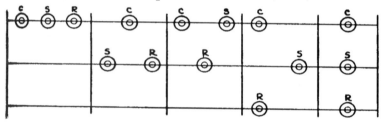

Fig. 8-7. Identification of compressor terminals.

RUN   START   COMMON

Fig. 8-8. Built-up terminals. These are on the obsolete twin-cylinder internal-mount H models.

Built-up Terminals

Some built-up terminals have screw- and nut-type terminals for attaching wires (Fig. 8-9). Others may have different arrangements. The pancake compressors built in 1953 and after have glass terminals that look something like those shown in Fig. 8-10. The terminal arrangement for S and C single-cylinder ISM (Internal Spring Mount) models resembles that shown in Fig. 8-11. Models J and PJ with twin-cylinder internal mount have a different terminal arrangement, similar to that shown in Fig. 8-12.

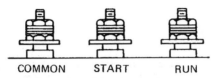

COMMON   START   RUN

Fig. 8-9. Built-up terminals. These are on all external-mount B and C twin-cylinder models and on F, PF, and CF four-cylinder external-mount models.

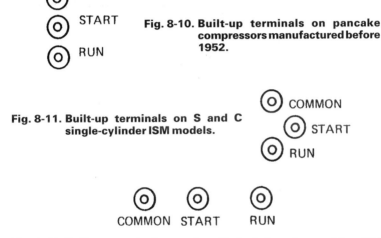

COMMON

START   **Fig. 8-10. Built-up terminals on pancake compressors manufactured before 1952.**

RUN

Fig. 8-11. Built-up terminals on S and C single-cylinder ISM models.

COMMON

START

RUN

COMMON   START   RUN

Fig. 8-12. Built-up terminals on twin-cylinder internal-mount J and PJ models.

(Figs. 8-7 through 8-12 courtesy of Tecumseh Products Company.)

Glass Quick-Connect Terminals

Figure 8-13 shows the quick connect terminals used on S and C single-cylinder ISM models. The AK and CL models also use this type of arrangement. Many of the CL models have the internal thermostat terminals located nearby.

Quick-connect glass terminals are also used on AU and AR air conditioning models. The AE air conditioning models also use glass quick connects. Models AB, AJ and AH also use glass quick connects, but notice how their arrangement of common, start and run varies from that shown in Fig. 8-13. Figure 8-14 shows how the AU, AR, AE, AB, AJ, and AH models terminate.

Glass terminals are also used on pancake-type compressors with P, R, AP and AR designations (Fig. 8-15). The T and AT models, as well as the AE refrigeration models, also use the glass terminals, but without the quick connect.

INTERNAL
THERMOSTAT
TERMINALS

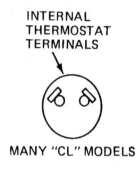

MANY "CL" MODELS

Fig. 8-13. Glass quick-connect terminals. These are used on S and C single-cylinder ISM models, as well as on AK and CL models. Note the location of the internal thermostat terminals on the CL models.

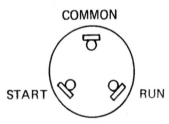

Fig. 8-14. Glass quick-connect terminals. This arrangement is found on AU, AE, AB, AJ, and AH models.

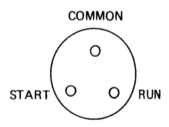

Fig. 8-15. Glass terminals. The pancake-type compressors P, R, AP, AR, and T, as well as AT models, use this terminal configuration.

(Figs. 8-13 and 8-15 courtesy of Tecumseh.)

Keep in mind that you should never solder any wire or wire termination to a compressor terminal. Heat applied to a terminal is liable to crack the glass terminal base or loosen the built-up terminals. This will, in turn, cause a refrigerant leak at the compressor.

Besides Tecumseh, there are other manufacturers of compressors for the air conditioning and refrigeration trade. One is the Americold Compressor Corporation. Two of the models made by them are the M series and the A series; Fig. 8-16 shows the terminations of these two compressor series.

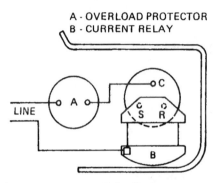

Fig. 8-16. Location of the terminals for the compressors and electrical connections on Americold compressors.

Starting Relay

The starting relay is usually mounted under the compressor terminal cover and consists essentially of a magnetically operated switch with starting contacts. The magnet coil of the relay is in series with the running winding of the compressor motor. When current is applied to the motor, the magnet coil raises the relay plunger and closes the starting contacts, thus connecting the starting winding in parallel with the running winding. As the motor approaches running speed, the current in the running winding and in the relay coil diminishes and the plunger *drops out*, which opens the starting winding circuit. The compressor then continues to operate on the running winding only.

To check for open relay starting contacts, short between the wires at the compressor terminals. If the compressor starts, the relay is defective. If the compressor does not start, check the relay magnet. To check the relay magnet coil, disconnect the power lead from the relay screw terminal. Short between the wires at the compressor terminals, and, at the same time, touch the wire terminal with the disconnected power lead. This bypasses the relay and connects the line directly to the compressor, but through the overload protector. If the compressor does not start, a check for open motor windings or electrical grounds can be made with an ohmmeter. Test between compressor terminals *common* and *run*, and between *common* and *start*.

CABINET LIGHT AND SWITCH

The cabinet light (or lights) is energized by the door-operated switch. If the cabinet light does not go on when the door is opened, it may be due to a burned-out light bulb, defective light switch or socket, or faulty wiring.

CONDENSER COOLING FAN

The condenser cooling fan (when used) is connected in parallel with the compressor. Therefore, if the compressor operates but the fan does not, the fan is either defective or disconnected. An excessive fan noise complaint may arise if one of the fan blades has been bent out of alignment. If any irregularity is noted, replace the fan blade. Another cause of excessive fan noise may be loose fan-bracket mounting screws. A typical fan assembly is shown in Fig. 8-17.

Replacing Controls

To replace controls on most models, proceed as follows:

1. Disconnect the electrical power cord.
2. Pull the control knob off the shaft.
3. Remove the control mounting screws and work the control out of the food-liner opening.

4. Disconnect the control leads from its terminal. When installing a new control, make sure that the capillary tube is firmly clamped against the stainless steel plate. If the control capillary tube is not located correctly, the control will not be able to sense the proper temperature, resulting in erratic refrigeration operation.

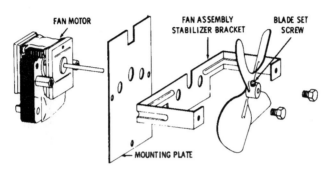

Fig. 8-17. Fan-motor and assembly components.

Calibrating Controls

If the cut-in and cutout temperature readings obtained by the use of a remote reading thermometer are not within 1 or 2° of those specified in the operating-temperature chart supplied by the manufacturer, recalibrate the control.

The temperature control may be calibrated to vary the cut-in and cutout temperatures by turning the range-adjustment screw. This screw, shown in Fig. 8-18, raises and lowers these temperatures by an equal amount. The procedure is as follows:

1. Turn the range adjustment screw right to raise and left to lower the cut-in and cutout temperatures (Fig. 8-18). Each one-quarter turn of the screw will raise or lower the temperatures approximately 2°. *Do not turn this screw more than one full turn as erratic operation may result beyond that point.*

2. Connect the line cord and recheck for proper cut-in and cutout temperatures.

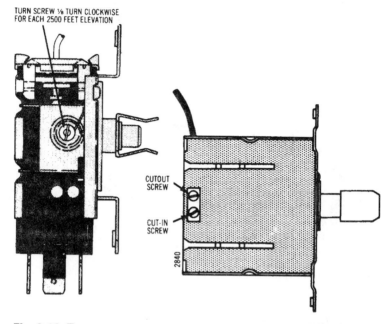

TURN SCREW ⅛ TURN CLOCKWISE
FOR EACH 2500 FEET ELEVATION

CUTOUT
SCREW

CUT-IN
SCREW

2840

Fig. 8-18. Temperature adjustment on Ranco and Cutler-Hammer temperature controls.

REPLACING SYSTEM COMPONENTS

When component replacements are necessary, careful checks must be made in order that identical parts are obtained from the manufacturers. Before actually opening a sealed refrigeration system, be sure that all the necessary tools are close at hand. The system should be opened only long enough to make the repair or component replacement. If the system is allowed to be open for an extended period of time, the motor winding in the compressor will pick up moisture, which will be very difficult to remove during compressor evacuation.

Compressor-Unit Replacements

When cutting into a system for compressor replacement, proceed as follows:

1. Pinch off the suction and discharge lines as close as possible to the compressor. This will prevent excess oil spilling when cutting the lines.
2. Clean the paint from the suction line, and cut the suction line close to the pinched area with a tube cutter. Place a fitting on the cabinet suction line, and plug it to prevent entrance of air and dirt.
3. Clean the paint from the discharge line at the pinched area. Then cut the discharge line close to the pinched area with a tube cutter. Install a fitting on the discharge line, and attach a compound gage with a proper fitting.
4. Remove the inoperative compressor by taking out the four spring clips holding it to the cabinet cross rails. Install a replacement compressor, making sure the rubber grommets from the inoperative compressor are remounted. The replacement compressor suction, discharge, and charging lines are equipped with flare nuts and plugs.
5. With a tube cutter, cut the compressor suction and discharge tubes as close to the flare nuts as possible, and install the proper fittings. Connect the compressor suction line to the cabinet suction line. Plug the compressor discharge line. *Note: Be sure to clean the paint from the tubing before cutting any lines.*
6. Remove the original drier, and install a replacement drier. Compression fittings are provided for the inlet to the drier. The output end of the drier connects directly to the capillary line and will have to be silver soldered.
7. Remove the plug from the compressor charging line, and connect a test can of refrigerant to the charging line via hoses and a manifold. Be sure to purge the lines.
8. Charge the system to 10 psig (pounds per square inch gage) as read on the condenser-line gage. Close all valves, and leak-test all connections.
9. Replace the test can with an exact measured charge of refrigerant.
10. Remove the plug from the compressor-discharge-line fitting. Apply power to the compressor, and evacuate the system to at least a 25-inch vacuum (as read on the condenser gage) for a minimum of 15 minutes. Then replace and tighten the

plug on the compressor-discharge-line fitting, and disconnect the power from the compressor.

11. Charge the unit to 2 psig as read on the condenser-line gage in order to break the vacuum in the unit.

12. Remove the gage from the condenser tubing and the plug from the compressor discharge tube. Then connect the fitting between the compressor discharge line and the condenser line. Make the connection immediately, while the unit still contains an atmosphere of refrigerant.

13. Begin to charge the unit, and apply power to the compressor. While the compressor is running, leak-test the high side.

14. After the system has been charged, pinch off the charging line and insert the plug previously removed back on the charging line. *Note: Leave the pinchoff tool on the line until the plug is reinstalled.*

15. Disconnect the power, permitting the low-side pressure to equalize. Leak-test once more, and recheck all joints and fittings. Repaint all tubing surfaces previously sanded.

Evaporator Replacement

To replace an evaporator in a sealed system, proceed as follows:

1. Disconnect the electrical line cord.

2. Cut off the suction line from the compressor close to the heat exchanger after the refrigerant charge is released. Place the fitting on the line, and plug the line to prevent entrance of air and dirt.

3. Clean the paint from the compressor discharge line about 2 inches beyond the joint, and cut the tubing with a tube cutter. Install the fittings on the lines. Plug the compressor discharge line, and attach a compound gage with the proper fitting to the discharge line going to the condenser.

4. Clean the paint from the compressor charging line, and, with a tube cutter, cut the line as close to the end as possible. Then install the fitting and plug on the open end of the line.

5. Remove the rear tube cover from the cabinet back to expose the heat exchanger tubing that is embedded in the foam insulation.

6. Cut the heat-exchanger line near the top of the cabinet back where it enters the liner. This will permit easier removal of the heat exchanger through the cabinet when removing the evaporator.

7. Remove the screws holding the evaporator to the top of the cabinet. Slowly pull out the evaporator and heat exchanger from the cabinet, as an assembly.

8. Take off the control cover and thermostat from the inoperative evaporator, and install the new evaporator.

9. Insert the heat exchanger and evaporator through the liner opening on the cabinet until the evaporator is properly positioned. Then remount the evaporator assembly and top cover.

10. At the rear of the cabinet, carefully reposition the heat exchanger.

11. Connect the cabinet suction line to the compressor suction line.

12. Evacuate and recharge the unit as outlined elsewhere in this chapter.

Drier Replacement

A new drier must be installed when any system component is replaced or whenever the system is opened. To replace the drier, proceed as follows:

1. Remove all paint and scale from the refrigerant lines for a distance of about 3 inches at both ends of the old drier (use steel wool or fine emery cloth).

2. Cut the lines approximately 1 inch from the old joints. To cut the capillary tube, score the walls with a knife or file. Make a uniform cut around the entire tube; then break it off.

3. Make an offset about 1 inch from the end of the capillary tube to prevent its penetrating too far into the new drier.

4. Immediately solder the new drier into place. Use silver solder and the proper flux.

Heat-Exchanger Replacement

To replace the heat exchanger, proceed as follows:

1. Purge the charge by cutting the process tube on the compressor, and install the service valve, leaving it open.
2. Remove the defective heat exchanger.
3. The capillary tube will have to be unsoldered from the evaporator inlet. Be careful not to restrict the evaporator inlet. Remove heat as soon as the solder liquefies.
4. Unsolder the suction line from the evaporator outlet; clean, and then cut the suction line about 4 inches from the compressor.
5. Unsolder the drier from the condenser outlet. Hold the outlet tube down, and wipe the solder off of the tube immediately after the drier is removed.
6. Swedge the compressor suction line so that the replacement suction line will fit inside.
7. Install the replacement heat exchanger, making all solder joints except on the condenser outlet.
8. Install the new drier on the condenser outlet.
9. Make an offset on the end of the capillary tube, and solder it in the outlet of the new drier.
10. Pull a vacuum for about 20 minutes; then add enough refrigerant to produce a pressure of 35 to 40 lb.
11. Leak-test the low side, then start the compressor and leak-test the high side.
12. Purge the charge, and continue with the vacuum as outlined.
13. Charge the system and reassemble the refrigerator, making sure to locate the system lines properly and seal where necessary.
14. Test-run the refrigerator to be sure it is operating properly.

Condenser Replacement

Most refrigerant condensers are made of copper-coated steel tubing. If a leak occurs, a piece of copper tube can be spliced in place, in which case the condenser does not need to be replaced. When a condenser is in need of replacement, proceed as follows:

1. Open the sealed system at the compressor process tube first. After the refrigerant charge is released, install a service valve on the process tube, leaving the valve open.

2. Before cutting the tubing lines going to the condenser, check the ends of the replacement-condenser tubing. If there is no excess tubing on the connector lines, unsolder the connector tubes from the old condenser.

3. Score and break the capillary tube about 1 inch from the old drier. Remove the old condenser, and install the replacement. Depending on the type of condenser to be replaced (fan-cooled or static), it might be preferable to make the joints on the replacement before installing it in its place (clean all tubing before soldering the joints).

4. After all other joints are made, make an offset in the capillary tube and install the new drier. *Note: When soldering these joints, make sure to leave the service valve on the compressor process tube open.*

5. Start evacuation procedures.

6. Refer to the evacuating, purging, and charging procedures as outlined elsewhere in this chapter.

7. After reassembling the refrigerator, make a test run to be sure that it is operating properly.

EVAPORATOR COOLING FAN

The evaporator cooling fan (when used) is designed to blow cold air over the evaporator compartment area to maintain an even temperature throughout. The air circulation also prevents frost accumulation on food packages and freezer racks. If the fan fails to operate or runs erratically, the reduced air circulation will cause unsatisfactory temperatures in the evaporator compartment.

Most troubles of this sort are usually due to defective wiring, a faulty fan-motor switch, or a bent motor shaft. The switch and fan motor may easily be checked out by means of an ohmmeter or the conventional test-lamp probes. If, after making the foregoing checks, the evaporator fan fails to operate, the trouble is either in the wiring or in the fan motor itself.

As a rule, refrigerator fan motors do not have a way to oil themselves. The bronze bearing has felt outside of it, which presumably oils the motor for life. However, this is often not the case. An overall statement is impossible to make, but if the steel portion around the bearing is enlarged, you can usually figure that there is a felt enclosure around the bearing. In this case, carefully drill a ⁹⁄₃₂-inch hole into the steel-enlarged portion, and, with an oil can, put in three or four drops of 20- or 30-weight automobile oil, being careful not to overoil. Place a little oil around the shaft while turning, and it will usually free up. These motors are of the high-impedance type and will stand a stall without burning up. Insert a small self-tapping screw into the hole that you have drilled to retain the oil.

TEMPERATURE-CONTROL SERVICE

There are generally two types of temperature-control devices, and their functions depend upon:

1. Temperature standard controls, serving solely as *on* and *off* switches for the compressor
2. Temperature combination controls, with automatic-defrost features that govern both defrost and compressor operation

Most problems resulting from temperature-control defects will show up in the refrigerator compartment or in both compartments, but never in the evaporator alone. Attention given to the refrigerator compartment and control adjustments is important because the evaporator compartment will react favorably to nearly all corrective measures. Unsatisfactory temperatures in the evaporator compartment only are usually the result of a poor door-gasket seal, poor air circulation in the evaporator or an inoperative fan, defrost heater wire, or defrost timer. All dual temps use a constant cut-in-type temperature control. Changing the control-knob setting will change the cutout temperature but will not affect the cut-in temperature.

A constant cut-in simply means that the control setting determines only the temperature at which the electrical contacts open, or cut out, interrupting power to the compressor. The temperature that causes the contacts to close always remains constant, regardless of control setting. The constant cut-in temperature is approximately 37°F. Therefore, whenever the temperature in the fresh-food section causes the control contacts to open, the frost that has accumulated on the cooling coil will begin to melt and drain off. All the frost will be dissipated since the control contacts will not close until the control capillary senses a temperature of 37°F, which is above the frost point. If defrosting of the cooling coil is not achieved, it is usually due to a high setting of the control dial that resulted in constant running of the compressor.

Checking Operating Temperatures

Cabinet and evaporator temperatures can be varied considerably by changing the control setting. If no change occurs, check the control cut-in and cutout temperatures. To check control-operating temperatures, proceed as follows:

1. Scrape away the frost on the inside of the evaporator adjacent to the feeler-tube thermal connection.
2. Using a few drops of water, freeze the bulb of an accurately calibrated remote-reading thermometer to the evaporator.
3. Set the control at *normal*. Close the cabinet door, and allow the compressor to run through two or three complete cycles.

REFRIGERATION TROUBLESHOOTING GUIDE

Symptoms and Possible Cause   *Possible Remedy*

Unit Dead, Will Not Run

(a) Blown fuse; rating of power source too low

(a) Check and replace if defective; check source with voltmeter (should check within 10% of 115 V).

(b) Broken service cord

(b) Check voltage at relay. If no voltage at relay, but voltage indicated at outlet, replace service cord or plug on cord.

Symptoms and Possible Cause   *Possible Remedy*

Unit Dead, Will Not Run (cont.)

(c) Inoperative thermostat

(c) Insert jumper across terminals of thermostat. If unit runs and connections are all tight, replace thermostat.

(d) Faulty capacitor

(d) Check capacitor using test cord and 150-W lightbulb. If bad or shows signs of leakage, replace capacitor.

(e) Inoperative relay

(e) Use starter cord and check unit. If unit runs with normal wattage and preceding conditions are correct, replace relay.

(f) Stalled unit

(f) Use starter cord and check unit. If unit runs with normal wattage and preceding conditions are correct, replace relay.

(g) Broken lead to compressor, timer, or thermostat

(g) Replace or repair broken leads.

Compressor Cycles on Overload Protector

(a) Voltage too high or too low

(a) Check line voltage-must be within 10% of name plate rating. Connect refrigerator to proper power source.

(b) Relay or overload protector inoperative

(b) Check continuity and replace if inoperative.

(c) Motor winding open

(c) Check continuity of motor windings. If open, unit must be replaced.

Symptoms and Possible Cause   *Possible Remedy*

Compressor Cycles on Overload Protector (cont.)

(d) Lead broken or connections loose

(d) Check for broken leads and loose terminal connections; make necessary corrections.

Unit Hums and Shuts Off

(a) Low voltage

(a) Check electrical source with volt-wattmeter. Under a load, voltage should be 115 V ± 10%. Check for possible use of extremely long extension cord or several appliances on same circuit.

(b) Faulty capacitor
(c) Inoperative relay
(d) Stalled compressor

(b) Replace capacitor.
(c) Replace relay.
(d) Check with test cord; if compressor will not start, replace.

Unit Runs Excessively

(a) Frequent door openings; high ambient temperature and humidity
(b) Poor door seal

(a) Advise user on proper location and use of refrigerator.
(b) Check to see if cabinet is level. Make necessary front and rear adjustments.

(c) Gasket not sealing

(c) Check door gasket. Realign door and, when indicated, replace door gasket.

(d) Interior light burns constantly

(d) Check light-switch operation. If light does not go out when door is closed, replace the light switch

Symptoms and Possible Cause *Possible Remedy*

Unit Runs Excessively (cont.)

(e) Insufficient air circulation

(e) Check position of refrigerator. The rear and sides must be several inches away from walls. Locate incorrect position.

(f) Gas leaking from unit

(f) Check unit for gas leak. If a leak is found, replace the unit.

(g) Loose connection of thermostat bulb to cooling coil

(g) Check connection and make necessary adjustment.

(h) Faulty thermostat

(h) Turn thermostat to *off* position. If unit continues to run, replace the thermostat.

(i) Refrigerant charge

(i) Too much or too little gas. Discharge, evacuate, and recharge with proper charge.

(j) Restriction or moisture

(j) Replace component where restriction is located. If moisture is suspected, replace drier-filter.

(k) Placing sudden load on unit

(k) Explain to user that heavy loading will cause long running time until temperatures are maintained. This running period may be several hours after, heavy loading of the cabinet.

High Noise Level

(a) Loose compressor mounting bolts

(a) Replace mounts. This type of noise usually

Symptoms and Possible Cause   Possible Remedy

High Noise Level (cont.)

	occurs during starting or stopping of the unit.
(b) Blower motor	(b) Check for motor wheel rubbing against housing; readjust. Replace motor if noise is excessive.
(c) Vibrating unit tubing	(c) Run hands over various lines. This can often help determine the location of the vibration. Gently reform tubing to eliminate vibration problem. Check discharge line from compressor to top of condenser to eliminate possible rubbing of line against condenser.
(d) Cabinet not level	(d) Check level, and, if necessary, make appropriate side-to-side and front-to-rear adjustments to level the cabinet.
(e) Location of cabinet	(e) An out-of-level floor may be a factor in causing noise. Also certain types of flooring transmit vibration more readily than others. Investigate type of floor construction where cabinet is located.
(f) Compressor	(f) Only after all external sources have been checked should the compressor be changed for noise.

Symptoms and Possible Cause   *Possible Remedy*

Food Compartment Temperatures Too Warm

(a) Inoperative thermostat

(a) Determine product temperatures. If slightly high, adjust control for longer running time. If control is discharged, broken, or if recording indicates switch settings are off, replace switch.

(b) Poor door seal

(b) Level cabinet. Adjust door seal; adjust tie rods in door.

(c) Repeated door openings or overloading of shelves in food compartment

(c) Instruct user regarding blocking normal air circulation in the cabinet.

(d) Inoperative fan switch

(d) Check leads and control. Replace if discharged, broken, or if other conditions would warrant replacing the control.

(e) Inoperative fan motors

(e) Replace motor.

(f) Lights stay on

(f) Adjust door to engage plunger. If switch is faulty, replace switch.

(g) Food compartment airflow control

(g) Check for obstruction in the passage from freezer to food compartment. Check for proper sealing around the control and bellows.

(h) Conventional refrigerator needs defrosting

(h) Excessive frost accumulation on freezer slows down heat transfer to refrigerant in the freezer. Excessive frost also resists normal airflow around

Symptoms and Possible Cause   Possible Remedy

Food Compartment
Temperatures Too Warm (cont.)

	the freezer. Suggest that user defrost more often.
(i) Improper bulb spacer	(i) Check for proper bulb spacer; use thicker spacer to lower switch cutoff point and lower cabinet temperature by increasing the running time.
(j) Seasonal control when used in closed position, or drop tray all the way back	(j) Change seasonal control or drip tray to the open position to allow more air circulation around in the freezer.

Food Compartment
Temperatures Too Cold

(a) Inoperative or erratic thermostat	(a) Turn control dial to highest setting and bypass thermostat. If unit starts and temperature goes down to normal level, replace the thermostat. Do not condemn the control until other factors are considered. If control does cycle the product regularly but foods are too cold, adjust the control warmer. If the unit does not cycle, replace the control.
(b) Very cold ambient temperature	(b) If in an extremely cold room, the freezer and cabinet temperatures tend

Symptoms and Possible Cause   Possible Remedy

Food Compartment
Temperatures Too Cold (cont.)

	to equalize, causing cabinet temperature to be below 32°F, refrigerator should be moved to a warmer location.
(c) Improper bulb spacer	(c) Use thinner bulb spacer to raise the effective switch cutoff point, and raise cabinet temperatures by reducing running time.

Freezing Compartment Too
Warm (Automatic Defrosting)

(a) Inoperative thermostat	(a) Note running time and freezer package temperatures; adjust control for longer running time. If control is broken or discharged, replace control.
(b) Poor door seal	(b) Level cabinet. Adjust door hinge wings. Check for interference from packages or foot pedal.
(c) Light stays on	(c) Adjust door to engage plunger. If switch is inoperative, replace switch.
(d) Blower motor inoperative	(d) Check to see that the blower motor is free. Check voltage to motor. Replace if inoperative.
(e) Inoperative timer	(e) Timer may not allow unit to go into defrost, causing

Symptoms and Possible Cause   Possible Remedy

Freezing Compartment Too Warm (Automatic Defrosting) (cont.)

	ice to build up on coil, which cuts down on the airflow. Replace timer if inoperative.
(f) Coil iced up	(f) Check drain heater, defrost heater, miter switch, and timer.

Water on Floor, Suction Lines Frosted

(a) Overcharge of refrigerant	(a) Frost back should stop after first couple of cycles. Purge slightly from high-side charging port. If too much has been purged, evacuate and recharge system.
(b) Freezer blower motor (when used) inoperative	(b) Check for voltage supply to motor. If none, check wiring. Remember this motor runs only when the compressor runs.

Oil on Floor

(a) Terminal stud leak	(a) Check with leak detector. Tighten stud. Evacuate and recharge with proper charge.
(b) Charging screw loose	(b) Check with leak detector. Tighten screw. Evacuate and recharge.
(c) Broken line	(c) Replace component or repair with line connec-

Symptoms and Possible Cause   *Possible Remedy*

Oil on Floor (cont.)

tor, if possible. Evacuate and recharge.

Frost in Freezer (Automatic Defrosting)

(a) Water spilled when ice-cube trays are placed in freezer

(a) Caution user that water spilled will not readily be removed. Tap bottom of cold-storage liner, and ice will break loose.

(b) Poor door seal

(b) Level cabinet. Adjust door hinge.

(c) Improper loading

(c) Check for interference from packages or door-opening pedal. Packages should not block airflow at the top in back of cold-storage liner.

(d) Inoperative freezer blower motor (when used)

(d) Check to see that blower wheel is free. Check voltage to motor. Replace if defective.

Excessive Moisture Inside Food Compartment (Temperature Normal)

(a) Poor door seal

(a) Level cabinet. Adjust door hinges. Adjust tie rods in door.

(b) Normal operation

(b) During humid weather, moisture will accumulate on cold surfaces within refrigerator, just as it accumulates on a glass containing an iced drink

Symptoms and Possible Cause   *Possible Remedy*

Excessive Moisture Inside Food Compartment (Temperature Normal) (cont.)

on a hot summer day. This is perfectly normal and is particularly noticeable when refrigerator door is opened frequently.

SUMMARY

A restricted capillary tube may be caused by moisture in the system, kinked tubing, or foreign particles lodged in the lines. Each of these conditions will cause similar symptoms to occur.

A kinked capillary tube will stop the flow of refrigerant, and the evaporator will be free from frost. The compressor will run continuously and eventually cycle on the overload protector.

A refrigerator system that is undercharged with refrigerants will produce various conditions, depending upon the degree of undercharge. During normal operation, a system that is fully charged will have frost covering the entire evaporator and accumulator. Any degree of undercharge or a gradual leak of refrigerant will be noticed first by the absence of frost on the accumulator. The compressor will run for long periods of time, and the food-compartment temperature may be colder than normal.

An accumulation of moisture or foreign particles may freeze or lodge in the evaporator tubing and cause a partial restriction at that point.

A defective compressor that is not pumping adequately will produce little cooling effect. The evaporator may be covered with a thin film of frost, but the evaporator temperature will not decrease to the cutout temperature of the control even with continuous running of the compressor.

Other component parts of the refrigerator may need attention. Each has its own characteristics, which must be understood if they are to be properly repaired or put into operating condition.

REVIEW QUESTIONS

1. What can cause a restricted capillary tube?
2. How do you detect an undercharged refrigerator?
3. How do you use an ohmmeter to test the electrical system?
4. What is the difference between the start and the run windings in a compressor?
5. How are capacitors connected on larger refrigerators?
6. What is the location of the starting relay?
7. Explain how the compressor unit is replaced.
8. Explain how to replace the heat exchanger.
9. What is the reason for the evaporator cooling fan?
10. What are the two types of temperature-control devices?
11. What could cause the unit to be dead and not run?
12. The unit hums and shuts off—what could cause this condition?
13. What causes a high noise level?
14. What are four possible troubles when the food-compartment temperature is too warm?
15. What are three possible causes of the food compartment being too cold?
16. What is the difference between a potential and a current relay?
17. Where is the potential relay most often used?
18. How are motor connections brought out of the hermetic shell of a compressor for connection to power lines?

CHAPTER 9

Household-Cabinet Defrosting Systems

Since the accumulation of frost on the evaporator has a direct relationship to the ambient temperature and humidity, the successful performance of any refrigerator depends upon maintaining cabinet and freezer temperatures within a range that will promote the most favorable conditions for keeping perishables and frozen foods. It should be noted that each time the refrigerator door is opened, some of the heavy cold air within the cabinet escapes and mixes with the warm room air. This process creates a low-pressure area within the cabinet, which draws in warm room air.

The rapid increase in cabinet air temperature results in an accumulation of excess moisture on the evaporator and cold surfaces within the refrigerator. An additional result of frequent door openings is that the temperature control will almost immediately energize the unit, which will run until the cabinet air temperature has been lowered to a degree corresponding to the setting of the temperature control switch. Naturally, with a greater number of door openings, the unit will operate with greater frequency and for longer periods of time, with a corresponding higher power consumption and increase in frost accumulation on the evaporator. One method of minimizing refrigerator door openings when preparing a meal is to remove all required food and dishes at one time, instead of opening the refrigerator door for each dish required. When this is done, door openings may be decreased as many as seven to ten times.

MANUAL DEFROSTING

Periodic defrosting is essential to the efficient and economical operation of any refrigeration system. The defrosting operation should be performed when the frost accumulation on the cooling unit becomes approximately ¼ inch thick; and, regardless of the thickness of the frost, the unit should be defrosted and the moisture removed once every two weeks (or more often if conditions require). If large quantities of moist food are stored or if the weather is exceptionally humid, it may be necessary to defrost more often.

Need for Defrosting

The need for defrosting is readily apparent if the following factors are considered: cabinet air temperatures are controlled by means of a temperature control, the control bulb of which is attached directly to the cooling unit and frequently insulated from the circulating air by a rubber insulator. The bulb is therefore directly affected by the temperature of the cooling unit but is affected only slightly by the temperature of the air in the cabinet. Since the cooling unit is the coldest spot inside the cabinet, moisture in the form of frost deposits on the outside surface of the cooling unit is increased considerably.

Position of Control or Switch—On refrigerators equipped with a defrosting switch, all that is necessary to defrost the unit is to turn the control to the "defrost" position (it is not necessary to turn it off or allow the door to remain open). Allow the switch to remain in this position until all the frost has melted. To defrost refrigerators not equipped with a defrosting switch, all that is necessary is to set the *on* and *off* switch to the "off" position.

Hot-Water Defrosting—In order to hasten the defrosting period, particularly in warm weather when it is desired to preserve ice cubes and allow only a small rise in cabinet air temperatures, pans or containers of hot water can be placed in the cooling unit after turning the control switch to the "off" position. The unit will then defrost in about 20 to 30 minutes.

Defrosting Tray—It is obvious that, before a defrosting operation, the defrosting tray should be empty and correctly located to receive the water from the melting ice of the cooling unit. After the defrosting is completed, the defrosting tray should be emptied, the interior of the refrigerator cleaned, and fresh water placed in the ice-cube trays.

AUTOMATIC DEFROSTING

Automatic defrosting is accomplished by providing an electric heater (or heaters) consisting of a conventional ohmic resistor encased in tubing, which, by various control methods, supplies the heat necessary to prevent excessive moisture accumulation on the evaporator. The defrost heater operates in a cyclical manner. The heater may be energized on cutout periods of the unit only, or it may be operated by a control-timer mechanism starting a defrost cycle once every 12 or 24 hours, as shown in Fig. 9-1.

The freezer and fresh-food compartments usually have a common drainage system. During a defrost cycle in the freezer compartment, the water runs down special drain tubes to a collector or pan under the cabinet, where it is evaporated into the room air.

Operation

There is no mystery or magic connected with the principle of automatic defrosting. In this connection, it should be noted that the various manufacturers of refrigerators employ different terms for automatic defrosting, such as *no-frost, frost-proof, frost-free*. A 12-hour system means a refrigerator defrosting system (Fig.

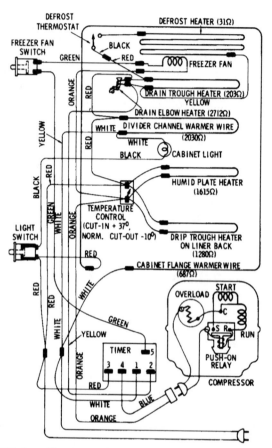

Fig. 9-1. Wiring diagram of an electric circuit used in a typical automatic-defrost refrigerator.

9–2) is used to automatically defrost the evaporator once every 12 hours. In a defrost system of this type, operation of the compressor is controlled by the temperature-control switch during both defrost and refrigeration cycles. Should the defrost-control switch be in a defrost position during a thermostat cycle (compressor off), the defrost cycle will not start until the temperature-control switch closes and starts the compressor.

The defrost controls usually employed are essentially single-pole, double-throw switching devices in which the switch arm is

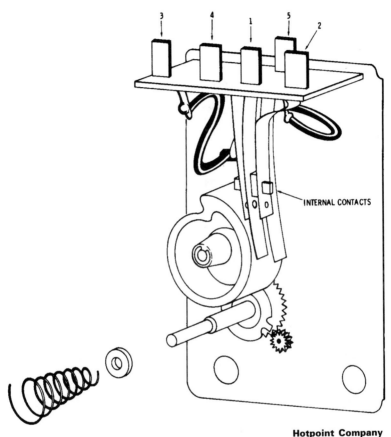

Hotpoint Company

Fig. 9-2. Defrost timer showing contact and terminal arrangement.

moved to the defrost position by an electric clock, as shown schematically in Fig. 9-3. The switch arm is returned to the normal position by a power element that is responsive to changes in temperature. Although there are several methods to accomplish automatic defrosting, depending upon the manufacturer of the refrigerating unit, a typical system operates as follows.

A small (usually 10W) liner heater operates during the *off* cycle of the compressor. This heater supplies a small amount of heat to the thermostat bulb, ensuring a reasonably short *off* cycle if the refrigerator is installed in a cold room or on an unheated back

293

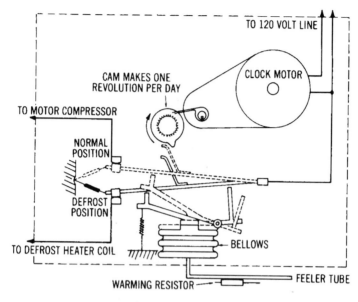

Fig. 9-3. Pictorial view of an electric-clock motor and contact arrangement in a typical automatic-defrost refrigerator.

porch. When the thermostat in the fresh-food compartment has energized the compressor, the refrigerant circulates through the cooling coil in the food compartment and passes through a restrictor into the cold coil in the freezer compartment. The refrigerant expands and evaporates, absorbing heat from the fresh-food and freezer compartments.

A fan in the freezer compartment circulates the air from the bottom, up through the coil, and out through the fan. This airflow pattern is unchanging and always goes from the bottom, up under the coil cover plate, through the coil, and out through the orifice in the coil-cover plate. Moisture is removed from the air as it passes through the coil and is deposited on the colder surface of the coil. Since all the moisture is deposited in the coil, there will be no frost on any other part of the freezer compartment or on any food packages stored in this area. Because frost cannot accumulate on the coil indefinitely, some provision must be made for its removal. This is done by means of an automatic defrost cycle.

The automatic defrost cycle for the freezer compartment is controlled by a timer with a 12-hour electrically driven movement. *The timer is running any time the cold control switch is closed.* The cold control switch also controls the compressor. Once the timer activates the switches on the end of its shaft (the timer is a motor), it removes the compressor from the circuit and the compressor stops running. However, the timer motor still has a complete circuit since the defrosting cycle started while the compressor was turned on. The timer had to be *on* since the cold control switch *controls* the compressor and timer motor. The only time that the timer motor operates independently of the compressor is during the defrost cycle. It has to keep running in order to cause the cams to do their work in opening and closing the time switches. Once the timer has completed its operation and the cams have gone back to normal position, the compressor is placed into the circuit once again. Since the defrost cycle interrupted the *on* cycle, the compressor is activated immediately to bring the coil down to the proper temperature as quickly as possible. There is, however, an additional 5-minute delay before the fan operation is resumed. This is to prevent the circulation of warm, moist air.

The only time the timer motor operates independently of the compressor, therefore, is during a defrost cycle. At this time (during defrosting) the compressor is *not* running. It was taken out of operation by terminals 1 and 4 being opened by the cam on the timer motor. However, only the cold control has to be closed in order for the timer motor to operate. And, since the cold control was on when the cycle started, it stays closed because the compressor is not running and no cold air is being produced to change the temperature inside where the control is located.

Keep in mind that the timer runs only when the cold control is *on*. This means that when the compressor is off, so is the timer motor—unless the cam-operated switches open and remove the compressor while the thermostat (or cold control) is still on, or closed. Hence, the timer motor runs only when the compressor does, for the sake of making sure that the cycle calls for defrosting in relation to the running time of the compressor (the exception being, of course, when the timer motor runs while the defrosting is taking place). If it is assumed that the compressor will run 50 percent of the time, there will be at least one automatic defrost a day in the freezer compartment. If the compressor should run

100 percent of the time (under extreme ambient conditions), there would be a maximum of two defrost cycles during a 24-hour period.

Defrost Sequence

The defrost sequence is as follows:

1. The timer turns off the compressor and the freezer-compartment fan. At the same time, it energizes the heater on the coil and the heater on the drain trough. At this point, the compressor and fans are not operating, but the heaters are activated.
2. The heater melts the frost on the coil, and the resulting water drains down into the trough through a tube in the bottom of the freezer compartment.
3. When the temperature at the bimetal thermodisk reaches approximately 65°F, the coil heater is turned off. The coil should be completely defrosted by this time. Approximately 10 minutes will elapse from the start of the defrost cycle to the time when the coil heater is cut off. If, for any reason, the bimetal disk should fail to function when the interior evaporator temperature reaches 65°F, the timer will cut off the coil heater after approximately 18 minutes from the start of the defrost cycle. The food will not be affected by the defrost cycle because the fan is off and warm air is not being circulated. The food packages are protected by the insulating cold air in the evaporator and the insulation afforded by the plate covering the cold coil.
4. The drain heater stays on for approximately 8 minutes after the coil heater has been shut off by the thermodisk and is cut off by the action of the timer. This 8-minute delay in cutting off the drain-trough heater is to ensure that all water is drained from the trough before the refrigeration cycle begins again.
5. Since the defrost cycle interrupted an *on* cycle, the compressor is activated immediately to bring the coil down to the proper temperature as quickly as possible. However, there is an additional 5-minute delay before fan operation is resumed in order to prevent the circulation of warm moist air.

Defrost-Timer Operation

Figures 9-4 to 9-6 show the action of a typical defrost timer during normal operation as well as the variation in the electrical circuit during the defrost cycle. Figure 9-4 indicates the circuit conditions during normal operation. One side of the power line goes to the timer motor and to contact terminal 1. The other side

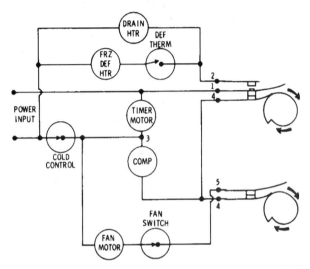

Hotpoint Company

Fig. 9-4. Electric circuit during normal operation.

of the power line runs through the cold control in the fresh-food compartment and to the other side of the timer motor. This same side of the power line also goes to one side of the compressor and then to one side of the fan motor. Contact terminal 1 makes contact with terminal 4, which goes to the other side of the compressor. Terminal 4 also makes contact with terminal 5, which goes to one side of the fan switch. When the fan-switch plunger is depressed and terminals 4 and 5 are making contact, the fan motor is energized.

The condition shown in Fig. 9-5 is the one that exists at the beginning of the defrost cycle. Terminals 1 and 4 have separated, which interrupts the power to the compressor. Terminals 4 and 5 are also separated, which interrupts the power to the fan switch, deenergizing the fan motor. Timer terminals 1 and 2 make contact, sending power to one side of the drain heater and to the cold-coil thermostat. The other side of the drain heater and one side of the coil heater connect to one side of the power line. This connection energizes the drain-trough heater. The coil heater is energized when the thermodisk closes.

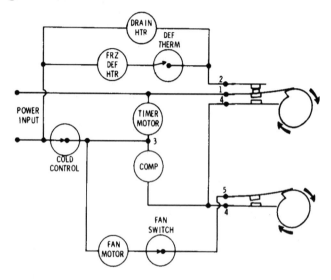

Hotpoint Company

Fig. 9-5. Electric circuit at start of defrost cycle.

Figure 9-6 shows the electrical situation after approximately 18 minutes of the defrost cycle have expired. Terminals 1 and 2 break contact and deenergize both the coil heater and the drain heater. Remember that, in normal operation, the cold-coil thermostat will deenergize the coil heater after approximately 10 minutes of the defrost cycle. Terminals 1 and 4 make contact and energize the compressor, while terminals 4 and 5 are still open to allow the 5-minute delay before the evaporator fan starts again.

After the 5-minute delay, terminals 4 and 5 make contact and the evaporator fan is energized once again. The action of the timer in breaking the circuit of the cold-coil heater is purely a mechanical one. If the cold-coil thermostat is operating properly, it will remove the coil heater from the circuit after approximately 10

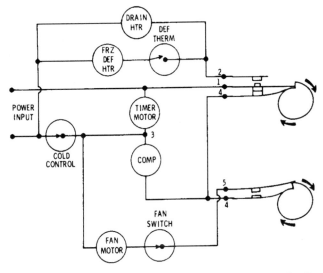

Fig. 9-6. Electric circuit near the end of defrost cycle.

minutes of the defrost cycle or when the temperature at the point of the cold-coil thermostat reaches either 55 or 65°F.

The schematic and pictorial wiring diagrams in Figs. 9-4 to 9-6 are representative of other makes of refrigeration. However, if in doubt and there is no schematic diagram pasted on the refrigerator, write the company that made the refrigerator for a schematic diagram of same.

DEFROST-THERMOSTAT TESTING

The freezer compartment defrost-thermostat contacts are usually set to a temperature change of about 15°F. Since a test might be made just after a freezer defrost cycle, it could appear that the defrost thermostat was inoperative because the contacts were still open. In order to make a more positive check, the following steps are recommended.

1. Disconnect the power cord.
2. Remove the breaker strip from the fresh-food compartment.

3. Connect the probes of a test lamp to the ends of the defrost-thermostat leads. Refer to the wiring diagram on the cabinet for the correct color code. (Needle-type probes are best for this test since they can be inserted into the connectors.)
4. Plug the test lamp into the wall receptacle. If the test lamp has a bright glow, the contacts are closed; but if the lamp has a dull glow, the contacts are open and additional testing is necessary.
5. It is important that the refrigerator line cord be disconnected; otherwise an erroneous reading will be made or the defrost and drain-trough heater could be energized. Points A and B in Fig. 9-7 show the circuit being tested and the points at which the test lamp is connected. If the initial test has shown the freezer thermostat contacts open, the serviceman should use a can of refrigerant for the next part of the test. Leave test lamp connected to points A and B.
6. Remove the food from the freezer compartment, and take off the freezer cover plate. Spray the defrost thermostat with a can of refrigerant for about 15 seconds. This should reduce the temperature to the vicinity of 15°F, thus closing the contacts. When the contacts close, a click will be heard and the test lamp will glow brightly. *If the contacts do not close after approximately three sprayings of refrigerant and the lamp retains its dull glow, then the serviceman has made a positive check and should replace the thermostat.* The serviceman, having tested the thermostat in the freezer and finding it operating normally, should test the freezer heater.

If you are accustomed to using a volt-ohmmeter and a clamp-on ammeter, these instruments are far superior to a test lamp. Note on the schematic wiring diagrams that at the top there are wattages given. These are highly important in troubleshooting domestic refrigerators as they indicate whether the unit is drawing the right amount of power, which can be expressed as

$$\text{Volts} \times \text{amperes} = \text{watts}$$

The ohmmeter can be used for testing continuity, but you must be certain that the cord to the power supply is unplugged or you might burn up the ohmmeter. If you use the voltmeter instead of the test lamp mentioned, you will obtain the same results with less danger to yourself or the components of the refrigerator. In using the volt-ohmmeter, always check to be sure that you have it set for ohms or the maximum voltage that you expect to encounter.

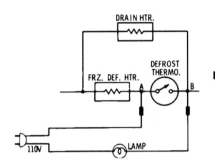

Fig. 9-7. Schematic diagram showing automatic-defrost thermostat connections.

Hotpoint Company

Parts of the Defrost System

The defrost system consists of only three functional components. These are the *defrost thermostat*, the *defrost heater* and the *defrost control*.

If there is trouble with the defrost system, the first thing to check is the defrost thermostat, which is a simple device. A bimetallic disc, pushing against a transfer pin, holds the contacts open (Fig. 9-16). As the evaporator temperature is lowered, the bimetal disc warps. The spring loaded contact arm pushes the transfer pin out of the way, allowing contacts to snap closed.

The thermostat design, incorporating the transfer pin, provides a fail-safe feature. In other words, when a defrost thermostat fails to function properly, it normally fails with the contacts open. It rarely fails with the contacts closed.

Defrost thermostats are precision built and tested to within ±6°F. of the specified limit. A unique characteristic of a bimetal disc is that its calibration is fixed and does not change. This provides reliability in excess of the life expectancy of the refrigerator or freezer. Continual life tests at the factory of 100,000 cycles, corresponding to 100 years, reveal that the calibration of a defrost thermostat does not drift out of tolerance (±6°F). A

301

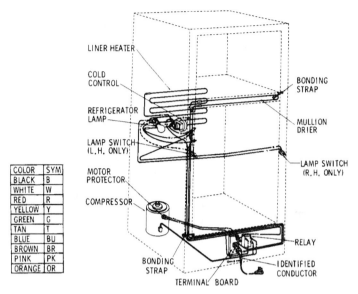

COLOR	SYM
BLACK	B
WHITE	W
RED	R
YELLOW	Y
GREEN	G
TAN	T
BLUE	BU
BROWN	BR
PINK	PK
ORANGE	OR

Frigidaire Corporation Div. of General Motors

Fig. 9-8. Pictorial wiring diagram of Model FCD-123T.

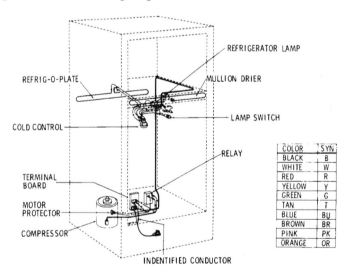

COLOR	SYM
BLACK	B
WHITE	W
RED	R
YELLOW	Y
GREEN	G
TAN	T
BLUE	BU
BROWN	BR
PINK	PK
ORANGE	OR

Frigidaire Corporation Div. of General Motors

Fig. 9-9. Pictorial wiring diagram of Model FGD-150T.

302

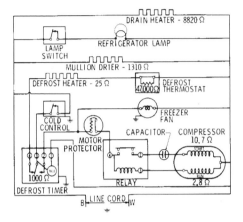

Frigidaire Corporation Div. of General Motors

Fig. 9-10. Schematic wiring diagram of Model FPI-152T.

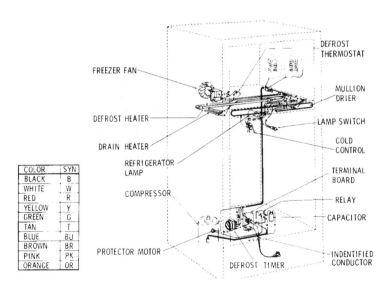

COLOR	SYN
BLACK	B
WHITE	W
RED	R
YELLOW	Y
GREEN	G
TAN	T
BLUE	BU
BROWN	BR
PINK	PK
ORANGE	OR

Frigidaire Corporation Div. of General Motors

Fig. 9-11. Pictorial wiring diagram of Model FPI-152T.

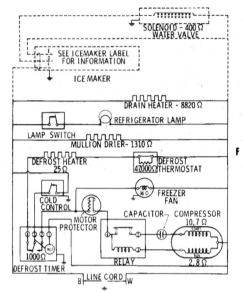

Frigidaire Corporation Div. of General Motors

Fig. 9-12. Schematic diagram of Model FPCI-152T-7.

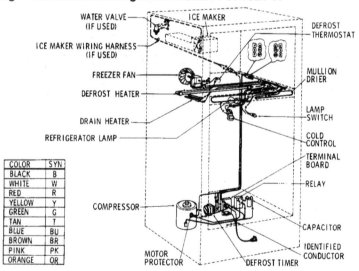

COLOR	SYN
BLACK	B
WHITE	W
RED	R
YELLOW	Y
GREEN	G
TAN	T
BLUE	BU
BROWN	BR
PINK	PK
ORANGE	OR

Frigidaire Corporation Div. of General Motors

Fig. 9-13. Pictorial wiring diagram of Model FPCI-152T-7.

304

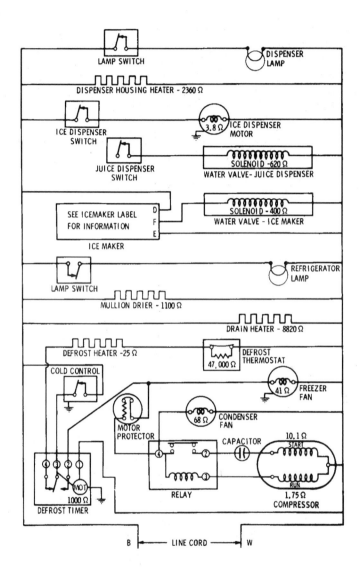

Frigidaire Corporation Div. of General Motors

Fig. 9-14. Schematic wiring diagram of Model FPF-200TI.

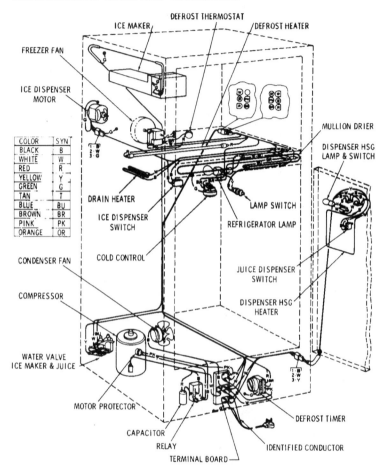

COLOR	SYN
BLACK	B
WHITE	W
RED	R
YELLOW	Y
GREEN	G
TAN	T
BLUE	BU
BROWN	BR
PINK	PK
ORANGE	OR

Frigidaire Corporation Div. of General Motors

Fig. 9-15. Pictorial wiring diagram of Model FPF-200TI.

slight "creep" of about 2°F occurs at about 28,000 cycles during the life of all defrost thermostats as the components "wear in." A defrost thermostat can, however, be incorrectly calibrated from the beginning. Therefore, if the defrost system has functioned properly for several months before failing, disregard the possibility of an incorrectly calibrated defrost thermostat. Do not suspect that the defrost thermostat has the wrong calibration unless residual ice is found in the evaporator.

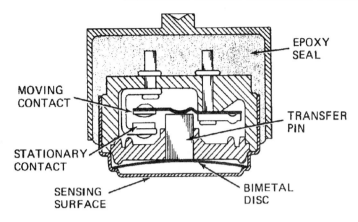

Fig. 9-16. A cutaway view of a defrost thermostat.

The only practical method for checking the defrost thermostat in the field is to test it for continuity. The contacts should be closed at all times, except during the later part of the defrost cycle and for the first ten minutes thereafter, when the compressor resumes operation. You can determine that the defrost thermostat contacts are closed on side-by-side models by feeling the mullion heater, which is in series with the defrost thermostat. Thus, if the mullion heater is warm, the thermostat contacts must be closed (Fig. 9-17).

Never replace a defrost thermostat unless it is known to be inoperative. Never substitute a defrost thermostat unless it is a recommended field correction or a factory authorized superse-dure. Under no circumstances should a defrost thermostat be bypassed, other than momentarily for testing purposes.

Check the defrost heater or heaters. Radiant (glass sheath) defrost heaters may be found on some GE and Hotpoint models. If the heaters are connected in series and one comes open, the whole arrangement will be inoperative. Figure 9-18 shows the defrost heater design.

Checking the circuit with an ohmmeter does not conclusively establish that the heaters are operative or inoperative. For example, the glass sheath may be broken, the glass darkened, the element coils bunched together, or the jumper lead between the heaters shorted to the ground. In this case, the ohmmeter may

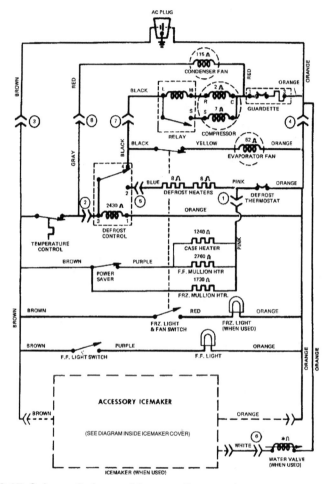

Fig. 9-17. Schematic for a refrigerator/freezer.

Fig. 9-18. Defrost heater encased in glass.

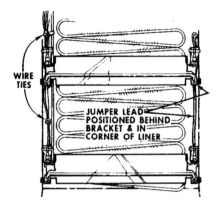

Fig. 9-19. Location of wires inside the walls of a refrigerator.

WIRE TIES

JUMPER LEAD POSITIONED BEHIND BRACKET & IN CORNER OF LINER

indicate the correct resistance value for the circuit. An infinity reading on the ohmmeter when it is applied to the heater circuit may be due to a detached wire at one heater terminal. When in doubt, these items should be visually inspected (Fig. 9-19).

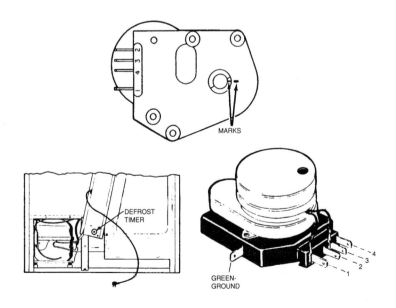

MARKS

DEFROST TIMER

GREEN-GROUND

Fig. 9-20. Defrost control in a refrigerator.

309

Where replacement defrost heaters are furnished in sets of two or three, always use the complete set—even though only one heater is actually inoperative. Never substitute defrost heaters unless it is a factory-authorized supersedure. When replacing defrost heaters, avoid handling the glass sheath. Handling leaves salt deposits, resulting in glass embrittlement. Rinse any fingerprints from the glass with water and dry with a clean paper towel.

Check the defrost control next (Fig. 9-20). The defrost control is an electromechanical device that is subject to electrical and mechanical failures. An ohmmeter can be used to check the electrical characteristics. The mechanical characteristics can be checked only by an operational test. To do this, manually advance the defrost control to the start of the defrost cycle (where the marks align in Fig. 9-20). Then, with the temperature control contacts closed, wait approximately five minutes for the control to advance itself automatically.

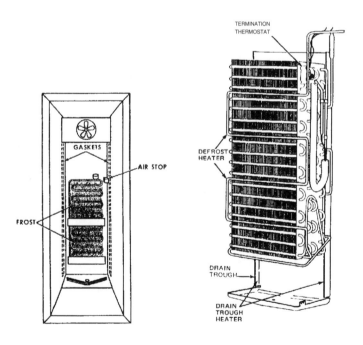

Fig. 9-21. Location of the evaporator coils inside a refrigerator.

When replacing a defrost control, check that the wiring is connected correctly and that the device will function properly. Remove all frost from the evaporator. Use a heat gun, if necessary. Partially melted frost remaining on the evaporator will refreeze. It will form a more solid mass of ice that may never be cleared by subsequent automatic defrosting. If in doubt, visually inspect the evaporator (Fig. 9-21).

Check the location and position of the coil cover gaskets and air stops. These items are necessary to direct the air flow across the evaporator. If coil cover gaskets and air stops are missing or mispositioned, moisture-laden air will bypass the evaporator. Frost accumulation will appear in unusual locations. This will stall the evaporator fan, frost the control console and cause defrosting problems.

Check for excessive heat leakage. Check that a door is not ajar due to a binding condition. Check that food packages have not been improperly placed so as to interfere with the door closing fully. Look for a poor door gasket seal caused by a racked cabinet, improper latch adjustment or torn door gaskets. If any of these conditions exist, more frost may accumulate than can be removed during the defrost cycle.

Consider the following when checking for the frost problem solution:

Room Ambient Temperature

Room temperature and especially humidity contribute to a defrosting problem. High ambient temperatures will indirectly impose a greater load on the refrigerator or freezer. This results in longer run time and greater frost load. High humidity, even with a low room temperature, can likewise result in an abnormal frost load.

Usage Conditions

Obviously, the heavier the usage, the longer the run time and greater the frost load. Under such conditions, more frost may be accumulated than can be removed during the defrost cycle. This results in residual ice.

Cabinet Sealing

Missing drain caps and leaks from torn door gaskets or at the tube seal can contribute to a defrost problem. If leaks in the outer case bottom seams are suspected, RTV® sealer can be applied to the perimeter of the bottom from underneath the cabinet.

Length of Heater Operation Time

Determine the length of time the defrost circuit is allowed to remain energized. As a general rule, with a heavily loaded evaporator, the heater should be energized for at least one-half of the defrost cycle time allotted.

Refrigerant Charge

A short refrigerant charge will allow greater-than-normal frost accumulation at the inlet of the evaporator and no frost at the outlet. This imbalance can result in incomplete defrosting of the evaporator. A lack of frost (cold mass) at the defrost thermostat will permit the thermostat to open sooner than normal.

SUMMARY

The successful performance of any refrigerator depends upon maintaining cabinet and freezer temperatures within a range that will promote the most favorable conditions for keeping perishables and frozen foods.

Periodic defrosting is essential to the efficient and economical operation of any refrigeration system. If large quantities of moist food are stored, or if the weather is exceptionally humid, it may be necessary to defrost more often.

Automatic defrosting is accomplished by providing an electric heater consisting of a conventional ohmic resistor encased in tubing, which, by various control methods, supplies the heat necessary to prevent excessive moisture accumulation on the evaporator. The heater may be energized on cutout period of the unit only, or it may be operated by a control-timer mechanism starting a defrost cycle once every 12 or 24 hours.

REVIEW QUESTIONS

1. Why is defrosting needed?
2. What is meant by manual defrosting?
3. How is manual defrosting accomplished?
4. How is automatic defrosting accomplished?
5. How often does defrosting take place in a refrigerator that is automatically defrosted?
6. How often should you defrost a manual-defrosting refrigerator?
7. List the steps in the defrost sequence.
8. Explain how the defrost timer operates.
9. How do you go about testing the defrost thermostat?
10. How is the ohmmeter used in testing the defrost circuit?
11. What are the three functional components of a defrost system?
12. What is the first thing to check if there is trouble with a defrost system in a refrigerator?
13. What is the temperature tolerance of a defrost thermostat?
14. How long are defrost thermostats expected to last?
15. Why doesn't an ohmmeter tell the whole story when there is electrical trouble in the defrost system?
16. Why don't you leave fingerprints on the heater glass enclosures?
17. How can defective door latches lead to defrost problems?
18. How can a racked cabinet cause problems with defrosting?
19. Why does heavy usage of a refrigerator or freezer create unusual defrost loading?
20. How does a short charge cause defrosting problems?

CHAPTER 10

Cabinet Maintenance and Repairs

Proper fit of the door is of the utmost importance in any refrigerator. The door seal should always be tested before a refrigerator is placed in operation or when servicing a refrigerator in the home. A poor door seal usually results in cabinet sweating inside, excessive frosting of the cooling unit, high percentage of running time, high power consumption, slow ice freezing, high cabinet air temperature, and possible sweating on the outside front of the cabinet. A typical door construction and seal assembly is shown in Fig. 10-1.

A good method of testing the tightness of a door seal is to placer a light inside the cabinet; this should be a strong flashlight. The light will be visible at any point where the door gasket does not make sufficient contact with the cabinet. Another method to locate imperfect door seals is by the use of a 0.003-inch-thick metal feeler. Locate the point of a poor seal by inserting the feeler

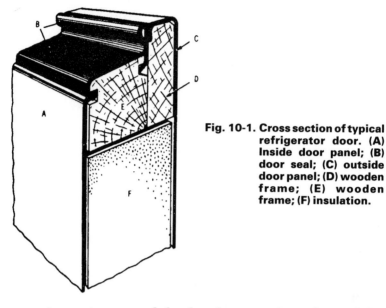

Fig. 10-1. Cross section of typical refrigerator door. (A) Inside door panel; (B) door seal; (C) outside door panel; (D) wooden frame; (E) wooden frame; (F) insulation.

at various points around the door between the gasket and the cabinet front with the door closed. A magnet-enclosed gasket seal is shown in Fig. 10-2.

If a poor seal is located, check the gasket first to see that it is not excessively worn. Check the hardware to see that it is not sprung

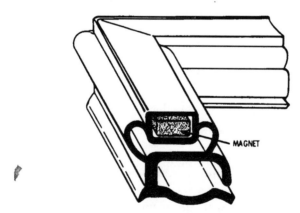

Whirlpool Corporation

Fig. 10-2. Cross section of typical magnetic-door gasket.

or worn and that the screws are tight. A poor seal can be corrected by rehanging the door, replacing the gasket, or properly adjusting or replacing the hardware. If, however, a poor seal is caused by the door being sprung out of line or by the front of the cabinet being out of line, the poor seal can be corrected as follows:

1. If the leak is on the hinge side of the door, open the door and hold it with the left hand. Then, with the heel of the right hand, strike all along the hinge edge until the door is sprung in sufficiently to tighten the seal.
2. If leakage is on the top-latch side, spring the door in at the top until contact is made along this side by opening the door, holding it at the bottom and pushing in at the top. Reverse this process if the leakage is at the bottom on the latch side.
3. To stop a leakage at the top or bottom of the door, spring the door in at the top or bottom by opening the door, holding one end and pushing on the end to be sprung in.
4. Place friction tape in back of the door gasket at any point where the gasket does not make a tight fit after the foregoing procedure has been followed. Be sure to fold the tape so that it does not show above the gasket.

DOOR REPLACEMENT

Doors are obviously replaceable by removing the hardware and transferring it to the new door, as shown in Fig. 10-3. Carefully center the door in the opening, and draw the hinge screws tight in rotation so that unequal pressure of the screws will not throw the door out of line.

Hinge Replacement

Hinges usually have a metal cover that conceals the screws. To remove this cover, insert a knife blade in the slot and pry out. The screws are now exposed and can be removed, and the hinge can be replaced. The hinge screws are embedded in heavy screw plates attached to the inside of the steel case, which prevents any sagging or distortion of the cabinet.

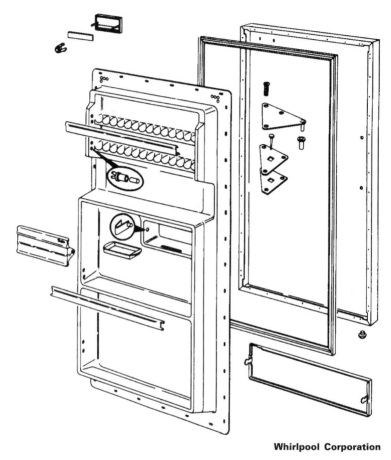

Whirlpool Corporation

Fig. 10-3. Exploded view of typical door parts.

Latch Replacement

First, remove the escutcheon that usually encloses the base of the handle by inserting a small screwdriver in the slot under the handle at the bottom of the escutcheon. Pry it off at the bottom, then lift it up and off. One type of escutcheon is removed by squeezing it in the middle of its plate to elongate it and then simply slipping it off. The defective latch can now be removed and replaced with a new one (Fig. 10-4).

Latch-Strike Adjustment

Three latch-tongue adjustments may be made to obtain the correct door closure and gasket seal. To make *up-and-down adjustments*, pencil a line on the cabinet flange just opposite the front edge of the latch mechanism nylon well (Fig. 10-5). Loosen the mounting screws, and adjust the latch tongue up or down until

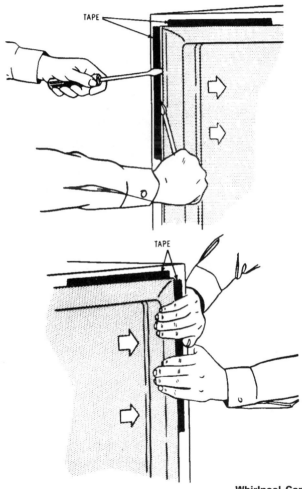

Whirlpool Corporation

Fig. 10-4. Replacement of food liner in a typical refrigerator door.

319

the lower point of the tongue is approximately ⅟₁₆ inch above the penciled line.

To make *in-and-out adjustments*, slowly close the door until a slight resistance is felt as the latch tongue contacts the rear edge of the mechanism nylon well. The distance between the door

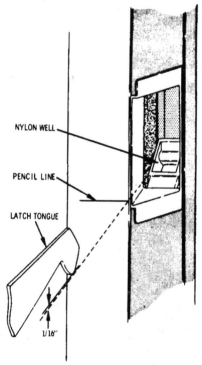

NYLON WELL

PENCIL LINE

LATCH TONGUE

1/16"

Fig. 10-5. Typical latch-tongue adjustment.

gasket and the cabinet flange should be approximately ¼ inch when the tongue touches the well. Only one distinct resistance should be noted as the door is closed.

If two contacts occur, the latch tongue is positioned too low on the door. Shims may be added to or removed from the latch tongue for in-and-out adjustments. To make *sideways adjustments*, hold the door in a partially open position, and measure the distance between the cabinet flange and the side of the latch tongue. Loosen the mounting screws, and adjust the tongue sideways until this distance is approximately ¼ inch.

320

This type of door was used on earlier models of refrigerators and freezers. The latching door has been entirely replaced in the industry with magnetic strips around the door. The magnetic door allows children to escape if trapped inside while playing in old refrigerators awaiting disposal. Many children suffocated when the latching door closed on them and they were unable to open it from inside. The magnetic strip also has the advantage of making a tighter seal all around the door frame.

Interior-Light Arrangement

Refrigerators are equipped with an automatic interior electric light, as shown in Fig. 10-6. The switch is operated by the door and

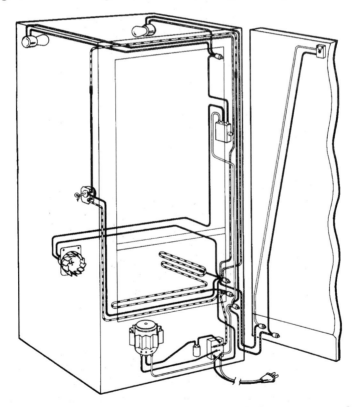

Fig. 10-6. Electrical system, including interior-light arrangement in a typical refrigerator.

is designed and wired to complete the lamp circuit and light the lamp when the door is opened. When the door is closed, the switch contacts open, and the light is extinguished.

Due to careful design of the switch and simplicity of the circuit, service complaints dealing with this part of the refrigerator are very infrequent. Trouble is most often found to be a burned-out lamp. Thus, if the lamp will not light, check for a burned-out or loosened lamp. If no trouble is found here, check the lightswitch plunger; it may be loose or stuck. Check all terminals of the light circuit for loose connections. Sometimes a broken wire may be responsible for the failure, although this is a rather remote possibility.

SUMMARY

The proper fit of the door is the most important thing in a refrigerator. The door seal should always be tested before a refrigerator is placed in operation or when servicing a unit in the home. A good method of testing the tightness of the seal is to place a flashlight inside and turn it on, close the door, and look for light leaks.

Hinge replacement is very critical since hinges support the weight of the door. The latch or the magnetic seal around the door is also important since the door must be held firmly closed to eliminate the migration of heat to the inside of the food compartments.

The door and the liners can be removed to be replaced after damage or because of imperfect fit. Shims may be added to or removed from the latch tongue for in-and-out adjustments.

Care should be taken to make sure the magnetic-door gasket is fitted properly.

REVIEW QUESTIONS

1. Why is the proper fit of a refrigerator door so important?
2. How do you check for tightness of a door?

3. How do you adjust the latch strike when the refrigerator has this type of catch?
4. How do you remove the roof liner on a typical refrigerator door?
5. What is the most frequent cause of trouble with the light in a refrigerator?

CHAPTER 11

Household Freezers

Household freezers are manufactured in various sizes to suit family requirements. The purpose of household freezers is to facilitate maximum food storage over a period of time. They are usually termed *horizontal* or *vertical* units, depending on individual requirements and installation space available. Household freezers are ordinarily manufactured in capacities of from 13 to 20 cu ft and equipped with hermetically-sealed compressor units.

CYCLE OF OPERATION

The operation of the refrigeration system can best be described with reference to the refrigerant-flow diagram shown in Fig. 11-1. The temperature of the freezer is regulated by a conventional temperature control, which starts the motor-compressor when the evaporator requires refrigeration and stops the motor-

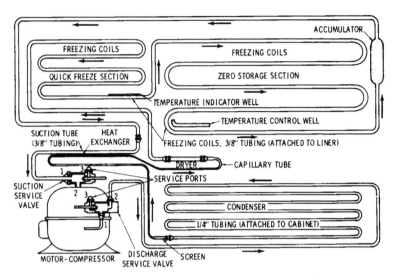

Fig. 11-1. Refrigerant-flow diagram of horizontal freezer.

compressor when the evaporator has been sufficiently cooled. When the temperature-control contacts close, the motor-compressor starts operating and reduces pressure in the evaporator.

The refrigerant evaporates, absorbing heat in the process, thus cooling the evaporator. The vapor is drawn through the suction line by the compressor, which compresses the refrigerant gas and forces it into the condenser under high pressure. The compression of the vapor causes the temperature to rise, permitting the condenser to transfer heat to the cabinet shell and into the air surrounding the cabinet, as shown in Fig. 11-2. As the compressed vapor gives up its heat, it condenses to a liquid in the condenser.

The capillary tube is made small in diameter in order to control the amount of liquid refrigerant forced through it from the condenser to the evaporator. For a short period of time after the motor-compressor shuts off, the pressure in the condenser continues to force refrigerant through the capillary tube until pressure is substantially equalized throughout the system. If the motor-compressor attempts to start immediately after it shuts off (before equalizing has occurred), it may stall and trip the overload

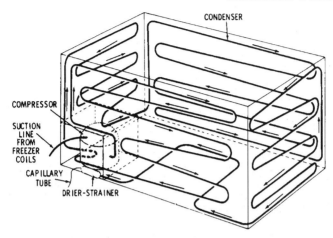

Fig. 11-2. Routing of the tubing in a typical horizontal freezer.

protector, thereby breaking the electrical circuit. This prevents the motor windings from burning out.

CHARACTERISTICS OF A CAPILLARY-TUBE SYSTEM

A refrigeration system that uses a *capillary tube* as the refrigerant control has certain characteristics and advantages. A capillary tube allows the liquid refrigerant to continue its flow from the condenser into the evaporator after the temperature control has shut off the motor-compressor. This effects a reduction in the condenser pressure during the *off* cycle and permits the motor-compressor to start easily against a relatively low pressure when refrigeration is again required.

The amount of refrigerant in a capillary-tube system is small but critical. There should be just enough refrigerant so that, during normal operation, evaporation will take place for the entire length of the freezing coil but not beyond. If there is *too little* refrigerant, there will be insufficient cooling of the freezing coils. If there is *too much* refrigerant, the excess liquid will flow into the suction line. Refrigerant in this line will cause it to frost during the *on* cycle, and the frost will melt and drip off during the *off* cycle.

ELECTRICAL SYSTEM

The hermetically-sealed motor-compressor requires a 60-hertz, 120V alternating current and has a capacity of from $\frac{1}{12}$ to $\frac{1}{3}$ hp, depending on the size of the total food-storing capacity of the freezer cabinet.

Fan Motor

The fan for cooling the motor-compressor in some models is mounted on the compressor housing. It is connected directly across the power supply to the motor-compressor after the temperature control and before the overload protector, as shown in Fig. 11-3. Connected in this manner, the fan will continue to cool the motor-compressor even if the overload protector has opened because of overheating. The fan will stop only when the temperature controls cut out.

Temperature Control and Signal Light

The temperature control, in addition to its conventional function, also controls the operation of the signal light, as noted in Fig. 11-3. The temperature control is an electrical switch controlled by the temperature of the control feeler tube. As the feeler tube gets warmer, the temperature control switches on and the motor-compressor starts. When the temperature of the feeler tube has been brought down to a predetermined temperature, the switch contacts open and the motor-compressor stops.

Thermostats

There are three commonly used types of thermostats on both upright and chest freezers (Figs. 11-4 through 11-6). Note the location of the adjustment screws on each thermostat.

The temperature at which the control opens (turns off the motor-compressor) is called the *cutout point*, and the temperature at which the control closes (turns on the motor-compressor) is called the *cut-in point*. Adjusting the temperature-control knob changes both the *cut-in* and *cutout* temperatures by varying the spring tension against which the diaphragm of the control exerts pressure. The design of the control provides adequate adjust-

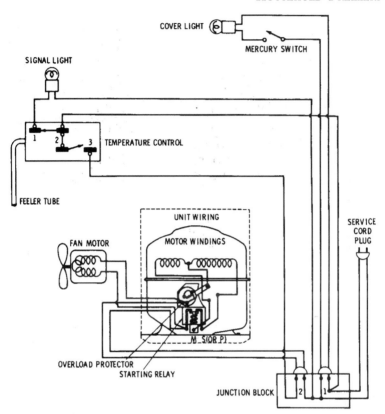

Fig. 11-3. Wiring diagram of electrical connections in a horizontal freezer.

ment to meet all normal requirements. For average use, the temperature-control knob should be set to the normal position, which is the warmest operating position. Since there is no "off" position, this makes it impossible to shut off the freezer accidentally.

The temperature control also operates the signal warning light. The signal circuit (inside the temperature control) will be closed, and the signal light located in the top part of the temperature-control escutcheon will be "on" when safe temperatures are being maintained in the food compartment. The expanding action of the control bellows (when the temperature in the food compartment rises 15°F or more above the cut-in setting of the control knob) opens the signal-circuit contacts and causes the signal light to go out when food-storage temperatures are unsafe.

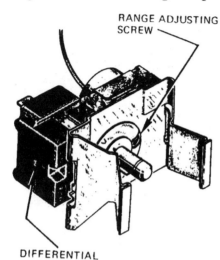

RANGE ADJUSTING SCREW

DIFFERENTIAL ADJUSTING SCREW

Fig. 11-4. G.E. thermostat.

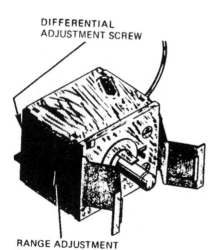

DIFFERENTIAL ADJUSTMENT SCREW

RANGE ADJUSTMENT SCREW

Fig. 11-5. Ranco thermostat.

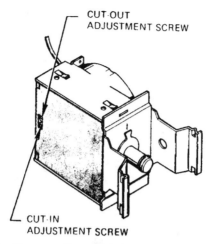

CUT-OUT
ADJUSTMENT SCREW

CUT-IN
ADJUSTMENT SCREW

Fig. 11-6. Cutler-Hammer thermostat.

The signal light will be "on" at all times, except as follows:

1. *Temperature control knob turned colder*—If the control knob is turned from a position near normal to a much colder position so that the cut-in point of the new control knob position is 15°F or more below the temperature at the control feeler tube, the signal light will go out. When the temperature is reduced to less than 15°F above the cut-in temperature of the new position, the signal light will come on.

2. *Large amounts of warm food in freezer*—If a large amount of warm food is put in the freezer, the light may go out. It is best to freeze smaller amounts of food at one time.

3. *Freezer door left open too long*—The signal light will go out if the freezer door is left open for an extended length of time.

4. *Blown fuse*—A blown fuse in the electrical supply (house circuit) will cause the signal light to go out.

5. *Service cord disconnected from wall outlet*—The signal light will go out if the service cord plug is disconnected from the wall outlet.

6. *Bulb (signal light) burned out.*

7. *Loose wire connections at control terminals.*

8. *Defective operation of refrigeration system*—If defective
operation of the refrigeration system causes the temperature in the food compartment to rise 15°F or more above the
cut-in point of the control-knob position, the signal light will
go out.

To obtain a clearer understanding of the temperature control
that regulates the operation of the motor-compressor and signal
light, refer to Figs. 11-7 to 11-9. During a normal off cycle of the
motor-compressor, the feeler tube of the temperature control is
cold and the bellows is contracted, allowing the spring to pull the
control arm down, as shown in Fig. 11-7. Contacts A and B are
open, and the motor-compressor is idle. Contacts C and D are
closed, and the signal light is on, indicating a safe food-storage
temperature.

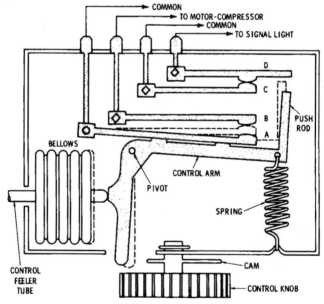

MOTOR-COMPRESSOR NOT RUNNING;
SIGNAL LIGHT ON
(FOOD STORAGE TEMPERATURES ARE SAFE)

**Fig. 11-7. Cutaway view of temperature control during normal *off*
cycle.**

During a normal *on* cycle of the motor-compressor, the bellows, control arm, and spring are in an intermediate position, as shown in Fig. 11-8. In this position, contacts A and B are closed

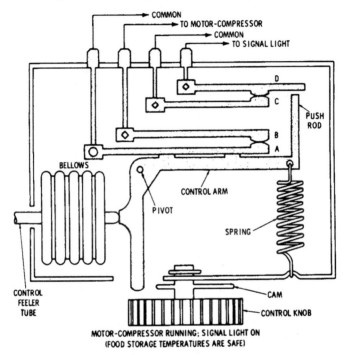

Fig. 11-8. Cutaway view of temperature control during normal *on* cycle.

so that the signal light is on and a safe food-storage temperature is indicated. If, for some reason, the temperature in the food compartment is 15°F or more above the cut-in temperature of the control, the bellows is expanded to a point that forces the right-hand end of the control arm upward, as shown in Fig. 11-9. Contacts A and B are closed, supplying power to the motor-compressor, but the push rod breaks contact at C and D and causes the signal circuit to open. The signal light will stay off, indicating an unsafe food-storage temperature, until the motorcompressor restores proper temperature in the food compartment. The bellows will then contract, and signal-circuit contacts C and D will close.

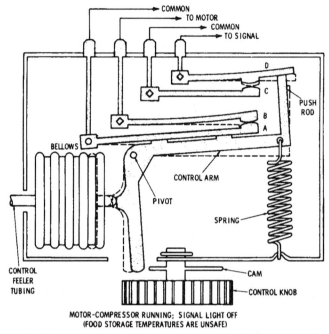

MOTOR-COMPRESSOR RUNNING; SIGNAL LIGHT OFF
(FOOD STORAGE TEMPERATURES ARE UNSAFE)

Fig. 11-9. Cutaway view of temperature control when signal light is off.

UPRIGHT FREEZERS

One basic refrigeration system is used on practically all upright freezers, the only major variation being the use of a fan-cooled condenser (Fig. 11-10) located beneath the cabinet on a few models. Other upright freezers use a static condenser (Fig. 11-11) located on the back of the cabinet. All units are of the conventional hermetically-sealed design. For service purposes, the compressor is considered as one part, and the freezer shelf and heat-exchanger assembly are considered as another and completely separate part.

Refrigeration System

The first few passes of the condenser tubing (static type, Fig. 11-11) form the oil-cooler loops that carry the partially cooled refrigerant gas back through the compressor. This lowers the operating temperature of the compressor and results in increased efficiency for the unit. The condenser releases the heat absorbed by the refrigerant and changes the hot gaseous refrigerant into a cool liquid refrigerant.

The capillary tube controls the flow of refrigerant to the freezer shelves. Part of the capillary tube is soldered to the suction line to form a heat exchanger. The heat from the capillary tube is transferred to the suction line, and the cold suction line further

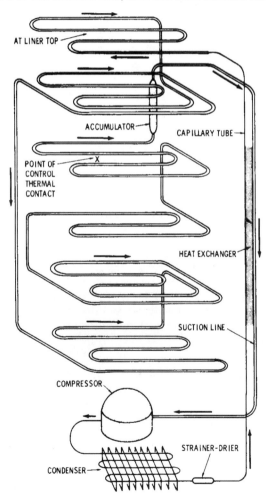

AT LINER TOP

ACCUMULATOR

CAPILLARY TUBE

POINT OF
CONTROL
THERMAL
CONTACT

X

HEAT EXCHANGER

SUCTION LINE

COMPRESSOR

STRAINER-DRIER

CONDENSER

Fig. 11-10. Refrigerant-flow diagram of upright household freezer with fan-cooled condenser.

cools the liquid refrigerant in the capillary tube. When the refrigerant leaves the capillary tube and enters the larger tubing of the freezer shelves, the sudden increase in tubing diameter forms a low-pressure area, and the temperature of the refrigerant drops rapidly as it changes to a mixture of liquid and gas. This cold mixture passes through the top shelf, ascending through each additional shelf until it reaches the accumulator.

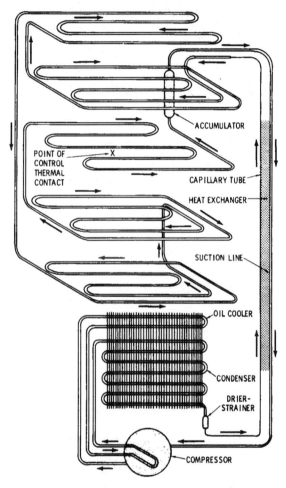

Fig. 11-11. Refrigerant-flow diagram of upright freezer with static condenser.

In the process of passing through the shelf tubing, the refrigerant absorbs heat from the storage area and is gradually changed from a liquid and gas mixture to a gas. The accumulator traps and evaporates any small amount of liquid refrigerant remaining in the low side, thereby preventing the liquid from entering the suction line to the compressor.

The refrigerant gas passes through the suction line to the compressor to be compressed and then repeats the cycle. The temperature-control and signal-light circuits are basically the same as those previously discussed in the section on horizontal freezers.

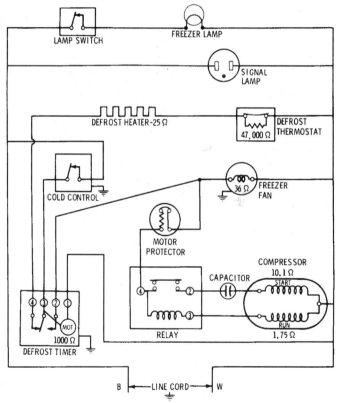

TOTAL RUNNING WATTS - MIN. 215 - MAX. 315

Frigidaire Corporation Div. of General Motors

Fig. 11-12. Schematic wiring diagram of Model FP-156V.

Electrical System of Automatic-Defrost Freezer

The electrical systems of automatic-defrost freezers are quite similar to the electrical systems of automatic-defrost refrigerators. Figure 11-12 shows a schematic wiring diagram, and Fig. 11-13 shows the same wiring in pictorial form. Service instruction charts have been worked out that will be helpful in locating and repairing faults likely to be encountered in day-to-day service work.

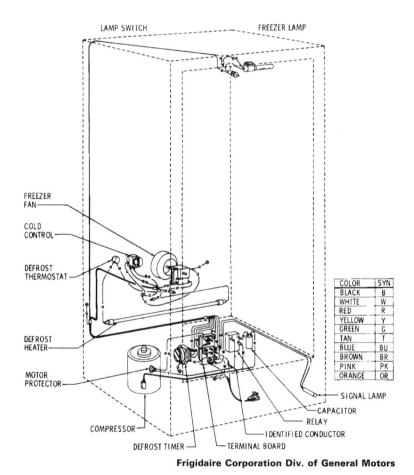

COLOR	SYN
BLACK	B
WHITE	W
RED	R
YELLOW	Y
GREEN	G
TAN	T
BLUE	BU
BROWN	BR
PINK	PK
ORANGE	OR

Frigidaire Corporation Div. of General Motors

Fig. 11-13. Pictorial wiring diagram of Model FP-156V.

Drainage Systems

Many defrost models have a defrost water drain and tube assembly for draining defrost water into a shallow pan. The drain tube is located behind the removable front grill (Fig. 11-14B). Chest models must be defrosted manually. A drain and tube assembly is located in the bottom left hand corner of the storage compartment (Fig. 11-14A).

Remove the drain plug from the inside bottom of the compartment. Place a shallow pan under the drain tube in front of the freezer and remove the cap. An alternate method is to insert at ½–in. male garden hose adapter into the drain tube and attach a garden hose. Remove the hose and adapter when defrosting is completed. Replace the drain plug and cap.

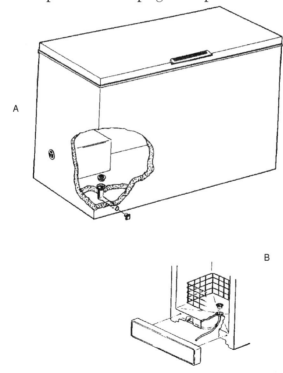

Fig. 11-14 (A). Location of the drain system for a chest freezer. (B) Location of the drain system for a manual defrost model upright freezer.

HOUSEHOLD FREEZER
TROUBLESHOOTING GUIDE

Symptom and Possible Cause   *Possible Remedy*

Compressor Does Not Run

(a) Low voltage

(a) Check with voltmeter. Voltage should be within 10% of that stamped on the compressor-motor.

(b) Defective compressor, relay, or overload protector

(b) Test the compressor. To test the compressor without any cabinet wiring in the circuit, unplug the compressor lead cord from the junction assembly and connect it directly to the electrical outlet (use a short extension cord if necessary). If the compressor starts, the trouble is not in the compressor, starting relay, or overload protector. Plug the lead cord into the junction assembly, and check the temperature control and all cabinet wiring for defects or poor connections. If the compressor does not start, check the overload protector, starting relay, and motor windings. To check an overload protector, short across its terminals. If the compressor starts, replace the overload with one having the same

Symptoms and Possible Cause   *Possible Remedy*

Compressor Does Not Run (cont.)

part number. If the compressor does not start, look for other trouble (low line voltage, defective starting relay, or defective compressor). To check a starting relay, pull it off of the compressor terminals. Remove the line-cord lead from the terminal on the relay and connect it directly to the run terminal of the compressor. Plug the line cord into a wall outlet, and, at the same timer, momentarily short between the run and start terminals. If the compressor starts, the relay is defective and must be replaced. If the compressor does not start, it is defective and must be replaced.

Signal Light Off

(a) Burned-out bulb
(b) Line cord unplugged
(c) Loose wire connection at control

(a) Replace.
(b) Plug into proper outlet.
(c) Reconnect looser wire at control.

Compressor Cycles on Overload Protection

(a) Low voltage

(a) See (a) under "Compressor Does Not Run."

341

Symptoms and Possible Cause   *Possible Remedy*

Compressor Cycles on Overload Protection (cont.)

(b) Defective starting relay or overload protector

(b) See (b) under "Compressor Does Not Run" for proper check.

(c) Poor air circulation around freezer

(c) Defective freezer or compressor-motor fan. Check and replace if necessary.

(d) Overcharged or restricted system

(d) An overcharged system may have a frostback condition on the suction liner. Trouble of this kind is usually recognized by water drippings when the compressor stops. Excessively long running time and a warmer than normal freezer compartment are other signs of an overcharged system. If a system is seriously overcharged, it must be evacuated and recharged.

Compressor Runs but Freezer Is Too Warm or Too Cold

(a) Improperly calibrated control

(a) The temperature control may be recalibrated by varying the cut-in and cutout temperature. Turn the calibration screw right to "raise" and left to "lower" the cut-in or cutout temperature.

(b) Poor gasket seal

(b) Check and repair as required.

Symptoms and Possible Cause   *Possible Remedy*

Compressor Runs Too Much or All the Time

(a) Incorrect control-knob setting

(b) Defective or improperly calibrated control

(c) Poor gasket seal

(a) Adjust temperature-control calibration screw.

(b) Replace defective control. Recalibrate control as noted in (a) in the preceding section

(c) Repair or replace as required.

Compressor Runs but Freezer Is Too Warm or Too Cold, or Compressor Runs Too Much or All the Time

(a) Undercharged or partially restricted system

(a) If the system is undercharged, the freezer will be colder than normal. An undercharged system must be purged, evacuated, and recharged with the proper amount of Freon. Before recharging, however, test for refrigerant leaks as instructed elsewhere in this book. Bent tubing, foreign matter, or moisture in the system may cause a partial restriction in the low-side tubing. This is usually indicated by frostfree tubing between the restriction and the capillary tube and by frost-covered tubing between the res-

Symptoms and Possible Cause   Possible Remedy

**Compressor Runs but Freezer Is
Too Warm or Too Cold, or
Compressor Runs Too Much or
All the Time (cont.)**

triction and the suction line. The restriction acts like a second capillary tube, increasing the pressure behind it (warming) and decreasing the pressure beyond it (cooling). Replace the component part if there is a partial restriction in the refrigerant tubing.

(b) Compressor not pumping adequately

(b) A compressor that is not pumping adequately will produce very little cooling effect. All cooling surfaces may be covered with a thin film of frost, but the temperature will not descend to the cutout temperature of the control, even with continuous running of the compressor. Place a hand on the accumulator surface for 2 or 3 seconds, and then examine the surface. If all frost is melted where touched, install gages and check operating pressures. If high-side pressures are lower than normal and low-side pressures are higher than

Symptoms and Possible Cause   Possible Remedy

**Compressor Runs but Freezer Is
Too Warm or Too Cold, or
Compressor Runs Too Much or
All the Time (cont.)**

normal, suspicions of defective compressors will be confirmed and compressors must be replaced.

Noisy Operation

(a) Tubing vibrating against compressor cabinet

(a) Adjust or fasten tubing.

(b) Loose compressor mounting; unlevel cabinet or weak floor

(b) Fasten compressor securely. Make sure the freezer is level by check ing, both from front to rear and from side to side, with a carpenter's level. An adjustable leveler is provided at each corner of the cabinet. Always level the freezer and check the gasket seal whenever the freezer is installed and again if the freezer is moved.

**Compressor with Capacitor in
Starting Winding Circuit
Keeps Tripping Overload
Switch**

(a) Capacitor defective

(a) Test with capacitor checker or a new capacitor. Replace if defective.

(b) Overload relay defective

(b) Replace overload relay.

SUMMARY

A refrigeration system used in refrigerators can best be described with the aid of its refrigerant flow. The temperature of the freezer is regulated by a conventional temperature control. This control starts the motor-compressor when the evaporator requires refrigeration and stops the compressor when the evaporator has been sufficiently cooled. When the temperature-control contacts close, the motor-compressor starts operating and reduces the pressure in the evaporator.

The refrigerant evaporates, absorbing heat in the process, thus cooling the evaporator. The vapor is drawn through the suction line by the compressor, which compresses the refrigerant gas and forces it into the condenser under high pressure. The compression of the vapor causes the temperature to rise, permitting the condenser to transfer heat to the cabinet shell and into the air surrounding the cabinet. As the compressed vapor gives up its heat, it condenses to a liquid in the condenser.

A refrigeration system that uses a capillary tube as the refrigerant control has certain characteristics and advantages. The amount of refrigerant in a capillary-tube system is small but critical. The hermetically-sealed motor-compressor requires alternating current to operate its $\frac{1}{12}$- to $\frac{1}{3}$-hp motor.

The fan for cooling the motor-compressor in some models is mounted on the compressor housing.

The temperature control, in addition to its conventional function, also controls the operation of the signal light. The temperature control is an electrical switch controlled by the temperature of the control feeler tube.

REVIEW QUESTIONS

1. What is a capillary tube?
2. How does the capillary tube control the flow of refrigerant?
3. What is a hermetically-sealed motor?
4. Where is the hermetically-sealed motor used?
5. What is the other function of the temperature-control switch?

6. What does an adjustment of the temperature-control knob do?
7. What happens to the refrigerant as it passes through the shelf tubing in the freezer?
8. How are the electrical systems of automatic-defrost freezers and automatic-defrost refrigerators alike?
9. What could cause the compressor not to run?
10. What could cause the compressor to cycle on overload protection?
11. What could cause the compressor to run but the freezer to be too warm or too cold?
12. What could cause the compressor to run too much or all of the time?
13. What causes noisy operation of the unit?
14. What causes the compressor that has a capacitor in the starting winding circuit to keep tripping the overload switch?

CHAPTER 12

Styles of Domestic Refrigerators and Food Arrangement

Refrigerators for household use come in a great number of styles and sizes. Some of the variations involve different cubic foot interior dimensions, single-door refrigerators with an interior door for the freezer compartment, and double-door units with one door for the refrigerator compartment and a separate door for the freezer compartment. They also come double wide, with a door on one side for the refrigerator compartment and another door on the other side for the freezer compartment. This last style sometimes has cold-water storage built into the door, with automatic fill and ice-cube dispenser. These features are usually optional.

Listed in Table 12-1 are some comparative figures of operation for different types of refrigerators. These are average figures, intended only to give some comparison as to operating costs. The frostless refrigerator, covered Chapter 7, will cost more to operate, of course, since heater coils are used in defrosting.

Table 12-1.

Refrigerator	Average Wattage (Est.)	kWh Annual
Refrigerator (12 cu ft)	241	728
Refrigerator (frostless— 12 cu ft)	321	1217
Refrigerator/freezer (14 cu ft)	326	1137
Refrigerator/freezer (frostless—14 cu ft)	615	1829

FOOD ARRANGEMENT

The proper arrangement of food is of the utmost importance if the best result of refrigeration is to be obtained. Generally, foods should be stored according to delicacy of flavor as well as their keeping qualities. Reference to Fig. 12-1 will indicate the proper food arrangement in a single-door refrigerator. The same relative arrangement should be maintained in a two-door refrigerator (Fig. 12-2). Food arrangement in apartment-type refrigerators is shown in Figs. 12-3 and 12-4. As noted in the figures, modern refrigerating units include a storage bin in the base of the cabinet to be used for dry vegetables that do not require refrigeration, such as potatoes, squash, dry onions, and turnips. The inner-door panel is used for storage of eggs, bottled foods, liquids, etc.

In this connection it should be noted that good air circulation is vital to efficient refrigeration and preservation of food. If air is restricted from circulating to all parts of the cabinet, food in the lower area will not be refrigerated sufficiently. It is therefore important that packages and other items of food be placed in the refrigerator in such a way as to allow sufficient air circulation. Large bulky items, especially square packages that are pushed against the rear wall of the cabinet, prevent cold air from circu-

Gibson Refrigerator Corporation

Fig. 12-1. Food arrangement in typical one-door-type refrigerator.

lating downward to the lower shelves and vegetable drawers. Space should be left between packages, and nothing should be placed too close to the cabinet walls. It is especially important that the refrigerator cabinet never be overloaded.

Placing hot foods or hot liquids in the refrigerator imposes an excessive load on the unit. Foods and liquids should be cooled to room temperature before they are placed in the refrigerator. Doing this will reduce the operating time of the unit.

DOOR OPENINGS

The user should be instructed that door openings should be kept at an absolute minimum. One way to do this is to remove all of the items from the refrigerator at one time when a meal is being prepared instead of opening the door each time something is needed. Hot, humid weather imposes a greater load on the refrigerator unit. During humid weather, moisture will accumu-

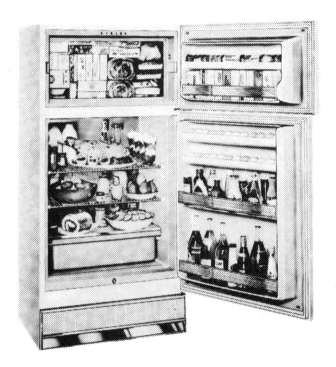

Fig. 12-2. Food arrangement in a typical two-door refrigerator with special freezer compartment.

late on the cold inner walls of the refrigerator cabinet. This is quite normal and is similar to the formation of moisture on the outside of a glass that is filled with a cold liquid. Although this condition is normal during periods of high humidity, it is annoying and can be minimized by opening the refrigerator door less frequently.

Gibson Refrigerator Corporation

Fig. 12-3. Method of arrangement for water-cooled refrigerator with a single-door.

WARM-WEATHER EFFECT

Warm-weather operation, particularly during summer months and in locations where air contains a considerable amount of moisture, will impose an additional effect on the refrigeration unit, known as *sweating*. When warm moisture-laden air comes in contact with the cold surfaces of the refrigerator, the moisture condenses, resulting in a condition that affects the inside surfaces of the refrigerator cabinet. This is a completely normal condition during warm, humid weather. The extent to which this sweating condition will take place is contingent upon the number of door openings, the length of time the door is left open, the temperature of the air outside the refrigerator, and the relative humidity.

Fig. 12-4. Food arrangement in a typical hotel-apartment refrigerator installation with an independent freezer unit.

A loose door seal will allow warm air to leak into the cabinet and cause excessive sweating during warm weather. If the door seal is good, the user must be informed that a certain amount of sweating is normal. Also, the user should be told that the condition can be kept to a minimum by following these recommended procedures:

1. Keep all liquids and moist food covered to prevent moisture from settling on the interior surfaces of the cabinet.
2. Keep the number of door openings to a minimum—*this is extremely important.*
3. Defrost often. On models that do not have the automatic-defrosting mechanism, it might be necessary to defrost as often as twice a week during warm, humid weather.

COOLING-COIL FROSTING

The refrigerator unit will run longer during very hot, humid weather and will cycle more frequently than in cooler weather. If the control knob is set *too high* under these conditions, the *off* cycle (on automatic-defrost units) may not be long enough to allow the cooling coil to defrost completely. In this event, the entire cooling coil will become heavily frosted within a very short period of time. This condition is a result of the inability of heat from the heating wires to penetrate the thick frost surface, and it will tend to increase the temperature in the fresh-food area.

The solution to this problem is to turn the thermostat to a low setting for a period of 12 hours or until the cooling coil has defrosted completely. Then the control knob should be kept at a lower setting to allow the cooling coil to defrost completely during any *off* cycle that might occur. Because each *on* cycle will begin with a cooling coil free from frost, better fresh-food temperatures will be obtained.

CLEANING

Regular cleaning of the refrigerator interior is of the utmost importance. A warm baking-soda solution is very good for cleaning the interior and will keep it fresh and sweet-smelling. In the absence of baking soda, warm sudsy water may be used. Plastic interior pans should also be cleaned in order to keep the food fresh and free from odors. Use only mild sudsy water (the same as that used for hand dishwashing) or a baking-soda solution. Exteriors should be washed regularly with clean sudsy water.

ODOR PREVENTION

In addition to regular cleaning, prompt attention to spillovers (especially milk) or an orange or lemon that has been overlooked is most important to maintaining a sweet-smelling refrigerator. When cleaning both the refrigerator and freezer sections, pay particular attention to the underside of the shelves and the moisture-pan and examine the crevices in the door seal, especially in humid climates.

Package any foods to be frozen in moisture- and vapor-proof wrapping paper. This is especially important in the case of fish, onions, and food containing garlic. It is also a good idea to rinse the hands in lemon water after handling fish and before handling the outside wrappings of fish packages. Charcoal canisters (available from hardware and department stores) can be placed in the refrigerator or food freezer to help in odor removal. Whenever a refrigerator is to be turned off for a few weeks (or when it is to be moved), it is important that it be thoroughly cleaned and dried. The door should be left ajar for air circulation, except during the actual move. A closed-up odor takes time to disappear after the refrigerator is cooled down and in use again. Swinging the door wide each timer it is opened will permit a greater change of air inside the refrigerator and will help speed up this process.

SUMMARY

The proper arrangement of food is of the utmost importance if the best result of refrigeration is to be obtained. Foods should be stored according to delicacy of flavor as well as their keeping qualities.

Space should be left between packages in a refrigerator so that cold air can circulate to all items in the food compartment. Placing hot foods or hot liquids in the refrigerator imposes an excessive load on the unit. Food and liquids should be cooled to room temperature before they are placed in the refrigerator.

Door openings on a refrigerator should be kept to a minimum. Hot, humid weather imposes a greater load on the refrigerator unit. During humid weather, moisture will accumulate on the cold inner walls of the refrigerator unit. This is quite normal and is similar to the moisture formation on the outside of a glass that is filled with a cold liquid.

A looser door seal will allow warm air to leak into the cabinet and cause excessive sweating during warm weather. If the door seal is good, the user must be informed that a certain amount of sweating is normal.

The refrigerator will run longer during very hot, humid weather. Regular cleaning of the refrigerator interior is of the greatest importance in order to prevent odors from forming and causing food damage.

REVIEW QUESTIONS

1. What is the difference in wattage used for a 12 cu ft refrigerator and that used by a 14 cu ft frost-free refrigerator?
2. Why is it important to properly arrange the food in a refrigerator?
3. Why is good air circulation inside a refrigerator important?
4. Why should packages inside a refrigerator be placed to allow space between each?
5. Why should hot liquids or solids not be placed into a refrigerator unit?
6. What does a loose door seal do in the way of increasing the defrost requirements of a refrigerator?
7. How does warm weather affect the operation of a refrigerator?
8. How often should you clean the inside of a refrigerator?
9. What do you do to prevent odors inside a refrigerator from being passed on to other food?
10. How does swinging the door wide open when opening the refrigerator aid in the elimination of odors?

CHAPTER 13

Installation Methods

Years ago, home-refrigeration systems required the cabinet to be located in one location and the refrigeration unit in another location, such as in the basement. Now, practically all home refrigeration is in a self-contained unit. A few instructions for the installation of self-contained units are appropriate. All that is required is to place and level the cabinet, remove the shipping bolts, test for leaks, and put the unit into service. The service is complete when the temperature-control dial is set in the desired position.

A method practiced by numerous distributors of household refrigerators is to uncrate the cabinet at the service shop. This is an excellent plan to follow inasmuch as it permits the unit to be operated and checked through its cycle. It also permits delivery of a tested unit and eliminates uncrating, with the accompanying disorder on the customer's premises. However, when this method is impractical, the following procedure is recommended:

1. Determine where the customer wants to locate the refrigerator, and then make an inspection with the customer, bearing in mind the following important factors that are necessary for an efficient and smoothly running refrigerator.
2. Perform the uncrating operations carefully so that the finish of the cabinet will not be scratched or marred.
3. Remove the shipping bolts and wood blocks that hold the condensing unit secure during shipment. Some cabinets are shipped with mounting springs or suspension parts in a separate package, usually accompanied by detailed instructions for this part of the installation. Regardless of the style of mounting, be certain that the unit floats freely on the spring or suspension mounting so that it will be able to absorb any vibration caused by the operation. Unpack and install the shelves and defrosting trays.
4. The back of the cabinet should be at least 3 inches from the wall, as in Fig. 13-1.

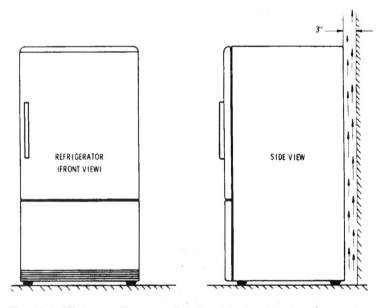

Fig. 13-1. Minimum distance to wall when installing self-contained refrigerator.

5. If the refrigerator is to be placed in a recessed wall or built-in cupboard, there must be an opening of at least 3 inches at the top and sides of the cabinet for proper ventilation, as shown in Fig. 13-2.
6. Be sure that the floor is solid. If not, it may be necessary to brace the floor directly beneath where the refrigerator is to be located.

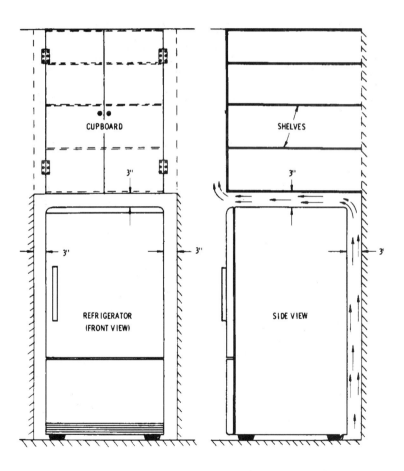

Fig. 13-2. Method of placing self-contained refrigerator in recessed wall space underneath a built-in cupboard. Dimensions shown are minimum for proper ventilation.

Leak Test

As a general rule, most self-contained units are shipped with the refrigerant confined in the condenser or liquid receiver. During shipment, it is possible that, due to strains at one or more joints, a leak develops. The unit should be tested for leaks before it is installed. Use an electronic hand-held leak detector or a halide gas leak detector.

INSTALLATION OF
ABSORPTION-TYPE REFRIGERATORS

Household refrigerators employing the absorption cycle do not differ in any important respect (with regard to location of units) from the units previously discussed in which the compression method of refrigeration is used. All units of this type require heat energy instead of mechanical energy to make the change in condition required in the refrigeration cycle. The heat energy required is usually obtained from a gas flame, with the gas piped from a convenient place (usually in the basement of the building).

There are two principal types of absorption units used, namely, *water-cooled* and *air-cooled* units. A typical refrigerator employing water-cooled condensers is shown in Fig. 13-3. This is an old unit and is no longer available. This type of unit usually necessitates tapping the water system of the house or apartment in which the installation is to be made. Although this is a rather simple problem, it should be done with care and in accordance with the Underwriter's Code or any local rules that affect such work. In addition, a pipe must be installed from the gas-supply line, which, in most locations, will require a permit from the gas company.

Cooling-Water-Line Installation

The purpose of the cooling water line is to furnish an uninterrupted supply of cool water to the condenser. After it is determined where the customer wants to locate the refrigerator, ascertain by means of a careful inspection that the location is suitable with respect to ventilation, accessibility, maintenance,

362

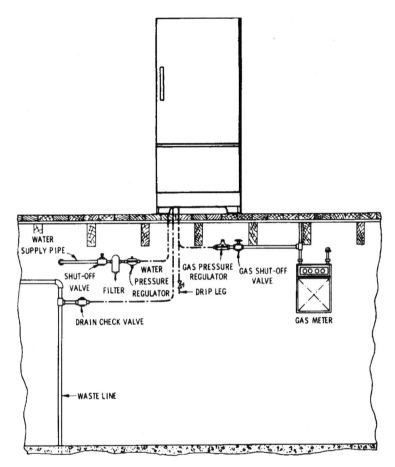

Fig. 13-3. Method of connection for a water-cooled refrigerator.

and repairs, etc.; then proceed. Make certain that the connections to the unit are made at a point in the water-supply line where full pressure and flow will be constantly maintained. Close the main valve in the water-supply line.

Make the extension from the house line with a ½-inch-diameter pipe to a point that is within approximately 10 feet of the unit. At this point, a shutoff valve, water filter, and pressure regulator should be installed, as shown in Fig. 13-3. The connection between the water pressure regulator and the unit should be made with copper tubing. When securing the lines, only brass fittings and nipples should be employed. The water line should be separated from the gas line to prevent condensation in the gas line.

Drain-Line Installation

The drain line is that part of the piping required to drain the heated water out of the condenser so that a constant flow of water in the system is maintained. This line is connected to the water-control-valve outlet of the unit and to a convenient drain outlet, or waste line. When the latter method is used, a check valve must be installed in the drain line to prevent water from backing up into the unit.

It is customary to install all connections as close to the unit as possible to facilitate the inspection and adjustment of the water-control valve. Vertical drops in the drain line should not exceeds to 6 feet since they may result in a partial vacuum in the water control valve and increase water consumption. An elimination of this tendency is sometimes accomplished by the installation of a larger pipe (½ or ⅝ inch) instead of the ⅜-inch standard tubing.

Gas-Line Installation

When installing a gas line, first turn off the main valve at the gas meter. The gas in the line is consumed by lighting a gas burner connected to the line. It is considered good practice to tap the refrigeration service main directly after the meter and ahead of any other gas appliances to ensure a steady and uninterrupted supply of gas.

The gas line usually consists of ½-inch black-iron pipe; this pipe should be extended to within approximately 10 feet of the unit, at which point the gas shutoff valve is installed in a closed position and the main gas valve is opened at the meter. All pilot lights that became extinguished when the gas was shut off should be lighted, and the gas jets that were opened to exhaust the gas from the line should be closed.

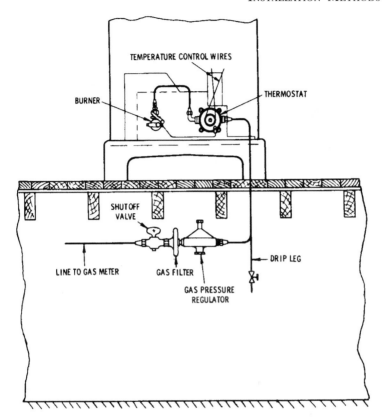

Fig. 13-4. Method of connection for an air-cooled refrigerator.

Gas Pressure Regulator

The gas regulator is now installed in the line next to the gas shutoff valve, as shown in Fig. 13-4. The function of this gas regulator is to automatically regulate the gas so it will maintain a constant burner pressure independent of any gas-line pressure fluctuations, thus ensuring a uniform rate of gas supply to the burner. The gas regulator should be installed at least 1 foot from the gas thermostat in a horizontal position and with the arrow on the regulator pointing in the direction of gas flow. The installation of the gas line is completed with copper tubing connected to the thermostatic gas control on the unit.

Traps in the line should be avoided since they have a tendency to become filled with condensation, thus interfering with the free flow of gas. The gas line would be insulated whenever it is in direct contact with cold walls or where it runs through cold rooms or refrigerated compartments. Before the installation is completed, a careful check for gas leaks as well as a thorough inspection of all joints should be made.

SUMMARY

A refrigerator should be uncrated and tested at the dealer's location before it is delivered to the customer. As a rule, self-contained units are shipped with the refrigerant confined in the condenser or liquid receiver. During shipment, it is possible that, due to strains at one or more joints, a leak develops.

Household refrigerators that use the absorption cycle do not differ a great deal from units employing the compression method of refrigeration. The outside appearance may be the same, but the absorption-type units must have a source of heat. In most instances, the source of heat is a natural gas heating element or burner.

The absorption-type unit has two principal types of cooling: air and water cooling.

The drain line is that part of the piping required to drain the heated water out of the condenser so that a constant flow of water in the system is maintained. This line is connected to the water-control-valve outlet of the unit and to a convenient drain outlet, or waste line. All connections should be installed as close to the refrigerator as possible.

The gas pressure regulator is installed in the line next to the gas shutoff valve. The gas regulator will automatically regulate the gas so that it will maintain a constant burner pressure independent of any gas-line-pressure fluctuations.

REVIEW QUESTIONS

1. Why do distributors of refrigerators check them out before they are delivered to the customer?

2. Why should a leak test be performed on a refrigerator after it has been delivered and installed?

3. Describe how the absorption-type refrigerator is installed differently from a unit employing compression-type refrigeration?

4. What is the purpose of a cooling-water line in an absorption-type refrigerator?

5. Why is a pressure regulator needed in the gas feed line of an absorption-type refrigerator?

6. Why should a refrigerator be located way away from the wall?

7. Why is it necessary to make sure the refrigerator has some space between the top of it and the bottom of the cabinet over it?

8. Why should a refrigerator be level?

9. Why shouldn't a refrigerator be placed on an unheated porch or in an unheated garage?

10. What does the modern refrigerator drain into when it defrosts automatically?

CHAPTER 14

Compressor Lubrication System

MOTOR LUBRICATION

In hermetically-sealed compressor units, the motor and compressor are lubricated by the splash system. This applies to smaller commercial units as well as domestic units. Lubrication of motor and fan bearings is usually accomplished by continuously feeding small quantities of oil through a piece of wicking or felt from a reservoir directly to the bearings. This makes it necessary to use an oil that is light enough in body to pass though the wick readily but heavy enough to form a proper film on the bearing surface at operating temperatures.

The oil must also have a low "pour point" so that it will flow freely to the bearings when starting up cold. The best oil will resist deterioration due to oxidation and polymerization, which develops stickiness and decreased lubrication. The use of properly refined oil will eliminate most of theses difficulties. When one considers the small amount required and the costly damage that may result from using an inferior oil, it can be seen that the best is the cheapest.

Grade-20W nondetergent automobile oil serves well for lubricating motor bearings of the bronze-sleeve type, or poured babbitt bearings. The small condenser-fan motors are usually supposed to be oiled for the life of the unit. Ordinarily they are single-bearing motors, with the bearing on the end opposite from the fan. If one of these fans has dry bearings and sticks, generally it will not burn the winding because it is a high-impedance-type motors. Should this happen, in many cases one may drill an ⅛-inch hole in the bearing cap and with an oil can relubricate the felt wick that is used to hold the oil for bearing lubrication. Then the shaft may be turned back and forth until oil gets into the bearing and frees the shaft. Use a self-tapping metal screw to close the drilled hole. In oiling motor bearings, do not overoil them because the oil will just run out and possibly get into the motor windings.

Many motors now come with ball bearings. Some of these, usually in smaller-size motors, are sealed on both sides and cannot be greased. If they get noisy, replace them with a direct-replacement bearing. On larger motors, the bearings are usually greasable. You may readily tell, as the bearing enclosure on the end bells will have two ⅛-inch pipe plugs. Ball bearings may be quickly ruined by using the wrong type of grease or by overgreasing them as they heat up. Remove one plug and replace it with an Alemite fitting. Use high-grade ball-bearing grease, which is a short fiber grease.

When you are ready to grease the bearings, remove the ⅛-inch pipe plug, which is usually about opposite to where you install the grease fitting. Now pump grease in until it starts to come out the plug hole. Stop adding grease, start the motor, and let it run for a little while until the grease stops coming out the plug hole. Replace the ⅛-inch pipe plug, and you have thus removed grease pressure on the ball bearing.

Set specific times for checking the motor lubrication; if an operator is in attendance, good times to grease are New Year's Day and the 4th of July.

COMPRESSOR LUBRICATION

There are so many types of refrigeration compressors that it is impossible to write specifically what will fit all requirements. Even when pressures and machines are quite similar in design, they are apt to require individual consideration for proper lubrication. Speed, clearance, temperature conditions, etc., have an important bearing, and the system of operation and the refrigerant used are the determining factors. Most oils are chosen for a given job as a result of actual tests. Therefore, servicemen in the field will profit by always using the oil specified by the manufacturer of the unit being serviced or an oil known to have the same specifications. Buy oils from reputable refiners only, preferably those who have had real experience in refrigeration problems.

Lubrication Methods

Compressors are lubricated in various ways, depending on their size and other factors. Compressors serving commercial refrigeration are lubricated by various method, such as *splash*, *semi-forced feed*, and *forced feed*.

Enclosed booster compressors are usually equipped with an internal forced-feed lubrication system that supplies oil under pressure to the main bearings, connecting-rod bearings, and compressor shaft seal. With reference to Fig. 14-1, the lubrication system consists of a rotary gear pump driven from the compressor crankshaft, the oil filter, and internal oil passages. The normal crankcase oil level covers the oil filter and the oil-pump suction to ensure that the pump is always primed and minimize the possibility of gas binding. The oil pump takes suction from the crankcase through the oil filter and discharges this oil vertically into a passage in the pump housing. From the pump, the flow of discharge oil divides—part flows to the pump end, and part flows through an internal pipe to the seal end of the compressor.

371

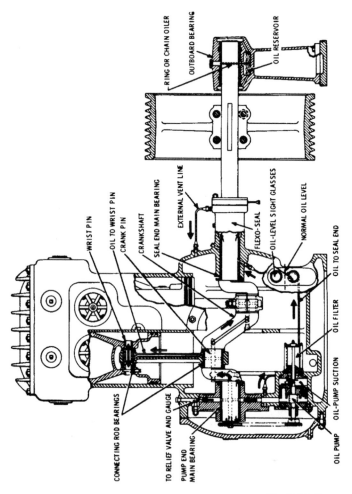

Fig. 14-1. Sectional view of a typical compressor showing lubrication details.

On two-cylinder compressors, part of the oil that flows to the pump end lubricates the pump-end main bearing, and the remainder flows into the crankshaft through drilled and piped passages to lubricate the connecting-rod bearings, crankpin, and wristpin. Oil passing through the bearings falls back into the crankcase. A relief valve and pressure gage are connected to the oil-pump discharge above the pump-end main bearing. The oil that flows to the seal end passes into the seal-end main bearings near the center. The oil that flows through the bearing toward the pump end falls directly into the crankcase. That oil passing through the bearing toward the seal end flows into the shaft-seal space to flood the space and lubricate the seal surfaces; it then returns to the crankcase through an external vent line above the shaft-seal housing.

Oil-Failure Switch—On large commercial compressors, to avoid operation at oil pressures below safe minimum, an oil-failure switch is usually furnished. The switch is activated by pressure difference between the crankcase pressure and the oil-pump discharge pressure. This switch opens and interrupts the control circuit to the compressor motor if the differential oil pressure falls below a safe limit.

Necessary Lubrication Precautions

Halocarbon refrigerants are miscible with oil, and, during operation, a certain amount of oil is carried through the system with the refrigerant. This is a normal condition, and the oil that leaves the crankcase when an idle system is started will soon begin to return as the system reaches normal operating temperatures and pressures. During shutdown periods, a certain amount of refrigerant is absorbed by the oil in the compressor crankcase. The actual concentration of refrigerant in the oil depends on the duration of the shutdown period, temperature of the oil, and pressure in the crankcase, the concentration increasing with lower oil temperature and higher standby crankcase pressures.

When the compressor is started, the crankcase pressure is rapidly lowered to the suction pressure, permitting the refrigerant to boil out of the oil. This causes the oil in the crankcase to foam, and, if the concentration of refrigerant is great enough, it is possible that sufficient oil may be carried out of the crankcase to cause temporary loss of oil. There are a number of methods of preventing too much loss of oil from the compressor after a shutdown.

To avoid excessive concentrations of refrigerant in the compressor crankcase oil during shutdown periods, a crankcase-oil heater is usually furnished. This heater is immersed in the crankcase oil and should be wired into the compressor-motor control circuit so that the heater is deenergized when the compressor is running and energized at all times while the compressor is idle.

Crankcase oil heaters cannot prevent liquid entering the crankcase during off cycles as the result of poor piping arrangements or leaking liquid-feed devices. However, they will maintain the crankcase oil at a temperature higher than other parts of the system, thus minimizing absorption of the refrigerant by the oil. Operation with this arrangement is as follows: whenever the temperature-control device opens the circuit or the manual control switch is opened for shutdown purposes, the crankcase heater is energized and the compressor keeps running until it cuts off on the low-pressure switch. The crankcase heater remains energized during the complete *off* cycle, and it is thus important that a continuous live circuit be made available to the heater during the *off* time. The compressor cannot start again until the temperature-control device or manual-control switch closes, regardless of the position of the low-pressure switch.

The use of a crankcase heater with single pumpout at the end of each operating cycle requires:

1. A liquid-line solenoid valve in the main line or in the branch to each evaporator.
2. Use of a relay or the compressor-motor starter auxiliary maintaining contact for obtaining a single pumpout operation before stopping the compressor.
3. A relay or auxiliary starter contact for energizing the crankcase heater during the compressor *off* cycle and deenergizing it during the compressor *on* cycle.
4. Electrical interlock of the refrigerant solenoid valve with the evaporator fan or chilled-water pump so that the refrigerant will be stopped when either the fan or pump is out of operation.
5. Electrical interlock of the refrigerant solenoid valve with safety devices, such as high-pressure cutout, oil-safety switch, and motor overloads, so that the refrigerant-flow valve will

close when the compressor stops through the action of one of these safety devices.

Domestic refrigerators and freezers are often placed in garages that are not heated. This is fine in the hot months; but when the temperature drops below the boiling point of the refrigerant, the refrigerant is absorbed into the oil and, upon startup, the oil will flow with the refrigerant to a degree, possibly lowering the level of oil in the compressor. If the home refrigerator or freezer is in such a cold spot, heat must be applied to the compressor so that the preceding results do not damage the compressor. This may be accomplished by a small lightbulb, placed so that it will not cause a fire. The proper method is to not place these refrigeration items where they will be subjected to extremely cold temperatures.

The crankcase heater is used primarily on larger types of compressors. With smaller compressors, the unit should be pumped down if it is to be shut off for a considerable length of time. This is accomplished by the installation of a vacuum gage on the suction side of the compressor, shutting off the compressor valve from the rest of the system and pumping the compressor crankcase down to about 1 lb of pressure. After accomplishing this, place tags on the equipment to explain what has been done so that whoever starts it again will know exactly what to do to the unit. This will also save loss of refrigerant through the shaft seal.

Compressors are sometimes equipped with flooded-type evaporators to prevent oil loss after a shutdown. It is recognized that neither automatic pumpdown control nor a single pumpout operation are practical in systems employing flooded evaporators unless suction-line solenoid valves are added to the system. Therefore, with flooded-type evaporators, the following control arrangements are considered satisfactory:

1. Manual operation as described in 2 and 3. No crankcase heaters are required.
2. Automatic control from temperature controllers or other devices, provided crankcase heaters are used and energized on the *off* cycles and the liquid solenoid valve closes whenever the compressor is stopped. Where water cooling of the compressor is employed, a solenoid valve in the water-supply line should close whenever the compressor is stopped.

3. Same as No. 2 except with the added precaution of a single pumpdown of the compressor for night or weekend shutdowns. This can be accomplished by manually closing the compressor-suction stop valve. If the single pumpout procedure is used, the compressor will automatically pump down once and then shut off on low-pressure cutout and stay off until manually restarted. This feature cannot be used if the low-pressure cutout also acts as the last capacity-control step.

SERVICING DATA

Before adding oil to a compressor, moisture must be reduced to an absolute minimum to avoid undesirable chemical reactions as well as the possible formation of ice and the resulting check of refrigerant flow. The oil should be dry when it is purchased, and it should be so handled that it will not have a chance to absorb moisture before it is put to work in a machine. Large users sometimes remove moisture by centrifugal or filter presses. Also, they may buy oil of lower quality (and, incidentally, lower price) and refine and filter this oil until it meets their specifications. This should not be attempted by anyone who does not have a well-equipped laboratory and competent chemical control.

The percentage of moisture in oil is related to its dielectric strength, which may be determined with suitable equipment. The oil should pass at least a 25- and preferably a 30-kV test. The latter is equivalent to approximately 0.01 percent, the former to about 0.03 percent, by weight of moisture. Do not purchase refrigeration oil in containers that are too large. When adding oil, pour what you feel you will need into a thoroughly cleaned and dried jar, and place the suction tube into the jar to draw the oil into the compressor. If you remove a filler plug to pour the oil directly into the compressor, thoroughly clean the area around the filler plug.

Acidity

An oil should not only be neutral when purchased but should remain so during storage and after being charged into a machine. In use, an oil may become acidic due to acids formed by the action of moisture on the refrigerant, as the result of oxidation of certain constituents of the oil or through formation of organic acids from saponifiable matter due to the action of oxygen and/or moisture. Obviously, prevention of the development of acidity during operation may be accomplished by keeping the system dry and using neutral oil that has a low oxidation number and is free from saponifiable matter.

Wax Separation

When methyl chloride, Freon-12 (and other Freons), and Carrene (methylene chloride) are mixed with oils in the amounts usually circulated (less than 1 to 10 percent or higher) in refrigerating systems, wax will separate upon sufficient cooling. This phenomenon is not to be confused with the possible separation of wax from pure oil. An oil may test wax-free and still separate wax on being mixed with the refrigerant. Due to the limited solubility of oil in liquid sulfur dioxide, wax separation does not present a problem.

Removal of Oil Sludges and Solids

The effects of the formation of an oil sludge in a refrigerating machine are obvious, but a few of the results are as follows:

1. Blocks oil passages and prevents proper lubrication
2. Packs behind piston rings, reducing compressor efficiency
3. Causes sluggish valve action, also reducing efficiency
4. Carries over to the expansion valve or float valve, causing sticking and failure to function properly
5. Remains in condenser coils and reduces cooling efficiency
6. Carries into the expansion coil with detrimental insulating effect

According to the Ansul Chemical Company, solids removed from refrigeration systems always have had an air of mystery about them. However, mystery always surrounds something that is not understood. Through continuing research, much has been clarified in the past few years, and it is important that the entire industry realize the simplicity of the problem and its solution. The various kinds of solids are easily classified as to both type and origin. They are:

1. Free metals
2. Corrosion
3. Metal oxides
4. Oil sludges
5. Ice in capillary tubes, expansion valves, or coils
6. All other extraneous solids

For convenience, these six types of solids, their origins and remedies, are listed in Table 14-1. The remedies for cleanup and recurrence of these refrigeration solids are simple and can be summed up this way: all are a case of filtration of solids, absorption by a desiccant of water and acid, and absorption of oil-decomposition products by the desiccant from the oil in the beginning stages of breakdown. In many cases, the oil breakdown progresses far enough to impair lubricity. It is difficult to tell when the breakdown has reached this point, and therefore an oil change is usually recommended when oil breakdown is involved.

Much money is spent to put controls and protective devices on equipment for the safety of the compressor. In many cases, these devices are never used. Often no means of protection against moisture and other contaminant troubles is installed. Yet moisture problems are far more prevalent than any of the problems that the common protective devices seek to overcome. This fact is proven by surveys that show that approximately 80 percent of all service calls are either directly due to moisture or traceable to it. Thus, a good filter-drier is the best insurance (Fig. 14-2).

In a clean, dry, acid-free system, there is nothing to filter. Moisture is the key to corrosion solids and oil sludges. If moisture levels are low enough, the acidity remains low and these trouble-

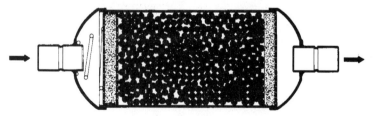

Virginia Chemicals

Fig. 14-2. Cutaway view of a filter-drier for the high side. The activated alumina desiccant beads provide high capacity for moisture, inorganic acids, organic acids, and oil breakdown materials.

Table 14-1. Suggested Methods for Removal of Refrigeration Solids

Type of Solid	Possible Cause	Possible Remedy
Free Metals: Powdered iron, bronze, or copper; chips of copper or solder	Compressor wear or incomplete cleaning of compressor body. Chips of copper may result from sawing copper lines or raw edges broken from flared joints. Origin of solder is obvious. Check for metallic iron with magnet. Corrosion solids do not respond.	Remove from system with filter-drier or plain filter.
Corrosion Solids: Chlorides of iron, copper, and aluminum; occasionally fluorides of those metal	Action of acid on these metals. Acid may result from moisture. oil breakdown, hermetic motor burnout, use of carbon tetrachloride or similar chlorinated solvents. or from heat breakdown of the refrigerant by soldering on lines containing refrigerant vapor.	Filter out solids with filter-drier. Remove acid and water with proper type of filter-drier.
Metal Oxides: Iron oxide and copper oxide	Iron oxide (iron rust) results either from allowing equipment to stand open to moist air or, most commonly, from liquid water circulating in the system along with the refrigerant. Copper oxide results from heating copper lines without passing an inlet gas, such as nitrogen, through the lines.	Remove from system with a good filter-drier. Water must be removed along with the iron oxide to prevent formation of more iron oxides and corrosion solids. Remove copper oxides with filter or filter-drier.
Oil Sludges: Vary in consistency from blown oil to dry powder or tarry mass	Result of heat in the presence of moisture and acid. May occur even if system is dry and acid free if compression ratio is too high — substantially above 10:1.	If oil is not too bad, it can be cleaned up by the proper activated alumna drier.
Ice In Capillary Tubes, Expansion Valves, or Coils	System contains more moisture than the refrigerant can hold in solution at the expansion valve, capillary tube, or coil temperature.	Dry system with good drier below freezeup and corrosion range.
All Other Extraneous Solids	Usually carelessness. Occasionally find desiccant or core sand, but this is very unusual.	Good filter or filter-drier.

some solids do not form. A factory-clean system fitted with the proper filter-drier at the factory and installed with care should operate indefinitely. The filter-drier is a means of prevention for the manufacturer and an all-around remedy for the service contractor.

SUMMARY

In hermetically-sealed compressor units, the motor and compressor are lubricated by the splash system.

Lubricating oil must have a low "pour point" so that it will flow freely to the bearings when starting up cold. Bronze sleeve bearings can be lubricated with Grade 20W oil as used in automobile engines. Many fans and single phase motors used in various locations in refrigeration systems are permanently lubricated when made and need no maintenance attention except under very hot or very cold ambient conditions above or below design specifications.

Many motors have ball bearings and are sealed. Ball bearing motors are noisier than sleeve bearing motors.

Compressors are lubricated in various ways, depending on their size and other factors. Compressors serving commercial refrigeration are lubricated by various methods, such as *splash, semi-forced feed*, and *forced feed*.

On two-cylinder compressors, part of the oil that flows to the pump end lubricates the pump-end main bearing, and the remainder flows into the crankshaft through drilled and piped passages to lubricate the connecting-rod bearings, crankpin and wristpin. On large commercial compressors, to avoid operation at oil pressures below safe minimum, an oil-failure switch is usually furnished.

Halocarbon refrigerants are miscible with oil, and, during operation, a certain amount of oil is carried through the system with the refrigerant.

To avoid excessive concentrations of refrigerant in the compressor crankcase oil during shutdown periods, a crankcase oil heater is usually furnished. The crankcase heater is used primarily on larger size compressors.

Compressors are sometimes equipped with flooded-type evaporators to prevent oil loss after a shutdown.

Before adding oil to a compressor, moisture must be reduced to an absolute minimum to avoid undesirable chemical reactions as well as the possible formation of ice and the resulting check of refrigerant flow. Oil should not only be neutral but should remain that way even during storage. The prevention of the development of acidity is important in any refrigeration system that uses oil for lubrication purposes.

An oil may test wax-free and still separate wax on being mixed with a refrigerant. Due to the limited solubility of oil in liquid sulfur dioxide, wax separation does not present a problem.

Much money is spent to put controls and protective devices on equipment for the safety of the compressor. Moisture is the key to corrosion solids and oil sludges. A clean system should operate indefinitely.

REVIEW QUESTIONS

1. Name various compressor lubrication methods.
2. What is the function of an oil-pressure switch?
3. State the operation of various compressor controls.
4. What precautions should be observed prior to adding oil toa compressor?
5. How does acidity affect refrigeration compressor lubrication?
6. What are the various methods used for removal of refrigeration solids?
7. What is meant by oil's *pour point*?
8. What does oxidation of oil cause?
9. Why is *non*detergent automobile oil specified?
10. What is meant by the splash method of lubrication?
11. What is a forced-feed lubrication system?
12. Why is an oil failure switch necessary?
13. What is a halocarbon refrigerant?
14. What does *miscible* mean?
15. Why is a crankcase heater needed in a compressor?

CHAPTER 15

Refrigeration Control Devices

In order to be of service, it is necessary that the refrigeration unit function without attention; that is, it must be made fully automatic in its operation. This automatic operation is obtained by the use of certain control devices, which connect and disconnect the compressor at periodic intervals as required to keep the stored food at the desired low temperatures.

Refrigerators are usually equipped with the conventional capillary tube for refrigerant control and a temperature-control switch for motor-compressor control. The motor and compressor in such systems are located in a common enclosure and are therefore termed a *sealed system*. Among the several advantages of the sealed system is its compactness, with an accompanying saving of space in the unit, in addition to the simplicity of the necessary control components.

Because of the numerous instances where older household refrigerators are still in use, it is important that the serviceman have a thorough knowledge of the operation of a few common types of older systems. This will enable him to understand others since they all work on the same basic principles. In general, these controls may be divided into two classes, namely, those employing a mechanical means for their operation, and those employing an electrical arrangement for their operation. Often, however, the two may be combined and the system made to function as a result of the combination. An automatic expansion valve (Fig. 15-1) is an example of a control that is purely mechanical, while a thermostatic type of control (a device for motor-compressor control) is an example of a control in the electrical class.

Although the function of a refrigerating system is the removal of heat and the consequent governing or maintaining of desired temperatures, many types of controls are employed other than temperature controls to complete the system. A great many controls have an indirect bearing on the temperature, although their primary purpose is to definitely operate some part of the system. The compression method of refrigeration has two main types of controls: refrigerant controls (liquid, vapor) and motor controls. Refrigerant controls are usually exercised by employment of the following: automatic expansion valve, thermostatic expansion valves, and capillary, or choke, tube. Motor controls are usually exercised by the use of pressure or thermostatic control.

AUTOMATIC EXPANSION VALVES

The automatic expansion valve is used on household evaporators of the dry type. It is located at the inlet to the evaporator and is used to control the flow of refrigerant from the receiver to the evaporator. The purpose of the expansion valve is to reduce the high-pressure liquid to a low-pressure liquid as it enters the cooling coil. When the pressure has been reduced by the compressor suction to a predetermined value, depending on the valve setting, the valve opens, admitting refrigerant to the evaporator. The admitted refrigerant evaporates almost immediately at a low temperature and is pumped away before admitting additional refrigerant to the evaporator.

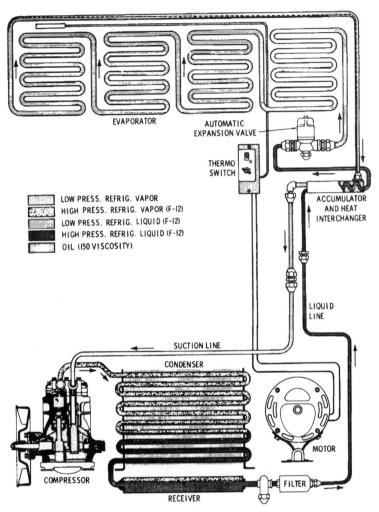

	LOW PRESS. REFRIG. VAPOR
	HIGH PRESS. REFRIG. VAPOR (F-12)
	LOW PRESS. REFRIG. LIQUID (F-12)
	HIGH PRESS. REFRIG. LIQUID (F-12)
	OIL (150 VISCOSITY)

Fig. 15-1. Refrigerant circuit and component arrangement for automatic-expansion-valve method of controlling refrigerant.

In reference to Fig. 15-2, which shows a typical automatic expansion valve, the operation is as follows: The condensing unit starts and reduces the evaporator back pressure to a point where the spring pressure forces the valve bellows, yoke, and needle downward until the valve is opened. This condition exists until

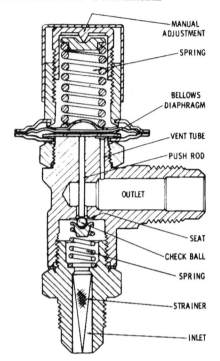

MANUAL ADJUSTMENT

SPRING

BELLOWS DIAPHRAGM

VENT TUBE

PUSH ROD

OUTLET

SEAT

CHECK BALL

SPRING

STRAINER

INLET

American-Standard Control Division

Fig. 15-2. Typical automatic expansion valve.

sufficient refrigerant is allowed to pass through the valve to raise the back pressure enough to overcome the spring tension, at which time the valve tends to close. Thus, during the time the condensing unit is in operation, a nearly constant pressure is maintained within the evaporator by the action of this valve. During the idle period, the rising evaporator pressure exerts a closing force, which holds the valve needle securely on its seat, thereby reducing the flow of refrigerant.

THERMOSTATIC EXPANSION VALVES

The thermostatic expansion valve differs from the automatic expansion valve in that its action is governed by the temperature, whereas the automatic expansion valve is governed entirely by pressure. In construction, it is similar to the automatic expansion valve except that it has, in addition, a thermostatic element. This thermostatic element is charged with a volatile substance (usually a refrigerant that is the same as that used in the refrigeration system in which the valve is connected). It functions according to the pressure-temperature relationship and is connected (clamped) to the evaporator suction line and to the bellows-operating diaphragm of the valve by a small sealed tube. A cutaway view of a thermostatic expansion valve is shown in Fig. 15-3.

The operation of the thermostatic expansion valve is governed by temperature. Thus, a rise in evaporator temperature will increase the temperature of the evaporated gas passing through the suction line to which the thermostatic-expansion-valve bulb is clamped. The bulb absorbs heat, and since its charge reacts in accordance with the pressure-temperature relations, the pressure tending to open the valve needle is increased and it opens proportionally.

Briefly, the greater the evaporator-gas-temperature rise in the evaporator, the wider the valve opens, and vice versa. The wider the valve opens, the greater is the percentage of coil flooding, which improves the heat transfer. It also causes the compressor to operate at a higher average suction pressure. Hence, an increased compressor capacity will also increase overall system capacity.

For the best performance of a thermostatic expansion valve, the power element bulb of the valve must be fastened to the suction line or drier coil as close to the evaporator outlet as possible. It is *extremely important* that the bulb be securely clamped to the suction line with metal clamps.

After the unit has stabilized, the adjustment of the valve is such that the frost line from the evaporator stops at or slightly beyond the bulb of the thermostatic expansion valve. This keeps liquid refrigerant from going back to the suction side of the compressor. The adjustment of this valve can be very touchy. A small turn of the adjustment valve either way can make quite a change in the refrigerator operation. A case in point: an automatic ice-flaking machine in a drugstore was not making and releasing ice flakes properly. The serviceman added refrigerant and checked it a couple of times, but it still did not do the job. Another man was called in, and with about one-half turn of the thermostatic

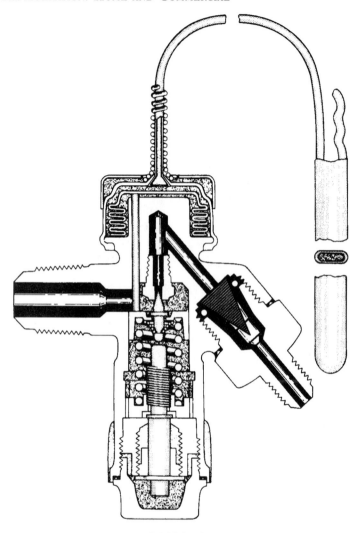

Frigidaire Corporation Div. of General Motors

Fig. 15-3. Typical thermostatic expansion valve.

expansion valve adjustment, the machine operated perfectly. However, it did take several tries at the valve setting and a little watching afterward to be assured that the expansion valve was operating properly.

Valve Selection

A thermostatic expansion valve selected or specified for any installation should be matched to the individual application. To do this is far more complex today than in years past. Valve selection involves several important factors that must be given careful consideration. They include:

1. Refrigerant used in system
2. Refrigeration load
3. Suction temperature
4. Liquid inlet temperature
5. Type of evaporator
6. Operating conditions

Many different refrigerants are used today. A valve to be selected for any given application should be designed specifically for the refrigerant in the particular system. Capacity of the valve must be matched to that of the evaporator. Manufacturers' literature indicates valve capacities covering a broad application range. The type of evaporator installed in the system dictates the type of expansion valve selected. If the evaporator has a considerable pressure drop, or is equipped with a distributor, an external equalizer valve must be used. Equally important, the type of evaporator and the type of installation should be considered in order to properly apply the various types of power element charged valves that are available.

Operation

A diagrammatical arrangement of the operation of a typical thermostatic expansion valve is given in Fig. 15-4, which shows that part of the refrigeration system in which the temperature and pressure are analyzed. Also shown is the cycle of action indicating the changes in the values of temperature and pressure at various points of the cycle as well as the direction of movement of the valve needle.

By following the oval cycle of action in a clockwise direction, it will be observed that when the temperature T_1 decreases at the thermostatic-power-element bulb at the evaporator outlet, the pressure P_1 decreases in the bulb and in the thermostatic-power

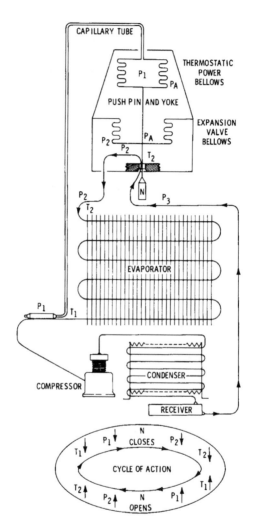

Fig. 15-4. Working principles of a thermostatic expansion valve.

bellows. The bellows then contracts, thus closing the needle valve N and decreasing the refrigerant pressure P_2 at the suction side of the valve and the inlet of the evaporator. The temperature T_2 then decreases due to the throttled action of N, causing the temperature T_1 at the outlet to the compressor to increase.

Just as soon as the temperature T_1 increases at the outlet and at the thermostatic bulb, the pressure P_1 increases in the bulb and the thermostatic-power bellows, causing the bellows to expand and the needle valve N to open. The refrigerant pressure P_2 and temperature T_2 increase at the suction side of the valve and at the inlet of the evaporator as a new supply of liquid refrigerant is admitted to the evaporator. As the refrigerant evaporates, the temperature T_1 decreases and the cycle of action repeats itself.

External Equalizer Application—In order to compensate for an excessive pressure drop through an evaporator, the thermostatic expansion valve must be of the external equalizer type, with the equalizer line connected either into the evaporator at a point beyond the greatest pressure drop or into the suction line at a point on the compressor side of the remote-bulb location. In general and as a rule of thumb, the equalizer line should be connected to the suction line at the evaporator outlet. By using the external equalizer type of thermostatic expansion valve with the equalizer line connected to the suction line; the true evaporator outlet pressure is imposed under the thermostatic-expansion-valve diaphragm. The operating pressures on the valve diaphragm are now free from any effect of pressure drop through the evaporator, and the thermostatic expansion valve will respond to the superheat of the refrigerant gas leaving the evaporator.

In general, for best evaporator performance, the thermostatic expansion valve should be applied as close to the evaporator as possible and in such a location as to make it easily accessible for adjustment and servicing. On pressure-drop and centrifugal-type distributors, apply the valve as close to the distributor as possible, or as shown in Fig. 15-5.

In multiple-valve applications, the external equalizing tubes should not be connected directly into the suction line or manifold. To secure maximum efficiency from the evaporator, connect the equalizing tube of each valve into the evaporator it controls just beyond the point of excessive pressure drop. This will ensure the suction pressure of only that evaporator being transmitted to the equalizing chamber of the valve, not the average suction pressure of all the evaporators.

Valve Selection—Thermostatic-expansion-valve size is determined by the Btu/hr or tons load requirement, pressure drop

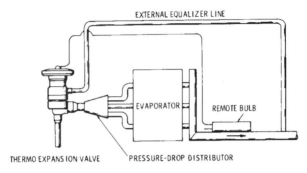

EXTERNAL EQUALIZER LINE

EVAPORATOR

REMOTE BULB

THERMO EXPANSION VALVE   PRESSURE-DROP DISTRIBUTOR

Alco Valve Company

Fig. 15-5. Thermostatic-expansion-valve application on a pressure-drop distributor feeding an evaporator.

across the valve, and evaporator temperature. It should not be assumed that the pressure drop across the thermostatic valve is equal to the difference between discharge and suction pressures at the compressor because this assumption will lead to incorrect sizing of the valve.

The pressure at the thermostatic valve outlet will be higher than the suction pressure indicated at the compressor due to frictional losses through the distribution header, evaporator tubes, suction-line fittings, and hand valves. The pressure at the thermostatic valve inlet will be lower than the discharge pressure indicated at the compressor due to frictional losses created by the length of the liquid line, valves, fittings, and possible vertical lift. The only exception to this is where the valve is located considerably below the receiver and the static head built up is more than enough to offset frictional losses. The liquid line should be properly sized, giving due consideration to its length plus the additional equivalent length of line due to the use of fittings and hand valves. When a vertical lift in the liquid line is necessary, an additional pressure drop due to loss in static head must be included.

Sizing of Expansion Valves—The selection and sizing of thermostatic expansion valves is of utmost importance for the best coil performance. Valve capacity must be at least equal to the coil-load rating and never more than twice that value. Any valve that is substantially oversized will tend to be erratic in operation, which will penalize both coil performance and rated capacity output. Liquid-line strainers should always be installed ahead of all thermostatic expansion valves.

Themostatic expansion valves are normally rated with Freon-12 refrigerant at 40°F evaporator temperature, 10°F superheat, and 60 psi differential (pressure at valve inlet minus pressure at valve outlet). For capacities of other differentials, or when used with other refrigerants such as Freon-22 and Freon-502, the valve-manufacturer's ratings must be consulted and closely followed in reference to capacity correction factors.

Although it is frequently assumed that when thermostatic expansion valves are used in low-temperature applications, some increased capacity results due to a higher pressure differential, this is not always true because of variations in valve design. It is always advisable under wide-range conditions to secure the valve-manufacturer's recommendations. As a further precautionary note, the power-element charges of all thermostatic expansion valves must be properly selected for the operating-temperature ranges and type of refrigerant used in the system.

Thermostatic expansion valves should be located as close as possible to the evaporator inlet and the bulb attached or inserted at a point where the refrigerant will not trap in the suction line. Keep the bulb away from the tees in common suction lines so that one valve will not affect any other valve. Externally equalized valves should be used on all multicircuited evaporators. In general, internally equalized valves are applied with single-circuited coils except when excessive drops are encountered.

How to Test a Thermostatic Expansion Valve

It is quite a simple matter to make a complete and accurate test of thermostatic expansion valves in the field. In most cases, the regular service kit contains all of the necessary equipment. The equipment required is as follows (Fig. 15-6):

1. Service drum full of Freon or methyl chloride (in the shop, a supply of clean dry air at 75 to 100 lb pressure can be used in place of the service drum): the service drum is used merely to supply pressure, and, for this reason, the refrigerant used does not have to conform with the valve being

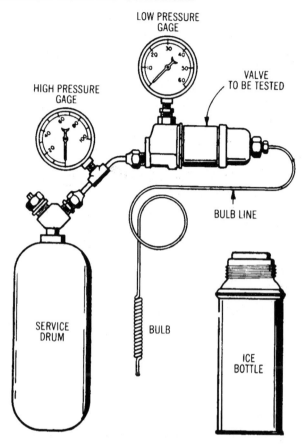

Fig. 15-6. Equipment needed to test a typical thermostatic expansion valve.

tested; in other words, a drum of Freon would be perfectly satisfactory for testing with sulfur dioxide, methyl chloride, or Freon valves.

2. High- and low-pressure gages: the low-pressure gage should be accurate and in good condition so that the pointer does not have too much lost motion. The high-pressure gage is not absolutely necessary but is recommended to show the pressure on the valve inlet.

3. Fittings and connections: these are required to complete the hookup, as shown in Fig. 15-6.

4. Finely crushed ice: one of the most convenient ways of carrying ice around is to keep it in a thermos bottle; otherwise, a milk bottle or other container is satisfactory. Whatever the container, it should be completely filled with crushed ice. Do not attempt to make this test with the container full of water and a little crushed ice floating around on top.

Procedure—Connect the valve, as shown in Fig. 15-6, with the low-pressure gage screwed loosely into the adapter on the expansion-valve outlet. The gage is screwed up loosely so as to provide a small amount of leakage through the threads. Proceed as follows:

1. Insert the bulb in the crushed ice.
2. Open the valve on the service drum, and be sure that the drum is warm enough to build up a pressure of at least 70 lb on the high-pressure gage connected in the line to the valve inlet.
3. The expansion valve can now be adjusted. The pressure on the outlet gage should be different for various refrigerants as follows:

Freon	22 lb
Methyl chloride	15 lb
Sulfur dioxide	3 lb

Note: Be sure to have a small amount of leakage through the gage connection while making this adjustment.

4, Tap the body of the valve lightly with a small wrench in order to determine if the valve is in operation. The needle of the gage should not jump more than 1 lb.
5. Now screw the gage up tight so as to stop the leakage through the threads and determine if the expansion valve closes off tightly. With a good valve, the pressure will increase a few pounds and then either stop or build up very slowly. With a leaking valve, the pressure will build up rapidly until it equals the inlet pressure.
6. Again loosen the gage so as to permit leakage through the threads, and then remove the feeler bulb from the crushed ice and warm it up with the hand or by putting it in water at

about room temperature. The pressure should increase rapidly, showing that the power element has not lost its charge. If the pressure does not increase when this is done, it is a sign that the power element is dead.

Note: With the new gas-charged expansion valves, the amount of charge in the power element is limited and the pressure will not build up above the specified pressure. This pressure is always marked on the power element and must be considered when testing gas-charged valves.

7. With high pressure showing on both gages, as outlined in step 6, the valve can be tested to determine if the body bellows leaks. This should be done by loosening the packing nut and using a halide leak detector or soap suds to detect the escape of gas. When making this test, it is important that the gage and other fittings are screwed up tight so as to eliminate leakage at other points.

Precautions—Be sure that the service drum has liquid in it and is warm enough to build up sufficient pressure. The high-pressure gage used (Fig. 15-6) will save a lot of trouble because it will show when there is not enough pressure on the inlet side of the valve. During the wintertime especially, the service drum may become cold and develop insufficient pressure to make a satisfactory test. Be sure that the thermos bottle or other container is full of finely crushed ice and does not have merely a little ice floating on top of the water.

CAPILLARY-TUBE SYSTEMS

The capillary-tube, or restrictor, system is presently in universal use on household refrigerators because of its simplicity of operation since it contains no valves or adjustments. Because of its nature, this type of system requires more accurate design to meet the particular requirements. The restrictor, in a sense, is a fixed control having no movable element responsive to load variations. Its element of variable control lies only in the natural variation of the factors affecting the rate of flow of the refrigerant. The most important of these are the pressure difference, the volume, and density of the refrigerant. The refrigerant circuit and component arrangement are shown in Fig. 15-7.

The positive force to push the refrigerant through the restrictor is the pressure differential between the inlet and outlet—the inlet being the condenser pressure, and the outlet being the evaporator pressure. Acting against this positive force is the resistance offered by the friction within the restrictor. The temperature of

FOR PURPOSES OF CLARITY, ILLUSTRATION DOES NOT SHOW
EXACT ROUTING, NUMBER OF PASSES OR PARTS LOCATION

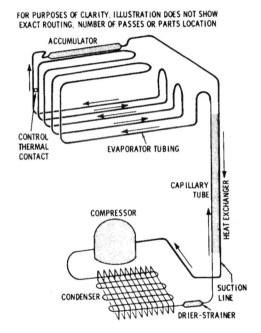

Fig. 15-7. Refrigerant circuit and component arrangement in a typical capillary and sealed household-refrigeration system.

the condensing medium has a greater effect on the pressure difference than the load change. The friction loss or pressure drop in the restrictor depends on the velocity, density, volume, and viscosity of the refrigerant and on the diameter and length of the restriction.

The velocity depends on the volume and density of the refrigerant delivered by the compressor. The volume and density depend on the temperature and pressure of the refrigerant and whether it is in a liquid or vapor state. The viscosity of the pure refrigerant changes very little. However, the presence of oil will affect the viscosity by increasing the pressure drop. The diameter and length of the restrictor are fixed quantities in any unit.

The restrictor, or capillary-tube, system accomplishes reduction in pressure from the condenser to the evaporator without the use of needle or pressure-reducing valves; instead it uses pressure drop or friction loss through a long small opening. With this system there is no valve to separate the high-pressure zone of the condensing unit from the low-pressure zone of the unit and evaporator. Therefore, the pressures through the system tend to equalize during the *off* cycle, being retarded only by the time required for the gas to pass through the small passage of the restrictor.

Capillary tubes, although not used to any great extent in commercial refrigeration applications, are employed extensively in domestic refrigeration and also in special refrigeration units such as water and milk coolers, ice cream cabinets, etc. The capillary tube (Fig. 15-8) consists of a very-small-diameter tube, the length of which depends on the size of the unit to be served, the refrigerant used, etc. This tube is usually soldered to the suction line between the condenser and evaporator to effect the necessary heat exchange. This device acts as a constant throttle, or restrictor, on the refrigerant; and due to its length and small diameter, it offers sufficient frictional resistance to the flow of

Fig. 15-8. Capillary tube showing the strainer connection.

refrigerant to build up the necessary head pressure to produce condensation of the gas.

If the two units (condenser and evaporator) were simply connected by a large tube, the pressure would rapidly adjust itself to the same value in both of them. Just as a small-diameter water pipe holds back water, allowing a pressure to be built up behind the water column with but a small rate of flow, similarly the small-diameter capillary tube holds back the liquid refrigerant, thus enabling a high pressure to be built up in the condenser during the operation of the compressor and at the same time permitting the refrigerant to flow slowly into the evaporator.

Due to the danger of foreign substances clogging the narrow passages in the tube, thus preventing the flow of refrigerant to the evaporator, a filter-drier is usually provided and connected between the condenser outlet and capillary-tube inlet.

Capillary-Tube Restrictions

Restrictions of the capillary tube may be caused by foreign particles lodged in the tube, moisture freezeup, or a bend or kink. A restriction or stoppage in the capillary will cause excess liquid refrigerant to collect in the condenser, resulting in excessive head pressure. This condition may be recognized by the fact that the frost level on the evaporator will be low while, at the same time, the condenser will be cool; the compressor will operate for a short period of time and then cycle on the overload. A starting and stopping of the unit several times will indicate if the stoppage is of a temporary or permanent nature. If the condition is of a permanent character, the indications are that the capillary tube is clogged and should be cleaned or, if necessary, replaced.

To clean the capillary tube, remove it from the system carefully, noting the direction in which the refrigerant flows through the coil. Then secure a drum of carbon dioxide and attach the coil to it in such a way that pressure will flow through it in the reverse direction. This will usually succeed in clearing the obstacle. If pressure is applied from and in the same direction as the refrigerant, the plug will probably be forced tightly into place and will be unremovable.

399

In many cases where the capillary hole in the tube is not very small, it is possible to straighten the tubing and run a piece of fine piano wire through it to clear the obstacle. The tube must be straightened; otherwise the wire cannot be pushed through. When moisture freezeup causes a restriction, it occurs at the outlet end of the capillary tube. Normally, a frost buildup can be noticed in this area. However, insulation wrapped around the tubing may conceal or limit the amount of frost formation. Expose the discharge end of the capillary and apply heat at this point. If there is enough head pressure, and if the restriction is caused by a moisture freezeup, a gurgling noise will be heard as the heat releases the refrigerant through the tubing. It is possible that this moisture will be picked up by the drier and cause no further trouble. However, if freezeup happens again, replace the heat exchanger or the drier, and purge, evacuate, and recharge.

If the trouble remains, change the heat exchanger. A kink in the capillary tube will have about the same symptoms as moisture freezeup except for the frost accumulation. Check the capillary tube along its entire length and, if possible, straighten the tube enough to relieve the restriction. Check the unit operation to see if the situation has been corrected. If the trouble still exists, replace the defective part.

Replacement of Capillary Tubes

Replacement of capillary tubes should be performed in the shop after discharging the unit. In replacing the capillary tube, make sure that the same length is employed and that the bore (inner diameter) is exactly the same as that of the old tube. To check the bore of the old tube, cut several sections from it and carefully fit small drill bits into the ends until one is found that fits snugly. The drill bit can then be measured with a micrometer so that the internal diameter of the tubing can be determined.

The capillary tube is usually soldered to the evaporator and filter connections with ordinary soft solder and may be removed with a heavy-duty soldering iron or a torch. The capillary tube must also be unsoldered from the suction line. Care must be taken when installing the new tube to see that the tube is clean where it is to be soldered. Care must also be taken to prevent too much solder from flowing into the connection, which would possibly close the end of the tube. After cleaning the ends of the tube with sandpaper or emery cloth and removing the pins from the ends of the tube, apply soldering acid and tin the tube at the point where the solder will eventually make the joint. Do not tin the tube all the way to the end as it would be practically impossible to install the tube and solder the connection without closing the ends with solder. Insert the tube 1½ inches into the evaporator fitting. Insert the other end of the tube as far as possible into the filter connection, and then pull the tube back out ⅛ inch. The tube may then be soldered to both connections and then clamped and soldered to the suction line in the same way as the original tube.

After changing the capillary tube, the unit should be baked out thoroughly and recharged. In case of tube breakage, a satisfactory field repair can be made by cutting off the old capillary tube several inches from where it enters the evaporator and joining the ends with a new capillary tube of exactly the same bore by means of a special capillary-tube union furnished by the manufacturer of the unit. This consists of a union with threaded openings at each end into which capillary tube nuts are fitted. The finished assembly should of course have the same length as the old capillary tube. This method makes it unnecessary to remove the evaporator plug when making repairs to the capillary tube. The expansion system is sometimes referred to as being the *dry system*, while the low-side float system is the *flooded system*. This is sometimes true, but not necessarily so, for the expansion valve may be used on a flooded system where the evaporator is so designed as to take the low-pressure vapor out of the top with the evaporator practically filled with liquid refrigerant.

DOMESTIC ELECTRICAL SYSTEMS

The electrical wiring in domestic electrical systems may differ considerably from unit to unit, depending on the protective feature of the motor-compressor as well as the defrosting circuit or circuits. All systems are made up of a cabinet-wiring harness, which includes the line cord, compressor leads, light switch,

control leads, and warmer wire leads (where used). In its simplest
form, the wiring components (Fig. 15-9) consist of the following:
motor-compressor, light switch, and temperature control. Some
refrigerator units use natural draft for condenser cooling, whereas
others use forced draft provided by a cooling fan usually located
underneath the cabinet.

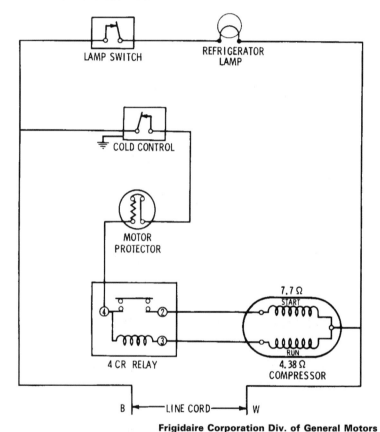

Frigidaire Corporation Div. of General Motors

Fig. 15-9. Schematic wiring diagram of Model D-100.

The functional requirements of the motor and compressor
devices are:

1. Starting of the unit
2. Stopping of the unit

3. Protecting the unit against overload caused either by high refrigerant pressure or excessive current drawn by the motor

The motor control mechanism in modern household refrigerators is actuated by a conventional thermostatic control switch. The length of the operating period of the compressor determines the quantity of the refrigerant pumped and amount of cooling produced. In order to function properly, the compressor must be started when the temperature of the evaporator has risen to a predetermined point and stopped when the temperature has dropped to another predetermined point, both of which must be at such levels as to maintain the desired storage-space temperature. In most household refrigerators this temperature is approximately 38°F.

The start-stop controls for the motor-compressor are operated directly by the temperature of the refrigerant in the evaporator. As the temperature of the refrigerant in the evaporator rises due to absorption of heat from the food load in the unit or from the water to be frozen, the pressure of the refrigerant rises. The rise in pressure caused by the rise in the refrigerant temperature is used to act through a small syphon bellows on a switch to start and stop the compressor. It is in this manner that the temperature in modern refrigerating units is automatically controlled to provide the desired cooling.

Compressor Wiring

Compressor-motor sizes vary with the cabinet capacity and are from $\frac{1}{12}$ to about $\frac{1}{3}$ hp. The compressor-motor employs a starting and running winding. The running winding is energized during the complete cycle of operation, whereas the starting winding is energized only during the starting period, when additional torque is required to overcome unit inertia.

Starting Relay

At the moment of starting, when the temperature control closes the electrical circuit, a surge of electric current passes through the running winding of the motor and through the relay coil. This energizes the starting-relay coil and pulls up the starting-relay armature, closing the starting-winding contacts. The current through the starting winding introduces a second out-of-phase, magnetic field in the stator and starts the motor. As the motor speed increases, the running winding current is reduced. At a predetermined condition, the running winding current, which is also the current through the starting-relay coil, drops to a value below that necessary to hold up the starting-relay armature. The armature drops, opens the starting winding contacts, and then takes the starting winding out of the circuit. The motor then continues to run on the running winding as an induction motor.

A schematic diagram of a household refrigerator circuit is shown in Fig. 15-9, and a pictorial wiring diagram of the same circuit is shown in Fig. 15-10. These are the simplest circuits you will probably come across. In Chapter 16, more complex diagrams will illustrate the later models with more equipment connected.

Overload Protector

In series with the motor windings is a bimetallic *overload protector*. Should the current in the motor windings increase to a dangerous value, the heat developed by passage of current through the bimetallic overload protector will cause it to deflect and open the controls. This breaks the circuit to the motor windings and stops the motor before any damage can occur. After an overload or a temperature rise has caused the bimetallic overload protector to break the circuit, the bimetallic strip cools and returns the contact to the closed position. The time required for the overload switch to reset varies with room temperature.

Temperature Control

The *temperature control*, or *thermostatic switch*, as the device is sometimes called, is the connecting link between the evaporator and the electrical system. Briefly, this control mechanism (Fig. 15-11) consists essentially of a thermostatic bulb clamped to the evaporator, together with a capillary tube and bellows. These three interconnected elements are assembled in one unit and

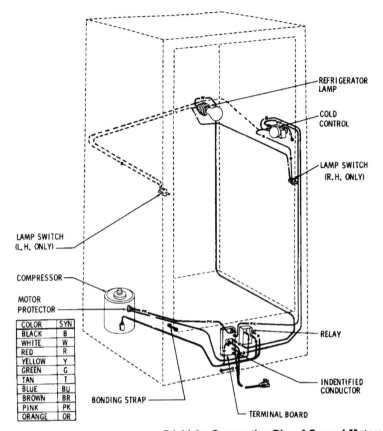

COLOR	SYN
BLACK	B
WHITE	W
RED	R
YELLOW	Y
GREEN	G
TAN	T
BLUE	BU
BROWN	BR
PINK	PK
ORANGE	OR

Frigidaire Corporation Div. of General Motors

Fig. 15-10. Pictorial wiring diagram of Model D-100.

charged with a few drops of refrigerant, such as sulfur dioxide or a similar highly volatile fluid.

As the temperature of the bulb increases, gas pressure in the bulb-bellows assembly increases, and the bellows pushes the operating shaft upward against the two spring pressures. The shaft operates the toggle, or snap mechanism. Consequently, the upward travel of the shaft finally pushes the toggle mechanism off center and the switch snaps closed, starting the compressor.

As the compressor runs, the control bulb is cooled, gradually reducing the pressure in the bulb-bellows system. This reduction

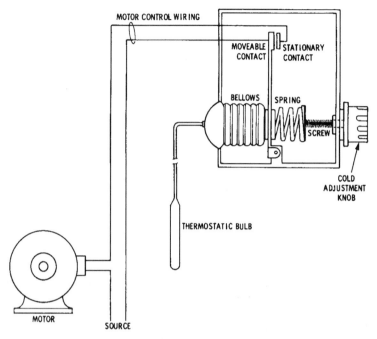

Fig. 15-11. Simplification of the working principles of a bulb-bellows system of temperature control.

of bellows pressure allows the spring to push the shaft slowly downward until it has finally traveled downward far enough to push the toggle mechanism off center in the opposite direction, snapping the switch open and stopping the compressor. The control bulb then slowly warms up until the motor again starts, and the cycle repeats itself. The thermostatic control is set to close the compressor circuit at a certain evaporator temperature and open the circuit at another predetermined temperature, and in this manner any desired cabinet temperature may be maintained.

TEMPERATURE CONTROL BY SUCTION PRESSURE

As previously mentioned, temperature may be controlled for commercial refrigerating units by means of thermostatic switches.

These are similar to domestic-type temperature control switches, but they usually operate at a lower voltage than that at which the motor operates. This is accomplished by the thermostat controlling a low-voltage relay to actuate the control switch for the compressor unit.

Pressure motor-control switches are most common in multiple and large commercial units. Those on the larger commercial units operate at a lower voltage than that supplied to the motor and actuate a relay to control the motor starting switch. Since the sizes of the motors used are variable, the control switches must operate at a lower voltage and current rating than the motor. The motor starting switch control pulls little current, so these switches can operate the magnetic coil that pulls the starting switch into start position.

The low-pressure, or suction, side of the refrigeration unit is an indicator of the temperature of the evaporator coil, and therefore one control works well regardless of the number of coils connected to it. This pressure control, installed in the suction line at the compressor compound valve, is actuated by a bellows with a pressure equal to the suction pressure, even though it may be at a minus or plus pressure. As the pressure builds up, the bellows pressure increases and the toggle switch is closed to start the compressor. As the pressure drops, the bellows pressure of the switch also drops, and at a preset point the contacts open and the compressor stops.

Another important control is a high-pressure cutout switch, which should be installed on compressors, especially the larger ones. This high-pressure cutout switch may be a separate unit from the low-pressure control, or it may be combined in the same enclosure.

Recording-chart mechanisms are quite often placed on refrigeration walk-in boxes and, more especially, on walk-in freezer boxes, so that the temperature variation for a 24-hour period will be recorded. This eliminates someone watching the temperature at frequent periods. A thermostat bulb is installed in the box, and a capillary tube leads through the wall to a clock-mechanism-operated chart. The recorder has a pen that will draw a line showing temperatures at all times during the 24-hour period.

SERVICING VALVES

If the valve does not appear to feed enough refrigerant to the evaporator, as evidenced by too low a suction pressure or too high a superheat, the trouble may be due to:

1. A dirty liquid screen.
2. Adjusting stem screwed in too far (improper superheat adjustment).
3. Too high a pressure drop through the evaporator.
4. Lack of refrigerant in the system (this is a very common source of trouble).
5. Vaporization of the refrigerant in the liquid line due to excessive pressure drop. This condition, which is not uncommon on Freon-12 installations, materially reduces the capacity of the valve. The remedy is to be sure the liquid line is of proper inside diameter in relation to its length, or else cool the liquid and gas mixture to recondense the gas by the use of an interchanger or subcooler.
6. The power-element bulb being located in a liquid trap in the suction line or evaporator.
7. Too low a head pressure.

If a valve appears to be feeding too much refrigerant, it may be caused by:

1. Adjustment stem screwed out too far.
2. Power-element bulb not making good contact with the suction line.
3. Dirt or scale on the valve seat.
4. Bulb placed in a location warmer than the refrigerated fixture.

If the suction pressure is lower than anticipated, this is no fault of the valve, provided the superheat of the vapor at the end of the evaporator is normal. The low suction pressure may be due to the load being light at the time or the condensing unit having more capacity than necessary. If a refrigerating system is not functioning properly, do not arbitrarily blame the valve. Many refrigerator control valves are needlessly replaced when the difficulty is due to improper installation of the valve or to some irregularity in another part of the entire system. Always thoroughly investigate the entire system, and make a definite analysis of the difficulty before attempting to rectify it. Be sure the functions of a thermostatic refrigerant control valve, its possibilities, and limitations are understood; and always be sure it is properly installed and adjusted before deciding to replace it.

SUMMARY

To obtain the necessary control of the refrigerant flow, various automatic control devices have been introduced. The purpose of these devices is to provide means by which the refrigerator machine is automatically connected and disconnected at certain periods to keep the food in storage at a required temperature. This freedom from attention has been obtained by the development of automatic controls to perform the various operations necessary for continuous production of low temperatures. As pointed out, control devices may operate on mechanical or electrical principles or may function as a combination of both.

It should be noted, however, that since it is not possible to include all the different control types in this chapter, the construction, operation, and adjustment of typical designs only are provided. Thus, the common types of liquid refrigerating metering devices such as float valves, capillaries, automatic and thermostatic expansion valves, including testing and adjustment methods, are fully described. Refrigeration control devices operating on electrical principles are fully explained in Chapter 16.

REVIEW QUESTIONS

1. What is the purpose of refrigeration control devices?
2. State the difference between mechanical and electrical control devices.
3. How do automatic controls operate?
4. What is the function of float valves in a refrigeration system?
5. Describe capillary tubes and their operation.

6. What is an automatic expansion valve?
7. How does an automatic expansion valve differ from a thermostatic expansion valve?
8. What is the function of an external equalizer?
9. Describe the selection and sizing of thermostatic expansion valves.
10. How may a thermostatic expansion valve be tested?

CHAPTER 16

The Electrical System

Because of the dependence upon an unfailing source of electricity for control and motor service, it is important that the serviceman or operator know the laws of electricity as well as the principles of motor operation. In this connection, it should be noted that equipment for motor starting, control, and protection varies with the type and service application. While motors are classified as either *alternating* (ac) or *direct current* (dc), the dc type will not be covered because the chances of finding dc motors used for refrigerators are quite remote.

It is not the intent of this one lesson to give a complete electrical course, as this is a subject unto itself. A rough coverage will be given, so that you will have an idea of the electrical operation of refrigeration units and control devices.

UNITS OF ELECTRICITY

In a dc system, the products of *volts* (V) multiplied by *amperes* (A) equals *watts* (W). Thus one watt is produced when one ampere flows under a pressure of one volt. This relationship is written:

$$\text{Amperes} \times \text{volts} = \text{watts}$$
or
$$IE = W$$

where I = current, amperes
E = voltage, volts
R = resistance, ohms

The relationship between current flow, resistance, and pressure is written:

$$\text{Amperes} \times \text{resistance} = \text{volts,}$$
or
$$IR = E$$

where I = current, amperes
R = resistance, ohms
E = pressure, volts

The unit of power is usually expressed in kilowatts (kW) or horsepower (hp). The relationship between horsepower, kilowatts, and watts is

$$1 \text{ kW} = 1000 \text{ W}$$

One horsepower is equal to 746 watts; therefore, to reduce *kilowatts* to *horsepower*, it is necessary to divide 1000 by 746:

$$1 \text{ kW} = \frac{1000}{746} = 1.34 \text{ hp}$$

From this we know that *kilowatts* may be converted into *horsepower* by multiplying by 1.34.

The formulas given here are fundamental only and may be rewritten to suit any particular condition or requirement, as shown in Table 16-1.

412

Table 16-1. Variations of Ohms Law*

Amperes	=	$\dfrac{\text{volts, } E}{\text{ohms, } R}$	=	$\dfrac{\text{watts, } P}{\text{volts, } E}$	=	$\sqrt{\dfrac{\text{watts, } P}{\text{ohms, } R}}$
Resistance	=	$\dfrac{\text{volts, } E}{\text{amperes, } I}$	=	$\dfrac{\text{watts, } P}{\text{amperes}^2, I^2}$	=	$\dfrac{\text{volts}^2, E^2}{\text{watts, } P}$
Volts	=	$\text{amperes} \times \text{ohms}$	=	$\dfrac{\text{watts, } P}{\text{amperes, } I}$	=	$\sqrt{\text{ohms, } R \times \text{watts, } P}$
Watts	=	$\dfrac{\text{volts}^2, E^2}{\text{ohms, } R}$	=	$\text{amperes}^2, I^2 \times \text{ohms, } R$	=	$\text{amperes, } R \times \text{volts, } E$

*P = power, measured in watts (W).

THE ELECTRIC CIRCUIT

A simple way to consider electricity is to use the example of water flowing through a pipe. In a manner similar to the flow of water in a pipe, electrical conductors are arranged to form a path for the flow of an electric current. The paths in which electric current flows are called *electric circuits*. Current, which is measured in amperes, is the movement of electrons (water) through a conductor (pipe).

In Fig. 16-1, a simple electric circuit is shown diagrammatically, consisting of a generator, motor, switch, and a series of wire conductors. Figure 16-2 shows a water system consisting of a

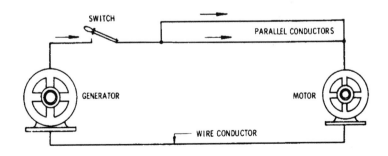

Fig. 16-1. Simple electric circuit consisting of a generator, motor, switch, and a series of wire conductors.

pump, water turbine, pipes, and valve, which corresponds to this electric circuit. The amount of water flowing through the pipe, expressed in gallons per minute, would correspond to the flow of current through the wire, expressed in amperes. The pressure of water in the pipe, expressed in pounds per square inch, would correspond to the pressure of electricity, or the voltage of the circuit, as it is usually expressed.

DIRECT CURRENT

The flow of electricity in a wire is often compared to the flow of water in a pipe. In a *direct current* (dc) circuit a voltmeter connected across the line and an ammeter connected in series with it (Fig. 16-3) will read the true voltage and current. The product of the two readings will be the power used, in *watts*.

Direct current flows in one direction, as indicated in Fig. 16-3. It flows from negative to positive (– to +). The only place that you will come across direct current in refrigeration will be in automobile air conditioning—the automobile uses a dc source.

SINGLE- AND THREE-PHASE CURRENT

A short explanation of single- and three-phase current is in order. One *hertz*, formerly known as cycles per second (cps), is one complete sine wave in one second (Fig. 16-4) and is 360 electrical degrees. The ordinary alternating current that we use is 60 cps and therefore is said to have a frequency of 60 hertz. There are two alternations in one cycle, and so 60-hertz current alternates 120 times per second. The term *cycles per second* is no longer used. This means that the voltage and current start at zero, go to maximum voltage in a positive direction, return to zero, go to maximum voltage in the negative direction, and then go back to zero. This takes place 60 times each second. Note that the direction of flow has alternated twice in one cycle. The values that you read on a voltmeter or an ammeter will be 0.707, or 70.7 percent of the peak voltage or current, and is the effectual value of the voltage or current.

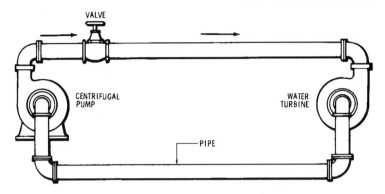

Fig. 16-2. Water system consisting of a pump, pipes, and valves, which correspond to the electric circuit.

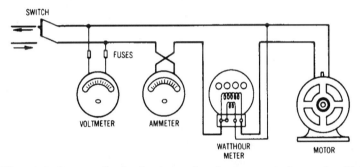

Fig. 16-3. One method of connecting instruments in a circuit to measure the voltage and current.

With direct current, the voltage and current are always in step with each other, and so the reading of a voltmeter multiplied by the reading of an ammeter will give the power in the dc circuit in watts. This is not always true with alternating current. Figure 16A shows a case where the load might be a resistive load, such as incandescent lighting. When this is true, the voltage and current will be *in phase*. In this case, voltmeter and ammeter readings could be used to arrive at the power, in watts.

Reactive loads (made up of winding, such as motors or ballasts for fluorescent lamps) produce self-inductance in the coils, causing a voltage in the opposite direction to the voltage from the

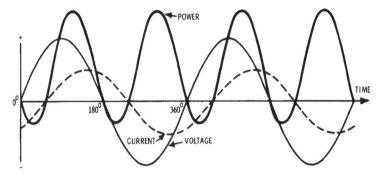

Fig. 16-4. Waveforms showing the relationship between current, voltage, and power in an ac circuit.

source, and causing the current in the windings to lag behind the applied voltage. Thus, the current reaches its maximum after the voltage has passed its maximum, which creates a *lagging power factor*. Capacitors will cause the current to lead the applied voltage, causing a leading power factor. This is shown in Fig. 16-5.

In an ac circuit, if you use a voltmeter and ammeter and multiply the results, you will not get watts but will get volt-amperes (VA). If at the same time you put a wattmeter in the circuit, it will read watts, and on a motor load the VA will be larger than the watts. This is due to the current lagging the voltage in time, resulting in a lagging power factor. The voltage will be the applied voltage, but the current will be larger than if the same load was on direct current since with the voltmeter and ammeter you

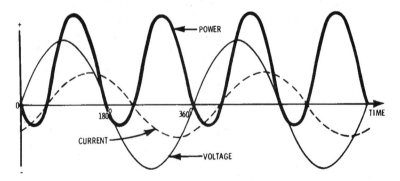

Fig. 16-5. Phase relationship between current and voltage. The voltage is leading the current.

reading apparent power, which in VA will be larger than the reading in watts on the wattmeter.

Thus, in figuring the size of conductors needed to supply a motor load, you have to use the full-load amperes on an ac circuit, which, as just stated, will be more than the full-load amperes on a dc motor fed from direct current. The difference between the apparent power (VA) and the real power (watts) is called *wattless power* and accomplishes nothing in pulling the load.

For induction motors, it is a lagging power factor, meaning current is lagging voltage. The power factor of a motor varies for different designs. It is a common error for the uninformed to assume that some motors are more efficient than others because the nameplate for one motor shows a lower current value than that of another motor of equal horsepower. Actually, the amount of power really drawn from the line in kilowatts can be the same. The motor having the higher current merely has a lower value of *power factor.* This will easily be found by the following example:

Example—Two single-phase ac motors A and B (Fig. 16-6), each having a capacity of 1 hp, have the data shown on their respective nameplates. The power taken from the line by the respective motors is:

Motor A (true power) = 110 V × 10 A × 0.8 (power factor)
                      = 880 Watts
Motor B (true power) = 110 V × 12.5 A × 0.64 (power factor)
                      = 880 Watts

The two motors using the same amount of power will cost the same to operate even though motor B takes more current than motor A. There is one thing to notice, however: the motor with

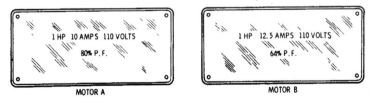

Fig. 16-6. Typical nameplate data.

the lowest power factor will have more current flowing, which will require a change in size of thermal element in the starter to provide equal protection. Also, larger wiring must be used to carry the heavier current.

Frequency and Slip

As explained previously, *frequency* is the number of hertz (cycles per second), and *alternations* is half the number of alternations alternating current makes per second. The frequency and number of poles in the stator of a motor will determine the speed of the motor on alternating current. The only motors that run pole for pole, or at synchronous speed, are synchronous and hysteresis motors, such as clock motors. On motors used in home refrigeration and air conditioning, motors run at a little less than synchronous speed, which is called *slip*, and regulates the power output of the motor. Thus a single- or three-phase motor at no load will run faster than at full load because as the load increases, there is more slip, causing the motor to pull more current to compensate for the additional load.

To find the synchronous speed of any ac motor, use the following equation:

$$\text{Motor rpm} = \frac{120 \times 60}{4}$$

Thus, for example,

$$\text{Motor rpm} = \frac{120 \times \text{frequency}}{\text{no. of motor poles}} = 1800 \text{ rpm}$$

Thus, all 4-pole, 60-hertz motors that operate at a maximum of 1800 rpm will usually be about 3 percent below this speed due to slippage, or 1750 rpm. A 4-pole, 50-hertz motor that operates at a maximum of 1500 rpm will usually be about 3 percent below this speed due to slippage, or 1450 rpm:

$$\text{Motor rpm} = \frac{120 \times 50}{4} = 1500 \text{ rpm}$$

Other examples at other frequencies can be worked out, remembering that the minimum number of poles is 2. An odd number of poles like 3, 5, 7, 9, etc., cannot be used. The only ac motors that run pole for pole, or, in other words, at synchronous speed, are synchronous motors.

Single-phase ac current is used mainly on domestic refrigeration and smaller commercial refrigeration.

Three-Phase Current

Most motors on larger commercial refrigerators are of the three-phase type. Figure 16-7 shows three-phase-voltage wave patterns. Notice that no two reach maximum at the same time, but that they are spaced 120 electrical degrees apart. On a motor, if you take an ammeter reading on one phase, this figure must be multiplied by 1.732 to get the total current that the motor is pulling (if you are determining horsepower). However, the ampere listing on the nameplate is the current per phase.

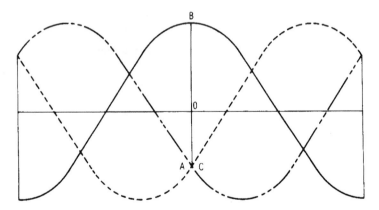

Fig. 16-7. Three-phase current, represented by sine waves that are 120° apart.

Transformer Connections

Figures 16-8 to 16-10 are schematic diagrams of transformer windings showing how single- and three-phase current is derived for service to buildings. Only the secondary side of the trans-

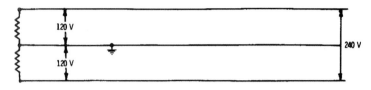

Fig. 16-8. Single-phase, 120/240-volt supply.

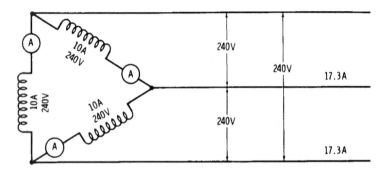

Fig. 16-9. Voltage and current in a delta transformer.

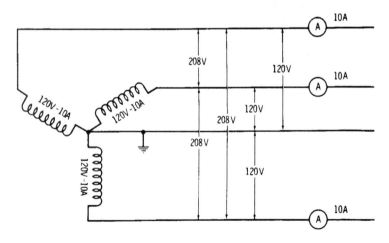

Fig. 16-10. Voltage and current in a wye transformer.

former is shown because this is where the power for refrigeration motors is derived.

Figure 16-8 shows the type of service that most homes receive. Figure 16-9 shows a closed delta-bank transformer, which can provide 120/240V single-phase as well as 240V three-phase service. Many commercial establishments receive this kind of service. Figure 16-10 shows a Wye three-phase transformer supply with 120/208V single-phase service available; also, 208V three-phase and, if necessary, 120V three-phase service are available. This type of service is very popular with commercial establishments. By using 277V transformers instead of the 120V transformers shown, we can get 277/480V single-phase or 480V three-phase service. This system is necessary for large-horsepower motors.

ENERGY

Energy is *the ability to do work.* Energy can neither be created nor destroyed. However, it may be converted from one form to another. The conversion of one form of energy to another is accompanied by some form of losses. *Potential energy* is stored energy, such as in a battery. *Kinetic energy* is energy in motion.

Since the rate of doing work is *power*, energy is equal to the product of power and time. For example, a 5-hp motor requires 2 hours to pump water necessary to fill an elevated storage tank; the energy represented by this work is 5×2 hours, or 10 horsepower-hours (hph). Now, a 10-hp motor could do the same work in half the time, but the energy used would still be 10 hph. Mechanical energy is measured in horsepower-hours. Electrical energy is measured in *watt-hours* (Wh), or *kilowatt-hours* (kWh). *One kilowatt-hour* equals *one thousand watt-hours.*

If power is known to remain constant on a circuit for a particular time interval, the energy can be determined by measuring the power with a wattmeter and multiplying this reading by the time in hours. For instance, if you measure the amount of energy consumed by one of the motors described previously during a period of 24 hours, assuming that the motors are running only 50 percent of the time mentioned and assuming a constant current, we obtain:

Energy consumed = 880 × 12 = 10,560 Wh (10.56 kWh)

In power systems, however, it is an exception for both the load and voltage on a circuit to remain constant for any great length of time. Consequently, the method of measuring energy just described is not practical. To be practical, the measuring instrument must be sensitive to all changes in power and its recording element must sum up the power demand continuously for small intervals of time. An instrument that does this may be called an *integrating wattmeter*, which is generally known as a *watt-hour meter* because its dials record the watt-hours (more commonly, kilowatt-hours) of energy passed through it. The *watt-hour meters* used in residences are smaller and in most cases have only one single-phase element.

MOTOR EFFICIENCY

The ratio of actual power delivered at the pulley or shaft and the power input at the terminals of a motor (computed in the same units) is called the *motor efficiency*. It is written:

$$\text{Efficiency} = \frac{\text{output}}{\text{input}}$$

As an example of calculating the efficiency of a single-phase ac motor, consider a motor with a nameplate containing the following data:

1 HP, 10 AMPS, 120 VOLTS, PF 80%

Since one horsepower equals 746 watts, substitution in our equation gives

$$\text{Efficiency} = \frac{746}{120 \times 10 \times 0.8} = 0.777 \text{ or } 77.7\% \text{ (approx)}$$

It will readily be observed that the electrical power, or the input to the motor, may conveniently be measured at the motor terminals and reduced to horsepower by the usual method of calculation. The mechanical power developed by the motor is not readily available since the nameplate gives only the average power that the motor will deliver under normal conditions.

The usual method adopted for calculating the mechanical power developed by a motor (and from which efficiency may be obtained, as described previously) is by means of a *Prony-brake test*. As shown in Fig. 16-11, the Prony brake consists of two blocks shaped to fit around the pulley of the motor. By means of two long bolts with handwheels at the top, the pressure of the blocks against the pulley can be varied. Two bars fastened to the top and bottom of the brake form an extension arm by which the torque of the motor is exerted upon the platform scale. The force factor *F* of the torque is measured as the net reading on the scale, in pounds; the radius factor *R* is measured in feet from the center of the shaft perpendicular to the bearing point on the scale.

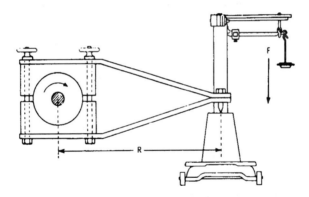

Fig. 16-11. Prony brake set up to test the electric motor.

The mechanical power of the motor equals its torque times the distance through which it operates per minute:

$$P = \frac{T \times 2 \times \text{rpm}}{33,000}$$

from which

$$T = \frac{P \times 5250}{\text{rpm}}$$

where P = power, hp

        T = torque (= F x R), lb-ft

        2π = 6.2832 (ratio of the circumference of a circle to the radius)

        rpm   = speed

        33,000 = no. of ft-lb/min = 1 hp

As an example, we may make a test of a 4-hp motor that has a force of 5.5 lb, a radius of 2.5 feet, and a speed of 1440 rpm. The power delivered will be

$$P = \frac{5.5 \times 2.5 \times 6.2832 \times 1440}{33,000} = 3.77 \text{ hp}$$

To find the efficiency, we must measure (by means of a wattmeter or other suitable method) the input to the motor at the same time that the output is determined. Both output and input are then converted to the same units (preferably watts), and the efficiency is found by the following formula:

$$P = \frac{T \times 2 = \text{rpm}}{33,000}$$

Other methods based upon the same principles but somewhat more convenient in measuring torque (or mechanical output of fractional-horsepower motors) are by means of a specially designed Prony brake, as shown in Fig. 16-12. In this case, the Prony brake differs from the one described previously only in construction. It requires a pulley, brake arm, and scale (may be either platform scale or spring balance). If a platform scale is used, be sure that the load is applied to the center of the platform. If a spring-balance scale is used, the pull must always be at right angles to the brake arm. In either case, the scale must have small enough graduations to accurately read torque on smaller-rated motors. The brake arm should be made so that the distance between the center of the pulley and contact point where the load is measured is exactly 12 inches. The scale reading will then be in pound-feet.

Before starting the test, make sure that the direction of rotation is such that the brake arm will be moved *against* the balance. In order to measure starting torque, clamp the arm to the pulley

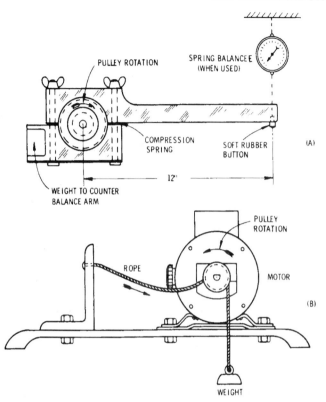

Fig. 16-12. Prony- and rope-brake methods of testing torque for fractional horsepower motors.

tight enough to allow the pulley to turn very slowly; then read the scale. To measure pull-in torque, release the clamp until the motor is just able to throw off the brushes and pull up to speed. Read the scale just as the brushes are leaving the commutator. The true pull-in torque is the highest scale reading, at which time the motor brushes will throw off and stay clear of the commutator.

PULLEY SPEED AND SIZES

On a motor, the driving pulley is called the *driver*, and the driven compressor pulley or flywheel is called the *driven*. In some cases a gear drive is encountered in which the total number of teeth must be counted. For gears, therefore, the number of teeth should be substituted whenever diameter occurs. By the use of a formula, the unknown can be easily found if the other three factors are known. For instance, if one diameter and the two speeds are known, the other diameter can be ascertained. In order to illustrate the formula, the diagrammatic layout in Fig. 16-13 will be helpful. The diameter and speed of both the driver and driven are given as a typical example.

The one important item to consider is that the load on the motor does not increase in direct proportion to the changes in speed of the compressor by changing pulley sizes. It is very easy to overload the motor by increasing its pulley size. Load checks on the current drawn by the motor should be taken after any pulley changes from those of the manufacturer's design.

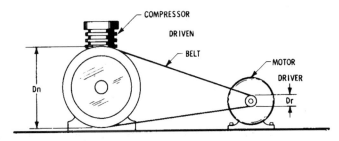

Fig. 16-13. Typical belt arrangement for a motor and compressor drive.

Calculation of Driver Diameter

Where the rpm of the driver, rpm of the driven, and diameter of the driven are known, the diameter of the driver may be found by multiplying the diameter of the driven by its rpm and dividing the product by the rpm of the driver. Thus, in accordance with the previous data, the problem may be written

$$\text{Diameter of driver} = \frac{\text{diameter of driven} \times \text{rpm of driven}}{\text{rpm of driver}}$$

or

$$Dr = \frac{Dn \times Rn}{Rr}$$

426

where Dn = diameter of driven pulley
$\quad\quad Dr$ = diameter of drive pulley
$\quad\quad Rn$ = rpm of driven pulley
$\quad\quad Rr$ = rpm of driver pulley

Example—

$$\frac{17 \times 200}{1700} = 2\text{-inch diameter}$$

Calculation of Driven Diameter

Where the rpm of the driven, rpm of the driver, and diameter of the driver are known, the diameter of the driven can be determined by multiplying the diameter of the driver by its rpm and dividing the product by the rpm of the driven. In accordance with this, we may write

$$\text{Diameter of driven} = \frac{\text{diameter of driver} \times \text{rpm of driver}}{\text{rpm of driven}}$$

or $\quad\quad Dn = \dfrac{Dr \times Rr}{Rn}$

Example—

$$\frac{2 \times 1700}{200} = 17\text{-inch diameter}$$

Calculation of Driver rpm

Where the diameter of the driver, diameter of the driven, and rpm of the driven are known, the rpm of the driver can be found by multiplying the diameter of the driven by its rpm and dividing the product by the diameter of the driver. This may be written

$$\text{rpm of driven} = \frac{\text{diameter of driver} \times \text{rpm of driver}}{\text{diameter of driven}}$$

or

$$Rn = \frac{Dr \times Rr}{Dn}$$

Example—

$$\frac{2 \times 1700}{17} = 200 \text{ rpm}$$

Where the diameter of the driven, diameter of the driver, and its rpm are known, the rpm of the driven can be found by multiplying the diameter of the driver by its rpm and dividing the product by the diameter of the driven. Therefore, we obtain

$$\text{rpm of driven} = \frac{\text{diameter of driven} \times \text{rpm of driver}}{\text{diameter of driver}}$$

or

$$Rr = \frac{Dn \times Rn}{Dr}$$

Example—

$$\frac{17 \times 200}{2} = 1700 \text{ rpm}$$

In the preceding calculations, no attention has been given to belt slip, which may take place. Actual tests bear out that the slip factor may amount to up to 3 percent, depending upon the size of pulleys, type of belt used, etc.

ALTERNATING-CURRENT MOTORS

Alternating-current (ac) motors are divided into two main classifications (depending on the type of power used): *polyphase* and *singe phase*. *Polyphase motors* are further divided into three classes:

1. Squirrel-cage induction
2. Wound-rotor induction
3. Synchronous

Single-phase motors, although somewhat less efficient than the polyphase type, are used mainly in applications where power requirements are small. Thus, fractional-horsepower motors are normally of the single-phase type. In integral-horsepower sizes, however, polyphase motors are commonly used.

Squirrel-Cage Motors

Squirrel-cage motors (Fig. 16-14) provide a variety of speed and torque characteristics. The speed of a squirrel-cage induction motor is nearly constant under normal load and voltage conditions but is dependent on the number of poles and frequency of the ac source. In the case of a squirrel-cage motor, however, the motor slows down when loaded an amount just sufficient to produce the increased current to meet the required torque.

The difference between synchronous speed and load speed is called the *slip* of the motor. This slip is usually expressed as a percentage of the synchronous speed. Since the amount of slip is dependent on the load, the greater the load, the greater the slip will be, i.e., the slower the motor will run. This slowing down of the motor, however, is very slight even at full load and amounts to from 1 to 4 percent of synchronous speed. Thus, the squirrel-cage motor is considered to be a constant-speed type. Certain designs of the squirrel-cage motor provide a high starting torque, and these types are used on refrigerator compressors and reciprocating pumps. Motors of this type used on highly pulsating loads are normally provided with flywheels.

Fig. 16-14. Schematic diagram of an across-the-line, non-reversible, single-speed, manually operated, three-phase, squirrel-cage induction motor.

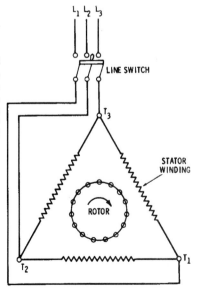

Wound-Rotor Motors

Wound-rotor motors (Fig. 16-15) are used for applications requiring high starting torque at a low starting current. A wound-rotor motor, along with its controller and resistance, can develop full load when starting with little more than full-load current. Motors of this type are provided with insulated phase windings on the rotor, and the terminal of each phase is connected to a collector ring. Stationary brushes riding on the collector rings are connected to adjustable resistors to provide a secondary winding by means of which the motor characteristics can be varied to suit various operation conditions.

Variations in operating speed are essential on many applications. It is often desirable to vary the operating speed of conveyors, compressors, pulverizers, stokers, etc., in order to meet varying production requirements. Wound-rotor motors, because of their adjustable and varying speeds, are ideally suited for such applications. However, if the torque required does not remain constant, the speed of the wound-rotor motor will vary over wide limits, a characteristic constituting one of the serious objections to the use of wound-rotor motors for obtaining reduced speeds.

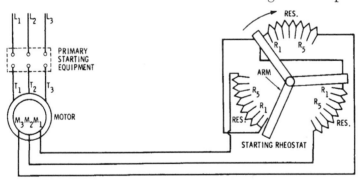

Fig. 16-15. Schematic diagram of the connection arrangement using a face-plate starter and speed regulator for a three-phase, wound-rotor motor.

Synchronous Motors

Synchronous motors are used for continuous-duty applications at a constant speed, such as fans, blowers, pumps, and compressors. The outstanding advantage of the synchronous motor is that its power factor can be changed to compensate for the low-power factor of other drives in the same location. When synchronous motors are employed exclusively as power-factor correction devices, they are termed *synchronous capacitors* because the effect on the power system is the same as that of a static capacitor. The power factor of a synchronous motor may be varied by adjustment of the field current. Thus, by adjusting the field current, the power factor may be varied in small steps, from a low *lagging* power factor to a low *leading* power factor. This makes it possible to vary the power factor over a considerable range and places the characteristics of the motor under the immediate control of the operator at all times.

Because of the higher efficiency possible with synchronous motors, they can be used advantageously on most loads where constant speed is required. Typical applications of high-speed synchronous motors (above 500 rpm) are such drives as centrifugal pumps, dc generators, belt-driven reciprocating compressors, fans, blowers, line shafts, centrifugal compressors, rubber and paper mills, etc.

The field of applications of low-speed synchronous motors (below 500 rpm) includes drives such as reciprocating compressors (largest field of use), Jordan engines, centrifugal and screw-type pumps, ball and tube mills, vacuum pumps, electroplating generators, line shafts, rubber and band mills, chippers, metal-rolling mills, etc. Synchronous motors are rarely used in sizes below 10 hp; their principal field of application is in sizes of 100 hp and larger.

Starting and Controlling Polyphase Motors

Polyphase induction motors may be started and controlled by several methods:

1. Directly across the line
2. By means of autotransformers
3. Using resistors or reactors in series with the stator winding
4. By means of the star-delta connection

Before discussing the various methods available for reducing starting current and improving line-voltage conditions, it is important to have a thorough knowledge of the effects of reduced-voltage starting on the motor as well as on the power system. Any method that reduces the starting current to the motor is accompanied by a reduction in starting torque. Therefore, it is essential to know something about the load-torque characteristics in determining if a given current limitation can be met. In other words, there are boundary conditions in which the permissible current to be taken from the line would not provide the needed output torque at the motor shaft necessary for the successful acceleration of its connected load. With all starting methods, the torque of a squirrel-cage motor varies as the square of the applied voltage at the motor terminals.

Starting Across-the-Line—This method of starting is generally the most economical but it is usually limited to motors of up to 5 hp because it requires a large starting current. With this method, the motor is connected directly to the full line voltage by means of a manually operated switch or magnetic contacter.

Starting with Autotransformers—In the case of the autotransformer type of starting, the current taken from the line varies as the square of the voltage applied to the motor terminals, and it is convenient to remember that the torque and line currents are reduced at the same rate. Thus, an autotransformer starter designed to apply 80 percent of the line voltage to the motor terminals will produce 64 percent of the torque that would have been developed if the motor had been started on full voltage. At the same time, the motor will draw 64 percent as much current from the line as would have been required for full voltage starting.

Starting with Resistors or Reactors—With resistor or reactor starting, the starting current varies directly with the voltage at the motor terminals because the resistor or reactor is in series with each line to the motor and must carry the same current that flows in each motor terminal. It is evident, therefore, that the resistor and reactor type of reduced-voltage starting requires more line current in amperes per unit of torque (in foot pounds) than does the autotransformer type. For example, if a motor connected to a loaded centrifugal pump is started with the 80 percent tap on the autotransformer, the initial torque is 64 percent. If, on the other hand, the motor were started with a primary resistor that limits the starting voltage to 65 percent of line voltage, the initial torque would only be 42.25 percent.

On some power systems, it is necessary to meet a restriction on the rate of current increase in starting. The rate of increase of current is determined to meet the conditions as they exist at that particular point on the system where the motor is started.

Star-Delta Starting—Some three-phase induction motors of the squirrel-cage type may be started by the star-delta method. This starting method is associated only with motors designed for their full power to be by a delta-connected three-phase winding. There must also be provided additional leads from the motor, which, when regrouped, will result in a star arrangement of the three-phase winding. Six main leads from the motor are required to accomplish the switch from a start-across-the-line (star connection) to a run-across-the-line (delta connection). The starting connection is always star since the voltage is $1/\sqrt{3}$, or 57.8 percent of the delta or line voltage.

From the foregoing data, it follows that this type of reduced-voltage starter (which is limited to 57.8 percent of the line voltage at starting) can be employed only where the motor has a light starting load. In all other applications, higher starting voltages are obtained through the use of resistors, reactors, or autotransformer, as previously outlined.

SINGLE-PHASE MOTORS

Single-phase motors, as their name implies, operate on only one phase. They are used extensively in fractional-horsepower sizes in most domestic and commercial applications. The advantages of using single-phase motors in small sizes are that they are less expensive to manufacture than other types and they eliminate the need for three-phase ac lines. Single-phase motors are used in fans, refrigerators, portable drills, grinders, etc.

A single-phase induction motor with only one stator winding and a cage rotor is like a three-phase induction motor with a cage rotor, except that the single-phase motor has no revolving magnetic field at the start and hence no starting torque. However, if the rotor is brought up to speed by external means, the induced currents in the rotor will cooperate with the stator currents to produce a revolving field, causing the rotor to continue to run in the direction in which it was started.

Several methods are used to provide the single-phase induction motor with a starting torque. These methods identify the motor as split-phase, capacitor, shaded-pole, repulsion, etc. Another class of single-phase motors is the ac series-wound universal type, which has electrical characteristics similar to dc series motors. Only the more commonly used types of single-phase motors are described. These include:

1. Split phase
2. Capacitor
3. Shaded pole
4. Repulsion start
5. AC series motor

Split-Phase Motors

The *split-phase motor* (Fig. 16-16) has a high-resistance auxiliary winding that remains in the circuit during the starting period but is disconnected through the action of a centrifugal switch as the motor comes up to speed. Under running conditions, it operates as a single-phase induction motor with only one winding in the circuit. The main winding is connected across the supply lines in the usual manner and has a low resistance and a high inductance. The starting, or auxiliary, winding, which is physically displaced in the stator from the main winding, has a high resistance and a low inductance. This physical displacement, in addition to the electrical phase displacement produced by the relative electrical resistance values in the two windings, produces a weak rotating field that is sufficient to provide a low starting torque.

After the motor has accelerated to 75 or 80 percent of its synchronous speed, a starting switch (usually centrifugally operated) opens its contacts to disconnect the starting winding. The function of the starting switch is to prevent the motor from drawing excessive current from the line and also to protect the starting winding from damage due to heating. The motor may be started

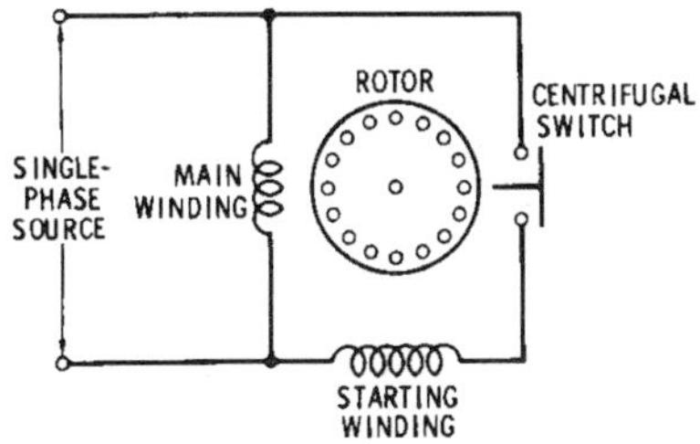

Fig. 16-16. Winding arrange-
ment of a split-phase
motor.

in either direction by reversing either the main or auxiliary winding.

Split-phase motors of the fractional-horsepower size are used in a variety of equipment, such as ventilating fans, oil burners, washers, and numerous other applications.

Capacitor Motors

The *capacitor motor* (Fig. 16-17) is a modified form of a split-phase motor, having a capacitor in series with the starting winding. The capacitor motor is manufactured in fractional-horsepower sizes and develops a high starting torque. Capacitor motors are used for constant-drive applications, such as fans, blowers, centrifugal pumps, and refrigerator compressors. During the starting period, a winding with a capacitor in series is connected to the motor circuit, and when the motor comes up to speed, a centrifugal switch cuts the capacitor and second winding out of the circuit.

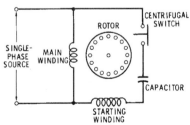

Fig. 16-17. Capacitor-start
motor-winding
diagram.

Capacitor motors are practically always dual-voltage motors and can be either 120 or 240V. They have replaced repulsion-induction motors practically 100 percent. Figure 16-18 shows the connections for both 120 and 240V.

435

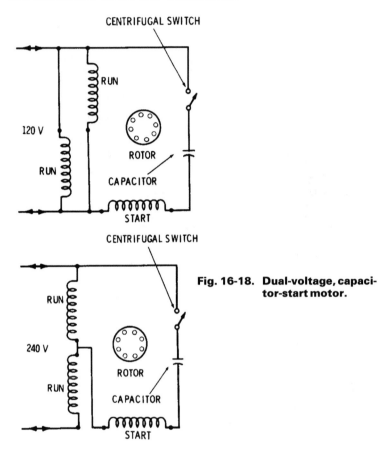

Fig. 16-18. Dual-voltage, capacitor-start motor.

Shaded-Pole Motors

The shaded-pole motor (Fig. 16-19) employs a salient-pole stator and a cage rotor. The projecting poles on the stator resemble those of a dc motor except that the entire magnetic circuit is laminated and a portion of each pole is split to accommodate a short-circuited copper strap called a *shading coil*. The shaded-pole motor is similar in operating characteristics to the split-phase motor and has the advantage of simple construction and low cost.

However, it has a low starting torque and low efficiency. It is manufactured in fractional-horsepower sizes and has found employment in a variety of household appliances, such as fans, blowers, electric clocks, hair dryers, and other applications requiring a low starting torque.

Most shaded-pole motors have only one edge of the pole split, and therefore the direction of rotation is not reversible. However, some shaded-pole motors have both leading and trailing pole tips split to accommodate shading coils. The leading-pole-tip shading coils form one series group, and the trailing-pole-tip shading coils form another series group. Only the shading coils in one group are simultaneously active, while those in the other group are on open circuit.

Fig 16-19. Schematic winding arrangement in a shaded pole motor.

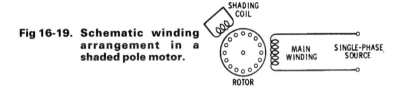

Repulsion-Start Motors

The *repulsion-start motor* (Fig. 16-20) has a form-wound rotor with commutator and brushes. The stator is laminated and contains a distributed single-phase rotor. In its simplest form, the stator resembles that of a single-phase motor. The motor is equipped with a centrifugal device, the function of which is to remove the brushes from the commutator and place a short-circuiting ring around the commutator as the motor comes up to speed. After short-circuiting the commutator, the motor operates with the characteristics of a single-phase induction motor.

Fig. 16-20(A). Schematic arrangement of windings in a repulsion-start induction motor.

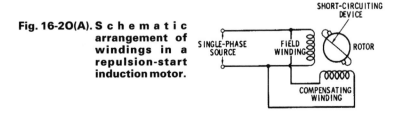

Fig. 16-20(B). A repulsion-induction motor with brushes and a wound rotor. Note how the brushes are located on the rotor and held in place with coiled springs.

MOTOR-COMPRESSOR CONTROLS

Two types of motor controls are used in the operation of commercial refrigeration plants, namely, *manual* and *automatic*. The manual control may be the ordinary snap and knife switches which must be operated by hand when the actuating mechanism is started or stopped. This type of control, although providing temperature control for a given set of conditions, will not provide automatic regulation. Also, with manual control, constant attention is required by the operator to maintain as near as possible the predetermined condition. In addition, the cost of operation is usually found to be excessive.

Automatically operated controls, on the other hand, provide all the features that are considered most desirable, both for temperature and for economy of operation. Here the operator or owner is relieved of all responsibility for maintaining the predetermined conditions, and consequently operating costs are kept at a minimum.

Contactors, Starters, and Relays

Most condensing units with motors larger than 1.5 hp have a starter or contactor; as a rule, these are furnished with the unit. Relays are a necessary part of many control and pilot circuits. These automatic switches are similar in design to contactors but are generally lighter in construction and carry a smaller amount of current.

Magnetic contactors are normally used for starting polyphase squirrel-cage and single-phase motors. Contactors may be connected at any convenient point in the main circuit between the fuses and the motor, and small control wires may be run between the contactor and the point of control.

Protection of the motor against prolonged overload is usually accomplished by time-limit overload relays, which are operative during both the starting and running periods. The action of the relays is delayed long enough to take care of heavy starting loads and momentary overloads without tripping.

The operation of a contactor, as applied to any individual installation, depends on two main parts—the operating coil and the thermal elements of the overload relay. Although the contactor proper is generally universal, it is necessary to have the proper coil and the proper thermal elements installed, depending on the line voltage, frequency, and protection desired.

Relays are furnished in a large number of types, the employment of which depends on the size of the condensing unit, power-source voltage, and other factors. As used on small single-phase motors, a relay is a necessary part of the motor-starting equipment. It consists essentially of a set of normally closed contacts that are in series with the motor-starting winding. The electromagnetic coil is in series with the auxiliary winding of the motor.

When the control contacts close, the motor starting and running winding are both energized. A fraction of a second later the motor comes up to speed, and sufficient voltage is induced in the auxiliary winding to cause current to flow through the relay coil. The magnetic force is sufficient to attract the spring-loaded armature, which mechanically opens the relay starting contacts. With the starting contacts open, the starting winding is out of the circuit, and the motor continues to run on only the running winding. When the control contacts open, power to the motor is interrupted. This power interruption causes the relay armature to close the starting contacts, placing the motor in a condition to start a new cycle when the control contacts again close.

Motor-Overload Protector

Motors for commercial condensing units are normally protected by a bimetallic switch operating on the thermo, or heating, principle. This is a built-in motor-overload protective device that limits the motor winding temperature to a safe value. In its simplest form, the switch, or motor protector, consists essentially of a bimetal switch mechanism permanently connected in series with the motor circuit, as shown in Fig. 16-21.

When the motor becomes overloaded or stalled, excessive heat is generated in the motor winding due to the heavy current produced by this condition. The protector located inside the motor is controlled by the motor current passing through it and the motor temperature. The bimetal element is calibrated to open the motor circuit when the temperature, as a result of an

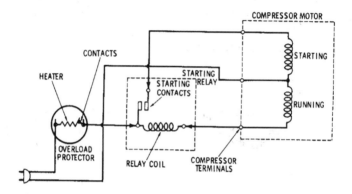

Fig. 16-21. Circuit arrangement of a typical domestic refrigerator.

excessive current, rises above a predetermined value. When the temperature decreases, the protector automatically resets and restores the motor circuit.

The advantage of the overload device obviously is that it will reduce service calls due to temporary overloads. When overload conditions occur, the protector will stop the motor for a short period of time to allow it to cool off, after which it will reset itself and allow the motor to run. This cycling may continue for some time until the load has been reduced to a point where the motor can safely handle it.

In this connection, it should be noted that servicing of motors equipped with built-in overload protectors must be handled with care since the compressor may be idle due to an overload and hence will start as soon as the motor cools off. This could result in a serious mishap to the operator or serviceman. To avoid such difficulties, open the electrical circuit by pulling the line plug or switch prior to any repair or servicing operation.

Motor-Winding Relays

A motor-winding relay is usually incorporated in single-phase motor compressor units. This relay is an electromagnetic device for making and breaking the electrical circuit to the starting winding. A set of normally closed contacts are in series with the motor starting winding (Fig. 16-21).

The electromagnetic coil is in series with the auxiliary winding of the motor. When the control contacts close, the motor starting and running windings are both energized. A fraction of a second later, the motor comes up to speed and sufficient voltage is induced in the auxiliary winding to cause current to flow through the relay coil. The magnetic force is sufficient to attract the spring-loaded armature, which mechanically opens the relay starting contacts. With the starting contacts open, the starting winding is out of the circuit and the motor continues to run on only the running winding. When the control contacts open, power to the motor is interrupted. This allows the relay armature to close the starting contacts. The motor is now ready to start a new cycle when the control contacts again close.

OIL-PRESSURE CONTROLS

Oil-pressure controls (Fig. 16-22) are designed to provide protection against breakdowns due to low lubrication oil pressure. Total oil pressure is the combination of crankcase pressure and the pressure generated by the oil pump. The net oil pressure available to circulate the oil to the bearing surfaces is the difference between the total oil pressure and the refrigerant pressure in the crankcase. This control measures that difference in pressures, which will be referred to as *net oil pressure*. A built-in time-delay switch, accurately compensated for ambient temperature, allows for pressure pickup on start and avoids nuisance shutdowns on pressure drops of short duration during the running cycle.

When the compressor is started, the time-delay switch is energized. If the net oil pressure does not build up to the cut-in point of the control within the required time limit, the time-delay switch trips to stop the compressor. If the net oil pressure does rise to the cut-in point within the required time after the compressor starts, the time-delay switch is automatically deenergized and the compressor continues to operate normally.

If the net oil pressure should drop below the cutout setting (scale pointer) during the running cycle, the time-delay switch is energized and, unless the net oil pressure returns to the cut-in point within the time-delay period, the compressor will be shut

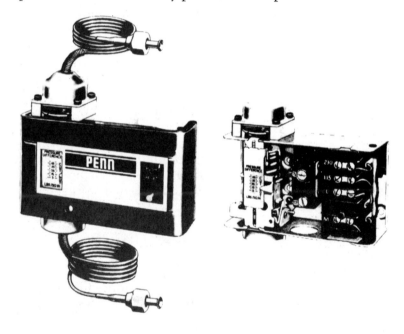

Fig. 16-22. Exterior and interior views of an oil pressure-failure control switch with a time-delay feature.

down. The compressor can never run more than the predetermined time on subnormal oil pressure. Fig. 16-23 is a schematic diagram of a typical compressor with an oil-pressure control connected.

Solenoid Valves

Solenoid valves are used on multiple installations and are electrically operated. A solenoid valve, when connected as shown in Fig. 16-24, remains open when current is supplied to it, and closes when the current is shut off. In general, solenoid valves are used to control the liquid-refrigerant flow into the expansion valve or the refrigerant-gas flow from the evaporator when the evaporator or fixture has reached the desired temperature.

The most common application of the solenoid valve is in the liquid line, which operates with a thermostat. With this hookup, the thermostat may be set for the desired temperature in the

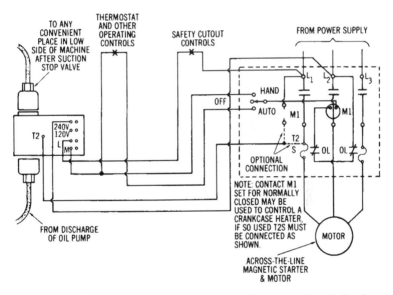

Fig. 16-23. Wiring diagram of an oil-protection control switch when used on a 240V power system.

fixture. When this temperature is reached, the thermostat will open the electrical circuit and shut off the current to the solenoid valve. The solenoid valve then closes and shuts off the refrigerant supply to the expansion valve. The condensing-unit operation should be controlled by the low-pressure switch. In other applications, where the evaporator is to be in operation for only a few hours each day, a manually operated snap switch may be used open and close the solenoid valve at the convenience of the operator.

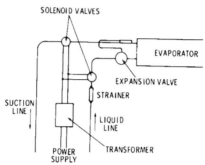

Fig. 16-24. Solenoid valve connected in suction and liquid evaporator lines.

TEMPERATURE CONTROLS

In modern condensing units, low-pressure control switches are being largely superseded by thermostatic control switches. Briefly, a thermostatic control consists of three main parts: a bulb, capillary tube, and power element (switch). The bulb is attached to the evaporator in such a manner as to ensure contact with the evaporator and contains a volatile liquid, usually a refrigerant such as sulfur dioxide or methyl chloride. The bulb is connected to the power element by means of a small capillary tube.

Referring to Fig. 16-25, as the evaporator temperature increases, the bulb temperature also increases, which raises the pressure of the thermostatic liquid vapor, causing the bellows to expand and thus actuating an electrical contact that closes the motor circuit starting the motor and compressor. As the evaporator temperature decreases, the bulb becomes colder and the pressure decreases to the point where the bellows contracts sufficiently to open the electrical contacts controlling the motor circuit. In this manner, the condensing unit is entirely automatic, producing just exactly the proper amount of refrigeration to meet any normal operating condition. A typical thermostatic control is shown in Fig. 16-26.

444

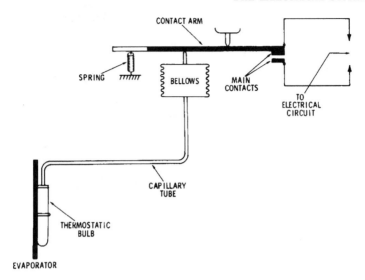

Fig. 16-25. Schematic diagram showing working principles of a thermostatic control switch.

Penn Controls, Inc.

Fig. 16-26. Exterior and interior views of a typical heavy-duty-refrigeration pressure control with a visible calibrated scale indicating cut-in and cutout settings. Control of this type is normally used with manual-starting compressor motors not integrally protected, thus making line starters unnecessary.

An automatic temperature-control system is generally operated by making and breaking an electric circuit or by opening and closing a compressed-air line. When using the electric thermostat, the temperature is regulated by controlling the operation of an electric motor or valve, and, when using the compressed-air thermostat, temperature regulation is obtained by actuating a compressed-air-operated motor or valve. Electrically operated temperature-control systems are used generally by manufacturers for practically all installations. However, compressed-air temperature-control systems have applications in extremely large central and multiple installations in close-temperature work and where a large amount of power is required for small control devices.

Bimetallic Thermostats

The so-called bimetallic type of thermostat operates as a function of expansion or contraction of metals due to temperature changes. Such thermostats are designed for control of heating and cooling in air-conditioning units, refrigeration storage rooms, greenhouses, fan coils, blast coils, etc.

The working principles of such a thermostat are shown in Fig. 16-27. As noted, two metals, each having a different coefficient of

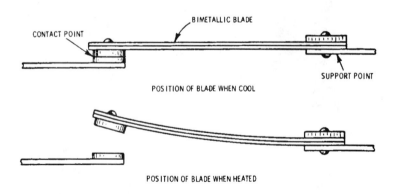

Fig. 16-27. Working principles of a simple bimetallic thermostat.

expansion, are welded together to form a bimetallic unit, or blade. With the blade securely anchored at one end, a circuit is formed and the two contact points are closed to the passage of an electric current. Because of the fact that an electric current produces heat in its passage through the bimetallic blade, the metals in the blade begin to expand, but at different rates. The metals are so arranged that the one which has a greater coefficient of expansion is placed at the bottom of the unit. After a certain time interval, the operating temperature is reached and the contact points become separated, thus disconnecting the appliance from its source of power. After a short period, the contact blade will again become sufficiently cooled off to cause the contact points to join, thus reestablishing the circuit and permitting the current to again actuate the circuit leading to the appliance. The foregoing cycle is repeated over and over again and, in this manner, prevents the temperature from rising too high or dropping too low. Figure 16-28 shows the construction features of this type of control.

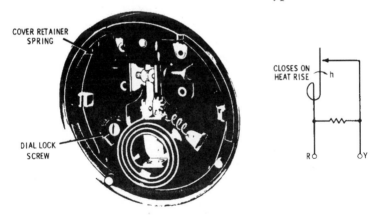

Penn Controls, Inc.

Fig. 16-28. Bimetallic thermostat with wiring hookup (low-voltage type).

DEFROST CONTROLS

Automatic defrosting is accomplished in several ways, the control method used depending on the type of refrigeration system, size and number of condensing units, and other factors.

Defrost-Timer Operation

In small and medium-size domestic applications, an automatic defrost-control clock may be set for a defrost cycle once every 24 hours or as often as deemed necessary. The defrosting is usually accomplished by providing one or more electric heaters, which, when energized by the action of the electric clock, provide the heating action necessary for complete defrosting.

The defrost controls, as usually employed, are essentially single-pole double-throw switching devices in which the switch arm is moved to the defrost position by an electric clock. The switch arm is returned to the normal position by a power element, which is responsive to changes in temperature. As the evaporator is warmed during a defrost period, the feeler tube of the defrost control is also warmed until it reaches the defrost cutout point of approximately 45°F. The defrost-control bellows then forces the switch arm to snap from the defrost position to the normal position (Fig. 16-29), thus starting the motor compressor.

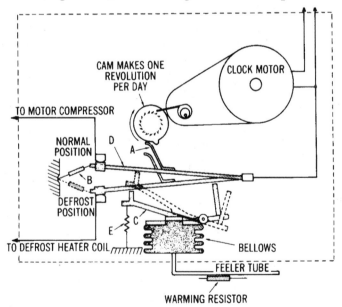

Fig. 16-29. Schematic arrangement of an electric-clock motor as employed in a 2-hour defrost control.

Another common method of automatic defrosting is the so-called defrost-cycle method in which the defrost cycle occurs during each compressor *off* cycle. In a defrost system of this type, the defrost heaters are connected across the thermostat switch terminals. When the thermostat switch is closed, the heaters are shunted out of the circuit. When the thermostat opens, the heaters are energized, completing the circuit through the overload relay and compressor.

Hot-Gas Defrosting

In any low-temperature room where the air is to be maintained below freezing, some adequate means for removing the accumulated frost from the cooling surface should be provided. As a result of recent developments, an improved hot-gas method of quick defrosting for direct-expansion, low-temperature evaporators is available. In order to apply this method, it is necessary to have two or more evaporators in the system because the hot gas required to defrost part of the system must be provided by the heat absorbed from the other cooling surface in a given system. The process of defrosting a plate bank with hot gas can be accomplished automatically by installing the proper controls.

ELECTRICAL-SYSTEM MAINTENANCE

The first thing to check when the unit will not run is the voltage at the compressor terminals. Be sure the contact points on the pressure control are closed. The voltage at the terminals should be within plus or minus 10 percent of the rated voltage shown on the compressor nameplate. If extreme low voltage exists, the cause may be inadequate wiring to the unit. Do not use 240V units on 208V circuits.

Starting Capacitors

For a quick check to determine whether a starting capacitor is still good, disconnect the capacitor lead wires and touch them for a second or two to line wires having the same approximate voltage as the voltage shown on the capacitor. *Caution: Never check a 120V capacitor on a 240V line, and do not hold a starting capacitor on the line more than 5 seconds.* Now touch the two capacitor wires together. If they spark on several attempts, the capacitor is good; if not, the capacitor is defective and should be replaced.

A more accurate means of checking a capacitor is to connect the two lead wires momentarily across the line and measure the current flow. The microfarad rating of the capacitor may be determined as follows:

$$\text{Microfarad rating} = \frac{2650 \times \text{amperes (for 60 hertz)}}{\text{volts}}$$

The reading thus obtained should be within 10 percent of the microfarad rating stamped on the capacitor. Running capacitors may be checked in the same manner as starting capacitors.

Overload Protectors

If there is no voltage at the compressor terminals, it is possible that the overload protector has tripped and is open. The protector can be checked by bridging the protector terminals with a jumper wire (which should contain a fuse for protection). If the compressor now starts, this indicates either a defective protector or an overload condition. Do not disconnect the overload protector and leave the machine operating without any protection.

Compressor Motors

The compressor motor may be checked by removing all wires from the compressor terminals, which are marked as follows: S for starting winding, R for running winding, C for common. Attach a two-wire test cord to the common and running terminals, and plug in the test cord to the power supply. Then momentarily touch the leads of an auxiliary starting capacitor, having the same microfarad rating and voltage rating as supplied with the unit, across the start and running terminals. If the motor does not start, it is probably burned out or has an open winding, or else the compressor is seized. If the motor does start, the trouble is in the electrical accessories.

Starting Relay (Voltage Type)

This type of relay is used on single-phase compressors. If the relay contacts fail to open, the motor will draw excessive current. It will cycle on the overload protector and may blow out the starting capacitor. This failure may be due to an open circuit in the relay coil or to welded contact points. The latter may be determined by disconnecting the power supply and manually trying to open the contact points. Sometimes a cracked relay base will prevent the contact points from opening sufficiently.

The relay holding coil may be checked by connecting a fused test cord to compressor terminals C and S and plugging the other end of the test cord into the power supply. Due to the fact that the relay requires more than line voltage to actuate, it may still not open, but by pushing the relay actuating arm with an insulated screwdriver, the contacts should remain open as long as the test current remains on. If not, the relay holding coil is faulty and the relay should be replaced by one of the same model.

Fan Motors

The fan motor should run whenever the compressor is running. If it does not, turn the fan by hand to see if it turns freely. If it turns freely and the lead wires are connected, it is defective and must be replaced. Fan motors are sometimes a source of noise, which can be caused by worn bearings, bent fan blades, etc. All fan motors having oiling cups should be oiled every six months. Sealed fan motors do not require oiling.

Wiring diagrams for various types of motors with protectors and starters are shown in Figs. 16-30 to 16-35.

COMPRESSOR-MOTOR REPLACEMENT

If the compressor motor burns out, the following procedure must be used to clean the refrigeration system of contamination resulting from the motor burnout:

1. Close the compressor service valves and remove the burned-out compressor.
2. Install the new compressor and a filter-drier in the suction line, using an adapter at the suction valve, if available, so

that gas passes through the filter-drier before entering the compressor. If an adapter is used, be sure that the hose between the filter-drier and compressor is clean. Use the largest-size filter-drier with openings matching the suction line.

3. Evacuate the compressor and the system cleaner with a good vacuum pump, break the vacuum with refrigerant, and reevacuate. Only the compressor and system cleaner need to be evacuated since the service valves have been closed, isolating the refrigerant in the system.

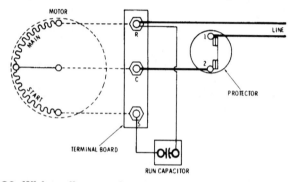

Fig. 16-30. Wiring diagram for a single-phase, permanent, split-capacitor motor with a two-terminal protector.

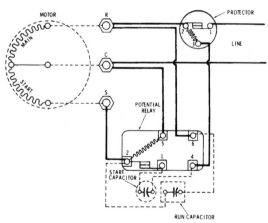

Fig. 16-31. Wiring diagram for single-phase, capacitor-start motor circuit using a potential relay and a three-terminal protector.

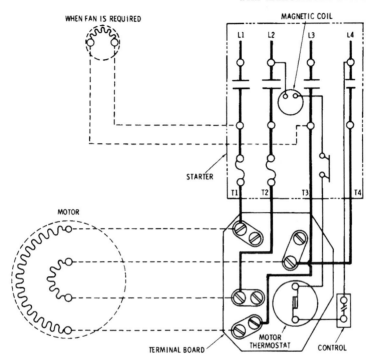

Fig. 16-32. Wiring diagram for a two-phase motor with magnetic starter and thermostat.

4. Open the compressor service valves, close the liquid-line valve, and pump down the system. Install an oversize filter-drier (at least one size larger than the normal selection size) in the liquid line, removing the old filter-drier if one exists.

5. Operate the system, using the usual starting procedure. Check the pressure drop across the system cleaner during the first hour of operation, and change the cores if it becomes excessive.

6. Take an oil sample in 8 to 24 hours. If oil tests free of acid and is clean, remove the system cleaner.

7. When the system cleaner is removed, replace the liquid-line filter-drier and install a sight glass with a moisture indicator in the liquid line.

8. In two weeks, recheck the color and acidity of the oil to see

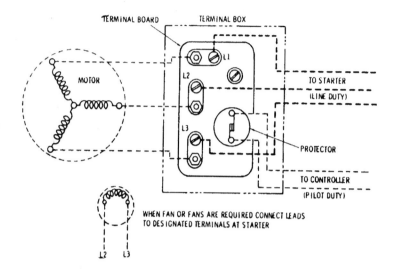

Fig. 16-33. Wiring diagram for a three-phase, star-connected motor.

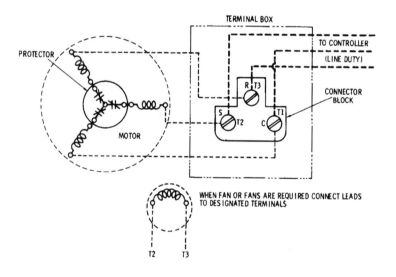

Fig. 16-34. Wiring diagram for a three-phase, star-connected motor with winding protector.

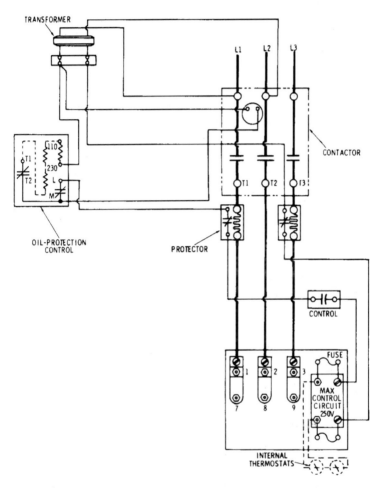

Fig. 16-35. Wiring diagram for across-the-line starting of a 550V compressor motor with oil-protection control.

if another change of the liquid-line filter-drier is necessary. Before the job is completed, it is essential that the oil be clear and acid-free. The sight glass with a moisture indicator will indicate if the filter-drier should be changed to reduce the moisture content of the system.

ELECTRICAL SYSTEM
TROUBLESHOOTING GUIDE

Symptoms and Possible Cause   Possible Remedy

Compressor Does Not Run

(a) Motor line open

(a) Close, start, or disconnect switch.

(b) Fuse blown

(b) Replace fuse.

(c) Tripped overload

(c) Reset overload relay.

(d) Control stuck open

(d) Repair or replace.

(e) Piston stuck

(e) Remove motor compressor head. Look for broken valve and jammed parts.

(f) Frozen compressor or motor bearings

(f) Repair or replace.

(g) Control off because of cold location

(g) Use thermostatic control, or move control to warmer location.

Unit Short-Cycles

(a) Control differential set too closely

(a) Widen differential.

(b) Discharge valve leaking

(b) Correct condition.

(c) Motor compressor overload cutting out

(c) Check for high-speed pressure, tight bearings, stuck pistons, clogged air- or water-cooled condenser, or water shutoff.

(d) Shortage of gas

(d) Repair leak and recharge.

(e) Leaky expansion valve

(e) Replace.

(f) Refrigerant overcharge

(f) Purge.

(g) Cycling on high-pressure cutout

(g) Check water supply.

456

Symptoms and Possible Cause   Possible Remedy

Compressor Will Not Start— Hums Intermittently (Cycling on Overload)

(a) Improperly wired

(a) Check wiring against diagram.

(b) Low line voltage

(b) Check main line voltage, and determine location of voltage drop.

(c) Open starting capacitor

(c) Replace starting capacitor.

(d) Relay contacts not closing

(d) Check by operating manually. Replace relay if necessary.

(e) Open circuit in starting winding

(e) Check stator leads. If leads okay, replace stator.

(f) Stator winding grounded

(f) Check stator leads. If leads okay, replace stator.

(g) High discharge pressure

(g) Eliminate cause of excessive pressure. Make sure discharge shutoff valve is open.

(h) Tight compressor

(h) Check oil level. Correct binding.

Compressor Starts but Motor Runs on Starting Winding

(a) Low line voltage

(a) Bring up voltage.

(b) Improperly wired

(b) Check wiring against diagram.

(c) Defective relay

(c) Check operation manually. Replace relay if defective.

(d) Running capacitor shorted

(d) Check by disconnecting running capacitor.

457

Symptoms and Possible Cause   Possible Remedy

Compressor Starts but Motor Runs on Starting Winding (cont.)

(e) Starting and running windings shorted

(e) Check resistances. Replace stator if defective.

(f) Starting capacitor weak

(f) Check capacitance. Replace if low.

(g) High discharge pressure

(g) Check discharge shutoff valve. Check pressure.

(h) Tight compressor

(h) Check oil level. Check binding.

Relay Burnout

(a) Low line voltage

(a) Increase voltage to not less than 10% under compressor motor rating.

(b) Excessive line voltage

(b) Reduce voltage to maximum of 10% of motor rating.

(c) Incorrect running capacitor

(c) Replace running capacitor with correct value capacitance.

(d) Short-cycling

(d) Reduce number of starts per hour.

(e) Incorrect relay mounting

(e) Mount relay in correct position.

(f) Relay vibrating

(f) Mount relay in rigid location.

(g) Incorrect relay

(g) Use relay properly selected for motor characteristics.

Starting-Capacitor Burnout

(a) Short-cycling

(a) Replace starting capacitor with series arrangement

Symptoms and Possible Cause   Possible Remedy

Starting-Capacitor Burnout (cont.)

	or reduce number of starts per hour to 20 or less.
(b) Prolonged operation on starting winding	(b) Reduce starting load (install suction-regulating valve). Increase voltage if low.
(c) Relay contacts sticking	(c) Clean contacts or replace relay.
(d) Improper capacitor	(d) Check for proper capacitor rating

Running-Capacitor Burnout

(a) High line voltage and light load	(a) Reduce voltage if over 10% excessive. Check voltage imposed on capacitor and select one equivalent to this in voltage rating.

Unit Operates Long or Continuously

(a) Shortage of gas	(a) Repair leak and recharge.
(b) Control contacts frozen	(b) Clean points or replace control.
(c) Dirty condenser	(c) Clean condenser.
(d) Location too warm	(d) Change to cooler location.
(e) Air in system	(e) Purge.
(f) Compressor inefficient	(f) Check valves and pistons.
(g) Expansion valve set too high	(g) Lower setting.
(h) Iced or dirty coil	(h) Defrost or clean.
(i) Cooling coils too small	(i) Add surface or replace.

459

Symptoms and Possible Cause   *Possible Remedy*

Unit Operates Long or Continuously (cont.)

(j) Expansion valve too small

(j) Raiser suction pressure with larger valve.

(k) Restricted or small gas lines

(k) Clear restriction or increase line size.

Head Pressure Too High

(a) Refrigerant overcharge
(b) Air in system
(c) Dirty condensor
(d) Unit location too hot
(e) Restricted water-cooled condenser
(f) Water shut off

(a) Purge.
(b) Purge.
(c) Clean.
(d) Relocate unit.
(e) Clean or replace condenser.
(f) Turn on water.

Head Pressure Too Low

(a) Refrigerant shortage
(b) Compressor suction or discharge valves inefficient

(c) Cold room or cold water

(a) Repair leak and recharger.
(b) Clean or replace leaky valve plates.

(c) None. Efficiency increased.

Noisy Unit

(a) Insufficient compressor oil
(b) Tubing rattle

(c) Mountings loose
(d) Oil slugging or refrigerant flooding back

(a) Add oil to proper level.

(b) Bend tubes away from contact.
(c) Tighten.
(d) Adjust oil level or refrigerant charge. Check expansion valve for leak or oversized orifice.

Symptoms and Possible Cause   *Possible Remedy*

Compressor Loses Oil

(a) Shortage of refrigerant.

(a) Repair leak and recharge.

(b) Evaporator design

(b) Changer oil to get higher gas velocity.

(c) Gas-oil ratio low

(c) Add 1 pint oil for each 10 lb of refrigerant added to factory charge.

(d) Plugged expansion valve or strainer

(d) Clean or replace.

(e) Oil trapping in liners

(e) Drain tubing toward compressor.

(f) Short-cycling

(f) Refer to "Unit Short-Cycles."

(g) Superheat too high at compressor suction

(g) Changer location of expansion valve bulb or adjust valve to return wet gas to compressor.

Frosted or Sweating Suction Line

(a) Expansion valve admitting excess refrigerant

(a) Adjust expansion valve.

Hot Liquid Line

(a) Shortage of refrigerant

(a) Repair leak and recharger.

(b) Expansion valve open too wide

(b) Adjust expansion valve.

Frosted Liquid Line

(a) Receiver shutoff valve partially closed or restricted

(a) Open valve or remove restriction.

(b) Restricted dehydrator or strainer

(b) Replace restricted part.

461

Symptoms and Possible Cause   Possible Remedy

Top Condenser Coils Cool When Unit Operates

(a) Refrigerant undercharge   (a) Repair leak and recharge.
(b) Compressor inefficient   (b) Check and correct
                                        compressor.

Unit on Vacuum; Frost on Expansion Valve Only

(a) Ice plugging expansion-   (a) Apply hot wet cloth to
     valve orifice   expansion valve. Moisture
                                        indicated by increase in
                                        suction pressure. Install
                                        drier.

(b) Plugged expansion-valve   (b) Clean strainer or replace
                                        strainer expansion valve.

SUMMARY

In this chapter, some of the most common electrical types of mechanical refrigeration and motor control devices have been presented. Various motor types, their controls, and operation characteristics have been presented first because the motor is the prime source of power in all compressor operations. In this connection, it should be noted that, although dc motors are seldom used in compressor drives, their classification and characteristics are included to assist the plant operator and others in distribution districts where only direct current is available.

The various methods of motor-compressor controls, including contactors, relays, and overload devices operating on the thermal, or heating, principles, provide the necessary information required in plant operation. Descriptions of temperature controls, such as thermostatic control switches and bimetallic thermostats, together with defrost methods and associated devices, are also given.

Electrical-system maintenance, giving details on capacitor function as used on single-phase ac motors and the checking methods to use in case of failure, has been given. The procedure for motor-compressor replacement will furnish information when a replacement due to burnout or other troubles becomes necessary. An electrical-system troubleshooting guide lists possible causes and remedies when the refrigerant system becomes inoperative.

REVIEW QUESTIONS

1. State the difference between the shunt and compound-wound motors.
2. Describe the starting arrangement in a shunt-wound motor.
3. What are the speed and torque characteristics of a squirrel-cage motor?
4. Describe the applications for wound-rotor motors.
5. Why are synchronous motors employed exclusively as power-factor correction devices termed synchronous capacitors?
6. Describe control methods for polyphase motors.
7. What are the starting devices used for single-phase motors?
8. Name five types of single-phase motors.
9. What are the advantages of using single-phase motors in the small sizes rather than polyphase motors?
10. Why is the starting torque greater in a capacitor motor than in a split-phase motor?
11. What are the disadvantages of shaded-pole motors?
12. Why is the repulsion-start induction motor so called?
13. Can an ac series motor operate on either alternating or direct current?
14. Why is automatic motor-compressor control desired instead of manual control?
15. What protection methods are used against overload on ac motors?
16. What is the function of time-limit overload relays in an ac motor circuit?
17. Describe and give the function of motor-winding relays.

18. What is meant by temperature control, and how does it operate?
19. Describe the operation of a bimetallic thermostat.
20. How does an automatic-defrost control operate?
21. What precautions should be observed in electrical-system maintenance?
22. What is the procedure to use in motor-compressor replacement?

CHAPTER 17

Commercial Refrigeration Principles

There are three primary methods used for producing mechanical refrigeration and consequently low temperatures: *use of natural ice; freezing mixtures;* and *vaporizing liquids.* The use of natural ice as a refrigeration medium is too well known to require any explanation. The freezing mixtures, usually called *brines,* generally consist of natural ice with the addition of calcium chloride, which lowers the freezing point of the mixture. The third method, vaporizing liquid, is most commonly used in mechanical refrigeration.

For any liquid, the temperature at which it boils or vaporizes is called its *boiling point.* For liquids having very low boiling points, it is not necessary that heat be supplied by any visible source because the heat in surrounding objects alone may be sufficiently high to cause boiling or vaporization. This is true of anhydrous ammonia, for example; its boiling point of –28°F is sufficiently low to cause it to boil violently when placed in an open vessel at room temperature. The absorption of heat by ammonia vaporization will cause the outside of the vessel to become heavily frosted by moisture condensed and frozen out of the air immediately surrounding it.

From this, it follows that, in order to produce refrigeration, a refrigeration plant would need only a supply of refrigerant, a regulating valve, and a coil of pipe in which the refrigerant would absorb heat from the substance in which the coil is immersed. Such a simple system would, however, be very uneconomical in that it would use refrigerants only once. The ability to recover refrigerants and use them over and over again is a continuous cycle employed in the compressor system of refrigeration.

The refrigerants primarily used in commercial systems are Freons and methyl chloride. Anhydrous ammonia is used in large installations where an operator is in attendance. Anhydrous ammonia is a hazardous material and should not be used where the commercial refrigeration equipment is unattended. It was used to a great extent in earlier commercial refrigeration facilities, especially in cold-storage plants and ice-making machines. Since it is classified as a hazardous material, it requires special treatment, even in the electrical wiring for operation of this equipment.

EFFECT OF PRESSURE ON REFRIGERANTS

As the boiling temperature of any liquid may be made to change with pressure, it is therefore an easy matter to cause the liquid refrigerant to boil at any desired temperature by placing it in a vessel where the required pressure may be obtained. The process of boiling, or vaporization, by which a liquid is changed to a vapor can be reversed; that is, the vapor can be reconverted into a liquid by employing pressure. This is called *condensation*. An increase in pressure, by raising the boiling point, will assist in the condensation of the vapor. Liquids used as refrigerants, by reason of their initial cost, must be recovered, and the process of condensation is employed for this purpose.

In Chapter 2 the pressures and temperatures of various refrigerants were given. These will vary slightly with altitude due to the difference in barometric pressures, but, for practical purposes, they are very helpful in refrigeration service.

ESSENTIAL PARTS OF A REFRIGERATION PLANT

The essential parts of a refrigeration plant are:

1. Compressor
2. Condenser
3. Receiver
4. Expansion valve
5. Evaporator coils
6. Suitable pipelines, shutoff valves, gages, thermostats, etc.

Since anhydrous ammonia was the refrigerant most commonly used in earlier times, it will be covered here as it is possible that you may still run across this type of system in commercial refrigeration service. The apparatus ordinarily used in an ammonia compression system is shown in Fig. 17-1. The compressor is merely a pump, which takes the ammonia vapor from the low-pressure side, compresses it, and delivers it to the high-pressure side. The *high-pressure* side of the plant comprises the compressor discharge piping, condenser, liquid receiver, and liquid piping between the liquid receiver and expansion valve. The *low-pressure* side comprises the piping leading from the expansion valve to the evaporator coil, evaporator coil, and piping leading from the evaporator coil to the compressor suction side. The pressure at the low side of the system remains practically constant and is indicated by the back-pressure gage at the suction side of the compressor.

In a refrigeration plant of this type, ammonia vapor superheated by the work of compression passes through a *condenser* where cooling water lowers its temperature until it condenses

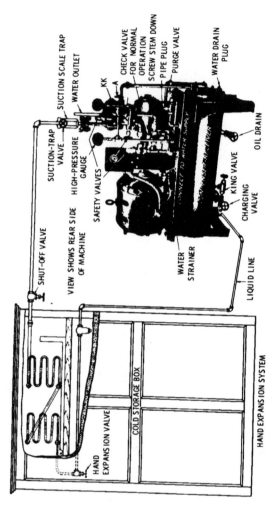

Fig. 17-1. Piping arrangement and accessories in a direct-drive ammonia refrigerating unit.

into liquid ammonia. The liquid leaving is subcooled to 82°F, a temperature that is slightly below that of condensation (86°F).

The *expansion valve* comprises the dividing point between the high- and the low-pressure sides of the system. The function of the expansion valve is to regulate the flow of ammonia from the high- to the low-pressure side. The heat absorbed from the refrigerator by the liquid ammonia in the evaporating coils causes the liquid to boil until vaporization is complete and vapor is slightly superheated. After the liquid refrigerant has produced its refrigerating effect in the evaporating coils, the gas is compressed to a suitable pressure. Ammonia gas, for example, when compressed to 154.5 psig, will condense at a temperature of 86°F. Cooling water of a lower temperature than 86°F is usually available for abstracting the latent heat from the vapor, thus converting it to a liquid (i.e., condensing it). To obtain this pressure of 154.5 psig this system requires the use of a compressor with some form of motive power and also a condenser in which the gas may be liquefied.

The regulating or expansion valve releases the liquid from the condenser or its accompanying liquid container or receiver and controls the pressure at which the evaporation and consequent refrigeration takers place. The expansion valve may be hand or automatically controlled, and its opening or closing allows more or less refrigerant to pass to the low-pressure side of the system. The liquid refrigerant in the low side is evaporated by the heat of the substance being cooled, and the resulting gas is pumped away by the compressor. To replace this, enough refrigerant is allowed to pass from the high side through the expansion valve to maintain a constant pressure and temperature in the low side. The compressor brings the low-pressure ammonia gas from the cold side, compresses it, and discharges it to the high-pressure side. This increase in pressure raisers the boiling or, in this case, condensing temperature of the gas so that comparatively warm cooling water readily condenses the compressed gas again to a liquid. This liquid ammonia is usually collected in a receiver, and from there it is passed through the expansion valve. The reduction in pressure lowers the boiling point of the ammonia, causing it to evaporate.

The actions of other refrigerants used in compression systems closely resemble that just described for ammonia. These systems (Fig. 17-2) operate fundamentally in the same manner, and are:

1. Evaporation of liquid at low pressure
2. Compression of gas to raiser its condensing temperatures

3. Condensation of gas at high pressure
4. Reduction of pressure on the liquid to lower its boiling temperature

Head pressures, back pressures, and working temperatures vary in accordance with the properties of each refrigerant and with conditions encountered in individual plants.

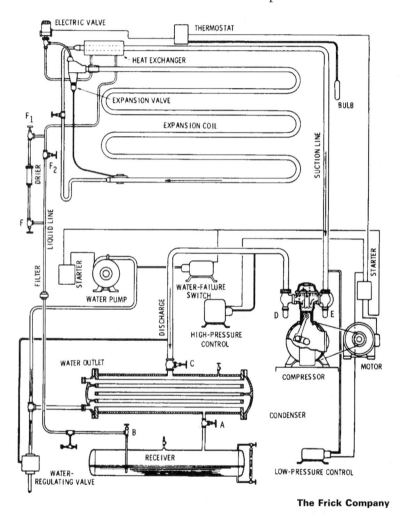

The Frick Company

Fig. 17-2. Standard refrigerating unit using Freon-12 or Freon-22.

CONDENSERS, FITTINGS, AND ACCESSORIES

It is important that a condensing unit of the proper size and loading point be selected. In making this selection, the following should be taken into consideration:

1. The capacity necessary to handle total application load without excessive running time or overload when ambient temperatures or water temperatures are at their peak
2. Selection of type unit, whether air- or water-cooled
3. Condensing unit location
4. Maximum summer air temperature
5. Maximum summer water temperature
6. Quality, quantity, pressure, and cost of condensing water
7. Total cost of operating an air-cooled unit as compared to a water-cooled unit
8. Refrigerant temperature range at which the condensing unit will be operated
9. Running time desirable
10. Availability of electrical supply of proper capacity and voltage

When necessary to locate the condensing unit above the evaporator, special care should be taken in sizing the suction line and installation of a trap at the base of suction line to ensure the return of the oil (which may circulate in the system) to the compressor.

In many instances the condenser is cooled by means of a fan on the driving motor, and the flywheel of the compressor has fan blades instead of spokes. The air is drawn through the condenser and blown across the compressor and motor of the unit. Where banks of units are installed (as in a supermarket) the separate units are mounted next to one another and in tiers, as shown in Fig. 17-3.

Where banks of refrigeration units are installed, as shown in Fig. 17-3, each unit has a set of louvers (Fig. 17-4) that open when the unit starts if the ambient temperature of the refrigeration unit room is too high. They will not open if the ambient temperature is cool; this is accomplished automatically and electrically.

Steffen's Market

Fig. 17-3. Bank of air-cooled supermarket compressors.

Steffen's Market

Fig. 17-4. Exterior view of on-demand louvers for air-cooled compressors.

COOLING-WATER REQUIREMENTS

An unfailing supply of suitable water is of the utmost importance in all refrigeration plants. By definition, water is a chemical compound of two gases, oxygen and hydrogen, in the proportion of 2 parts by weight of hydrogen to 16 parts by weight of oxygen, at a normal pressure of 14.7 psi. At 62°F 1 U.S. gallon of water (231 cu in) weighs 8.33 lb, or 8⅓ lb, for ordinary calculations.

Water cooling and refrigeration are interrelated since, without water, commercial refrigeration would not be possible. The primary considerations when planning a refrigeration system are the temperature, chemical composition, and cost of the required water supply. Condenser water may be obtained from city mains, open ponds, rivers, lakes, wells, or recirculated water. In cases where the water temperature is high due to exposure to the sun or where it is desired to use the water over again, expediency requires artificial methods of water cooling for condenser use. Atmospheric cooling is usually obtained by one of the following methods: *cooling ponds, spray cooling ponds,* and *cooling towers.*

Cooling Ponds

The cooling pond is usually resorted to when the location of the plant is such that a natural body of water is available nearby. Here, the water is cooled by being forced into surface contact with the air, and the amount of cooling will depend on the relative temperature and humidity of the air. Any available pond or lake may be used as a means of cooling recirculated water without the use of sprays, provided it is large enough for the purpose. The heated water should be delivered as far from the suction connections as possible. Since the cooled water tends to sink due to its greater density, the best results will be achieved when the suction connection is applied to the bottom of the pond.

Spray Cooling Ponds

The spray cooling pond differs from the ordinary cooling pond in that means are introduced to accelerate the cooling effect. This is achieved essentially by properly designed sprays or nozzles, which break the water into fine drops, but not into a mist, since, to be effective, the drops must fall back into the pond and not be carried away. The number of nozzles required depends on the size of the pond, i.e., the water requirement of the plant. In average practice, the nozzles are located from 3 to 8 feet above the water level and in horizontal rows from 10 to 16 inches apart. Such an arrangement is shown in Fig. 17-5.

The pressure at the inlet to the spray system may be from 5 to 12 psi. Here, as in the cooling pond, the final temperature of the water depends on air temperature and humidity. In most cases, however, reduction of water temperature from 10 to 15° may be obtained with a single spraying.

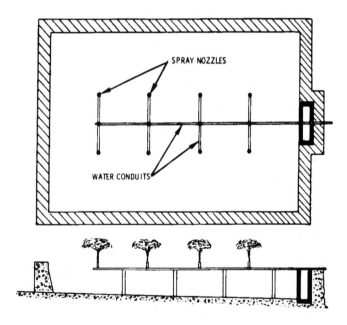

Fig. 17-5. Spray-nozzle arrangement in a typical spray cooling pond.

Cooling Towers

Water-cooling towers are perhaps the most common method of obtaining water cooling and have been in general use for a considerable time. Water-cooling towers are divided into two classes, depending on their design: *open* and *closed* towers.

Open Towers—The open tower depends for its functioning on a number of platforms or drops for breaking up, or atomizing, the water, or on filming water over large surfaces exposed to the dry atmosphere. The closed tower, also called a *fan tower*, has much less exposed surfaces and depends for its functioning on the circulation of a large volume of dry air through falling water for its cooling effect.

The open types consist generally of drip pans installed at regular intervals, with space between to allow free access of the air; and each pan has holes for the equal distribution of the water. Shavings, boards, mineral wool, tile, and even slate have been used with some success in the pans in this type of tower, as well as in the closed type. The question is largely one of installation expense and consideration of the deterioration factor. Almost any material or device that provides satisfactory distribution and separation of the water along with adequate retardation is satisfactory for the purpose.

Such a tower (Fig. 17-6) is open at the sides and depends upon the natural air circulation of the atmosphere. Its efficiency varies with the velocity of the wind, humidity, and temperature as well as the design of the tower for separation and retardation. The deterioration factor in this type is quite large since the destructive effect of air and water under these conditions is considerable.

Dimensions of Cooling Towers—A cooling tower for a 50-ton plant is approximately 12 × 6 feet at the bottom and 24 feet high. Sufficient air could be supplied by two 8-foot blowers, which would consume about 8 hp.

Efficiency of Cooling Towers—The efficiency of a cooling tower is the number of degrees to which it can cool the water. Very seldom can the water be cooled much below 75°F in the summer. The higher the temperature of the initial water, however, the more efficient the tower will be in its operation. Condenser water from steam condensers is furnished at a temperature ranging from 165 down to 80°F, and the operation of the towers under these circumstances is very efficient. Refrigerating plants have a range of temperature depending simply on the pressure maintained in the condenser and seldom rising above

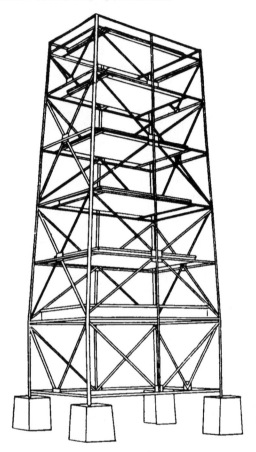

Fig. 17-6. Structural arrangement of an open cooling tower. In a water-cooling system of this type, the water is discharged into the top pan and, after passing through all the various pans, returns to the condenser.

120°F for the initial temperature in the cooling tower. The evaporation of the water in the cooling tower must be resupplied, of course, which represents a certain loss. In refrigerating plants, the loss of the water is from 5 to 15 percent for each circulation.

Closed Tower—The closed tower is practically identical with the open tower, with one important modification: the walls of the tower are enclosed and air is supplied at the bottom and forced upward throughout the tower by means of a fan assembly, such as the one in Fig. 17-7. The air supply under these circumstances can be varied by mechanical means and the resulting cool effect made practically independent of temperature and humidity variations of the outside atmosphere. Efficient operation of the tower is further independent of the existence of winds.

Fig. 17-7. Roof-top packaged cooling unit.

When considering the installation of a cooling tower, there may be no need for fans, except in cases where a larger floorspace is not available. Fans require considerable installation expense and a continuous expense for running and repairs. In the majority of plants, there is always plenty of space either in the yard, on top of a roof, or over the condensers for a sufficiently large cooling tower without having to install a fan. If such a tower is properly constructed, it will give the same results as towers with fans and can be installed at a low cost.

The amount of water used is an important item unless there is an unlimited supply. If water is purchased, it should be passed over a cooling tower and used again and again. The cooling tower will pay for itself in a short time in the amount of water saved. In installations where reliability is a matter of prime importance and cost of installation a matter of minor significance, the closed tower is invariably chosen. The majority of large power plants use closed towers.

Cooling-Tower Control

Temperature controls for refrigerating service are designed to maintain adequate head pressure with evaporative condensers and cooling towers. Low refrigerant head pressure, caused by abnormally low cooling-water temperature, reduces the capacity of the refrigeration system.

Two systems of control for mechanical and atmospheric draft towers and evaporative condensers are shown in Figs. 17-8 and 17-9. As noted in Fig. 17-8, the control opens the contacts on a temperature drop. These contacts are wired in series with the fan motor (or to the pilot of a fan-motor controller), stopping the fan when the cooling-water temperature falls to a predetermined minimum value corresponding to the minimum head pressure

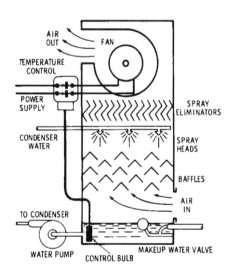

Penn Controls, Inc.

Fig. 17-8. Wiring diagram of a control for a cooling tower with force-draft cooling.

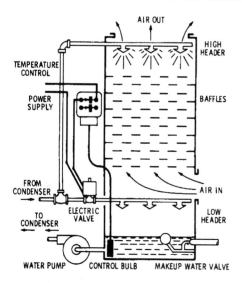

Fig. 17-9. Wiring diagram of a control for a cooling tower with atmospheric-draft cooling.

for proper operation. In the control system shown in Fig. 17-9, the contacts close on a temperature drop and are wired in series with a normally closed motorized or solenoid valve, opening the valve when low cooling temperature occurs. The cooling water then flows through a low header in the atmospheric tower, reducing its cooling effect.

COOLING-SYSTEM COMPONENTS

Head-Pressure Control

In applying refrigerant condensers, it is well to take advantage of a lower-than-design condensing temperature whenever this is possible to achieve the effects of more refrigeration capacity from the system, lower brake horsepower requirements per ton of refrigeration, and lower compressor discharge temperatures. At the same time, it is often necessary to take precautions that the condensing pressure does not drop down too low during periods of low temperature of the water or outside air. This is particularly true on jobs that have higher internal or process loads throughout the year and also on systems having compressor capacity control. The principal problem arising from abnormally low condensing pressures is the reduction of the pressure difference available across the expansion valve to the point where it is unable to feed sufficient refrigerant to the evaporator to handler the cooling load.

Water-Cooled Condensers

With water-cooled condensers, head-pressure controls are used for the dual purpose of holding the condensing pressure up and conserving water.

Water-Regulating Valves

The control commonly used with water-cooled condensers is the condenser water-regulating valve. This valve is usually located in the supply water line to the condenser. It is a throttling-type valve and responds to a difference between condensing pressure and a predetermined shutoff pressure. The higher this difference, the more the valve opens up, and vice versa. A pilot connection to the top of the condenser transmits condensing pressure to the valve, and the valve assumes an opening position according to this pressure. The shutoff pressure of the valve must be set slightly higher than the highest refrigerant pressure expected in the condenser when the system is not in operation to make sure the valve will close and not pass water during *off* cycles. This pressure will be slightly higher than the saturation pressure of the refrigerant at the highest ambient temperature expected around the condenser.

Condenser water-regulating valves are usually sized to pass the design quantity of water about 25 to 30 lb difference between design condensing pressure and valve-shutoff pressure. In cooling-tower application, a simple bypass can also be used to maintain condensing temperature, employing a manual valve or an automatic valve responsive to head-pressure change. The valve can be either the two- or three-way type.

Evaporative Condensers

When installed outdoors, the unit should be located so that there is a free flow of air around it. A minimum clearance of twice the width of the unit should be provided at the air inlet and air outlet. Care should be taken that the flow of the discharge air is not deflected back to the inlet. Short-circuiting in this fashion can materially reduce the cooling capacity. Special consideration should be given to short-circuiting on multiple installations. It is very desirable to position the unit so that the prevailing summer wind will blow into the inlet. It is also desirable to take precautions against a strong wind blowing into the outlet, which can reduce the capacity and might overload the fan motor. A windbreak erected about 6 feet from the air discharge is often useful in preventing this. In some outdoor locations, it may be desirable to place a coarse screen over the air inlet to exclude large objects, such as birds, paper, and other debris.

If the unit is installed indoors, careful consideration must be given to supplying and removing an adequate amount of air from the space. If the discharge air is allowed to recirculate, the wet-bulb temperature will rise and capacity will be reduced. It is usually possible to omit a duct to the air inlet. The unit can draw air from the room, but provision should be made to supply an adequate amount of replacement fresh air by special blower units when required.

The piping to and from the unit should be installed in accordance with good practice, using pipe sizes at least as large as the fittings on the unit. It is desirable to use vibration eliminators on all lines to and from the units. The units should be installed so that the top cover can be lifted for maintenance. The makeup water line should be connected to the float valve. If it is desired to use the makeup line to fill the system, it may be desirable to make the pipe size larger than the float-valve-connection size.

In some locations, it is permissible to let the overflow drain on the roof; if this is not desired, the overflow should be piped to a sewer. A bleed, or blowdown, connection should be incorporated in the piping. This should be installed in the discharge pipe from the pump at a high enough level so that the bleed flows only when the pump is operating. If the unit is installed outdoors and must operate in winter, the installation should be made using an indoor storage tank in a heated space.

Air-Cooled Condensers

With air-cooled condensers (Fig. 17-10), the pressure-stabilizing method is often employed as a means of obtaining pressure control. This is achieved by backing liquid refrigerant up in the coil to reduce the effective condensing surface. The basic principle of the pressure stabilizer is simple and unique: it is a heat-transfer device that transfers heat from the hot-gas discharge of the compressor to the subcooled liquid leaving the condenser.

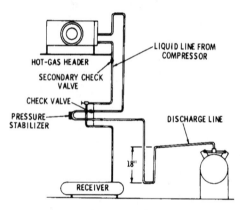

Fig. 17-10. Air-cooled condenser arrangement with a hot-gas header and pressure stabilizer.

This heat exchange is controlled by the regulating valve installed between the condenser and the receiver. This valve is set at the desired operating pressure and throttles from the open to the closed position as the head pressure drops. The throttling action of the valve forces liquid to flow through the heat-exchanger section of the pressure stabilizer. The heat picked up by the liquid raises the receiver pressure and prevents further flow of liquid from the condenser, which causes flooding of the condenser and, by reducing the effective surface, maintains satisfactory discharge and receiver pressures. This ensures a solid column of liquid at the expansion valve at a pressure which guarantees satisfactory operation of the system.

The pressure stabilizer is provided with a spring-loaded check valve in the heat-transfer liquid section which remains closed during warm-weather operation to insure against liquid refrigerant reheating.

Driers

Moisture in a refrigeration system is one of the main causes of trouble, particularly to the compressor, resulting in the formation of acids that may attack the motor insulation, valves, and valve plate. Some driers are most effective when located in the refrigerated space, although driers are now on the market that manufacturers claim are not affected by temperature. The decision of whether or not to use a permanent drier on the higher temperature systems is one of judgment. In the case of package-type air conditioners and water chillers, it may not be necessary if proper dehydration techniques have been followed at the factory, no refrigerant connections are broken in the field, and any refrigerant added is at a sufficiently low moisture content. If there is any doubt that the system is initially free of moisture or that the system can be maintained dry in operation, then a drier is indicated to protect the compressor and prevent freezing at the liquid-feed device.

In the case of hermetic compressors, there is an additional reason for keeping moisture out of the system: the motor windings are exposed to the refrigerant gas, and excessive moisture can cause a breakdown of the motor-winding insulation, possibly resulting in both burnout of the motor and distribution of the products of decomposition throughout the refrigeration system. A full-flow drier is usually recommended in hermetic compressor systems to keep the system dry and prevent the products of decomposition from getting into the evaporator in the event of a motor burnout.

Side-outlet driers are preferred since the drying element can be replaced without breaking any refrigerant connections. The drier is best located in a bypass of the liquid line near the liquid receiver. The drier may be mounted horizontally, as shown in Fig.

17-11, but should never be mounted vertically with the flange on top since any loose material would then fall into the line when the drying element is removed.

A three-valve bypass should be used, as shown, to provide a means for isolating the drier for servicing and when employing open-type compressors to allow only partial refrigerant flow through the drier to reduce pressure drop. In systems using hermetic compressors, it is preferable to have all refrigerant going through the drier at all times for the reasons outlined in the preceding paragraphs.

Reliable moisture indicators are now on the market for installation in refrigerant liquid lines. These devices will indicate whether or not a system has excessive moisture in it. They are valuable additions to the system as an indicator of when the drier cartridge should be replaced.

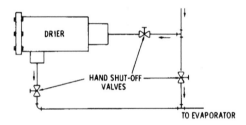

Fig. 17-11. Drier with piping connections.

Fittings and Valves

One of the most important considerations when installing a refrigerating plant is to make sure that the piping system is absolutely tight since even the slightest leakage is dangerous. It is important to remember that the refrigeration cycle is a closed one with the compressor in series in order to obtain continuity. Tightness is important, especially with ammonia, although other refrigerants are hardly less demanding in this respect. Because pressures are almost always above the pressure of the atmosphere, the possibility of air leaking into the system is slight; but the piping and fittings must be of a material to comply with the various codes in existence, and the joints must be designed for the particular application and system performance.

Referring back to Fig. 17-2, you will seer more valves shown than you will find on an ordinary commercial system. Valves D and E on the compressor are compound valves, as explained earlier in Figs. 6-1 and 6-2. These two valves are always present, and valve B at the discharge of the receiver tank will always be found. The other valves shown in Fig. 17-2 are convenient but not always installed.

Should you wish to pump the system down to change an expansion valve, you would close valve B, place a compound gage on valve E by backing the valve stem all of the way out, and place a pressure gage on valve D by backing the valve stem all of the way out. Then crack both valves by backing the stem out slightly. Purge the air out by loosening the nut where the line attaches to each gage. Now start the compressor. If the compressor is controlled by a pressure-actuated switch on the low side, this switch will have to be blocked closed, or it will open when the low side reaches the preset point. Stay in attendance with the compressor when doing this. If you hear it scrubbing oil from the crankcase (this will be very noticeable), stop the compressor and, proceeding slowly, turn the compressor on and off by hand. This operation will draw the refrigerant back from the closed valve B on the receiver tank, back through the compressor, and put it in the receiver tank.

Draw a vacuum on the low-side gage, and then shut the machine off. If the vacuum holds, crack the valve on the receiver tank slightly to allow liquid to enter the evaporator, watching the low-pressure gage. When this gage shows about 1 lb of pressure, shut off valve B. You will want to have a very slight pressure on the system, so that when you open the system at any point, such as to change the drier-filter or replace the expansion valve, there will be refrigerant pressure, however slight, to blow out and not suck air into the system. You may now replace the drier or expansion valve. Before connecting the discharge side of either of these, crack valve B, slightly allowing refrigerant to pass through and escape into the air to purge out any air in them. Then attach the discharge side tubing to the component(s) you replaced. Open valve B all of the way, checking for leaks at the points where you were working.

Valves F, F_1, and F_2 are handy to have, but expensive. Thus, if you wish to change the drier, close valves F and F_1, which are ordinarily open, and open valve F_2, which is normally closed. Remove the drier, remembering that the high-side pressure is on it, so crack the connections slowly until the pressure is released. Remove the drier, put in a new drier, connecting the side toward valve F first; then crack valve F slightly to purge the air out, and connect the side of the drier on the valve F_1 side. Close valve F_2, open valves F and F_1, and operate the compressor, checking the pressures at the gages on the compressor, especially the high-side gage for abnormal pressure. The system should be again ready to operate. However, check for leaks when under pressure.

Ordinarily, you will not have valves C and A on the condenser. These two valves are what you might call luxuries—they are nice, but expensive. If you needed to replace the flapper valves in the compressor head or the shaft seal, you would proceed as explained for the drier replacement. If the flapper valves in the compressor head are to be replaced, remove the head slowly as there will be more pressure here. As you tighten the head again, crack valve B to purge the air out of the head. Close valve B, and tighten the head-bolts. Put a little pressure on the head by cracking valve B and checking for leaks. You are again ready to operate the compressor.

Without valves C and A on the condenser, if the condenser leaks, you will lose the refrigerant anyway, and so you should not be too concerned. Replace the condenser, and pump down the system with a vacuum pump, as explained previously, drying out the system, then recharge. A condenser problem is very seldom encountered.

Heat Exchangers

Although sometimes a subject of controversy in high-temperature application, it is generally agreed that in medium-and low-temperature refrigeration systems, heat exchangers (when properly applied) perform the following functions of utmost importance to overall system performance:

1. Subcooling the liquid refrigerant entering the thermal expansion valve reduces the flash gas load at the evaporator

inlet. It likewise increases the heat-content difference of the refrigerant during its evaporating phase, which produces more useful work in the evaporator.

2. In the process of subcooling, the heat extracted from the liquid refrigerant is transferred to the suction gas, thereby ensuring a dry suction return to the compressor at an entering superheat level which will produce the best possible volumetric efficiencies for the refrigerant used.

3. The increase in suction-line temperature will likewise reduce the possibility of sweating.

4. The use of a heat exchanger will permit more open adjustment of the thermal expansion valve without the risk of serious floodback of liquid to the compressor under light or variable load conditions. At the same time it ensures maximum utilization of the evaporator surface.

The advantages of a heat exchanger may be cancelled out if its use increases the suction-line pressure drop by an excessive amount. On a theoretical heat exchanger without pressure drop, designed to superheat the suction gas to 65°F from a normal 20°F suction temperature, it is entirely possible to obtain as high as a 16 percent increase in the overall efficiency of the machine, as compared to a system without a heat exchanger. There are many factors involved, however, to obtain such a figure, and, for practical purposes, it may readily be said that the overall efficiency of a system may be increased 8 to 10 percent by the proper adaptation of a heat exchanger.

In all low-temperature applications, it is more important to correctly size and properly apply heat exchangers. Their selection must be based on accurate performance ratings checked out against the calculated design loads involved for each evaporator or for the entire system. Likewise, care must be taken to ensure that both liquid and suction connections are properly sized in order to reduce entrance and exit losses to a minimum.

Balancing the System

The application of air cooling to the low sides in refrigeration work is basically an air-conditioning problem. It so happens that the refrigeration industry has certain standards that make the adaptation fairly easy, as compared to the adaptation of a coil to air conditioning work. In the proper application of a coil, it is most important to make sure that there is a balance between the condensing unit and the low side. The temperature difference between the air to be passed over the coil and the refrigerant temperature in the coil is the point that determines the conditions which will be maintained in a given cooler.

For the general purpose of refrigerators involving meats, vegetables, and dairy products, it is common procedure to balance the low side to the condensing unit at a 15°F temperature difference; that is, they are balanced to maintain a temperature difference of 15°F between the refrigerant in the coil and the air temperature. It has been learned by experience that, if this is done, one may expect to maintain 80 to 85 percent relative humidity in a cooler, which is a good range. A coil that is selected for a wide temperature difference will therefore maintain a lower relative humidity in service, whereas one that is selected for too close a temperature difference will produce relative humidities that are higher than required for practical operation, and surface sliming may result on stored meat products during winter periods when loads are reduced. It has been determined over years of practice that coils should be selected to balance between 12 and 15°F temperature difference for general-purpose refrigerated storage coolers.

On straight vegetable coolers, where higher humidities are desired, the coil should be selected to balance the compressor at a 10°F temperature difference because this will produce an average relative humidity of 90 percent within the refrigerated space. The same recommendation applies to florists' display boxes, and, in both cases, the maintenance of a high relative humidity in long-term storage is beneficial, whereas some exception with reference to meat products was noted previously. On low-temperature units, if one stops to consider that the amount of dehumidification is in proportion to the temperature difference, it is obvious that the closer the temperature difference, the less frost accumulation. It is recommended that coils for low-temperature work be selected to balance the condensing unit at a 10° temperature difference or less.

REFRIGERATION-SYSTEM OPERATION

It is essential to thoroughly clean the compressor before placing it in service, and the best time for cleaning is just prior to the initial start of the system. The crankcase cover should be removed and the inside of the crankcase thoroughly cleaned out. Only clean lint-free rags should be used for this purpose. If waste is used, lint and threads will stick to the casting surfaces.

Before starting the compressor, make certain that the crankcase is properly filled with a good grade of machine oil as recommended by the manufacturer of the machine. The initial oil charge may be filled through the crankcase opening to a sufficiently high level for satisfactory operation. Although compressors are usually given a factory test before shipment, a no-load run of from 10 to 20 hours (depending on the size of the compressor) is recommended. Remove the head covers and discharge valves, and pour oil on top of the piston as close to the cylinder walls as possible to ensure initial lubrication to the cylinder walls. Rotate the compressor by hand to make sure everything is free; then start the motor and compressor and run the unit at idle to ensure that every part is properly lubricated before the unit goes into actual service. Before starting the compressor, it is preferable to oil prime the gear pump by removing the small pipe plug on top of the strainer and pouring in oil. This oil will run back into the pump and seal the gears.

When satisfied that all parts are properly lubricated and that there is no undue heating, the unit should be stopped and the discharge-valve and cylinder-head covers replaced, making sure that the cylinder-head cover gaskets are in place and that they are properly tightened. Before starting the plant, however, it is important to test the system for leakage at a pressure equal to the maximum discharge pressure at which the system is expected to operate. This will not apply to all commercial refrigeration because many compressor units do not have oil pumps.

Starting the Plant

Some refrigeration units come with a refrigerant charge in the receiver tank, and others do not. The system may come with a slight charge of an inert gas, such as nitrogen, to ward off moisture-laden air. Irrespective, the complete system, i.e., new lines, evaporator coils, etc., must be pumped down with a vacuum pump to get rid of all air and moisture. As a general rule, the compressor will not be charged with refrigerant in the receiver tank, and so we will proceed from that angle:

1. Open all valves, installing a compound-gage system on the high and low sides. Connect the vacuum pump, as described previously, and pump the system down until no more bubbles are coming out through the oil container, into which the discharge of the vacuum pump was inserted. Continue pumping after this for about 1 hour.

2. Back the suction valve all the way out, shutting off the opening to the compressor and also to the discharge line of the compressor (suction line). This closes the connection where the vacuum pump was connected.

3. Shut off the vacuum pump and disconnect it. Connect a cylinder of refrigerant to where you disconnected the vacuum pump from the compressor, and crack the refrigerant cylinder valve slightly to purge the line from it to the compressor.

4. You are now ready to charge the system.

5. With the refrigerant cylinder upright, start the compressor, and screw the compressor valve in slightly, then open the refrigerant cylinder valve. Control the gas going in by the valve on the compressor, being sure that the discharge valve on the compressor is screwed in just slightly so that the high-side pressure may be checked.

6. The vacuum pump should have indicated any leaks, but do not rely on this. When you have a few pounds of pressure on the entire system, stop this operation, backing out the suction valve to shut off the gas to the system from the refrigerant tank.

7. Test for leaks, as explained previously. If no leaks are found, continue charging the system by cracking the suction valve, turning it in a turn or two.

8. Do not turn the refrigerant tank upside down because you will scrub oil from the compressor. Leave this tank upright so that you receive only gas into the system.

9. There are two methods used to tell when the proper charge is reached: one is that the receiver tank has a sight glass on its end to tell you the liquid level in the receiver; the other is a sight glass in the liquid line from the receiver tank.

10. If there is a sight glass approximately one-half the way up on the end of the receiver tank, shut off the refrigerant when this point is reached.

11. If the sight glass in the liquid is used, you will see bubbles in the sight glass until pure liquid refrigerant is flowing, at which time you will have a clear sight glass and know that the system is receiving liquid refrigerant into the evaporator.

12. At this point stop adding gas and let the system operate for a considerable time to be sure that you have enough refrigerant.

13. If, after a time, bubbles appear in the sight glass, you know that you do not have quite enough refrigerant in the system.

14. All this time watch the high-side gage. If the pressure is abnormally high, you have not removed all the air from the system; if this is the case, shut the system off, and in a little while bleed some refrigerant from the high-side valve. If the system was properly cleared of air by the vacuum pump, this problem will not occur.

15. If all systems seem to be operating properly, watch the evaporator for freeze-over. It should soon freeze over completely.

16. If the frost goes back beyond the bulb attachment to the suction line, adjust the expansion valve to bring the frost back to the bulb. This is a slow process, and you must not hurry it.

Stopping the Plant

If the plant is to be out of operation for some period of time, it is well to pump the system down, which is accomplished as follows:

1. Shut valve B at the receiver tank in Fig. 17-2.
2. Install a low-pressure gage that reads inches of mercury on the low side of the compressor.
3. Start the compressor. You may have to block the low-

pressure cutoff switch to keep the compressor running. Pump a slight vacuum, and then stop the compressor for a little while. If there is still a vacuum in the system, crack the valve B on the receiver tank (Fig. 17-2) slightly until 1 or 2 lb of refrigerant is in the system, then shut off all valves. Be sure to tag the compressor plainly, so that someone else will know what has been done.

The pressures on the low-side gage for turn-on and shutoff will materially aid in setting the temperatures desired because most commercial refrigeration systems are controlled by a low-side pressure switch.

COMPRESSOR TROUBLESHOOTING GUIDE

Symptoms and Possible Cause   *Suggested Remedy*

High Head Pressure at Discharge

(a) Air or noncondensable gas in system

(b) Inlet water warm
(c) Insufficient water

(d) Consenser clogged
(e) Too much liquid in condenser-receiver

(a) Purge from condenser. Increase water flow. Check pump and flow lines.

(b) Increase water flow.
(c) Clogged water pump strainer or pipe line; clean out.

(d) Clean condenser tubes.
(e) Remove excess into service drum.

Low Head Pressure at Discharge

(a) Charge of refrigerant in system too low
(b) Discharge valve leaks slightly

(a) Add refrigerant from service drum.
(b) Repair or replace.

Symptoms and Possible Cause   *Suggested Remedy*

Low Head Pressure at Discharge (cont.)

(c) Too much cold water

(c) Throttle outlet water valve to condenser.

(d) Faulty thermal expansion valve

(d) Adjust or replace expansion valve.

High Suction Pressure

(a) Overfeeding of expansion valves

(a) Adjust valve to higher superheat.

(b) Leaking compressor suction valves

(b) Remover compressor head and examine valves.

Low Suction Pressure

(a) Insufficient refrigerant in system

(a) Add from service drum.

(b) Restriction in liquid lines

(b) Remover and clean strainer.

(c) Stopped suction strainer

(c) Remove and clean carefully.

(d) Underfeeding of expansion valves

(d) Adjust valve to decrease superheat.

Compressor Stops from High-Pressure Cutout

(a) High-pressure cutout setting incorrect

(a) Make correct setting.

(b) Overcharge of refrigerant

(b) Remove excess into service drum.

(c) Water failure

(c) Investigate as to why water failure switch did not function.

(d) Pump failure

(d) Check over pump and motor.

Symptoms and Possible Cause   *Suggested Remedy*

Compressor Stops from High-Pressure Cutout (cont.)

(e) Stoppage in water line

(e) Check water strainer and valve settings.

Compressor Short-Cycles on Low-Pressure Cutout

(a) Coils covered with excess frost

(a) Defrost coils.

(b) Clogged screens in liquid lines

(b) Clean screens.

(c) Improper setting of back-pressure valves

(c) Regulate properly.

(d) Also see causes under "Low Suction Pressure."

Compressor Does Not Start

(a) Overload has tripped

(a) Push reset.

(b) High-pressure cutout open

(b) Push reset.

(c) Low charge of refrigerant

(c) Repair leak. Recharge system.

(d) Low-pressure cutout open

(d) See "Low Suction Pressure."

(e) Electric valve closed

(e) Check thermostat, which may be in the "off" position.

(f) Water-failure switch open

(f) Check water supply.

Compressor Runs Continuously

(a) Shortage of refrigerant

(a) Charge additional refrigerrant from service drum.

Symptoms and Possible Cause   *Suggested Remedy*

**Compressor Runs
Continuously (cont.)**

(b) Compressor valve   (b) Check for leaks.
    leakage
(c) Cylinders overheated   (c) Repair or replace valves.

SUMMARY

The principles of commercial refrigeration are based on circulation of refrigerants through the various components comprising the refrigeration plant. To understand how the system operates, it is necessary to study its essential parts.

REVIEW QUESTIONS

1. Name three principal methods used for producing mechanical refrigeration.
2. What is the effect of pressure on refrigerants?
3. Name the difference between a direct and indirect expansion system.
4. What are the essential parts of a refrigeration plant?
5. Name the components of the high- and low-pressure side of a refrigeration system.
6. What is the function of the expansion valve, and where is it located in the system?
7. What precautions should be observed prior to installation of condensers, fittings, and accessories?
8. What is the function of driers in the refrigeration system?
9. What is the function of heat exchangers?
10. What are the adopted standards in balancing a refrigeration system?
11. What precautions should be observed in the operation of a refrigeration system?

CHAPTER 18

Brine Systems

Refrigeration is distributed by several methods. In direct expansion systems, the volatile refrigerant is allowed to expand and evaporate in a pipe or coil placed in the space to be cooled where the refrigerant absorbs its latent heat of evaporation from the material to be cooled. This method of refrigeration is used in small cold-storage rooms, constant-temperature rooms, freezer rooms, and other locations where possible losses due to the leakage of refrigerant would be low.

INDIRECT REFRIGERATION SYSTEM

In the indirect system, some refrigeration medium, such as *brine*, is cooled down by direct expansion of the refrigerant and is then pumped through the material or space to be cooled, where it absorbs its sensible heat. Brine systems are used to advantage in large installations, where the danger due to leakage of the large amount of refrigerant present is important, and in locations with fluctuating temperatures. For small refrigerated rooms or spaces where it is desired to operate the refrigerating machine only part of each day, the brine coils are supplemented by holdover or congealing tanks.

Holdover Tanks

A *holdover tank* is a steel tank containing strong brine in which direct expansion coils are immersed. During the period of operation of the refrigeration machine, the brine is cooled down and is capable of absorbing heat during the shutdown period, the amount depending upon the quantity of brine, its specific heat, and the temperature head.

Congealing Tanks

Congealing tanks serve the same general purpose as holdover tanks but operate on a different principle. Instead of a strong - brine solution, they contain a comparatively weak brine solution, which freezes, or congeals, to a slushy mass of crystals during the period of operation. This mass of congealed brine, in addition to its sensible heat, is capable of absorbing heat equivalent to its latent heat of fusion. For the same refrigerating effect, congealing tanks can be made much smaller than holdover tanks for the reason that advantage is taken of the latent heat fusion of the brine. Thus, there is a considerable saving in construction and maintenance cost for this type of system.

BRINE

In addition to acting as simple carrying medium for refrigeration, a satisfactory brine should possess certain other properties desirable from an operating standpoint. Briefly, it should have a low enough freezing point to remain liquid at the lowest temperature encountered at any point in the refrigeration system. It

498

should be as noncorrosive as possible and should not be subject to precipitation when contaminated with the refrigerant through accidental leakage. The specific heat of the brine should be as high as possible in order to avoid the necessity of handling excessively large volumes.

Of the two materials commonly used for the making of brine, sodium chloride and calcium chloride, the latter is much to be preferred because it fulfills all the foregoing requirements in that it is less corrosive on equipment and is capable of furnishing brines having a much lower freezing point than sodium chloride. For practical purposes, the lowest freezing point possible with sodium (common salt) brines is 0°F, while with the calcium brines, temperature may safely run as low as –58°F.

Calcium Chloride

The content of actual calcium chloride ($CaCl_2$) in commercial solid form is 73 to 75%, and in flake form it is 77 to 80%, the balance being principally water of crystallization with a little salt (NaCl). Magnesium chloride is absent, thus precluding the danger of corrosion from mixed chlorides. About 0.1% of lime [$Ca(OH)_2$] is contained in the product; the alkalinity that this contributes to the brine is desirable for protection of iron and steel apparatus against corrosion.

Calcium chloride dissolves readily in water; in fact, it has such an attraction for water that when exposed to air under ordinary conditions it absorbs atmospheric moisture until the calcium chloride is dissolved. Its solutions are relatively stable and, when made up to suitable strength, remain liquid at temperatures far below the lowest temperatures required in commercial refrigeration practice. The properties and freezing points of calcium chloride brines of various strengths are given in Table 18-1. The indicated freezing point of any brine is the temperature at which crystals begin to form in it; it does not mean that the brine solidifies completely at that temperature.

In the refrigeration industry, the specific-gravity hydrometer is generally used to determine the strength of brine. In Table 18-1 corresponding salimeter and Baume hydrometer readings (in degrees) are given, in addition to specific gravities.

499

Table 18-1. Properties of Calcium Chloride Brine

Density of Brine at 60°F		% Anhydrous Calcium Chloride	Freezing Point °F	Weight of Brine at 60°F		Pounds Dow 73-75% Solid Calcium Chloride		Pounds Dow 77-80% Flake Calcium Chloride	
Specific gravity	Baume scale			Pounds gallon	Pounds cu ft	per gallon of solution at 60°F	per cu ft of solution at 60°F	per gallon of solution at 60°F	per cu ft of solution at 60°F
1.00	1.0	0.0	32.0	8.34	62.4	0.00	0.0	0.00	0.0
1.02	2.8	2.3	30.2	8.50	63.7	0.26	2.0	0.25	1.8
1.04	5.6	4.7	28.0	8.67	64.9	0.55	4.1	0.51	3.8
1.06	8.2	7.0	25.9	8.84	66.2	0.84	6.3	0.78	5.9
1.08	10.7	9.2	23.4	9.00	67.4	1.13	8.4	1.06	7.9
1.10	13.2	11.4	20.3	9.17	68.7	1.42	10.6	1.33	9.9
1.12	15.5	13.5	16.5	9.34	69.9	1.70	12.8	1.59	11.9
1.14	17.8	15.6	12.0	9.50	71.2	2.00	15.0	1.87	14.0
1.16	20.0	17.6	7.0	9.67	72.4	2.30	17.2	2.15	16.1
1.18	22.1	19.5	1.4	9.83	73.7	2.60	19.4	2.43	18.2
1.20	24.2	21.5	− 5.8	10.01	74.9	2.90	21.7	2.72	20.3
1.22	26.2	23.3	−13.2	10.17	76.2	3.20	24.0	3.00	22.5
1.24	28.1	25.1	−21.3	10.33	77.4	3.51	26.2	3.29	24.6
1.26	29.9	26.9	−30.8	10.50	78.7	3.83	28.6	3.58	26.8
1.28	31.7	28.7	−44.1	10.66	79.9	4.14	31.0	3.88	29.0
1.29	32.6	29.6	−59.8	10.75	80.5	4.31	32.2	4.03	30.2
1.30	33.5	30.5	−41.8	10.84	81.2	4.47	33.4	4.19	31.3
1.32	35.2	32.2	−16.6	11.01	82.37	4.80	35.9	4.50	33.6
1.34	36.8	34.0	4.1	11.18	83.62	5.14	38.4	4.81	36.0
1.36	38.4	35.7	21.6	11.34	84.86	5.48	41.0	5.14	38.4
1.38	39.9	37.4	37.0	11.51	86.11	5.83	43.6	5.46	40.8
1.40	41.4	39.1	50.4	11.68	87.36	6.18	46.2	5.79	43.3
1.42	42.9	40.8	60.6	11.84	88.61	6.36	47.6	5.96	44.6

Initial Charges

The solid form was formerly the only type of calcium chloride on the market. It was delivered in steel drums, and, in order to put it in convenient size for handling, it was necessary to break up the contents of the drum by pounding it about its circumference. After this, the end of the drum was cut off with an axe and the calcium chloride was dumped into the brine tank, where it dissolved in the water. A more convenient form of calcium chloride is now available. This is flake calcium chloride, which, as its name indicates, is a loose flake material that does away with all the labor involved in preparing the solid type for solution. By the use of flakes, it is necessary only to pour the chloride from the drum or the convenient and easily handled 100-lb moisture-proof bag into the brine tank, which previously has been partially filled with water. If the calcium chloride is placed directly in front of one of the brine agitators, it will soon go into solution.

Practically any water may be used to make up brine, but for new installations, excessively hard water should be avoided if possible. In making up brine for new installations, it will be noticed that when calcium chloride is dissolved in water, considerable heat is developed. Warm alkaline brines readily attack zinc or galvanized coatings, and so the best procedure is to dissolve the calcium chloride and cool the brine in a tank that contains no zinc-coated metal. In case no extra tank is available, the ice cans should be removed from the freezing tank before making up the brine.

The quantity of calcium chloride needed to charge a refrigeration system at a given strength is determined by the volume of brine needed. The capacity of a system is the sum of the volume of the brine tank plus the volume of the brine coils, coolers, auxiliary tanks, etc., less the brine displacement of ice cans, expansion coils, etc. Having determined the net capacity of the system in cubic feet, the amount of calcium chloride needed may be calculated by consulting Table 18-1, which shows the strength of brines and their corresponding freezing points. The brine should be of such a strength that it will not freeze at the lowest temperature in the system. The proper strength of calcium chloride brine should have a freezing point of from 10 to 15°F below the minimum brine temperature prevailing at the point of ammonia expansion.

To prepare the brine, find in the table the pounds of calcium chloride per cubic foot of brine corresponding to that freezing point and multiply by the net capacity of the system in cubic feet. The result is the pounds of calcium chloride required.

Foaming of Brine

The presence of foam on the surface of a refrigeration brine is often troublesome. It is due to air being given off by the brine or to a recent formation of a fine colloidal precipitate. Weak brines are capable of holding more air in solution than are strong brines; therefore, it is advisable to keep the brine up to strength at all times. When a weak brine is strengthened by the addition of calcium chloride, the air is forced out of solution in the form of tiny bubbles. The presence of insoluble corrosion particles tends to stabilize the foam, or froth, thus formed.

When calcium chloride has not recently been added to a brine, and froth, or foam, appears, it is an indication that corrosion is taking place or that an excess of air is being introduced into the brine. Steps should be taken at once to ascertain the cause and to prevent its continuance. The user of an electric light rather that an open flame for investigating the foaming condition of a brine is recommended because hydrogen gas might be present. The hydrogen is a by-product of electrolytic corrosion.

BRINE-SYSTEM CORROSION

All metals used in construction of ice plants are subject to *corrosion*, which is most commonly an oxidation of the metal surface caused by oxygen in the air, in absorbed air found in water, or in water solutions. In the case of calcium chloride brine, the dissolved oxygen should be kept to a minimum by eliminating, as much as possible, contact of the brine with air. Agitation and rapid circulation at such areas of contact should be avoided. Care should be taken that the brine lines discharge under the surface of the liquid at all times since any dropping or splashing of brine in the air tends to increase the amount of oxygen dissolved by the liquid. Naturally, brines used in open systems, such as ice tanks, are likely to develop corrosive tendencies more easily than brines used in closed systems. Nevertheless, care in design and operation of all brine refrigerating systems so as to minimize the amount and time of contact with air will reduce the attendant oxygen absorption and help appreciably in reducing corrosion.

Other factors that affect absorption of oxygen in refrigerating brines are the density and temperature of the solution. Weaker brines dissolve more oxygen; therefore, brines always should be kept at full operating strength. Although higher-temperature brines, in themselves, dissolve slightly less oxygen than colder ones, the reaction between the oxygen and the hydrogen film is speeded up a great deal at warmer temperatures. Hence, in off-seasons of the year or under other conditions when refrigerating capacity is to be reduced, it is desirable to keep the -temperature of brine about 30°F in any inactive units.

The second major cause of corrosion taking place in refrigerating brines is due to atoms of hydrogen combining on the surface of the metal to form molecules of the gas, which escape as bubbles. Such action is largely a function of the acidity or alkalinity of the brine.

Corrosion Retarders

Numerous rust and corrosion preventive compounds are on the market. Many have sodium silicate or "water glass" as a base. These are sometimes useful in clarifying sodium chloride brines, but there seems to be no authoritative data available that show a lessening of corrosive effects resulting from their use. Sodium silicate, when added to calcium chloride brines, causes a precipitate of calcium silicate to form, which only adds to whatever sediment is already present.

The corrosion committee of the A.S.R.E (American Society of Refrigerating Engineers), in their latest report, recommended the user of from 100 to 150 lb of sodium chromate per 1000 cu ft of calcium brine. The commercial grade of sodium dichromate is usually employed for this purpose, but in order to change the dichromate to the normal chromate, it is necessary to add 27 lb of caustic soda for each 100 lb of dichromate of soda added. Both materials should be added in the form of strong solutions in order that they may be mixed quickly into the brine. The degree of protection afforded by this treatment is stated to be from 80 to 90 percent. In other words, the corrosion is calculated to be from 10 to 20 percent of what unavoidably occurs in an untreated brine system.

pH Values

The letters pH denote the negative logarithm of the concentration of the hydrogen ion in grains per liter. Because of the fact that this scale is based on a logarithmic equation, a variation of one unit indicates a tenfold variation in the actual acidity or alkalinity of the solution. A value of 7 on the pH scale is the neutral point; higher values indicate alkalinity, and lower values acidity. Thus, for example, a solution having a pH value of 9 is ten times as alkaline as one having a pH value of 8; or, on the acid side, a pH of 5 indicates ten times the acidity shown by a pH of 6.

One method of determining the pH value of a solution is by using a *universal indicator solution*. A few drops of the solution added to a test tube half full of brine will show a color indicating the pH value of the sample. The best reading will be obtained by gently shaking the test tube just enough to spread the color about 1/2 inch from the top of the sample. If the mixture is shaken too much, the color will be diffused and it will be difficult to accurately identify the shade. Table 18-2 lists the pH values indicated by the main division of the various colors.

If a simpler test is desired, a phenolphthalein indicator solution may be used. This solution, containing 5 grains of phenolphthalein crystals per liter of denatured alcohol, may be obtained from any druggist. A few drops added to a test tube partially filled with brine will be colorless if the brine is neutral and will turn white if the brine is acid. If the brine is alkaline, the addition of phenolphthalein indicator will produce a red color—pink if slightly alkaline and deeper red if more highly alkaline (above a pH of 10, no further color change will be noted).

A properly treated brine should be sufficiently alkaline to produce a slight pink color when the phenolphthalein test is made. This test can often be used to advantage in checking results obtained with a universal indicator solution because, due to the green color of chrome-treated brine, pH values between 7.0 and 9.0 are often not readily distinguishable.

Table 18-2. pH Color Table

Condition	Color	pH Value
Acid	Red	4.0 or lower
Acid	Orange	5.5
Acid	Yellow	6.5
Neutral	Greenish yellow	7.0
Alkaline	Green	8.0
Alkaline	Green	9.5
Alkaline	Violet	10.0 or higher

Acidity

Sodium dichromate, when dissolved in water or brine to retard corrosion, makes the solution acid. Before treating with dichromate, the acidity or alkalinity of the brine should be determined by one of the methods described. If the brine is acid or neutral (a pH of 7.0 or lower), sufficient caustic soda should be added to increase the pH value of the brine to 8.0 or to produce an alkaline reaction if the phenolphthalein test is used.

If the untreated brine has the proper pH value, the acidifying effect of the dichromate may be neutralized by adding 27 lb of 76% commercial flake caustic soda for each 100 lb of sodium dichromate used. Caustic soda must be thoroughly dissolved in warm water before it is added to the brine. An alternative method of neutralizing acid brine is to suspend a bag of slaked lime in the freezing tank until enough has dissolved in the brine to bring the pH value to the desired point.

When treating a brine with sodium dichromate, do not add the entire amount at one time; use approximately one-fourth of the desired amount and then test for alkalinity, adding caustic soda if needed before proceeding with the dichromate treatment.

Alkalinity

In order to most effectively retard corrosion, it is desirable to maintain refrigerating brine in a slightly alkaline condition (pH

505

7.5 to 8.0). Alkaline brines are generally less corrosive than neutral or acid brines, although with high alkalinity, the activity may increase. Calcium chloride brines are usually highly alkaline, and, in some cases, the normal dosage of sodium dichromate may not reduce this alkalinity to the most desirable point. Such brines can be brought to a more nearly neutral condition by adding an excess of sodium dichromate or by adding chromic, acetic, or hydrochloric acid. The first course is preferable, as there is less chance damaging the equipment due to carelessness or accident, although acid may be used if desired. The direct use of acid has the advantages of producing a relatively larger pH depression per pound of material added and is less expensive.

The only common cause for a brine becoming alkaline in service is leakage of ammonia where it can be absorbed by the brine. An increase in alkalinity does not prove the presence of ammonia, however, and for this reason a direct test should be made periodically. Naturally, if ammonia is found to be present, the leak should be located and promptly repaired. If there is any considerable amount of ammonia in the brine, arrange a temporary aerating system to eliminate as much of it as possible by heat before using chemicals. The alkalinity may then be reduced to the recommended pH value by the addition of sodium dichromate or either of the acids mentioned.

Testing for Ammonia

In sodium chloride (table salt) brine or in water, traces of ammonia may be detected by the user of Nessler's solution in a strong concentrate of caustic soda, which producers a brown precipitate or brownish-yellow (reddish-brown) coloration, depending on the amount of ammonia present. The reaction is produced when a few drops of the solution are added to a small sample (1 or 2 ounces) of brine. This test cannot be made satisfactorily with calcium chloride brine. When the test is made on sodium chloride brine, no precipitate will be formed if ammonia is not present; reddish-brown precipitate will be formed if ammonia is present. If the test is made on a calcium chloride brine, a white precipitate of calcium hydroxide will be formed if no ammonia is present; if ammonia is present, this precipitate will be marked by a reddish-brown color and will appear very similar to that produced by a salt brine. The reactions are quite definite and the distinction very easy to make.

The procedure depends upon whether or not the original sodium brine has been treated with dichromate to keep down corrosion. Begin by pouring a sample of the brine into one of two test tubes until the tube is about half full. If the brine has been treated as recommended and contains dichromate, the orange or yellow color caused by the dichromate must first be removed by adding a reagent of barium salt, a drop at a time, until a heavy orange precipitate forms. When this solid sinks to the bottom of the tube, the liquid above it will be colorless. Do not add a large excess of the reagent; add only enough to precipitate all the dichromate and obtain a clear brine after the solids settle. Carefully pour off some of the top portion of the treated brine into the second test tube, being careful not to disturb the sediment, which should remain.

PREPARING NEW BRINE

When new brine is to be prepared for use in a freezing tank, great care must be exercised to avoid excessive damage to the galvanizing on the ice cans. This is particularly true when new cans are involved; older cans will usually have a slight coating of the products of corrosion that will protect them to some extent. In any case, the sodium dichromate should be added to the brine with the salt or calcium chloride before the brine comes in contact with the ice cans. Both salt and calcium chloride brines are very active electrically, and their activity increases as the temperature is increased. A brine that would not seriously damage the galvanizing at a temperature of 15°F may be active enough to strip much of the galvanizing from new cans when brought to a temperature of 60 or 70°F. The brine should be mixed outside of the freezing tank, cooled by the addition of ice or by other means to a temperature of 40°F or lower, and then pumped into the freezing tank.

An alternative method is to put the cans in position in the freezing tank, fill the tank with cold water to a height somewhat below the normal brine level, and apply refrigeration until the water temperature is below 40°F. Be careful not to freeze up the coils or cooler. With the agitators running, slowly add the salt, caustic soda, and sodium dichromate. When calcium chloride is used, it must be added very slowly and the temperature of the solution checked frequently because this salt produces relatively large quantities of heat as it goes into solution.

SUMMARY

Because the brine system is of considerable importance in indirect refrigerating applications, we have discussed it extensively. In present-day installations, calcium chloride is used in preference to sodium chloride. Thus, the properties of calcium chloride have been fully tabulated.

The problem of corrosion and treatment methods has been given in considerable detail. The determination of acidity or alkalinity in a brine solution by means of pH measurements is also discussed. Finally, the preparation of new brine and ice cans is given.

REVIEW QUESTIONS

1. Name the various methods of refrigeration distribution.
2. What are the chemical constituents of brine?
3. Name the chemical formula of calcium chloride.
4. What is the purpose of congealing tanks in the brine system?
5. What is the proper strength of calcium chloride brine for a freezing point of from 10 to 15°F?
6. How can the presence of foam on the surface of a refrigeration brine be avoided?
7. What methods are used to lessen brine-system corrosion?
8. Define pH values in determining the actual acidity or alkalinity of a brine solution.
9. What methods are used in testing for ammonia in the brine system?
10. Explain the methods used when preparing new brine to avoid excessive damage to the galvanizing of ice cans.

CHAPTER 19

Ice-Making Systems

The manufacturer of artificial ice will always occupy a large sector of the refrigeration industry. The ice-making industry is presently making ice by several different processes, depending on whether distilled- or raw-water ice is desired. Production of artificial ice takes two different forms, depending on requirements: *can ice* and *plate ice*.

Can Ice

In the can system, the water is placed in galvanized iron cans or molds immersed in a brine tank, which is kept cool by ammonia expansion coils. In this system, unless means are introduced to prevent it, air and other impurities have a tendency to collect in a core in the center of the can. Making ice from distilled water will eliminate this trouble, but, due to its high cost, it is not widely used. At the present time, the tendency is to use raw water and to agitate it in order to eliminate the air and impurities. This agitation is usually accomplished by special agitation equipment, a refrigerant air jet, or special air piping.

Plate Ice

In the plate system (which is still used but is not being installed in new plants to any large extent), hollow pans through which cold brine or ammonia circulates are immersed in a tank of water until ice 8 to 12 inches thick is formed. The plate is arranged so as to allow the liquid ammonia to feed into it and the gas to return to the compressor in the usual manner.

AIR PIPING

Air piping, when used, must be carefully installed according to the manufacturer's specifications. The air to the blower is usually drawn from under the framework and returned directly from the blower to the laterals without any further conditioning. The only exception is in cold weather when it sometimes becomes necessary to take the air from a warm room or heat the air slightly. The temperature of the air in the laterals must be kept from 46 to 50°F in order to prevent freezing of the moisture in them or of ice on the drop pipes. Keep the air pressure as low as possible—usually from 1¼ to 1½ lb is sufficient. In exceptional cases, 2 lb may be necessary.

One outstanding feature of the air tube is that the spring on the bracket is made to enter the lifting hole in the can, thus centering the tube. In addition, this spring clamps the bracket and keeps the tube in the center of the can even though the can might tip lengthwise in the tank. The tube can swing crosswise in the can, but the can itself will not tip crosswise. The swinging action prevents freezing-in of the tube until the block is entirely closed. When placing the tube in the can, squeeze the spring down and start the end of the spring in the hole in the can, and then slip the bracket over the edge of the can; the tube will be accurately placed without further adjustment.

Be careful, when putting the fitting into the drop pipe, to leave it sitting at an angle that ensures free swinging of the tube. The secret of good ice is in keeping the tube centered in the can. When pumping cores, be sure to wash the inside of the ice block and dropline thoroughly so as to remove any deposits that may have started to collect. Have the core spotlessly clean before refilling with fresh water. Pull the ice regularly, a definite number of blocks per hour, on the hour. Work from one end of the tank to the other, drawing out every other row.

ICE CANS

Commercial sizes of ice cans vary with the weight of ice cakes required. The sizes shown in Table 19-1 are in accordance with standards adopted by the Ice Machine Builders Association of the United States.

Table 19-1. **Standard Welded and Riveted Ice Cans**

Rated Capacity (lb of ice)	Inside Dimensions (inches)			Depth of Immersion (inches)	Displacement (cu ft)
	Top	Bottom	Height		
50	5 × 12	4½ × 11½	31	28	0.9
50	6 × 10	5½ × 9½	31	28	0.9
50	8 × 8	7½ × 7½	31	28	1.0
60	5 × 14	4½ × 13½	31	28	1.1
100	8 × 16	7¼ × 15¼	31	28	2.0
200	11½ × 22½	10½ × 21½	31	28	4.0
200	14½ × 14½	13½ × 13½	35	31	3.5
300	11 × 22	10 × 21	48	46	6.0
300	11½ × 22½	10½ × 21½	44	40	5.7
400	11½ × 22½	10½ × 21½	57	52	7.4

Ice Removal

To remove the ice, the cans are lifted out of the brine and sprayed with, or dipped in, warm water. This loosens the ice so that when the can is inclined on its side, the cake slides out. The cans are usually tapered to facilitate ice removal.

Filling

The ice cans are usually filled by means of a filler device that is so constructed as to automatically shut off the water supply when the can is filled to the proper height. The filler is inserted in the can, and the water is turned on. As the can fills, a floating ball rises until the can is filled to the right depth, when the ball automatically closes a valve. An automatic ice-can filling system of this type is shown in Fig. 19-1. This can filler reduces tank labor by filling a considerable number of cans at one time through the operation of single lever. The life of the ice cans is increased by this method of handling as compared to the usual careless and injurious procedure followed when filling one can at a time.

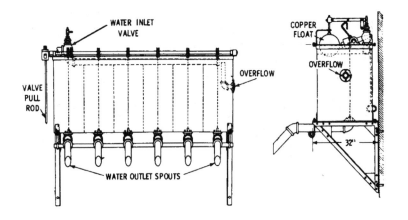

Fig. 19-1. Automatic can filler.

A pressure-type can filler is shown in Fig. 19-2. In this system, the combination valves immediately under the pressure tanks will open when the operating lever is pulled forward to fill the ice cans. Observe that the water-supply valve, which fills the press sure tanks through the header at the bottom, is closed. Also make sure that both the outlet and inlet valves are fully closed when the operating lever is in the intermediate position halfway between the emptying and refilling positions. After the ice cans are filled, the lever is normally pushed all the way by the operator, allowing the combination valves to open, and opening the valve on the water-supply line.

The linkage to the water-supply valve must be set with sufficient looseness or free play so that the bottom supply valve does

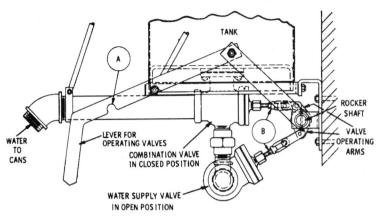

Fig.19-2. Operating levers and valves on a Frick pressure-type can filler.

not begin to open until the combination valves above have closed completely. Adjust the linkage so that all the combination valves have equal openings. A check valve at the base of each of the combination valves makes all the water in the pressure tank above it run into the ice can which that particular tank is filling. The valves are opened by the lever and links but are closed solely by the pressure of the springs inside the valve bodies. Keep the stem packing nuts just tight enough to prevent leakage and loose enough to permit the stems to slide freely.

INFERIOR ICE

Cloudy or milky ice is usually caused by the presence of air. It may be due to insufficient reboiling, overworking the reboiler or, more likely, an insufficient supply of steam to the distilled-water condenser, in which case the rapid steam condensation causes a vacuum, with the result that air is drawn in and mixed with the water.

A white core in the ice cake is usually caused by carbonate of lime and magnesia in the water. In many cases, it comes from overworking the boiler, carrying too much water, and not "blowing off" often enough.

513

FREEZING TANKS

Freezing tanks are made of such material as wood, steel, or concrete. Wooden tanks have a relatively short life and are subject to leaks. For this reason, freezing tanks made of steel coated with waterproof paint are preferable. Tanks made of reinforced concrete are also recommended as superior to those made of wood. The freezing tank (Fig. 19-3) contains direct-expansion freezing coils equally distributed throughout the tank and submerged in brine. The tank is provided with a suitable hardwood frame for supporting the ice cans and a propeller or agitator for keeping the brine in motion. The brine in the tank acts as a medium of contact only; the ammonia evaporating in the freezing coils extracts the heat from the brine, which again absorbs the heat from the water in the cans, thereby freezing it.

The tank itself should not be any larger than necessary to hold the cans, coils, and agitator. About 2 inches should be left

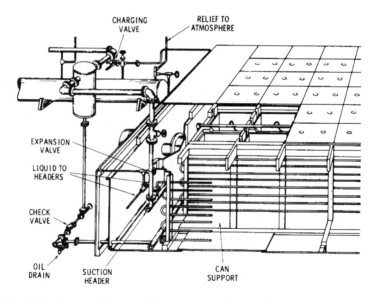

Fig. 19-3. Arrangement of the accumulator and connections in a loose-can tank using flat coils between each row of cans.

between the molds and 3 inches between the pipes and the molds. Insulation of the freezing tank is accomplished by using 12 to 18 inches of good insulating material on each of the sides and not less than 12 inches under the bottom.

The brine temperature should be maintained at 10 to 20°F, and the back pressure in the ammonia coils from 20 to 28 psi, which is equivalent to a temperature of 5 to 15°F in the coils.

CAN DUMPS

When harvesting a large number of cans in one lift, the can dump takes the form of a dump board, as shown in Fig. 19-4. Grids holding four to six cans are frequently handled with one dump. Such dumps are constructed of heavy steel channels lined with wood timbers to protect the can from wear. A dip tank is always used in connection with dumps handling more than one can at a time. The arrangement of cans is shown in Fig. 19-5.

Single-can dumps (Fig. 19-6) are furnished with or without sprinkler pipes for thawing the ice loose. When equipped with sprinkler pipes, the dump is usually so arranged that the water supply to the sprinkler pipes is automatically turned on when the can is in the dumping position and turned off as soon as the dump returns to its normal position. The weights are so distributed that the dump returns to or near its normal position by gravity after the ice has left the cans. When harvesting several cans at one time, the dump can be provided with hydraulic power, which acts through a piston to tilt and return the dump. Suitable pans are usually installed to catch the drip from the dump and protect the floor.

PLATE-ICE SYSTEMS

Plate ice is made by one of several methods:

1. Direct-expansion plate system
2. Direct-expansion plate system using "still" brine
3. Brine-cell plate system

515

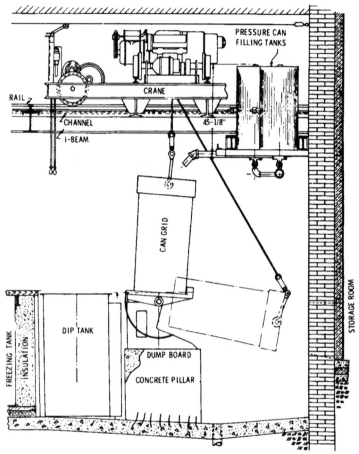

Fig. 19-4. Arrangement of the dip tank, dump board, crane, and can filler for a group-lift plant.

4. Brine-coil plate system
5. Block system

Direct-Expansion Plate System—The *direct-expansion plate system* is the simplest in construction and consists of direct-expansion zigzag coils with ⅛-inch plates of iron bolted or riveted in place. The thawing off of the face of the ice is accomplished by

516

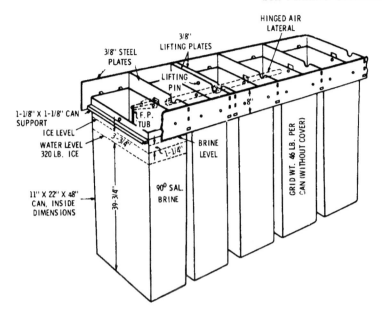

Fig. 19-5. Method of securing the ice cans in a can grid for a group lift.

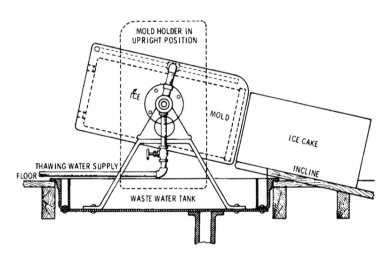

Fig. 19-6. Single-can dumper with iron stand and dump box. The control valve of the thawing pipe in this case is connected to one of the trunnions of the dump box.

turning the hot ammonia gas from the machine directly into the tank coils.

Direct-Expansion Plate System with "Still" Brine—The *direct-expansion plate system with still brine*, known as the *Smith plate*, has coils immersed in a brine solution contained in a water- and brine-tight cell. Thawing off is accomplished by turning hot gas into the coils.

Brine-Cell Plate System—The *brine-cell plate system* consists of a tightly caulked and riveted cell or tank about 4 inches thick, provided with proper bulkheads or distributing pipes to give an even distribution of brine throughout the plate. The thawing off of the face of the ice is accomplished by circulating warm brine through the plate.

Brine-Coil Plate System—The *brine-coil plate system* is similar to the direct-expansion plate system except that brine is circulated through the coils instead of ammonia. Thawing off is accomplished by means of warm brine circulated through the coils.

Block System—In the *block system*, the ice is formed directly on the coils through which either ammonia or brine is circulated. After tempering, the ice is cut off in blocks the full depth of the plate by means of steam cutters, which are guided through the ice close to the coils.

Harvesting—The method used in harvesting plate ice is similar in all of the previously enumerated systems except for harvesting block ice. One method consists of using hollow lifting rods and thawing them out with steam. Another method uses solid rods that are cut out when cutting up the ice. Still another method uses chains which are slipped around the cake when it floats up in the tank. The common size for plate ice is 16 × 8 ft 11 in. thick, although other sizes are sometimes made.

The arrangement of plate-ice freezing tanks does not differ materially from that of can ice, except that they are divided into compartments of the required size and number. Each compartment is furnished with one or more plates on which the ice is frozen. The compartments are made of sufficient size to hold a supply of ice for one day so that each compartment may be emptied, cleaned, and refilled every 24 hours (less if demand requires.)

Freezing Methods

The process of freezing ice in plate plants is principally as follows: several vertical hollow iron walls are built in a larger tank. The tank is filled with pure well water so that the iron walls are entirely submerged. The hollow iron walls are placed parallel to each other at a distance of 2 to 3 feet. The freezing fluid, consisting either of cold brine or ammonia, is passed through the hollow walls, with the result that the water will freeze on the outside of the walls. The water is kept in agitation by means of either a propeller, pump, or compressed air, so that the water is kept continually on the move carrying the air with it to prevent it from being frozen in the ice. After the ice is frozen on the walls to the required thickness, the freezing fluid is shut off from the walls and a warm fluid passed through instead until the ice is loosened and taken out of the tank.

The plate walls differ in construction with the type of freezing fluid used. Thus, if cold brine is used, the brine has to be cooled in a separate refrigerator from which it is pumped through the walls and back to the refrigerator. This is the same method used for cooling rooms by means of brine pipes. The plate walls in such a case are generally constructed of iron. Because of the expansion and contraction caused by warm or cold brine being passed alternately through the hollow walls, it is very difficult to keep them tight.

Where ammonia gas is used as the freezing fluid, the walls are built up of expansion pipes, which are connected at each end by return bends to make one continuous zigzag coil. To obtain an even surface, the coils are covered with thin iron plates, with the ice frozen on the outside. To loosen the ice when thick enough, the cold ammonia gas is shut off and hot gas is passed through the pipes. Agitation is carried on by means of air jets located midway between the plates, sometimes in the center, sometimes 3 or 4 feet from one end, and sometimes at both ends of the plates.

Cutting

Cutting of the plates into cakes is accomplished in various ways, such as by steam cutters, power saws, and hand plows. In the block system, however, where the ice is cut off the plate in the tank, the cakes are removed by means of a light crane and hoist and are then divided into the required sizes with an axe or bar.

AUTOMATIC ICE MAKERS

Automatic ice makers produce fragmented ice for processing, shipping, and protection of a variety of food products. This ice is manufactured in various sizes and in quantities of up to 46 tons every 24 hours. As noted in Fig. 19-7, the ice maker consists basically of a *freezing section*, a *harvesting section*, and a *control panel*.

The *freezing section* includes a supporting frame, freezing tubes with liquid and gas headers, accumulator and float control valve, drip pan, oil and liquid traps, water-distribution system, automatically operated suction, liquid, and hot-gas valves, the necessary solenoid and stop valves, and a galvanized sheet-steel tube enclosure.

The *harvesting section* consists of a supporting base, ice-breaker assembly and motor, ice sizer, water drain pan with float valve and recirculating pump and motor, baffle plates, and water grid.

The *control panel* includes a timer, terminal board, relay, restarter mechanism, ice-selector switch, and pressure gages.

Operational Features

The accumulator float control valve feeds liquid refrigerant to the stainless-steel freezing tubes. Water is directed against the tops of the cold tubes. The water flowing over the tubes is frozen into solid, clear shells of ice. The cool excess water is collected in

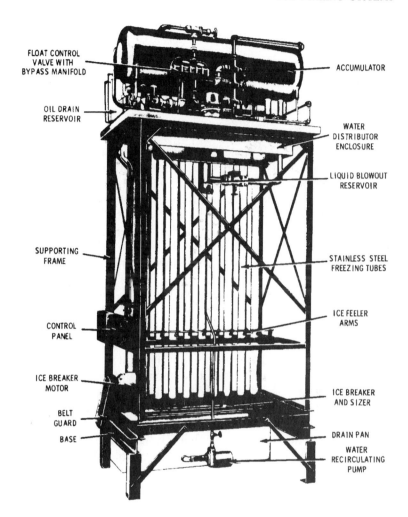

Fig. 19-7. Typical shell ice maker (steel enclosure removed).

a pan at the base of the unit for recirculation. A float valve maintains the correct water level in the pan. When the ice reaches a predetermined thickness (after approximately 11 minutes), discharge gas is admitted to the tubes to force out the cold ammonia. The warm gas causes the ice to loosen (in less than 1 minute) and slider into the ice breaker. Ice thickness may be varied from ⅛ to ½ inch on standard units by regulating the duration of the freezing time.

DRY ICE

Dry ice is merely solid carbon dioxide, which has a temperature of approximately –110°F at normal atmospheric pressure. Carbon dioxide is produced either by heating certain limestones or by treating the stone with acid. Another method of production that is widely used commercially is a fermentation process; still another is burning coke. Carbon dioxide is also developed in the petroleum industry as a by-product of hydrogen production and is found in blast-furnace gases. Other larger potential sources are the fields of natural gas or oil that produce carbon dioxide as a be-product.

Because dry ice is so excessively cold, it is necessary to develop means to control its refrigerating effect. The most common methods employed for its control are: (1) insulating it from the objects or space to be cooled, and (2) using and controlling a secondary or intermediate refrigerant. In recent years, dry ice has found a constantly increasing field of usage. In transportation of goods that must be held at low temperatures, more payload can be carried because of the lighter weight of dry ice compared to ice-and-salt mixtures. Deterioration of equipment due to melting mixtures is avoided. The principal disadvantage is its rather high cost, which is being reduced, however, by development of more economical methods of manufacture. Another problem has been temperature control, but solutions are being found and several methods for proper control and economical use have been developed.

A recently devised plan for using dry ice efficiently for cooling small compartments, truck bodies, and refrigerator cars consists essentially of a dry-ice compartment or bunker surrounded by a jacket. Bunker and jacket are heat-insulated from the space to be cooled in the car, truck body, or other container but are constructed so that heat can easily pass from the fluid in the jacket to the dry ice in the bunker. From this jacket, cooling coils extend into the upper part of the space to be cooled. Temperature within the cooling coils and therefore in the refrigerated space is controlled at will by a thermostatic valve, which regulates the flow of a secondary refrigerant in the cooling coils. This secondary refrigerant is a liquid with a high coefficient of thermal expansion, which circulates by gravity flow down through the bunker jacket and up through the cooling coils. The large variation in density causes fairly rapid circulation.

A modification of this system is one in which dry ice is placed in an insulated metal container surrounded by a coil through which a liquid with a low freezing point circulates. The liquid circuit has a valve in it controlled by a thermostat in the refrigerated compartment. When refrigeration is needed, the thermostat opens the valve and allows the liquid to circulate. The liquid in the coil surrounding the dry-ice chamber gives up heat to the ice, causing it to vaporize. The vapor, in escaping, actuates a diaphragm-type pump that circulates the liquid. When no refrigeration is required, the thermostat closes the valve, the resistance stops the pump, the liquid stops circulating, and the evaporation of the ice is very much retarded. This system is applicable to compartments of almost any size, and quite a wide variety of temperatures can be maintained.

ICE-MAKING CALCULATIONS

Capacity

Ice-making capacity is usually equal to about 50 to 70 percent of the refrigeration capacity, as expressed in tons of refrigeration per day. Of necessity, however, such operating conditions as initial temperature of the water supply, room temperature, and effectiveness of insulation will influence this ratio. If heat-leakage losses are known, the ice-making capacity can be closely estimated. The following example illustrates the method.

Example— Assume that the water supply has a temperature of 60°F and the temperature in the brine tank is 12°F. Calculate the ice-making capacity, assuming a 20 percent heat-leakage loss.

Solution—If it is remembered that the specific heat of water is

1 and that of ice 0.5, our calculation will be as follows: To lower the temperature of the water to the freezing point, the Btu/lb is 60 – 32 = 28 Btu. To freeze 1 lb of water without change in temperature requires 144 Btu. To reduce the temperature of 1 lb of ice to 12°F, (32 – 12) 0.5 = 10 Btu.

The foregoing add up to 182 Btu/lb of ice made. To this must be added the heat-leakage losses of 20 percent, that is, 1.2 × 182, or 218.4 Btu. Total refrigeration effect to manufacture 1 ton of ice is accordingly 2000 × 218.4 = 436,800 Btu. Since 1 ton of refrigeration = 288,000 Btu, it follows that the ice-making capacity is

$$\frac{288,000}{436,800} = 0.659$$

or 65.9 percent of refrigerating capacity.

Heat Losses

Heat losses in ice making usually vary from less than 10 percent in the best installations up to a possible 50 percent under poor conditions. When the ice production and refrigeration rating of the machine is known, the heat losses may be calculated as in the following example.

Example—Assume an ice plant is producing 10 tons of ice daily from a refrigerating compressor rated at 15 tons refrigerating capacity, other conditions being equal to those given in the previous example. What will be the heat losses?

Solution—The total heat required to be removed per pound of ice, exclusive of heat losses is 182 Btu. Heat per ton of ice = 182 × 2000, or 364,000 Btu. A 15-ton plant has the capacity to withdraw 288,000 × 15, or 4,320,000 Btu. Consequently, the capacity to produce ice exclusive of heat losses is

$$\frac{4,320,000}{364,000} = 11.9 \text{ tons of ice (approx.)}$$

The number of tons of ice due to heat losses is 11.9 – 10 = 1.9. The percent of heat loss is therefore 1.9 ÷ 11.9 = 0.16, or 16 percent (approximately).

Freezing Time

The time required for freezing ice depends upon the temperature of the brine and thickness of the cake. With brine at about 15°F, it will take approximately 50 hours to freeze 11½ inches of ice. The longer time a given thickness is allowed to freeze, the better the quality of the ice will be. The time for freezing may be calculated mathematically if it is remembered that, for different thicknesses, the time required is proportional to the square of the thickness. In other words, 8-inch ice will require four times as long to freeze as will 4-inch ice; and 11-inch ice will require approximately seven and a half times as long to freeze as 4-inch ice, etc.

The following empirical formula has been used to determine the freezing time for ice in cans:

$$\text{Freezing time (in hours)} = \frac{At^2}{32 - T}$$

where t = width of ice at the top (narrow way)
T = temperature of brine, °F
A = constant (usually taken as 7)

Example— If it is assumed that the brine temperature is 14°F and a standard 300-lb can with inside dimensions of 11 × 22 inches at the top is used, calculate the hours of freezing time required.

Solution—A substitution of values in the freezing formula gives

$$\frac{7 \times 11^2}{32 - 14} = 47 \text{ hours (approx.)}$$

ICE STORAGE

Manufactured ice is stored very much in the same manner as natural ice, insulation being the most important factor. Strips of lath, sawdust, hay, straw, and (in the South) rice chaff are used for packing it. The space allowed per ton of ice should be about 50 cu ft. Manufactured ice is usually shipped in cars, where it is packed and insulated the same as when put in storage. Ice-storage warehouses are usually equipped with a cooled anteroom, thus obviating the necessity of frequent opening of the main-room door. A certain amount of ice can be kept in the anteroom as may be found necessary.

SUMMARY

Detailed methods of ice making have been fully presented since the manufacture of artificial ice will always be necessary to fill a constant need in the various fields of food preservation. Modern methods of ice making, such as plant layout and component features, are dealt with in the first part of the chapter, and automatic ice-making units are presented in the latter part.

Dry ice and its importance in the protection of perishable foods during transportation, together with control methods, have been discussed. Finally, various calculations provide information on the determination of plant size and time requirements when certain other factors are known.

REVIEW QUESTIONS

1. Name two different forms for the production of artificial ice.
2. What is the function of air piping in an ice-making plant?
3. What methods are used for filling ice cans?
4. How can cloudy or milky ice be avoided?
5. What types of materials are used in freezing tanks?
6. Name the several methods used for production of plate ice.
7. Name the method used in harvesting plate ice.
8. What are the freezing methods used in plate ice?
9. What are automatic ice makers?
10. What are the operational features of automatic ice makers?
11. What does dry ice consist of?
12. How may heat losses be calculated?
13. What is the method of storing manufactured ice?

CHAPTER 20

Supermarket and
Grocery Refrigeration

Modern supermarket and grocery refrigeration generally consists of various prefabricated units designed to furnish the necessary cooling effect for a great variety of perishable foods, each unit being individually temperature-controlled to prevent spoilage. Depending upon size and other requirements, the various refrigerated units usually consist of the following: *display cases*, *walk-in coolers*, and *storage freezers*.

Placement of the various units depends upon such factors as available floorspace and nonperishable food placement, but, in general, the refrigerated units should be placed in locations to provide the maximum facility for display and customer convenience.

DISPLAY-CASE ARRANGEMENT

Display cases, as the name implies, are used for display of perishable products. In the display of meat, for example, it is generally said by merchants that display cases are made to *sell* meats, not *keep* them.

If the customer has access to the product, it must be wrapped and priced. In wrapping, absorbent papers are placed under the meat to absorb blood, which is found to leak from the cut meats. The wrapping keeps the meat from the air and thus tends to stop discoloring. Also, the wrapping will show the weight, price per pound, and the product price. Customers like to pick out the piece of meat that they want without interference from the butcher. With this in mind, most meat counters are not enclosed. The refrigerant temperature is lower than the room temperature, and cold air tends to flow downward, so the meat cases may be open for the customer to pick out the piece of meat that he or she wishes to buy. A typical example of such a meat display counter is shown in Fig. 20-1.

The meats thus displayed are presented to the customer, and what is not sold one day will be in very presentable condition for the next day's sales because it is thoroughly wrapped in transparent covering. The meat case must be well illuminated, with lighting having the proper color-temperature to bring out the right color of the meat. At night, many markets have covers that are drawn over the tops of meat cases to also keep the meat at proper temperatures, but with modern designs, this is not too much of a problem.

The sides and the back are well insulated, as are the ends of the display case. The front is made of Thermoglass, so that the meat is clearly visible to the buying public. A great variety of open and closed meat cases are presently available. The most recent development are vertical shelf units for cold cuts, bacon, ham, etc. Regardless of the case design, however, the basic refrigeration requirements remain the same.

Because of the increase in self-service stores and supermarkets, open display cases, similar to the one in Fig. 20-2, are generally used in preference to the closed type, although closed cases are still used in many locations, particularly where higher-than-

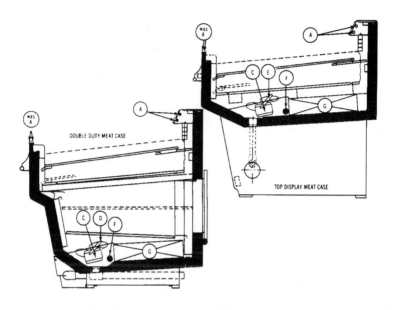

DOUBLE DUTY MEAT CASE

TOP DISPLAY MEAT CASE

Operational Parts List	8' Case	12' Case
A. LO-WATT ANTI-SWEAT WIRE Rub Rail formerly had this wire—don't replace	25 ohm/ft.	11 ohm/ft
B. HI-WATT ANTI-SWEAT WIRE Rear Riser formerly had this wire—replace with A	8 ohm/ft.	4 ohm/ft.
C. FAN MOTOR (2) 2 Fans each for 8 and 12' cases	5 Watt	5 Watt
D. FAN BLADES (Double Duty Case) (2) 3 Blades	7.75-18°	7.75-32°
E. FAN BLADES (Top Display Case) (2) 3 Blades	6-27°	7.75-18°
F. (OPTIONAL) DEFROST HEATER HEATUBE	76" 1600W	124" 2400W
G. DEFROST LIMIT SWITCH L 50-2 (Open 50°, Close @ 30°)	Klixon	Klixon
H. EXPANSION VALVE 1 Ton Sporlan	GF-1C	GF-1C

Tyler Refrigeration Corporation

Fig. 20-1. Typical supermarket meat-display counter.

529

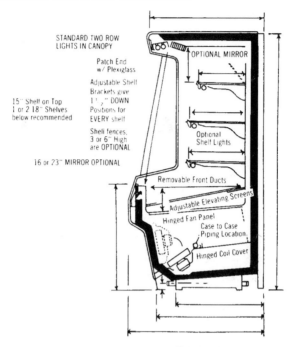

STANDARD TWO ROW
LIGHTS IN CANOPY

Patch End
w/ Plexiglass

Adjustable Shelf
Brackets give
1 1/2" DOWN
Positions for
EVERY shelf

15" Shelf on Top
1 or 2 18" Shelves
below recommended

Shelf fences,
3 or 6" High
are OPTIONAL

16 or 23" MIRROR OPTIONAL

OPTIONAL MIRROR

Optional
Shelf Lights

Removable Front Ducts

Adjustable Elevating Screens

Hinged Fan Panel

Case to Case
Piping Location.

Hinged Coil Cover

Tyler Refrigeration Corporation.

Fig. 20-2. Open, multishelf cold-meats display case.

average incomes allow for greater consumption of more expensive meat. Objective studies have indicated that, in general, refrigeration performance is reasonably equal in both the open and closed types. The installation and adjustments made by service personnel, however, are not usually comparable; hence there is a possibility of a great variance in performance in any given make.

OPEN DISPLAY CASES

Open display cases keep the product cool by circulating chilled air from a coil. Double-duty cases feature a lower storage, which furnishes a place to store packaged cuts prior to display. Reach-in or reach-through storage refrigerators may also be used to supply needed storage space.

Multishelf cases have proven very popular because they offer so much more display area in relation to floorspace. Prepackaged items, such as sandwich meats, are often equipped with holes or hooks so that they can be hung from metal pins. There is a dual gain in this vertical display because light intensities are diminished on the hanging package surface, which reduces fading. Light does fade processed meats, such as ham and bacon, but does not affect fresh meat by itself.

All open cases are affected by outside air turbulence caused by heating/cooling ducts, open doors, windows, etc. These drafts should be eliminated by directing ducts away from all open cases and by keeping outside doors closed.

The open, shelved display cases are also used for frozen food. This type of display case is shown in Fig. 20-3. Dairy products and fresh produce cases are also of the open-type display, where the customer has access to the products. Figure 20-4 shows a dairy-products display case, and Fig. 20-5 illustrates a fresh-produce display case-both with convenient access for the customer.

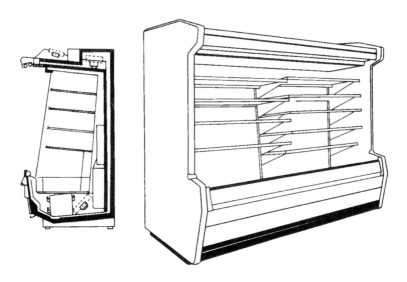

Tyler Refrigeration Corporation

Fig. 20-3. Open, shelved frozen-food case.

531

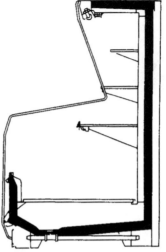

Tyler Refrigeration Corporation.

Fig. 20-4. Open-faced dairy-products case.

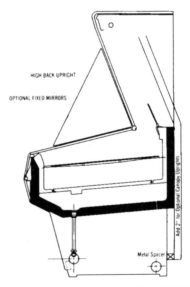

Tyler Refrigeration Corporation.

Fig. 20-5. Easy-access produce case.

DISPLAY-CASE LIGHTING

One of the major sources of heat in a display case is *radiant heat*. Some of it comes from warm ceilings over the cases, but most of it comes from one thing the store manager could not do without, namely, lighting. Brightly lit stores contribute greatly to the sales of meat products. Thus, foot-candle (fc) readings of 75 to 100 are commonly used in supermarkets and other food stores.

In order to highlight the meat display, more lights are employed. Therefore, it is not unusual to have 100 to 150 fc of light at the display level. The quantity of light does not affect fresh meat. The undesirable effect comers from the infrared (heat) rays in the light, which penetrate the cold blanket of air, warming the packages in the same manner as winter-time sunbathers stay warm, although the air outsider their shelters may be below freezing.

One method that quickly and dramatically illustrates the effect of radiant heat is the wrapping of a single cut of meat in aluminum foil. Within 24 hours it will be found that, because the aluminum rejects the radiant heat by reflection, the temperature of the wrapped package will very closely approach the temperature of the air within the display case. The temperature of adjacent conventionally wrapped packages can rise substantially, depending on the infrared radiators that are present.

Fluorescent lamps distribute the light more evenly than incandescent lamps. Deluxe cool white or deluxe warm white fluorescent lamps will give truer reds and flesh colors, with less infrared, than incandescent lamps. The lighting should be 100 fc or better.

RADIANT-HEAT EFFECT

Radiant-heat transfer can be a major problem in food storage. This is particularly true in supermarkets where the self-service open meat cases are built with the maximum airflow permissible to hold ambient air-mixing losses to a minimum. This airflow will remove the heat from the products at a fairly rapid rate, but if the rate of radiant-heat input exceeds the rate of heat withdrawn by the case air, then the product temperature will rise. This can be demonstrated by wrapping a piece of meat in aluminum foil and placing it in the center of any open meat case. In a few hours, this particular cut will have a temperature considerably lower than its neighbor in conventional wrapping material—a temperature fairly close to the temperature of the surrounding case air. However, no one can sell meat in aluminum foil.

If attempts are made to run the case colder or if night covers are used, the product will freeze because the air temperature in open meat cases is normally kept as close to the freezing temperature of red meat as possible. The product adjacent to the source of cold air will obviously be somewhat colder than that at the return-air opening because the case air warms up by absorbing heat from the product as it passes across the case. Thus, lowering the entering-air temperature will lower the return-air temperature and the products temperature at that point, but it will also threaten the product at the entering side with freezing. Similarly, night covers that alleviate the input of radiant heat at night will bring the meat temperature much closer to the case-air temperature so that a cabinet running at optimum temperature levels during the day will freeze the meat at night.

It seems the only way to combat the problem is to eliminate or lessen the intensity of the heat source or find a way to change the heat-absorbing rate of the meat. This can only be done by the architects, store planners, and all others involved in the creation of a food market.

Lighting is also an important transmitter of radiant heat. Thus, for example, the surface temperature of electric bulbs can run quite high (over 2000°F for incandescent lamps) and, although small in area, can contribute heavily to the load. This problem can be lessened by using diffusers and lenses that will cut down the intensity of the light and also provide a screen between the meat and high-temperature light source. The effect of light on the color of meat is actually a separate and distinct problem. In many cases, the product temperature may be excellent, but the meat may change color due to overlighting.

There are many other sources of radiant heat. However, for this discussion they are not as important as the ceiling and lights.

There are many new heat sources to be watched. Among these are the now popular barbecue and bakery departments. They are potentially radiators that can add radiant heat to perishable products. Also, solar heat through windows in some cases has become a most important factor. It must be concluded that meat, a high profit maker and one of the most perishable items in the food stores, deserves the utmost care in the control of its environment, which can be controlled only by careful preplanning and the user of good construction techniques.

CLOSED MEAT-DISPLAY CASES

Closed meat-display cases differ somewhat in operation from the open type. For example, a butcher who has become accustomed to a three-day meat life in an open display case with a minimum of product attention now finds it necessary to constantly groom the display for maximum appeal. The unwrapped product is much more subject to contamination from a variety of sources. The butcher may also be accustomed to open-meat-case temperatures of 28 to 34°F. Attempting to maintain these temperatures in a closed case results in coils icing up and in a very poor meat surface. Drying also occurs.

Coil Arrangement

The remaining problem is the application of refrigeration. Since the cold air will not rise higher than the coil from which it originates, it is necessary to have the coils as high as possible in the case. Visibility enters into this problem because it is necessary to let the customer see as much of the interior of the case as possible. Thus, the top containing the coil must be as narrow as possible, which reduces the volume of the top coil; therefore, refrigeration must come from some other source. This involves introducing another coil lower down-either under the shelf, against the front, or embedded in the floor or walls.

This problem has been solved differently for different-shaped cases. The principal thing to be observed is that the circulation from the top coil is not directed immediately on the stock (which would tend to dry out the display on the top shelf) but is dispersed into the case, keeping up the relative humidity and preserving the original bright color of the stock display. In all cases the lower coils are arranged to cool the stock, partly by circulation, but mostly by radiation, which contributes to the same conditions just mentioned.

MULTIPLE CONNECTION

Two or more cases are often connected to one refrigeration machine, as shown in Fig. 20-6. Great care must be exercised in this type of installation to satisfy all the conditions that may have to be fulfilled. It cannot be too strongly emphasized that unless the fixtures chosen are suitable for multiple connection, one fixture may operate too warm and another too cold, or the refrigeration machine may run all the time, resulting in spoilage losses and excessive power bills.

Some installations have temperature requirements such that individual machines should be used for each fixture. There are,

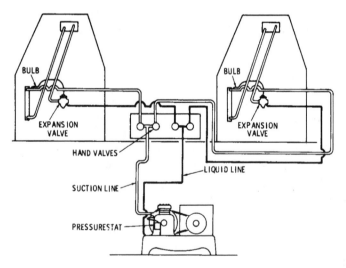

Steffen's Market

Fig. 20-6. Typical multiple connection.

however, many installations where conditions are suitable for multiple connections. In general, the requirements for multiple operation are:

1. All multiple-connected fixtures must be in good repair.
2. Service on all fixtures must be uniform.
3. The coils must be balanced to operate at the same back pressure; in other words, all fixtures must give desired temperatures while operating at the same back pressure during the same running time.
4. The fixtures should not be too far away from the refrigeration machine. For best operation, the machine should be located as close to one fixture as another.
5. Three fixtures can be multiply connected, but the chances of successful operation are decreased as the number of fixtures is increased.

If conditions on a job will fulfill all these requirements, the fixtures may be satisfactorily multiply connected.

DISPLAY-CASE MAINTENANCE

Regardless of the type, make, or age of a meat case, one practice that will add to the display life of all meat is a regular weekly cleaning of the case interior. This is standard practice for many chain stores and is recommended by most case manufacturers. The product should be removed, the machine shut down, and every screen, tray, and interior surface scrubbed down with a mild soap (detergent) or bicarbonate of soda solution. *Never use ammonia—the fumes will discolor meat rapidly.* Water should be used sparingly, and none should get on electrical parts. Use of a hose should be limited to flushing the drain periodically.

INSTALLATION

When installing refrigeration equipment in supermarkets and grocery stores, it is of the utmost importance that the liquid and suction lines be of ample size in order to avoid pressure drops.

Refrigerant piping systems should be designed to accomplish the following:

1. Ensure proper refrigerant feed to the evaporators.
2. Prevent excessive amounts of lubricating oil from being trapped in any part of the system.
3. Protect the compressor from loss of lubricating oil at all times.
4. Prevent liquid refrigerant from entering the compressor.
5. Maintain a clean and dry system.

Pressure drop in liquid lines is not as critical as it is in suction and discharge lines. The important factors to remember are that the pressure drop should not be so great as to cause gas formation in the line or result in insufficient liquid pressure at the liquid-feed device.

Suction Lines

In laying out the suction lines, the following should be adhered to:

1. The suction line should be sized for a practical pressure drop at full load.
2. It should be designed to return oil from the evaporator to the compressor under minimum-load conditions.
3. It should be designed to prevent liquid from draining into the compressor during shutdown.
4. It should be designed so that oil from an active evaporator does not drain into an idle evaporator.

Most refrigerating piping systems contain a suction riser, either by reason of the relative location of the evaporator and compressor or because the riser is introduced, for the purpose of minimizing the possibility of liquid draining from the evaporator into the compressor during compressor *off* cycles. Oil circulating in the system can be returned up gas risers only by entrainment with the returning gas.

Dehydrator

A permanent liquid-line dehydrator should be fitted into the line at the time of the installation. In the event that the dehydrating agent becomes saturated, it should be replaced. Dehydrators will eliminate moisture, which may cause sticking expansion valves and internal corrosion of the various parts.

Oil Separators

An oil separator will greatly improve the operation of the system but is not to be considered a service necessity.

Controls

In most installations of this type, display cases and walk-in coolers do not operate at the same temperature. Where the chill room is also connected to the same condensing unit, it will be necessary to add more controls. When using a forced-air evaporator, experience has shown that an evaporator regulating valve and a snap-action valve are necessary. The evaporator regulating valve will maintain the operating pressure in the evaporator at or above a given point, while the snap-action valve will permit the evaporator to cycle and thus obtain defrosting. Check valves must be used in the suction line leading to each low-temperature evaporator, consisting of either one large check valve installed in the main low-temperature suction line outside the low-temperature fixture or an individual check valve after each bank of freezing plates.

Air Circulation

The results obtained with mechanical refrigeration depend largely on air circulation within the fixture. The importance of providing proper air circulation in supermarket display cases and cooling rooms cannot be too strongly emphasized. Air circulation is necessary, not only for the proper distribution of refrigeration, but also to assist in the removal of gases and odors produced by stored-food products.

A fixture contains a cooling unit or heat absorber, which, in the case of mechanical refrigeration, is the evaporator. As the heat is absorbed from the air in contact with the evaporator, it is cooled, becomes heavier, and falls. In order that refrigeration may take place, the cold air that falls must be circulated around the stored food. It is in this manner that the cold air absorbs the heat in the food product and fixture. It expands, rises, and again comes into contact with the evaporator. This cycle is repeated over and over, the speed of the airflow depending on the relative temperature of the food and fixture.

Baffles

To assist and direct the circulating air, certain guides, termed *baffles*, and cooling-unit decks are constructed and installed within the refrigerator. They normally divide the fixture into two parts: the evaporator compartment and the food compartment. Circulation depends primarily on the existence of a temperature difference between the evaporator and the food compartments. To ensure maintenance of this temperature difference, the best practice is to insulate the evaporator deck with 1 to 1½ inches of corkboard (or its equivalent) for refrigerant temperatures below 20°F. Unless an evaporator deck is properly insulated, the temperature on both sides will be very nearly equal. Circulation under these circumstances will be very sluggish and often inadequate to maintain the desired temperatures in the food compartment. Foods may become slimy, and the evaporator deck may drip water into the food compartment due to condensation on the bottom of the deck.

MECHANICAL CENTERS

A recent simplification in supermarket and grocery-store refrigeration consists of a complete refrigeration-machinery package to serve an entire food store. The main advantages claimed for such prefabricated condensing units are that they cost less and, with proper planning, remove the machinery room from the selling area, thus offering additional display facilities.

Mechanical centers of this type (Fig. 20-7) are manufactured to suit various requirements. Horsepower sizes for both zero- and normal-temperature applications are available; and although rooms have been built to accommodate units down to ½ hp, 30-hp remote condensing units have also been designed. Each unit is completely prewired, including circuit protection and prepiping to a specified terminal area. The machinery unit, being

Sherer-Gillett Company

Fig. 20-7. Prewired mechanical center consisting of three condensing units of 10, 3, and 1 hp, respectively.

factory preassembled, may be placed in an appropriate remote location, such as on the roof or on a slab adjacent to the building or food store.

COMBINED SYSTEMS

Avoiding the use of compressors primarily designed for air conditioning eliminates the hazard of using unproven compressors for critical low-temperature requirements of a food store. Combined-system technique avoids electrical damper controls and shuns gadgets. It uses electrical defrost with no moving parts and proven reliability. Figure 20-8 shows a typical prewired condensing unit.

Sherer-Gillett Company

Fig. 20-8. 10-hp prewired condensing unit.

Of significant importance to food operators is the availability of a food-protection alarm system. This protecting device is an electronic sensor imitating the temperature reaction of food and can precisely determine the time a warning should be initiated. It is specifically designed to warn under precise conditions but avoids nuisance calls. The combined system, along with a mechanical center, employs principles familiar to the mechanic serving the food store, avoiding the need for special maintenance training. Thus, the legal requirements for employment of a stationary engineer will be eliminated in most cases.

DISPLAY-CASE DEFROSTING

The accepted defrosting method used in commercial display is almost universally the *direct-electric method*. Basically, this system consists of a resistance-type electric heater and an electrically operated time switch to initiate the defrost. Several methods of termination are used, the particular one chosen depending on the design of the refrigeration system and, to some extent, on the customers' preference.

Straight-Time Method

The straight-time system is controlled entirely by a clock. The length of defrost is determined by past experience with similar systems or by adjustment after installation. This system, while reliable, has the disadvantage of making no allowance for changing conditions of service load, humidity, or other variables. To ensure proper operation at all times, the length of defrost must be set for the most severe conditions, which may result in excessively long defrost periods much of the time.

The advantages of the straight-time defrost are simplicity and low cost, although cost would not seem so serious a consideration in relation to the cost of the total system. The electrical diagram of this type of system is shown in Fig. 20-9.

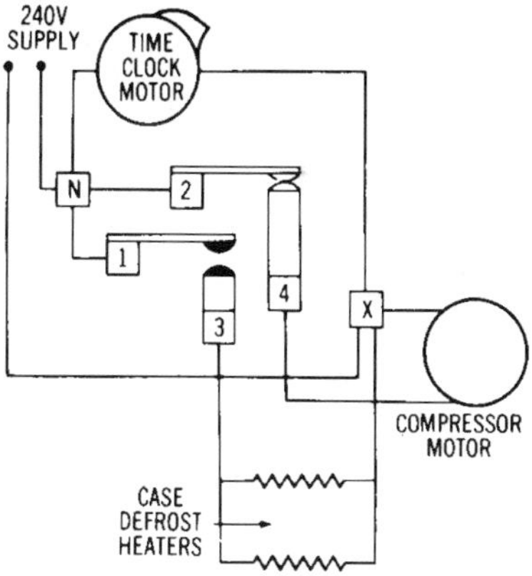

Fig. 20-9. Straight-time defrost and control diagram.

Pressure Method

The pressure method (Fig. 20-10) depends on the rise of the low-side refrigerant pressure as the coil temperature rises during defrost. These are popularly termed *TP controls*. At a predetermined pressure (most often around 45 psig for low-temperature

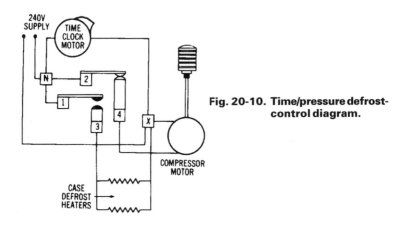

Fig. 20-10. Time/pressure defrost-control diagram.

systems and 40 psig for medium-temperature systems), the defrost is terminated by a pressure-actuated switch. Pressure-terminated controls are usually provided with a "fail-safe" time termination in case the pressure switch fails to operate.

The pressure-terminated control has the advantage of adjusting automatically to the differing conditions of service load and humidity and results in defrosts only of such duration as are necessary to clear the coil of frost. Its application should be limited to systems where the temperature/pressure relationship of the coil is accurately sensed by the control because of the basic characteristic that the pressure in a system will usually respond to the area of coldest temperature. Hence, cold suction lines or cold machines (if they are colder than the coil or coils), in effect, will be the controlling point. Faulty operation will often result when used on systems where part of the system is at a lower temperature than the termination temperature of the coil during defrost.

Temperature Method

Termination on temperature rise depends on a thermostat in the refrigerator to be defrosted, usually located on the coil. Termination thermostats are mostly of the nonadjustable type and are often set for a temperature of around 50°F. Temperature-terminated controls have all the advantages of the pressure-terminated types and the additional advantage that they are not dependent on accurate reflection of pressures due to temperature at remote parts of the system. The cost of these controls is somewhat higher, but, in most instances, the additional advantages will be worth the extra cost. A diagram of a typical temperature-terminated control is shown in Fig. 20-11.

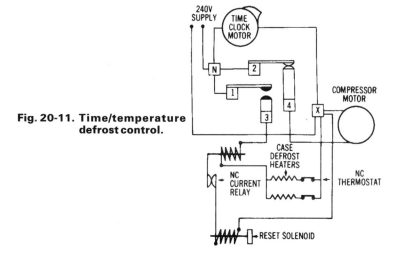

Fig. 20-11. Time/temperature defrost control.

The foregoing discussion exemplifies the versatility of the electric method of defrosting and one of the numerous reasons for its acceptance and popularity. No matter what type of control is used, the length of defrost is dependent on the voltage at the heater. The amount of heat supplied by an electric heater varies directly as the square of the voltage. Assume a typical defrost heater delivers 2400W at 240V. This same heater with the supply voltage reduced to 208V will deliver only 2140W, which does not represent a serious reduction but is worthy of note.

The current draw of the heater is also related to the voltage. Citing the same example as before, at 240V the heater will draw 10A; and at 208V, 10.3A. These differences may or may not be important in choosing the size of the supply wire. Supply voltage obviously has some effect on the length of the defrost period. As voltage is reduced, some increases in length of defrost result. Generally, the increase is not significant and is more than compensated for by the usual safety factors in the settings.

DRAINAGE REQUIREMENTS

An important item in refrigeration for display cases is the drain trap and associated piping. Drain water from the freezer coils should be conducted out of the refrigerator without the possibility of allowing warm air to enter. To accomplish this, a trap should be installed in the drain conductor. A drip space should be provided between the end of the drain pipe and the sewer to prevent sewer gas from backing up into the refrigerated space.

A typical installation is shown in Fig. 20-12. Notice that all drain pipe and fittings must have a diameter of at least 1 inch. Each section of pipe that is connected directly to any walk-in cooler must have an adequate slope, with ¼ in. being considered the minimum. The drain trap must *not* be located closer than 12 inches to the outlet from the cooler case.

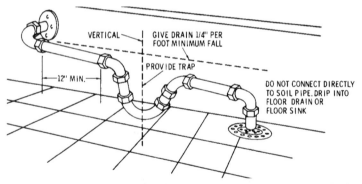

Fig. 20-12. Typical drain-installation diagram. All drain pipe and fittings must have at least 1 inch inside diameter. Every line connected directly to a display case must have a minimum slope of ¼ in/ft. Drain lines should never be connected directly to the soil pipe, but should be allowed to drip into the floor drain or an open sink. On low-temperature equipment, a trap must be installed at least 12 inches from the fixture.

WALK-IN COOLERS

Walk-in coolers, as the name implies, consist of a refrigerating unit provided with a suitable walk-in door to permit storing and preservation of larger amounts of perishable foods. Refrigerating units of this type may be purchased from various manufacturers, complete with all the necessary accessories. Formerly, walk-in coolers were made of oak, maple, or other hardwood exteriors and with a spruce lining. The insulation usually consisted of 3 to 4 inches of cork placed between the inside and outside surfaces. At the present time, however, they are made of light steel-clad Fiberglas-insulated panels. Units of this type (Fig. 20-13) can be erected easily and put into operation simply by connecting wiring to the required electrical source. Large refrigerators of this type are usually equipped with an air conditioning and humidifying unit.

Sherer-Gillett Company

Fig. 20-13. Cutaway view of a Sherer-Gillett metal walk-in cooler.

Floor Arrangement

Sectional prefabricated walk-in coolers are furnished with or without an insulated floor. In the floorless units, it is estimated that about 1.3 Btu of heat per hour per square foot can be expected to come in from the ground. Thus, floor losses are considered negligible. If the unit is installed on a concrete floor over a space, insulation is necessary to prevent sweating beneath the cooler. This insulation can be a sprayed-on type (where equipment is available to apply it), or a sheet of styrene can be cut and fastened to the area with mastic.

A level entry is becoming more important as the use of hand and electric trucks increases. The advantage of recessed floors and the convenience of a level entry afforded by a floorless cooler can also be had by recessing a sectional insulated floor. The supplier should be informed when the order is placed so that the door can be properly fitted.

An insulated concrete floor (on the ground) can be constructed for a low-temperature (freezer) cooler, but full consideration of all the facts must be taken. The only perfect insulation is a 100 percent vacuum, and this is not fully obtainable even in a thermos bottle or in the laboratory. Regardless of the thickness of the insulation in the floor under a low-temperature cooler, a freeze zone will gradually extend into the ground to the extent that the heat of the earth permits. If the subsoil is or becomes wet, the inevitable icing and expansion will take place, and the floor is likely to heave (especially in clay subsoil). If the freezer is located along an outside wall, the danger of heaving and subsequent damage to the footings and walls is an added possibility in areas of severe winters.

The only sure way to prevent a freeze zone from reaching moisture is to provide a source of heat to act as a barrier. One method, shown in Fig. 20-14, is to lay a pipe grid below 4 inches of styrene foam insulation, and circulate a nonfreezing liquid (permanent antifreeze, brine, etc.) through it. The heat is furnished from the unit condenser through a heat exchanger. Another practical method is that of laying 4-inch drain tile in rows and connecting the rows to plenum chambers so that air can be circulated under the floor. The air can make an open circuit through the freezing room above. Small freezers generally do not have a freeze zone deep enough to cause trouble. The necessity of a heat barrier depends chiefly on the nature of the subsoil and size of the freezer.

A tamped dry sand or gravel fill will make a good base on which to build an insulated floor. Some soil conditions may warrant a concrete floor below the insulation.

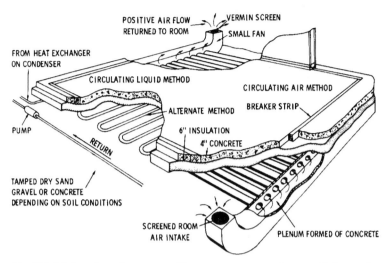

POSITIVE AIR FLOW
RETURNED TO ROOM

VERMIN SCREEN

SMALL FAN

FROM HEAT EXCHANGER
ON CONDENSER

CIRCULATING LIQUID METHOD

CIRCULATING AIR METHOD

ALTERNATE METHOD

BREAKER STRIP

PUMP

6" INSULATION

4" CONCRETE

RETURN

TAMPED DRY SAND
GRAVEL OR CONCRETE
DEPENDING ON SOIL CONDITIONS

SCREENED ROOM
AIR INTAKE

PLENUM FORMED OF CONCRETE

Fig. 20-14. Insulated concrete-floor arrangement for a walk-in cooler showing plenum chambers for air circulation.

Walls

The simplest form of sectional walk-in coolers—four plain walls with a top and a door—will fill some customer requirements. However, there is a great variety of standard sections up to 10½ feet high to make up just about any possible floor plan. Rolling cold conveyors can be fitted through special wall sections to facilitate the processing and packaging of meats and other perishables. Glass-door display fronts (Fig. 20-15) bring the customer close to the refrigerated items, and the newest open-display multishelf cases, attached to the cooler, place tempting food items at the customer's finger tips. Thus, the walk-in cooler can become an effective selling tool as well as a holding area.

A freezer and cooler may share a common wall, saving the price of one wall. On the freezer side of the wall, temperatures may be 0 to –20°F, while on the cooler side they may be 40°F at 90% relative humidity. The wall will be cooler than 40°F, and since the dew point for the air at this condition is 37°F, the wall surface will sweat or frost. There will be less sweating if a thicker insulation is used in the freezer walls, but additional insulation will not

549

Sherer-Gillett Company

Fig. 20-15. Exterior view of a sectional walk-in cooler permitting additional units to be installed according to requirements.

eliminate the condition entirely. It is better to place a cooler against an erected freezer rather than vice versa. If there is a door through the common wall, sweating or frosting will be more noticeable. This condensed moisture usually is no great problem, but it is good to be aware of it.

If separate coolers and/or freezers are to be adjacent, it is good to maintain a space between when installing to allow air circulation. This will prevent moisture from collecting and damaging the sections. Clear spans are possible up to 20 inches in coolers and up to 16 inches in freezers. Widths greater than these are attained with I beams. The tops are self-supporting, but extra weight, such as oversize coils or overhead meat tracks, must be fastened to supports above the top by means of rods. In addition to meat tracks, accessories such as meat rails, shelving, and floor and platter racks are also available from the manufacturer.

Wholesale Storage

Inasmuch as the wholesale storage of meat is generally in larger pieces—quarters or halves, or even whole animals—and the time of storage is longer, both the temperature and the relative humidity should be lower. Wholesale-storage temperatures are generally kept right on the edge of freezing. Because the rooms are larger, forced convection is readily applicable since the circulation can be more readily dispersed, cooling the room without blowing directly on the stock. Common practice, however, is still largely in favor of steel pipes or cold plates, the frost being brushed off instead of melted as a means of defrosting. The general practice in wholesale storage is to separate the refrigeration department into a number of rooms, each designed for a different and specific purpose, storage of cured meats, storage of carcasses, etc.

Frozen-Meat Storage

In the present stage of development, frozen-meat storage is a special problem. Bacteria are not killed at low temperatures. Their growth, however, is greatly reduced. Loss of color is also reduced as the temperature is lowered. In general, it may be said that the temperatures required are the lowest economically obtainable, generally around –10F°, but –20°F is better. It is necessary to provide a vestibule for entering such low-temperature storage rooms to prevent the inrush of warm air, both for economy and to prevent the deposit of moisture on the stock and coils.

Generally, it may be said that steel-pipe coils or plate coils composing the shelves on which the stock is piled are good practice. Small storage amounts are generally best in chests rather than refrigerators with swinging doors because the cold air tends to lie in a chest when the lid is open but spills out when a vertical door is swung. The application resembles ice cream hardening rooms, and the same general rules will apply.

MEAT PREPARATION

Chilling Fresh-Killed Meat

The chilling of fresh-killed meat presents difficulties, particularly in the retail market where a certain amount of killing is done and no particular provision is made for separating the warm fresh-killed meat from the storage stock. In larger markets, a separate chill room is always provided. It is best to do this in every case if economically possible because as long as the animal heat is present, the fresh-killed carcass steams like one's breath when introduced into a refrigerated atmosphere. This steam circulates and condenses on every cold thing in its path, thus wetting the surfaces of all the stored meat that happens to be in the same room.

If the amount of fresh-killed meat is small as compared to the capacity of the refrigeration and the size of the room in which it is introduced, this moisture will be absorbed by the air circulation carried to the coils and down the drain. On the average, it may be said that if the carcass is allowed to hang outside for a few hours, a single beef or a couple of hogs can be hung in a 6×8-foot market cooler if heavy-duty coils have been installed. If four or five animals are to be killed at one time, however, a separate chilling room should be provided, and it should be cleared of storage stock when a new batch of carcasses is put in.

Forced circulation coils are especially adaptable to meat-chilling rooms because they have the ability to take the heat out of the meat rapidly without excessive dehydration due to the moisture given off by the warm carcasses. When the machine equipment for coolers in which fresh-killed meat will be chilled is selected, the hot-meat load should be added to the normal-heat load of the cooler.

Packaged Meat

By the time packaged meat is put on display, it has usually come a long way. Before being packaged, it has been cooled, warmed, transferred two or three times, and then cut and maybe chopped or ground. By this time, the forces of loss have certainly gained some headway. The amount of heat gain depends to a great extent on the treatment that the product has received prior to this step. It may be stated that any harm that has occurred to the product prior to being put on display cannot be overcome simply because it is put in a pretty package.

Each properly wrapped cut will maintain its own humidity and is protected from further bacterial and mold spore contamination. Therefore, when the packaged meat is placed in the open-display case, it is subject primarily to temperature. One authority states that the ideal meat temperature for minimum bloom life is from 32 to 34°F. From 34 to 38°F is a semidanger zone, speeding up oxygenation and hastening fading, but not to the point where troubles are created if the store turnover is sufficiently high. If the meat temperature rises above 38°F, the fading becomes so rapid that the wrapped portions can become discolored and unsalable in well under 8 hours. These temperatures also may have been reached at some time before the cut was displayed. This is the reason much stress is placed on quick transfer of packages, use of cool cutting rooms and/or package conveyors, and minimizing other contacts that warm the meat.

SUMMARY

In this chapter an attempt has been made to furnish detailed information about modern methods of food preservation as presently practiced, as well as display-case arrangements, placing of condensing units, and related items. Since the radiant-heat effect is a major problem in the field of food preservation in storage, the methods used for air circulation, insulation, and other factors to prevent heat penetration have been discussed in detail.

Because of space consideration, mechanical centers housing complete equipment units (usually in remote locations) have been presented. These self-contained units provide a late development in the field of food preservation. Additional features, such as defrosting methods, controls, drainage requirements, walk-in coolers and floor arrangements, meat preparation, etc., conclude the chapter.

REVIEW QUESTIONS

1. Name three methods for the preservation of foods in supermarkets and grocery stores.
2. What is the display-case arrangement?
3. What are the advantages in using open display cases in supermarkets?
4. State the number of foot-candles of light used in display cases for meat.

5. What are the sources of radiant heat and their effect on food-storage display cases?
6. What is the proper coil arrangement in meat display cases?
7. What precautions should be observed in the multiple connection of condensing units?
8. How may pressure drops be avoided in refrigeration lines?
9. Name methods of control used in supermarket refrigeration.
10. What is the advantage of mechanical centers?
11. What are the accepted defrosting methods used in commercial display cases?
12. What provision is made for drainage for display cases?
13. What doers a walk-in cooler consist of?
14. What type of insulation should be provided for walk-in coolers?
15. State the temperature requirements for frozen-meat storage.
16. Give the methods of meat refrigeration prior to storage.

CHAPTER 21

Locker Plants

A technique of food preservation that has become very popular during the last decade is that of storing food in special freezing lockers. One general method of doing this in common use at the present time is by means of community lockers. The community locker or locker plants are usually equipped very much in the same manner as large refrigerating plants, but they have special facilities, such as locker rooms, quick-freeze rooms, cutting rooms, aging rooms, chill rooms, and customer rooms.

LOCKER-PLANT CONSTRUCTION

There are many types of locker plants in operation, and additional plants are designed to suit requirements of the particular location and trade under consideration. If an attempt is made to classify theses types of refrigeration plants, most of them can be grouped as follows:

1. Plants equipped exclusively for locker rental with all the related facilities
2. Plants combining locker rental with a retail establishment, such as a grocery or butcher store
3. Plants combining locker-rental service with certain manufacturing processes, such as slaughtering, curing of meats, and freezing of fruits and vegetables
4. Locker storage rooms or buildings attached to refrigeration plants, such as ice plants and creameries

The primary consideration when planning a locker plant is to determine whether a plant of any type will prove economically successful in the location under consideration. This requires a thorough study of local factors relating not only to the prices and proximity to production of food, but also to consumer habits, including the sale of meats, fresh foods, and commercially frozen food.

DESIGN CONSIDERATIONS

Every locker storage plant presents individual engineering problems, but each should be designed with economy of space and location of facilities previously determined as a primary consideration. In addition, the probability of expansion must be kept in mind to make the smallest possible additional investment when required. Since refrigeration space is the most expensive, it is necessary that the part of the plant that is lined with insulation and equipped with coils be constructed with the smallest amount of space possible. However, there are certain minimum requirements that must be considered. Thus, if a certain minimum requirement for insulation or floorspace is exceeded, it may become impossible to operate economically.

The floor plan of a properly designed and equipped locker storage plant is shown in Fig. 21-1 and should include the following features:

1. A *chill room* with an air circulation of ample capacity
2. A well-equipped *processing room* (not refrigerated) in

which to clean, cut, wrap, and label products to be frozen and stored

3. A suitable *quick-freeze room* held at a temperature of –5 to –15°F

4. A *locker room* with a closely controlled temperature

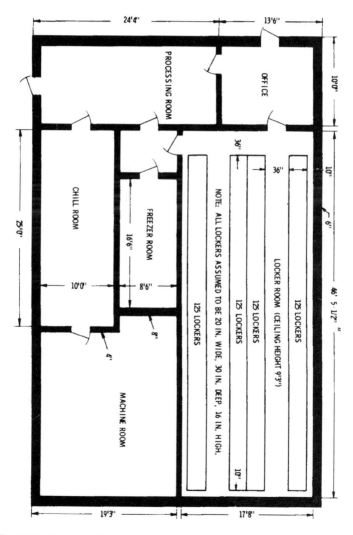

Fig. 21-1. Layout of rooms in a typical locker plant.

5. Refrigerating machinery of adequate capacity that is automatically controlled

In addition to these, such additional facilities as a customer room, aging room, pickling room, vegetable-storage room, and other auxiliary refrigerated space should be considered.

Chill Room

The function of the chill room is to remove the animal heat from the meat. The temperature is held at 30 to 38°F. Frequently, a chill room and an aging room are combined in one. The main advantage in having a separate aging room is the ability to maintain an even temperature. Thus, when warm products are placed in the chill room, the temperature of the room will tend to increase a certain amount, depending, of course, on the amount of the heat load and frequency of operation.

The size of the chill room depends on the size of the plant. It is considered good practice to allow ½ sq ft per locker. Thus, a 300-locker plant should have about 150 sq ft of floorspace. The height of the ceiling should be sufficient to accommodate meat rails or overhead tracks. However, if no meat tracks are used, 10 feet will be a good average height.

Processing Room

The processing room (sometimes termed the *cutting room*) is where the cutting, cleaning, wrapping, and labeling of products take place. This room is insulated but not refrigerated. The usual equipment for the room consists of regular butcher tools, such as cutting blocks, grinding machines, saws, knives, and scales. If the processing room is part of a large plant, a smokehouse for curing ham and bacon and a pickling room for processing corned beef are usually included.

Quick-Freeze Room

The quick-freeze room, as the name implies, is where the meats, fruits, and vegetables are quickly frozen before being placed in the locker compartment. The quick-freeze room is held at a temperature of –5 to –15°F° In this room, the floorspace allowance should be approximately ⅕ sq ft per locker. Thus, a 300-locker plant should allow about 60 sq ft of floorspace for its quick-freeze room. The height of the ceiling usually averages 9 to 10 feet. The time required for quick freezing ranges from 3 to 6 hours.

Locker Room

The locker room is the heart of the locker-plant system. The size of this room depends on the dimensions of the individual lockers and the amount of space allowed between the locker rows. Locker sizes have now become more or less standard, although each manufacturer has his own individual design intended to make his equipment the most attractive to the user and plant operator.

Metal is used almost exclusively in locker construction. Most lockers today are either $20 \times 30 \times 15$ inches or $20 \times 30 \times 17$ inches. Some are of the rolling-drawer type, and others are of the straight-compartment type with locking doors. Lockers are usually supplied with either padlocks or locks built into the door proper in order that the customer may be reasonably surer that his storage space is safer. The plant operator always has a master key for all lockers, which enables him to place frozen food in lockers after processing.

The aisles between the locker rows should be from 32 to 38 inches wide. There should be an allowance of approximately 2 feet above the lockers for the cooling coils or units. It has been found convenient to maker lockers not more than six tiers high. The insider finish of the locker room should be of odor-free material. This is important since in long storage, even a slight trace of odor will taint the food. Such finishes as hard cement plaster, mastic, or an odor-free wood, such as spruce, are most commonly used.

INSULATION REQUIREMENTS

The arrangement of any plant should be such that it reduces the required insulation to a minimum and yet has the correct amount of insulation in each room to ensure the most economical operation possible. Refrigeration doors should be located so as to permit the least amount of heat from entering the refrigerated rooms from the outside. The quick-freeze room should be planned with as little outside area as possible and with the door opening into either the locker room or anteroom. This design will reduce the service load on this section and prevent excessive condensation on floor and walls of adjacent high-temperature rooms.

One of the primary considerations to be dealt with when designing a cold-storage or refrigeration plant will be the type of insulation material. Several materials are commonly used to prevent heat leakage through walls, floors, and ceilings. Cork has long been used for low-temperature insulation because it has a high resistance to heat flow. This property is largely due to the fact that cork is a cellular material, and each of the cells contains air. Air is one of the best known insulators, and any material that contains small air cells is a good insulator. The more thoroughly the air is confined, the better are the insulating characteristics. Many so-called dead-air spaces actually allow circulation of air to the extent that their value is reduced.

Insulating materials can be divided into two general classes: those that have both insulating value and structural strength, and those that have insulating value only. Certain insulating materials, such as cork and rock cork, are available in board or block form and have sufficient structural strength to be used for self-supporting partition walls. Other materials, such as shavings, sawdust, flakewood, redwood-bark fiber, and quilt materials must be held in place by an additional structure. When erecting any kind of insulation, the recommendations of the manufacturers should be carefully followed. Figure 21-2 shows a method of insulation used for masonry walls, while Fig. 21-3 shows how a concrete ceiling can be insulated.

The most important consideration in any cold-storage insulation is moisture. Every possible effort should be made to build cold-storage walls airtight; if they are not airtight, they will not be moisture-proof. Moisture tends to travel through a wall from the warm to the cold side. It enters the wall on the warm side as a vapor, but as it penetrates the wall, its temperature drops, and when the dew point is reached, condensation begins. If the temperature is below freezing, the moisture freezes and not only

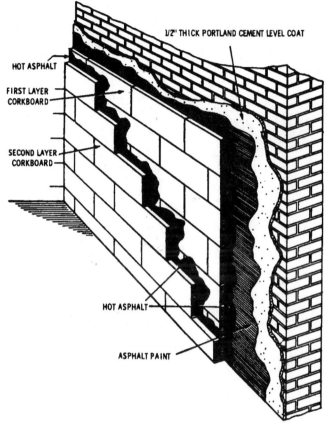

HOT ASPHALT

1/2" THICK PORTLAND CEMENT LEVEL COAT

FIRST LAYER CORKBOARD

SECOND LAYER CORKBOARD

HOT ASPHALT

ASPHALT PAINT

Fig. 21-2. Method of cold-storage insulation in masonry-wall construction.

reduces the resistance of the wall to heat flow, but also causes rapid deterioration.

EQUIPMENT AND CONTROLS

A locker refrigerating plant, in order to function effectively, must incorporate certain temperature controls and other devices in addition to the condensing unit, piping valves, and evaporator

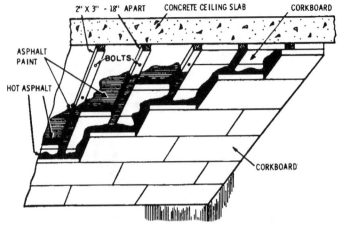

Fig. 21-3. Method of cold-storage insulation as applied to concrete ceilings.

coils. Control equipment has been developed to a high state of effectiveness, and several companies manufacture such equipment exclusively. The arrangement of equipment in a typical locker plant is shown in Fig. 21-4.

Evaporators

Although there are many forms of evaporators in use, the plate-type evaporator has found a constantly increasing use in locker-plant installations. A vacuum plate-type evaporator installation is shown in Fig. 21-5. Evaporators of this type are constructed especially for low-temperature applications and are made of steel tubing that has been flattened to give a greater area of contact with the plate; these tubes are encased in a welded steel shell. After fabrication, a vacuum is drawn on the plate, which ensures a positive contact between the tubing and the plate. All joints are welded, and the outside surfaces of the plate are sprayed with aluminum paint or molten zinc to provide a lasting and durable finish. The use of zinc-finished plates is recommended where they are above freezing temperatures (32°F).

Standard-size locker-room plates are 12 inches wide with four passes of tubing inside; they are usually available in lengths of 6, 7, 9, and 12 feet. Standard-size freezer-room plates are 22 inches

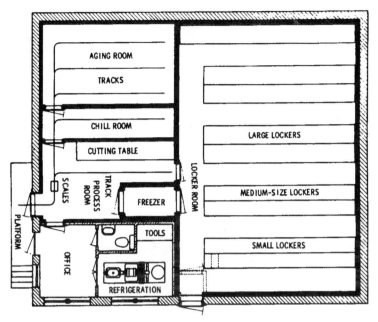

Fig. 21-4. Arrangement of equipment in a typical locker plant.

wide with eight passes of tubing; they are available in 4-, 5-, 6-, and 9-foot lengths. The refrigerant tubing on the outsider of all plates in which the refrigerant lines are connected is of ⅝-inch copper.

Locker-room plates are usually packed separately by the manufacturer and must be installed in banks by the installer. The plates in the banks may be connected together in series or parallel, as desired. With a series hookup, the number of plates per bank must be limited to the amount of tubing that can be handled successfully by one expansion valve. Experience shows that four 12-inch × 12-foot plates, or its equivalent in tubing, are the limit per bank in a series hookup.

With a parallel hookup, it is recommended that not more than six 12-inch plates be connected to one manifold, although there have been instances where eight plates operated successfully. Always use one expansion valve for each bank of plates. The two methods of assembling plates are shown in Fig. 21-6. After the

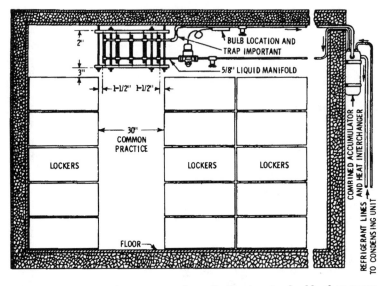

Fig. 21-5. Details of evaporator installation in a typical locker room.

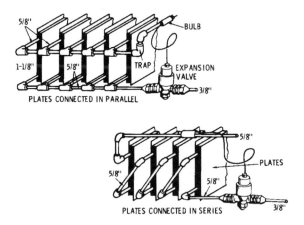

Fig. 21-6. Method of connecting evaporators.

plates are assembled, they should be suspended from the ceiling and provision made to raise or lower each corner of the bank to permit leveling, thus ensuring uniform frosting.

Freezer plates are always connected in series and installed as shelves in the freezer section (Fig. 21-7). A suitable rack should be made locally to support the plates. This rack may be of galvanized iron pipe or angle iron. The direction of expansion through the freezer plates should be from the bottom upward, using a suitable expansion valve. No more than four 6-foot × 22-inch plates should be used on one expansion valve. The spacing of the plates will depend on the type of products being frozen. For example, if packages the size of 2½-gallon ice cream containers are to be frozen, then the plates must be spaced accordingly to accommodate them. For small packages of meat, the plates can be spaced much closer. The common practice is to install the plates with varied spacings to accommodate any size package.

Liquid and Suction Lines

Liquid and suction lines must be of ample size to avoid pressure drop. The suction line leaving each bank of locker-room plates must be brought up above the top of the plates in order to form a trap. The expansion-valve bulb should be located at a point above the suction manifold (or on a series hookup above the plate outlet) to ensure flooding the tubes in the top plate (Fig. 21-6).

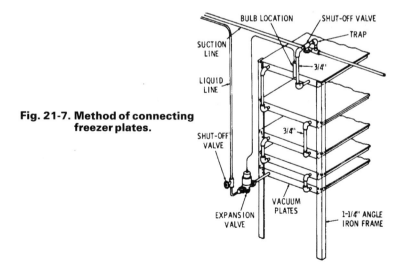

Fig. 21-7. Method of connecting freezer plates.

565

It is advisable to install a drier coil on each bank of plates in order to allow the expansion valve to be properly adjusted without excessive frost back. This will also permit each expansion valve to operate independently of the others without spillover, which would affect all the other expansion valves. In some instances, suction lines may be long enough within the room to accomplish this without a drier coil. The suction line from the freezer plates should also be installed so as to permit each expansion valve to function independently. The connection into the main suction line should be made so as to prevent oil-clogging due to faulty oil return (Fig. 21-7).

Hand Valves

Hand valves are not a necessity but will aid in servicing the equipment. Common practice is to install a hand shutoff valve ahead of each expansion valve and another one at the end of each bank of plates (Fig. 21-7).

Accumulators

Accumulators should be used on every installation. A combination accumulator and heat interchanger should be installed outside the low-temperature fixture but as close to the fixture as possible. Its purpose is to permit complete flooding of all plates with refrigerant without permitting the liquid refrigerant to enter the crankcase. It will also level out the usual intermittent oil return that is common on low-temperature evaporators.

Dehydrators

A permanent liquid-line dehydrator should be installed at the time of the original installation. In the event the dehydrating agent becomes saturated, it should be replaced. Dehydrators will eliminate moisture, which may cause sticking expansion valves and internal corrosion of various parts.

Oil Separators

An oil separator will greatly improve the operation of a locker system, but it is not to be considered a service necessity.

Controls

In most locker installations, the freezer and locker-room plates operate at the same refrigerant temperature, allowing products to freeze in the desired time unless unusual freezing demands are placed on the freezing-room plates. However, there are times, especially during cold weather, when the condensing unit does not operate a sufficient length of time to give the desired freezing. In any case, it will be necessary to control the operation of the condensing unit by a cold-control knob installed on the low-pressure switch. This will enable the user to change the switch settings to obtain the desired freezing time. The locker room in this instance must be controlled by a room thermostat operating a liquid-line solenoid valve (Fig. 21-8). Receiver capacities must be sufficient to hold the amount of refrigerant pumped out when the liquid-line solenoid closes.

Where the chill room is also connected to the same condensing unit, it will be necessary to add additional controls. When using

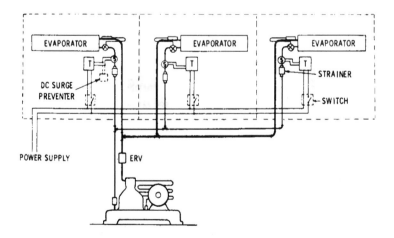

Fig. 21-8. Schematic arrangement of the multiple installation of solenoid valves in liquid lines: S, a solenoid valve; T, thermostat; ERV, evaporator regulating valve. When using solenoid valves on dc circuits of 64V or higher, it is necessary to use a surge preventer to properly protect the solenoid windings against the reactive surges set up when a dc circuit is opened.

a forced-air evaporator, experience has shown that an evaporator regulating valve and a snap-action valve are necessary. The evaporator regulating valve will maintain the operating pressure in the evaporator at or above a given point, while the snap-action valve will allow the evaporator to cycle and thus obtain defrosting.

Check valves must be used in the suction line leading to each low-temperature evaporator—either one large check valve installed in the main low-temperature suction line outside the low-temperature fixture, or an individual check valve after each bank of plates. The second recommendation is somewhat less desirable than the first, as it allows refrigerant to condense in the length of suction line in the locker room. Unless a check valve is used, the relatively warm gas from the chill-room evaporator will condense in the colder evaporators and keep the low-pressure switch from operating.

SUMMARY

Locker plants provide yet another method of food preservation. These are usually equipped in the same manner as large refrigerating plants, except that they are provided with locker rooms and special customer rooms in addition to the customary cutting, chill, and quick-freeze rooms. Construction and design methods, together with complete locker-plant layouts, have been fully described, including insulation requirements, controls, and such accessories as valves, accumulators, and oil separators as part of the condensing-unit operation.

REVIEW QUESTIONS

1. Describe the construction of community locker plants.
2. What is the floor plan arrangement in locker plants?
3. What are the functions of the chilled processing and quick-freeze rooms in locker plants?
4. State the approximate dimensions of individual lockers.
5. Describe the insulation requirements in locker plants.
6. What is the control arrangement in locker plants?
7. Describe the arrangement of liquid and suction lines.
8. Name accessories used in locker-plant operation.

CHAPTER 22

Special Refrigeration Applications

WATER COOLERS

Water coolers are manufactured in many sizes to suit various requirements. Water cooling equipment is divided into two classes: *bottle coolers* and *pressure or tap coolers*.

Bottle Coolers

The bottle type of water cooler differs from the pressure type mainly in that the former does not require any plumbing connection since the bottled water (usually distilled) is delivered to the customer's premises as required, whereas the pressure type depends on tap water connected to the cooler by means of a water supply line.

Bottle coolers (Fig. 22-1) lend themselves to office use, particu-

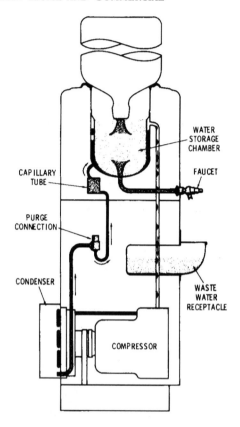

Fig. 22-1. Bottle water cooler with condensing unit and associated connections.

larly in temporary locations, since no plumbing is required. The only requirements are an electric source for the condensing unit and the availability of distilled water, which is usually delivered in 5-gallon glass bottles.

Condensing Unit—The condensing unit normally used in bottle coolers is of the hermetic or sealed rotary type, with a 120V, 60-hertz motor of approximately $\frac{1}{12}$ hp, Most condensing units of this type are equipped with a natural-draft finned-type condenser, which is so arranged and baffled that efficient circulation is ensured.

Cooling Chamber—With reference to Fig. 22-2, the evaporator consists of two concentric shells soldered together at the top and bottom. The liquid refrigerant enters the space between the shells at the bottom, and the suction line is connected at the top of the outside shell. A restrictor-type refrigerant control is located in the insulation of the cooler, and the thermostatic switch-bulb well is soldered to the evaporator. The inner shell forms the water-storage tank of ½-gallon capacity, and all parts of this cooling chamber are hot-tinned to prevent corrosion.

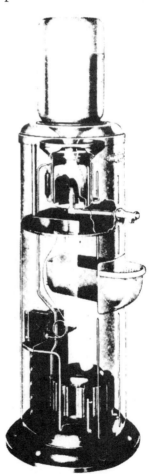

Fig. 22-2. Phantom view of a typical bottle water cooler.

Controls—Operation of the condensing unit is controlled by a thermostatic switch and cold control for temperature adjustment by the user. Motor protective relays are normally incorporated in the condensing unit.

Pressure Water Coolers

The pressure-type water cooler differs from the bottle cooler mainly in that the water is supplied from the building water lines as fast as it is used, whereas in the bottle cooler the water must be replenished from time to time. As shown in Fig. 22-3, the individual pressure is a self-contained, complete refrigerating

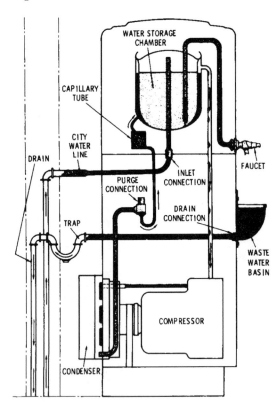

Fig. 22-3. Pressure water cooler showing the water and refrigerant connections.

system and water-cooling unit. It is equipped with all the necessary fittings for dispensing drinking water as well as catching and disposing of wastewater. This type of water cooler is usually equipped with a compact fractional-horsepower compressor (from ⅛ to ⅓ hp) and has capabilities ranging from 3 to 19 gal/hr.

Precooler—A precooler is a device in which the incoming water receives partial cooling from the wastewater, thus decreasing the quantity of heat to be removed by the compressor. The amount of heat removed by the precooler depends on the volume of wastewater as compared to the water drawn from the cooler. The amount of heat that the precooler will remove also depends on the length of the tubing comprising the precooler. In various designs of precoolers and water coolers, it may be said the precooler increases the capacity of the unit from 25 to 75 percent.

Remote Bubbler—There are times when an individual water cooler is located so that a proportion of those drinking from it are in another room. This means that they have a long way to go for a drink or that they do not get the amount of water they should have. To avoid this, individual water coolers designed for remote-bubbler connection are supplied by various manufacturers.

The water cooler should be located at the point where the greatest demand for water exists and the remote fountain (bubbler) run to the area of lesser demand. This run should never exceed a total of 15 feet. The pipe running from the water-cooler unit to the remote bubbler should be brass and no larger than ⅜ inch and preferably ¼ inch. The pipe must be insulated for ice water with cork or its equivalent.

Plumbing Connections—If the water inlet valve is brought through the floor, a shutoff valve should be installed under the flooring. After installing the shutoff valve, bring the water inlet line through the floor and up through the hole in the base of the cooler, where a reducing connector should be installed for connection to the copper tubing in the cooling unit.

To facilitate the installation of the drain circuit, a union should be installed in the pipe so that the union is accessible through the panel opening in the rear of the cabinet. The pipe should then be continued through the hole in the cooler base, through the floor to a trap, and then to the main drain. If water connections cannot be brought through the floor but must run through the wall, it may be necessary to bring the piping through the cabinet grilling in the rear of the cooler rather than through the cabinet shell. This will permit removal of the panel without disconnecting the water lines.

To install the drain line when the piping is to be brought out the back of the cooler, first install the connection by means of a nipple leading into the drip pan. A street ell should then be screwed into the elbow, and, by proper adjustment of the elbow, the street ell can be brought out through the service-panel opening.

Cycle of Operation for Typical Bottle Water Cooler

On a bottle-type model shown in Fig. 22-1, drinking water is supplied from an inverted bottle that rests on the top of the water cooler. The neck of this bottle extends down into a watertight extension above the cooling chamber. When water is drawn from the chamber, the water level falls below the outlet of the bottle, allowing air to enter the bottle and water to flow out until the water level covers the bottle opening.

Water from the bottle passes down to the cooling chamber through a baffle, which directs the incoming water along the refrigerated sides of the cooling chamber. Cooled water is drawn from the bottom of the cooling chamber and passes through a tube to the faucet at the front of the cooler. Wastewater from the faucet is caught in a receptacle that has a drain tube running from the chamber outlet to a connection at the rear of the water cooler, where it is sealed with a special safety plug.

Cycle of Operation for Typical Pressure Water Cooler

On pressure models, city water enters the cooler through the inlet connection at the rear of the cooler. Here the water passes through the precooling jacket surrounding the drain tube, where it is precooled by the wastewater in the drain. The precooled water from this jacket enters the cooling chamber at the top. A baffle inside the cooling chamber directs the incoming water down the sides of the cooling chamber close to the refrigerated walls. The outlet water tube runs close to the bottom of the cooling chamber where the coolest water lodges. The cooled water passes through this tube to the self-closing-valve pressure regulator and bubbler on some models, and to the faucet on other models.

Air that finds its way into the cooling chamber is released through a small hole in the outlet tube near the top of the chamber. The glass filler connection is connected in the outlet line between the cooling chamber and the self-closing valve. Water for a remote bubbler passes through a tube leading from the bottom of the water-cooling chamber to a connection at the rear of the water cooler. This connection is sealed with a special safety plug when a remote bubbler is not used. From the bubbler or faucet, wastewater passers through the catch-basin strainer to a short length of drain tubing leading directly to a drain connection at the rear of the cooler.

Remote Water Chillers

Water chillers for water fountains may be of the individual tube, where the compressor is located within the unit, or they may consist of devices that have the drinking fountain in one location and the refrigeration unit in another. In the latter case, they are called *remote water chillers*. These come in all sizes, and the size is usually the gallons per hour (gph) rating. They may be able to deliver 5 to 24 gph, 29 to 38 gph, or even more.

Figure 22-4 shows the 5- to 24-gph unit that is wall-mounted and has a recessed drinking fountain. It provides precooled water with modern fountain styling and can be concealed under a counter, between or within walls, or high on a wall or column (Fig. 22-5).

Typical installation when used with multiple-outlet circulating systems is shown in Fig. 22-6. A circulating system is recommended when multiple outlets are to be served and where it is not possible to install the chiller close to the cold-water outlet(s). In determining the number of fountains that can be hooked up to one chiller, be certain that the cooler capacity is adequate or not less than the total number of gallons needed for the location. It is recommended that all outlets be no more than 20 feet from the

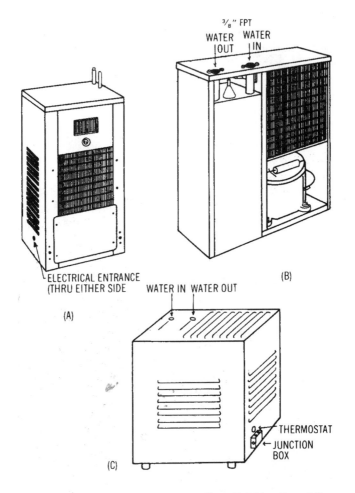

Haws Drinking Faucet Company

Fig. 22-4. Remote water chillers of the 5- to 24-gal. capacity.

chiller and all cold-water piping be covered with sponge rubber or ice-water-type insulation of adequate thickness.

Figure 22-6 shows a typical installation with one remote chiller serving several outlets; Fig. 22-7 shows how a typical installation

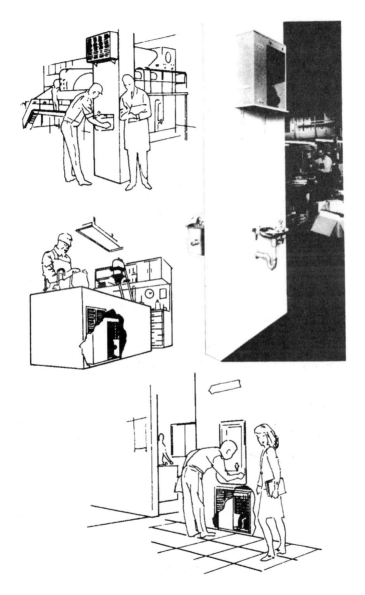

Haws Drinking Faucet Company

Fig. 22-5. Typical installations of remote water chillers.

with two remote chillers is connected in parallel. These coolers may be positioned side by side or as shown. Piping must be connected as shown no matter how the chillers are located. Circulating pumps must be installed in the return or supply line to the chiller, not in the cold-water-outlet line.

Figure 22-8 shows the 29- to 38-gph-size chillers. The horizontal model is recommended for locations where air doers not circulate freely, where room temperatures approach 100°F, or where excessive dust and lint are present. You must make provisions for the disposal of wastewater from the condenser.

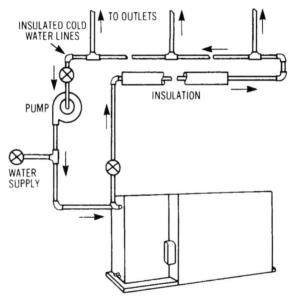

Haws Drinking Faucet Company

Fig. 22-6. Typical installation of a one-unit remote chiller.

These units operate with a hermetically-sealed refrigeration unit. The operation is similar to any refrigeration unit of similar design. The only variations are in the hook-up of the water supply and its return. Wastewater must also be eliminated into the sewer at the fountain location.

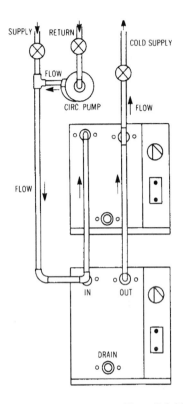

SUPPLY RETURN COLD SUPPLY

FLOW

CIRC PUMP FLOW

FLOW

FLOW

IN OUT

DRAIN

Haws Drinking Faucet Company

Fig. 22-7. Typical installation of a two-unit remote chiller connected in parallel.

ICE CREAM REFRIGERATION

Refrigeration employed in ice cream dispensing, as used in stores, consists of special prefabricated freezer cabinets of steel-shell construction and usually finished in a white painted finish. Sometimes the freezer is mounted on a stand only; but more often, a unit is made up of a storage cabinet and a hardening cabinet along with the freezer. The insulation (Fig. 22-9) usually consists of approximately 4 inches of corkboard on the sides, ends,

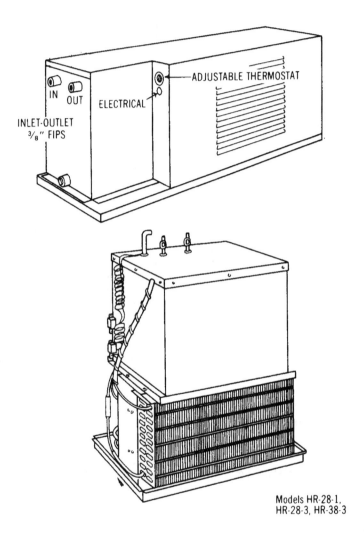

Models HR-28-1,
HR-28-3, HR-38-3

Haws Drinking Faucet Company

Fig. 22-8. Horizontal (A) and vertical (B) remote water chillers. This is the larger capacity unit with a capability of delivering 29 to 38 gal/hr.

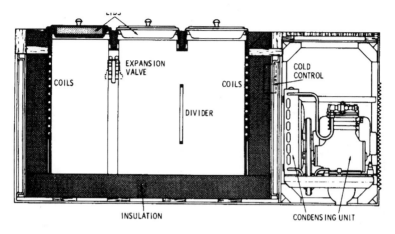

Fig. 22-9. Cross-sectional view of a typical ice cream cabinet.

and bottom supported by a steel framework. This insulation in a well-designed cabinet is encased in a steel shell and sealed against infiltration of moisture by an asphalt compound. The lids are commonly of the hinged type, having a soft rubber lip on the outer edge, which forms a tight seal against the soft rubber insulating collar at the top of each compartment.

The freezer capacity is usually 1, 2½, or 5 gallons per batch and is usually built integral with a hardening and storage space for 40, 60, or 80 gallons, depending on individual requirements. In the combination-plant setup, the hardening space is provided as an integral part of the unit. Often, it is necessary to supply one or two extra hardening cabinets for additional hardening and storage space. The load can be figured on these individual units exactly as in the large plants and summed up to give the total hardening load.

Condensing Units

Depending on the particular requirements, the condensing unit may be obtained as a part of the cabinet or mounted in a remote location. As a general rule, counter freezers are applicable for use with Freon-12 refrigerants. However, some of the old ammonia-type brine counter freezers are still in use. Although condensing-unit sizes may vary, Table 22-1 shows the approximate compressor-size requirements for various ice cream cabinet freezers.

Table 22-1. Freezer Requirements

Freezer, gal	Hardening Space, gal	gal/hr	Daily Capacity, gal	Approx. Compressor Size, hp
2½	None	12½	Continuous	1
2½	40	12½	40	1
2½	60	12½	60	1½
5	None	20	Continuous	1½
5	40	20	40	1½
5	60	25	60	2

Note: Compressors are to run at –20°F suction and 100°F condensing pressure on hardening cabinets.

It must be understood that Table 22-1 was prepared by one manufacturer for his own equipment and is offered only as an example. Different freezer and cabinet constructions may result in slightly different machine recommendations. A typical condenser unit used in ice-cream cabinets is shown in Fig. 22-10.

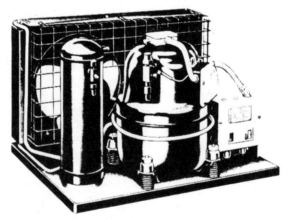

Tecumseh Products Company

Fig. 22-10. Exterior view of an air-cooled condensing unit suitable for remote operation of an ice cream cabinet.

Control Methods

Controls employed consist of the usual automatic expansion valve to control the amount of refrigerant entering the evaporator. This valve is normally insulated with a rubber and cork composition cover and installed in the control compartment. In addition, most units provide a combination accumulator and heat interchanger for improved expansion-valve operation and higher overall efficiency. Cold-control adjustment is made by a conventional thermostatic switch, which, in most cases, contains the "on" and "off" feature and motor protector element.

Multiplexing Ice Cream Cabinets

Multiplexing remote-control ice cream cabinets have been found satisfactory where the user knows which cabinet will be subjected to the heaviest service or require the coldest temperature. However, where the operating conditions cannot be determined with any degree of certainty, a method has been employed that is very flexible in operation. This method, shown in Fig. 22-11, employs a temperature-regulating valve on all cabinets with condensing-unit operations controlled by the low-pressure switch. The automatic expansion valves are still used to control the refrigerant to each cabinet. It is suggested that manifolds be

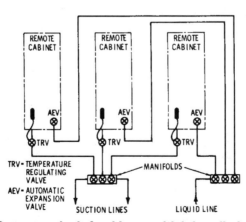

Fig. 22-11. Correct method of making a multiple installation of two or more ice cream cabinets on one compressor.

used because they permit servicing any one cabinet without materially affecting the operation of the other cabinets connected in the system.

BOTTLE AND CAN BEVERAGE COOLERS

The day of the chest-type beverage cooler is past; the present-day trend is toward upright automatic dispensers, which are coin-operated. These dispensers are of the dry type, with a fan circulating the cold air. They are self-contained, operating on 120V, 60-hertz ac power. The compressor and motor in self-contained bottler and can coolers are usually of the hermetic type with the condenser, fan, fan motor, shutoff valves, check valve, and relay all located in the machine compartment and assembled on a unit base.

Some reach-in units are used where beer and liquors are cooled in liquor stores. The refrigeration equipment on these is very similar to the grocery-type units, and servicing is the same.

MILK COOLERS

There are three general types of coolers used to cool milk on the farm: the *aerator type*, the *vat type*, and the *tube cooler*.

Aerator Type

Aeration is the cooling of milk by allowing it to flow by gravity over a surface cooler through which either cold water or brine is pumped to reduce the temperature. Principally, it consists of a series of tubes located one above the other so that the milk can be distributed over the top tube and allowed to drip from one tube to the other and be cooled by surface contact. Usually, this aerator is in two general sections: the top section, which uses water for a cooling medium, and the bottom section, usually supplied with refrigeration.

In most localities, water between 70 and 80°F may be obtained from city mains or a cooling tower and is used to cool the pasteurized milk from an initial temperature of 145°F down to some point between 75 and 80°F, depending entirely on the temperature and quantity of water available. Milk then enters the bottom section of the aerator, which is supplied with refrigeration, either directly in the direct expansion, or indirectly as in the brine or sweet-water type of aerator.

Vat Type

The *vat type* of pasteurizer and cooler combined, known generally as the vat type of pasteurizer, consists of a large porcelain or glass-lined vat containing a helical coil, which is submerged and rotating during the processing. During pasteurizing, either hot water or steam is circulated through this coil to supply the heat necessary to pasteurize the milk. Immediately after the pasteurizing is accomplished, tap water is circulated through the coil to bring the milk down to a point between 75 and 85°F, after which chilled brine or fresh water is circulated to furnish further reduction down to 40 or 45°F.

This type of equipment has an advantage in that the milk is handled only once, making it unnecessary to put it through a separate pasteurizer, precooler, and aerator. The chief disadvantage with this type of equipment lies in its high initial cost. Some slight difficulty is also sometimes experienced in cleaning it. The calculation of the refrigeration loads for the two setups is practically identical.

Tube Cooler

The third type of cooling equipment consists of a *concentric tube cooler* wherein the milk passes through a center tube while water for precooling and final chilling is passed through the angular space around the center tube within the outsider tube. Here, again, the refrigerator-load calculations are identical to those described previously.

Condensing Unit

The *condensing unit* (Fig. 22-12) is normally of the hermetic preassembled type, the size of the motor depending on the requirements. The condensing unit of this type of cooler is usually mounted on top of the cooling cabinet. In some instances,

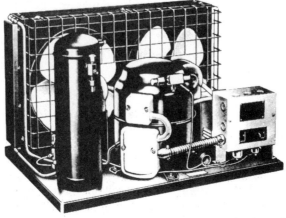

Fig. 22-12. Exterior view of a 2-hp, air-cooled, hermetic-type condensing unit for bulk-milk cooler units.

however, it may be desirable to install a milk cooler with the condensing unit at some location other than on top of the cabinet. In other instances, the milk-cooling tank may require two units to meet the heavy load conditions or to obtain the proper water circulation for a pulldown in the required time.

Agitator Unit

The *agitator*, as the name implies, provides the necessary water circulation by impeller action. The agitator assembly is designed for placement in the cooler tank through a hole cut in the top of the stationary lid. Since the agitator is independent of the condensing unit, it is driven by a small ac motor, usually about 1/10 hp, operating on 120V, 60-hertz, single-phase current, and is normally equipped with a wound-coil (magnetic) reset-type of overload protector.

SERVICING AND REPAIR

It is very important that every serviceman be able to properly diagnose any difficulty in special refrigeration units. As a general rule, service complaints will fall under three general headings:

1. *Unit Does Not Run*—The failure of a unit to run when plugged into an electrical outlet may be due to one of several reasons, which must be determined by the serviceman before changing the unit.

   a. The source of power should be checked with a test cord.
   b. Check for broken wires in the lead-in cord.
   c. Check the "off" and "on" switch to make sure the switch is in the "on" position.
   d. The proper type of power must be used.
   e. Check for a defective thermostatic switch that is not making contact at this point in the circuit.
   f. Check the motor-protector relay.
   g. Test the capacitor to determine if it is functioning properly.

   If all the foregoing points have been checked very carefully and still the unit refuses to run, then the unit must be changed.

2. *Unit Runs but Does Not Refrigerate*—This condition may occur just after installation if the cabinets have been stored in an extremely cold place or exposed to low temperatures during delivery and do not have time to warm up after being installed. It has happened, in some instances, that the cabinet was installed in a location where low outside temperatures were encountered. If this is not the cause of the unit failure, it will be necessary to change the unit.

3. *Unit Does Not Refrigerate Properly*—This condition may be due to one of several causes, which must be diagnosed very carefully by the serviceman before assuming the unit to be defective. Any one of the following may be at fault:

   a. Unit is not properly charged.
   b. Temporary low line voltage.
   c. Defective switch.
   d. Dirty condenser.
   e. Condenser fan motor does not operate.

Leaks in the System

In the event that any welded part of the condensing unit develops a leak, it will be necessary to change the unit. The leak will usually be discovered through the presence of oil around the point at which the leak develops. It must not be assumed, however, that the presence of oil on any part of the unit is an indication of a leak. An actual refrigerant leak may be discovered with a detector. A shortage of refrigerant in the system may result in partial frosting of the evaporator, which should be checked with the cold control set in the coldest position. If the entire evaporator doers not cool when the control is set in this position, the unit may be short of refrigerant.

Electrical Tests

Testing a motor in a sealed condensing unit to determine why it does not operate becomes a very simple process if the correct procedure is followed. Each test made should be one of a series of eliminations to determine what part of the system is defective. By checking other parts of the wiring system before checking the unit itself, a great deal of time can be saved since, in most cases, the trouble will be in the wiring or controls rather than in the unit.

In order to make a complete electrical test on electrical outlets and on the unit itself, it is advisable to make a test cord in the manner shown in Fig. 22-13. By connecting the black and white

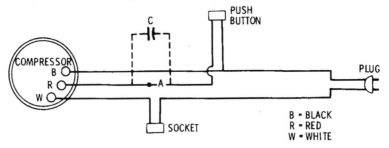

Fig. 22-13. Standard test cord.

terminal clips together and placing a lightbulb in the socket, the cord may be used to check the wall outlet into which the unit is connected. By connecting the white and red terminal clips, this

588

same test may be made by depressing the push button. This will serve as a test to make certain the push button is in working order. This test cord should have a capacitor installed in the red lead to the push button if the compressor is a capacitor-start-type unit. When these tests have been completed and it is known that current is being supplied to the unit, the next step is to check the three wires on the base of the compressor unit. Pull the plug from the wall receptacle, and carefully examine the nuts which hold the wires in position. Then try each wire to be sure it is held firmly in place, as a loose wire may keep the unit from operating. Test the thermostatic switch to determine whether or not contact is being made at that point. Turn the cold-control knob several times. If this fails to start the unit, then short across the thermostatic switch terminals on the switch. To do this, it will be necessary to remove the switch cover from the top of the switch. If the unit starts, it is an indication that the thermostatic switch is not operating properly and must be repaired or replaced. After the thermostatic switch has been checked and if the trouble is not located, it will be necessary to determine whether the trouble is in the motor, motor-protector relay, or capacitor.

Capacitor Test

The capacitor must be checked before testing the unit itself. This is done in the following manner:

1. Disconnect the capacitor wires from the motor-protector relay.
2. Connect these two wires to the black and white terminals of the test cord.
3. Put a 150W lightbulb in the receptacle on the test cord, and plug it into an outlet.

If the 150W bulb does not light, it is an indication that the capacitor has an *open circuit* and must be replaced. However, if the bulb does light, it is not an indication that the capacitor is perfect. This must be checked further by shorting across the two terminals of the capacitor with a screwdriver with an insulated handle. If the brilliancy of the light changes—that is, if the lightbulb burns brighter when the terminals of the capacitor are shorted—it indicates that the capacitor is in proper operating condition. A decided sparking of the terminal will also be noticed when the terminals are shorted. If the brilliancy of the bulb does not change, it is an indication that the capacitor has an internal short circuit and must be replaced.

Replacing the Motor-Protector Relay

If the motor-protective relay is found defective during the preceding test, replace it. The motor-protective relay is usually accessible for replacement from the rear of the cabinet. Remove the nuts holding the relay in position, and remove the lead clamp. Notice the position of the electrical leads before replacing them. Place the electrical leads on the replacement relay, and check them for correct connections against the wiring diagram.

MILK PASTEURIZATION

Although the farm dairy is not primarily interested in pasteurization of milk, it will be well to bring you up to date on pasteurization methods. The newest and most accepted method of pasteurization is what is called the high-temperature-short-time method, which involves temperatures of 155 to 185°F. Ordinary milk for consumption is usually raised to a temperature of 161°F for a minimum of 15 seconds and then quickly brought down to a temperature of 40 to 45°F. This is often accomplished by running the hot milk slowly over a series of horizontal refrigerated tubes, thus allowing the removal of the higher temperature until the desired final temperature is reached.

Simple Load Calculations

Assume that the processing plant handles 800 gallons of milk daily. To cool the milk from a pasteurizing temperature of 161 to 40°F, the following assumptions are made:

1. The specific gravity of milk is 1.03.
2. The weight of milk is 8.6 lb/gal.
3. The weight of water is 8.34 lb/gal.
4. The specific heat of milk is 0.95.

The equation for heat load is

$$H = WS\,(t_1 - t_2)$$

where H = Btu required
W = weight of fluid being cooled
S = specific heat Btu/lb cooled
t_1 = higher temperature
t_2 = lower temperature

Substituting values in the preceding equation, we obtain

$$H = 800 \times 8.6 \times 0.95(161 - 40) = 790,856 \text{ Btu}$$

If it is assumed that ample water is available, it is entirely feasible to cool milk with water to 80°F. Consequently, the heat to be removed by water is

$$H = \frac{790,856 \times 81}{121} = 529,416 \text{ Btu}$$

If the water used for cooling is allowed to rise 10°F, the amount of water necessary is

$$\frac{529,416}{10 \times 8.34} = 6348 \text{ gallons}$$

If a 4-hour period is allowed for the cooling, the amount of water necessary per hour is 6348 ÷ 4, or 1587 gallons. If 25 percent of the water is allowed for losses, the gal/min requirement of water finally is

$$\frac{1587 \times 1.25}{60} = 33 \text{ gal/min (approx.)}$$

To cool milk from 80 to 40°F, the amount of heat to be removed is

$$H = 800 \times 8.6 \times 0.95(80 - 40) = 261,440 \text{ Btu}$$

The preceding figures, of course, will not apply to the home dairyman, as he is not interested in the pasteurization of the milk. What he is interested in is the size of condensing units and tank capabilities to take care of two milkings a day. The manufacturer

591

of the tank and condensing unit will arrive at the sizes of the equipment, given the gallons per day that are to be processed.

SUMMARY

Special refrigeration applications covered include water coolers, ice cream cabinets, bottle and can beverage coolers, and milk coolers. In each of the foregoing, condensing-unit equipment and control features have received particular attention.

Insulation requirements, plumbing methods, and multiple-condensing-unit operation are fully described, in addition to a service and repair section. The latter part of the chapter provides instructions for using various testing methods to correct operation difficulties.

REVIEW QUESTIONS

1. Name the various classes of water coolers.
2. Describe the difference between bottle and pressure water coolers.
3. What is the arrangement of the plumbing connections in pressure water coolers?
4. Describe the cycle of operation in bottle and pressure water coolers.
5. State the arrangement of the refrigerating components in ice cream dispensing cabinets.
6. What are the control methods used in ice cream refrigeration?
7. What is meant by multiplexing ice cream cabinets?
8. Describe bottle and can beverage cooler cabinets.
9. What are the three general types of milk coolers used on the farm?
10. Describe the servicing and repair methods used for special refrigeration units.

CHAPTER 23

Cold-Storage Practice

The primary reason for cold storage of perishable foods is its preservation from decay and spoilage, which will impair its usage. Refrigeration not only has saved incalculable quantities of meat, fish, fruits, eggs, milk, and cream from spoilage, but has played an important part in changing the diet of the world. No longer are the inhabitants of one hemisphere, country, or locality dependent on local foodstuffs, but they may draw on the entire world as a source. This has made possible great developments in agriculture and livestock-raising in countries far distant from potential markets and has permitted the utilization of the great productivity of the tropics in supplying fruits and foods for other zones.

Nearly all of the industries involved in the preparation of food and drinks make large use of artificial ice or mechanical refrigeration. For example, the dairy industry finds that, for precooling milk and cream and in the manufacturer of butter and ice cream, refrigeration is indispensable.

QUICK FREEZING

There are several reasons for the popularity of the *quick-freezing method* for preserving fruits and vegetables. Perhaps the most important is the timer and labor saved as compared to the usual home-canning method. Other reasons are the preservation of flavor and color so as to approach very nearly the fresh product. Some farm products that may be frozen and stored are berries, peaches, apricots, plums, cherries, citrus and tomato juices, asparagus, peas, lima beans, snap beans, corn, cauliflower, broccoli, spinach, brussels sprouts, and squash. When handling farm products on a large scale, it is important that the planting supervision of the crops in the field is timed so that a properly ripened and matured product will flow in an even manner to the freezing plant, and thus avoiding larger peak-freezing loads, which the plant may not be equipped to handler.

In general, it may be said that the greater the speed of the freezing, the better the products will be in texture and taste upon thawing. At present, locker-plant freezing is sometimes of one variety and sometimes of another. It is proper, however, to quick-freeze everything going into a locker. The usual apparatus for quick freezing in a locker plant is a series of cold plates or shelves through which air may or may not be blown.

The temperature of the room in which meats, fruits, or vegetables are quickly frozen before they are placed in the locker varies from –5 to –15°F. Most items are individually wrapped in small packages to enable quick penetration of the cold. The time required for quick freezing ranges from 3 to 6 hours.

Frozen foods have become a very large industry, so much so that one may purchase most any item in the frozen-foods department of a supermarket. Prepared dinners, fish, vegetables, juices, ice cream, and most any other item that one should wish for can be purchased frozen. Frozen foods, however, should not be thawed and then refrozen.

FOOD PREPARATION FOR QUICK FREEZING

Blanching

The method of preparing most vegetables for quick freezing is known as *blanching*. This procedure consists in partly cooking and softening the vegetables for freezing; it aids in cleansing and produces a nearly natural color. The main purpose, however, is to kill bacteria and mold. Once killed, bacteria and mold do not multiply at freezer or storage temperatures, but they should be killed by blanching before freezing.

Blanching is done by placing the vegetables in boiling water the length of time specified for different varieties. After blanching, they should be placed immediately in cold water to remove the heat as rapidly as possible. It is also highly important that products for freezing be placed in the freezer as soon after blanching and packaging as possible. Fruit jars, jelly glasses, and waxed cups are generally satisfactory for the packaging of frozen products. Waxed containers are the most satisfactory for most purposes. Class containers should have the lids loose during the freezing period to prevent breakage. In filling any container, care should be used not to fill to the top. An air space should be allowed for expansion during freezing.

Sugar and Salt Packing

Sugar can be used in two ways—dry or in solution. When used dry, the sugar should be fairly well distributed over the product. This method serves partly to control oxidation when applied to strawberries, red raspberries, and loganberries and is particularly effective with fruits such as apricots and peaches.

Sugar solutions are made hot and then cooled before being applied to the fruit. Densities are determined on the basis of weight. To make a 50% density solution, for instance, use 5 lb sugar and 5 lb water; while to make a 60% solution, use 6 lb sugar and 4 lb water. The solutions must be cooled before pouring over the fruit. Salt solutions are also made on a weight basis.

Preparation of Fruits

Fruits are packed either drained or in syrup. The use of syrup is advantageous in most cases because it protects the fruit from the air.

Blackberries—Wash and sort the berries. For the best results, pack the fruit with sugar 3:1, or with a 50% syrup solution. Paraffined cups are recommended.

Cranberries—Select the riper and more colored berries; sort and wash. Barely cover the fruit with a 50% syrup. Cranberries may be frozen, however, without sugar or syrup. Paraffined cups are recommended.

Loganberries—The loganberry is well adapted to freezing, with ripe, firm berries being best. Wash the fruit carefully and pack in paraffined cups, glass jars, etc. Add sugar at the rate of 2:1, taking care to cover the fruit well. If syrup is used, a 50 to 60% solution is best.

Black Raspberries—Wash, sort, and pack in containers, covering with sugar 3:1. If syrup is used, a 40 to 50% solution is satisfactory.

Red Raspberries—Wash, sort, and pack. Sugar 3:1, or use a 50% syrup solution.

Strawberries—The fruit should be picked when well colored and ripe but not soft. Cap and wash, and then pack in containers with a 3:1 sugar ratio, or a 60% syrup solution. Wax cups may be used.

Youngberries—The youngberry is a mild-flavored large berry that is very well adapted to freezing. After washing and sorting, pack 3:1 with sugar, or use a 30 to 40% syrup solution.

Black Cherries—Use only well-ripened fruit that has been washed and sorted. Stems may be left on or removed. The syrup pack is recommended for best results, using a 40 to 50% solution.

Sour Cherries—Use only bright red, tree-ripened fruit with a slightly acid taste. Wash, stem, pit, and sort carefully; then pack the fruit in waxed cups, using sugar 5:1 or a cold syrup of 60%.

Apricots—Apricots are in the best condition for freezing when firm and ripe, showing good color and maturity. Soft fruit is to be avoided because freezing contributes to a loss of firmness. Keep the fruit cool and handle it quickly. Avoid bruising. Wash the fruit carefully; then halve and pit. Peeling is not necessary-in fact, the skins help to hold the halves more firmly together. If peeling is desired, however, the fruit can be dipped in boiling water and subjected to steam or lye, as practiced in commercial canning. If lye is used, the fruit should be rinsed in a weak citric-acid bath. A syrup pack is recommended for this fruit due to its rapid oxidation. A 40 to 50% solution is satisfactory.

Figs—Wash and sort carefully, removing the stem up to the base of the fig. Pack the fruit without peeling in paraffined cups. Use the dry pack or syrup with a density of 35%. In most cases, the syrup pack will be found better.

Grapes—Many of the popular varieties of grapes can be frozen. Tokay, concord, muscat, and others are suitable. Maturity is essential to obtain full flavor. Wash, sort, and stem carefully, placing the fruit in glass jars or wax containers. Cover the grapes with a syrup of 40% density.

Peaches—When selecting peaches, choose only those of high quality with predominant flavor. The fruit should be firm and ripe. Handle it quickly to prevent oxidation. Sliced peaches are very much better than halves but require more care. A 50% syrup pack is recommended. They should be placed in the quick freezer as soon as possible.

Preparation of Vegetables

Vegetables are packed either well drained or in brine. The use of brine is recommended in most cases.

Asparagus—Frozen green asparagus, when properly handled, are satisfactory. Careful sorting is essential to obtain good, succulent and tender stalks. Prepare and pack quickly to avoid shrinkage or shriveling. Blanch for 2 or 3 minutes in boiling water, and then chill quickly in cold water. Pack and seal in airtight containers without further treatment. If a brine pack is desired, containers that do not seal airtight may be used. A 2% brine is preferred.

Beans—Green or waxed beans are very satisfactory to quickfreeze. Wash and blanch for 2 or 3 minutes. Dip in cold water, drain and chill quickly, and then pack in glass jars or waxed containers.

Lima Beans—Lima beans should be harvested while still young and tender, shelled, blanched in boiling water for 2 or 3 minutes, and then cooled in water. Drain well and pack in glass or waxed containers.

Broccoli—Broccoli has been frozen with unusual success, the product never losing its characteristic fresh green color. Use only the tender stalks with compact heads, sorting carefully, and cutting back the stems to the part that is tender. Blanch in boiling water 3 or 4 minutes and cool in fresh rinse water. Drain and pack in glass or waxed cups or boxes.

Cauliflower—Cauliflower can be frozen after being carefully trimmed. Remove all the green leaves, cut the larger flowerets apart, and then soak for a short time in a weak brine solution. Blanch in boiling water for 2 or 3 minutes, cool and drain promptly, and pack in glass or waxed containers. Freeze in either a 2% brine solution or as a dry pack.

Sweet Corn—Obtain corn of proper maturity—in the full milk stage. Blanch in boiling water for not more than 6 minutes; then cool in fresh water. Wrap each ear in wax or parchment paper, and pack in tin cans or waxed boxes. For cut corn, blanch as outlined and cut from the cob. Use care not to cut or scrape the cob. Pack the cut corn in glass or in waxed cups or boxes, covering with a 2% brine solution.

Mushrooms—The small button-sized mushrooms are best for freezing. Care must be taken in handling so as not to damage the caps lest discoloration appear. Sort for size, if necessary, and wash carefully. Blanch in boiling water for 2 to 4 minutes (depending on the size of the mushrooms) and cool rapidly in water. Pack in glass or waxed cups, preferably with a 2% brine solution.

Peas—Obtain peas that are absolutely fresh from the vine; otherwise they will be of a high starch content. Hull and wash. Blanch in boiling water for 1 minute, cooling quickly in plenty of cold water. The peas can be packed dry or in 2% brine. Pack in glass jars, parchment-lined wax boxes, or waxed cups.

Spinach—Although somewhat difficult to handle, spinach makes a very good frozen product. Care must be taken to see that the spinach is not too advanced in maturity. It should be well washed to remove all sand and grit. Blanch in boiling water for 2 to 2½ minutes. Rinse well in cold water, drain, and pack without added liquid in glass jars, parchment-lined waxed boxes, or in waxed cups. Table 23-1 lists the storage temperatures for various foods and is supplied through courtesy of The Frick Company, Waynesboro, Pa.

Table 23-1. Condensed List of Storage Temperatures

Products	How Packed	Temp., °F
Almonds	In shells	34-38
Apples	Barrels	30-32
Apricots	Baskets	40-45
Beans	Green, bushel baskets	36-40
Berries	Short periods, 3 or 4 days	36-40
Butter	Boxes or crates	38-40
Cabbage	Bulk	32-36
Cabbage	Boxes or crates	32-33
Calves	Fresh, held in chill	30-36
Calves	Fresh, short time	36-38
Candy	Bulk or boxed	60-65
Canteloupes	Box or crate, 3 weeks	34-36
Carrots	Bushel crates	34-40
Cherries	Fresh, in boxes or crates	36-40
Cherries	Frozen	0-5
Cheese	Cream, Limburger, etc.	38-40
Condensed milk	Bulk	36-40
Corn	Green, bushel crates	36-40
Corn	Dried, sacks	40-45
Cream	Fresh, 40-qt cans	32-36
Cream	Condensed	36-40
Currants	In boxes	36-40
Dates	Cured, boxes	40-45
Eggs	Crates, 30 dozen	28-30
Figs	Dried	40-45
Fish	Fresh, short periods	25-30
Fish	Frozen, regular storing	0-10
Fish	Dried, in boxes or barrels	35-40
Flour	Any kind, from cereals	35-40
Fruit	Usual boxes or baskets	35-40
Fruits and vegetables	Dried	40-45
Furs	Coats, for summer season	25-30
Furs	Rugs of animals, or stuffed	25-30
Game	Frozen	0-10
Game	In cooler, short period	25-30
Grapes	Large baskets	34-40
Grapefruit	Boxes	35-40

Table 23-1. Condensed List of Storage Temperatures (Cont'd)

Products	How Packed	Temp., °F
Hams, etc.	Cured	35-40
Hominy	Boxes	35-40
Lard	Barrels or cans	36-40
Lemons, limes	Boxes	35-40
Lettuce	Bushel crates	35-37
Maple sugar	Boxes or syrup	40-45
Meat	Frozen	0- 5
Meat	In chill	30-32
Meat	In cooler	36-38
Meat	Cured or smoked	40-45
Milk	Fresh, 40-qt cans or bottles	38-40
Milk	Condensed	36-40
Molasses	Barrels	40-45
Mutton	Fresh, short period	30-32
Nuts	In shells	35-40
Oils, lard, cottonseed, etc.	Barrels or cans	35-40
Oleomargarine	Boxes	25-30
Olive oil	Barrels	35-40
Onions	Sacks	32-36
Oranges	Boxes	35-40
Oysters	In shells, sack	36-42
Oysters	In tubs or cans	30-34
Peaches, pears	Bushel baskets	35-40
Peas	Green, bushel crates	36-40
Pineapple, plums	Boxes	40-45
Poultry	In freezer	0-10
Poultry	Cooler, short period	36-38
Potatoes	Sacks	34-38
Rice	Sacks	40-45
Skins	Dried	25-30
Skins	Uncured	30-35
Strawberries	Qt boxes, crated	36-40
Vegetables	In barrels or crates	36-40
Watermelons	Short period	34-36
Wines	Barrels	40-45

SUMMARY

Cold storage, its purpose for storage of perishable food, and the methods used have been fully described in this chapter. Quick freezing and the preparation of food, including blanching, have been discussed in detail. The preparation and packaging of common fruits and vegetables prior to freezing have been given considerable space, and it is hoped that this important feature will be helpful for the individual home owner as well as lockerplant operators and others interested in modern cold-storage practice.

REVIEW QUESTIONS

1. Describe quick-freezing methods for fruits and vegetables.
2. What are the preparations necessary for quick freezing?
3. In general, how does the speed of the quick-freezing method affect the product?
4. What should be the temperature of fruits and vegetables prior to food-locker placement?
5. What is meant by the term *blanching*, and what is the procedure?
6. Describe the reasons for adding salt and sugar solutions to various products.
7. What types of containers are used for storage of quickfrozen foods?
8. Why is brine used in the preparation of vegetables for quick freezing?
9. Enumerate the several vegetables that are successfully stored by the quick-freezing method.

CHAPTER 24

Fans and Blowers

Fans and blowers are used in refrigeration and air conditioning plants to provide cooling and airflow. Depending on the particular application and construction principles, they are sometimes termed exhausters or propeller fans. Fans may be classified according to the construction as *centrifugal*, or *radial flow*, and *axial flow*.

In the centrifugal fan, the air flows radially through the impeller within a cylinder, or ring. Fan performance may be stated in various ways, the air volume per unit time, total pressure, static pressure, speed, and power input being the most important. The terms, as defined by the National Association of Fan Manufacturers, are as follows:

1. The *volume* handled by a fan is the number of cubic feet of air per minute expressed as fan-outlet conditions.
2. The *total pressure* of a fan is the rise of pressure from fan inlet to fan outlet.

3. The *velocity pressure* of a fan is the pressure corresponding to the average velocity determination from the volume of airflow at the fan-outlet area.
4. The *static pressure* of a fan is the total pressure diminished by the fan velocity pressure.
5. The *power output* of a fan is expressed in horsepower and is based on fan volume and fan total pressure.
6. The *power input* to a fan is expressed in horsepower and is measured horsepower delivered to the fan shaft.
7. The *mechanical efficiency* of a fan is the ratio of power output to power input.
8. The *static efficiency* of a fan is the mechanical efficiency multiplied by the ratio of static pressure to the total pressure.
9. The *fan-outlet area* is the inside area of the fan outlet.
10. The *fan-inlet area* is the inside area of the inlet collar.

With respect to the foregoing, it should be noted that while the total pressure truly represents the actual pressure developed by the fan, the static pressure may best represent the useful pressure for overcoming resistance.

PROPELLER FANS

A propeller fan, as noted in Fig. 24-1, consists of a propeller or disk-type wheel within a mounting ring or plate, and includes the mechanical supports for either a belt drive or direct connection. Propeller fans move the air in an axial direction by propulsive force in an action similar to that of a ship's propeller. Fans of this type are usually termed *exhaust*, or *ventilating*, fans. When propeller fans are made with flat or nearly flat blades, they are sometimes termed *disk fans*.

CENTRIFUGAL FANS

A centrifugal fan (Fig. 24-2) consists of a fan rotor or wheel with a scroll-type housing and includes drive-mechanism supports for

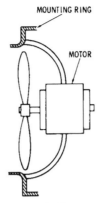

Fig. 24-1. Mounting arrangement of a propeller fan.

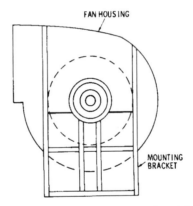

Fig. 24-2. Mounting arrangement of a centrifugal fan.

either belt drive or direct connection. This type of fan may have an inlet on one or both sides. Those with one inlet are usually termed *exhausters*, while those with two inlets are called *blowers*.

FAN CHARACTERISTICS

Fans and blowers may operate with equal efficiency at any speed that their mechanical strength allows. The efficiency, however, varies with the different orifice area but not according to any fixed law for all designs. The performance curves of a 24-inch centrifugal fan are shown in Fig. 24-3. When designing or estimating the performance of any fan, such performance characteristics are the basis for selection of the motor and associated controls.

POWER REQUIREMENTS

When determining the horsepower requirements for a fan or blower, it is essential to remember that for a certain discharge orifice the power required varies directly as the cube of the speed. The horsepower requirements of a centrifugal fan generally decrease with a decrease in the area of the discharge orifice if the speed remains constant. The horsepower requirements of a propeller fan increase as the area of the discharge orifice decreases if the speed remains unchanged.

MOTOR TYPES

Fans used in refrigeration and air conditioning are usually driven by electric motors. Small-size fans in the higher-speed ranges are usually equipped with direct-connected motors, whereas larger-size fans and those operated at lower speeds are generally V-belt driven.

When selecting a motor for fan operation, it is advisable to select the next larger standard size than the fan requires. However, it should be kept in mind that direct-connected fans do not require as great a safety factor as belted units. It is desirable to employ the belt drive when the required fan speed or horsepower requirement is in doubt since a change in pulley size is relatively inexpensive if an error is made.

Directly connected small fans for various applications are usually driven by a single-phase ac motor of the split-phase, capacitor, or shaded-pole type. The capacitor motor is more efficient electrically and is used in districts where there are current limitations. Such motors, however, are usually arranged to operate at only one speed. In such cases, where it is necessary to vary the air volume or pressure of the fan or blower, the throttling of air by a damper installation is usually made. In large installations, such as when mechanical draft fans are required, various drive methods are used, such as:

1. A slip-ring motor to vary the speed
2. A constant-speed, directly connected motor, which, by means of movable guide vanes in the fan inlet, serves to regulate the pressure and air volume

FAN SELECTION

Most often, the service determines the type of fan to use. When operation occurs with little or no resistance, particularly when no duct system is required, the propeller fan is commonly used because of its simplicity and economy in operation. When a duct system is involved, a centrifugal or axial type of fan is usually employed. In general, centrifugal and axial fans are comparable with respect to sound level, but the axial fans are somewhat lighter

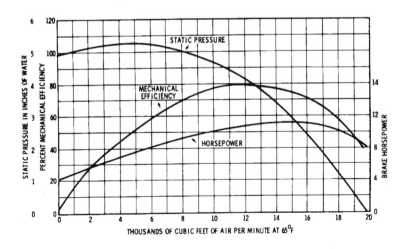

Fig. 24-3. Performance characteristics of a 24-inch centrifugal fan with backwardly inclined blades when running at 1600 rpm.

and require considerably less space. The following information is usually required for proper fan selection:

1. Capacity requirement in cubic feet per minute
2. Static pressure or system resistance
3. Type of application or service
4. Mounting arrangement of system
5. Sound level or use of space to be served
6. Nature of load and available drive

The various fan manufacturers generally supply tables of characteristics, which usually show a wide range of operating particulars for each fan size. The usual tabulated data include static pressure, outlet velocity, revolutions per minute, brake horsepower, and tip, or peripheral, speed.

SUMMARY

Fans and blowers are used in refrigeration and air conditioning plants to provide cooling and airflow. Fans may be classified as either centrifugal or radial flow or axial flow.

The centrifugal fan makes the air flow radially through the impeller within a cylinder or ring. The fan performance may be stated as volume of air per unit time, total pressure, static pressure, speed, or power input to the motor.

A propeller fan consists of a propeller or disk-type wheel within a mounting ring or plate and includes the mechanical supports for either a belt drive or direct connection. When propeller fans are made with flat or nearly flat blades, they are sometimes referred to as disk fans.

Fans and blowers may operate with equal efficiency at any speed that their mechanical strength allows. The efficiency, however, varies with the different orifice area but not according to any fixed law for all designs. There are certain performance curves furnished with each fan by the manufacturer.

The horsepower requirements for a centrifugal fan generally decrease with a decrease in the area of the discharge orifice if the speed remains constant. The horsepower requirements of a propeller fan increase as the area of the discharge orifice decreases if the speed remains unchanged.

REVIEW QUESTIONS

1. Describe the function of fans and blowers in a refrigeration system.
2. Give two classifications of fans employed in a refrigeration system.
3. Describe terms as defined by the National Association of Fan Manufacturers.
4. What is a propeller fan, and how does it operate?
5. How is the efficiency of a fan affected by the different orifice areas?
6. What are the horsepower requirements for a fan?
7. What drive methods are employed in fan operation?
8. Describe the various types of motors employed in fan operation.
9. What criteria are used in fan selection?

CHAPTER 25

Refrigeration Piping

The success of any refrigeration plant depends largely on the proper design of the refrigeration piping and a thorough understanding of the necessary accessories and their functions in the system. In sizing refrigerant lines, it is necessary to consider the optimum sizes with respect to economics, friction losses, and oil return. It is desirable to have line sizes as small as possible from the cost standpoint. On the other hand, suction- and discharge-line pressure drops cause a loss of compressor capacity, and excessive liquid-line pressure drops may cause flashing of the liquid refrigerant with consequent faulty expansion-valve operation.

Refrigerant piping systems, to operate successfully, should satisfy the following:

1. Proper refrigerant feed to the evaporators should be ensured.

2. Refrigerant lines should be of sufficient size to prevent an excessive pressure drop.
3. An excessive amount of lubricating oil should be prevented from being trapped in any part of the system.
4. Liquid refrigerant should be prevented from entering the compressor at all timers.

PRESSURE-DROP CONSIDERATIONS

Pressure drop in liquid lines is not as critical as it is in the suction and discharge lines. The important thing to remember is that the pressure drop should not be so great as to cause gas formation in the liquid line and/or insufficient liquid pressure at the liquid-feed device. A system should normally be designed so that the pressure drop due to friction in the liquid line is not greater than that corresponding to 1 to 2° changer in saturation temperature. Friction pressure drops in the liquid line include the drop in accessories, such as the solenoid valve, strainer-drier, and hand valves, as well as in the actual pipe and fittings from the receiver outlet to the refrigerant feed device at the evaporator.

Friction pressure drop in the suction line means a loss in system capacity because it forces the compressor to operate at a lower suction pressure to maintain the desired evaporating temperature in the coil. It is usually standard practice to size the suction line to have a pressure drop due to friction not any greater than the equivalent of a 1 to 2° change in saturation temperature.

LIQUID REFRIGERANT LINES

The liquid lines do not generally present any design problems. Refrigeration oil is sufficiently miscible with commonly used refrigerants in the liquid form to assure adequate mixture and positive oil return. The following factors should be considered when designating liquid lines:

1. The liquid lines, including the interconnected valves and accessories, must be of sufficient size to prevent excessive pressure drops.

2. When interconnecting condensing units with condenser receivers or evaporative condensers, the liquid lines from each unit should be brought into a common liquid line, as shown in Fig. 25-1.

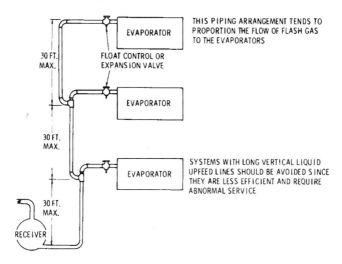

Fig. 25-1. Piping diagram showing liquid riser connections when evaporators are located at different levels above the receiver.

3. Each unit should join the common liquid line as far below the receivers as possible, with a minimum of 2 feet preferred. The common liquid line should rise to the ceiling of the machine room. The added head of liquid is provided to prevent, as far as possible, hot gas blowing back from the receivers.

4. All liquid lines from the receivers to the common line should have equal pressure drops in order to provide, as nearly as possible, equal liquid flow and prevent the blowing of gas.

5. Remove all liquid-line filters from the condensing units, and install them in parallel in the common liquid line at the ceiling level.

6. Hot gas blowing from the receivers can be condensed in reasonable quantities by liquid subcoolers, as specified for

613

the regular condensing units, having a minimum lift of 60 feet at 80°F condensing medium temperature.

7. Interconnect all the liquid receivers of the evaporative condensers above the liquid level to equalize the gas pressure.
8. The common and interconnecting liquid line should have an area equal to the sum of the areas of the individual lines.
9. Install a hand shutoff valve in the liquid line from each receiver.

Where a reduction in pipe size is necessary in order to provide sufficient gas velocity to entrain oil up the vertical risers at partial loads, greater pressure drops will be imposed at full load. These can usually be compensated for by oversizing the horizontal and downcomer lines to keep the total pressure drop within the desired limits.

INTERCONNECTION OF SUCTION LINES

When designing suction lines, the following important considerations should be observed:

1. The lines should be of sufficient capacity to prevent any considerable pressure drop at full load.
2. In multiple-unit installations, all suction lines should be brought to a common manifold at the compressor, as shown in Fig. 25-2.
3. The pressure drop between each compressor and main suction line should be the same in order to ensure a proportionate amount of refrigerant gas to each compressor, as well as a proper return of oil to each compressor.
4. Equal pipe lengths, sizes, and spacings should be provided.
5. All manifolds should be level.
6. The inlet and outlet pipes should be staggered.
7. Never connect branch lines at a cross or tee.
8. A common manifold should have an area equal to the sum of the areas of the individual suction lines.
9. The suction lines should be designed so as to prevent liquid from draining into the compressor during shutdown of the refrigeration system.

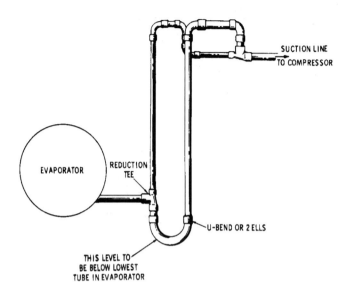

Fig. 25-2. Double-suction-riser arrangement providing oil return at minimum load.

DISCHARGE LINES

The hot-gas loop accomplishes two functions in that it prevents gas that may condense in the hot-gas line from draining back into the heads of the compressor during the *off* cycles, and prevents oil leaving one compressor from draining down into the head of an idle machine. It is important to reduce the pressure loss in hot-gas lines because losses in these lines increase the required compressor horsepower per ton of refrigeration and decrease the compressor capacity. The pressure drop is kept to a minimum by sizing the lines generously to avoid friction losses but still making sure that refrigerant line velocities are sufficient to entrain and carry along oil at all load conditions. In addition, the following pointers should be observed:

1. The compressor hot-gas discharge lines should be connected as shown in Fig. 25-3.

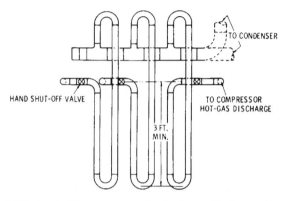

Fig. 25-3. Method of interconnecting hot-gas discharge lines.

2. The maximum length of the risers to the horizontal manifold should not exceed 6 feet.
3. The manifold size should be at least equal to the size of the common hot-gas line to the evaporative condenser.
4. If water-cooled condensers are interconnected, the hot-gas manifolds should be at least equal to the size of the discharge of the largest compressor.
5. If evaporative condensers are interconnected, a single gas line should be run to the evaporative condensers, and the same type of manifold provided at the compressors should be installed.
6. Always stagger and install the piping at the condensers.
7. When the condensers are above the compressors, install a loop having a minimum depth of 3 feet in the hot-gas main line.
8. Install a hand shutoff valve in the hot-gas line at each compressor.

WATER VALVES

The water-regulating valve is the control used with water-cooled condensers. When installing water valves, the following should be observed:

1. The condenser water for interconnected compressor condensers should be applied from a common water line.
2. Single automatic water valves or multiple valves in parallel (Fig. 25-4) should be installed in the common water line.
3. Pressure-control tubing from the water valves should be connected to a common line, which, in turn, should be connected to one of the receivers or to the common liquid line.

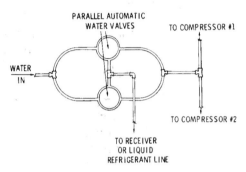

Fig. 25-4. Method of interconnecting water valves.

MULTIPLE-UNIT INSTALLATION

Multiple compressors operating in parallel must be carefully piped to ensure proper operation. The suction piping at parallel compressors should be designed so that all compressors run at the same suction pressure and oil is returned in equal proportions to the running compressors. All suction lines should be brought into a common suction header in order to return the oil to each crankcase as uniformly as possible.

The suction header should be run above the level of the compressor suction inlets so that oil can drain into the compressors by gravity. The header should not be below the compressor suction inlets because it can become an oil trap. Branch suction lines to the compressors should be taken off from the side of the header. Care should be taken to make sure that the return mains from the evaporators are not connected into the suction header so as to form crosses with the branch suction lines to the compressors. The suction header should be run full size along its entire length. The horizontal takeoffs to the various compressors should be the same size as the suction header. No reduction should be made in the branch suction lines to the compressors until the vertical drop is reached.

Figure 25-5 shows the suction and hot-gas header arrangements for two compressors operating in parallel. Takeoffs to each compressor from the common suction header should be horizontal and from the side to ensure equal distribution of oil and prevent accumulating liquid refrigerant in an idle compressor in case of slopover.

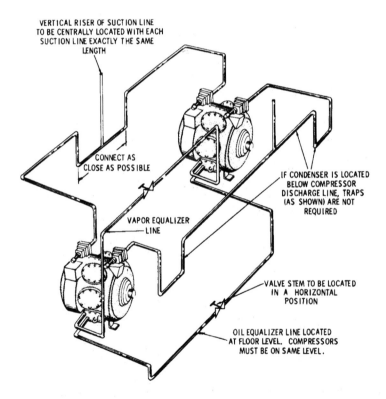

VERTICAL RISER OF SUCTION LINE
TO BE CENTRALLY LOCATED WITH EACH
SUCTION LINE EXACTLY THE SAME
LENGTH

CONNECT AS
CLOSE AS POSSIBLE

IF CONDENSER IS LOCATED
BELOW COMPRESSOR
DISCHARGE LINE, TRAPS
(AS SHOWN) ARE NOT
REQUIRED

VAPOR EQUALIZER
LINE

VALVE STEM TO BE LOCATED
IN A HORIZONTAL
POSITION

OIL EQUALIZER LINE LOCATED
AT FLOOR LEVEL. COMPRESSORS
MUST BE ON SAME LEVEL.

Fig. 25-5. Connections for the suction and hot-gas headers in a multiple compressor installation.

Discharge Piping

Hot-gas piping at paralleled compressors should be arranged to prevent excessive vibration and to prevent oil and liquid refrigerant from draining back to the head of any of the compressors. The branch hot-gas lines from the compressors should be connected into a common header. This hot-gas header should be run at some level below that of the compressor discharge connections. For convenience, it is often at the floor. This is equivalent to the hot-gas loop for the single compressor. The hot-gas loop accomplishes two functions:

1. It prevents gas that may condense in the hot-gas line during the *off* cycles from draining back onto the heads of the compressors, thus eliminating compressor damage from this potential hazard.
2. It prevents the oil leaving one compressor from draining down onto the head of an idle machine.

The solid lines in Fig. 25-6 show the correct hot-gas piping from two paralleled compressors. Each broken line has a muffler for reducing hot-gas pulsations, and each line drops down into the hot-gas header at the floor before rising and going to the condenser. The piping shown by the broken lines in Fig. 25-6 is incorrect and can result in three kinds of trouble:

1. The use of a *bull-headed tee* at the junction of the two branches and the main will greatly increase the gas turbulence and pressure drop at this point. Hammering in the line may also result.

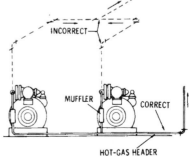

Fig. 25-6. Correct and incorrect hot-gas connection when compressors are connected in parallel.

2. Any gas condensed in the discharge riser during an *off* cycle can run down and accumulate on the heads of the compressors, causing possible valve breakage on startup.

3. Oil leaving one operating compressor will be thrown into the branch line of the other compressor (which may be idle). This situation will cause the oil to drain down onto the head of the idle compressor.

Interconnecting Piping

In addition to properly designing the suction and hot-gas piping at paralleled compressors, there are also certain other piping connections required between compressors and condensing units for proper parallel operation. These piping connections are shown in Fig. 25-7.

When two or more compressors are to be interconnected, they should be placed on foundations such that all oil-equalizer tappings are exactly level. An interconnecting oil-equalization line should connect all of the crankcases in order to maintain uniform oil levels and adequate lubrication in all compressors. The oil equalizer may be run level with the tappings, or, for convenient access to the compressors, it may be run at the floor. Under no conditions should the equalizer be run at a level higher than that of the tappings.

For the oil-equalizer line to work successfully, it is necessary to equalize the crankcase pressures. This is accomplished through the use of a gas equalizer line that is installed above the oil level. This line may be run to provide headroom or run level with the tappings on the compressors. It should be piped so that the oil or condensed vapor will not be trapped. Both lines should be the same size as the tappings on the largest compressor and should be valved so that any one machine can be taken out for repair without shutting down the entire system. Neither line should be run directly from one crankcase into another without forming some sort of U bend or hairpin to absorb vibration.

Condensing-Pressure Equalization

When multiple condensing units are interconnected as shown in Fig. 25-7, it is also necessary to equalize the pressure in the

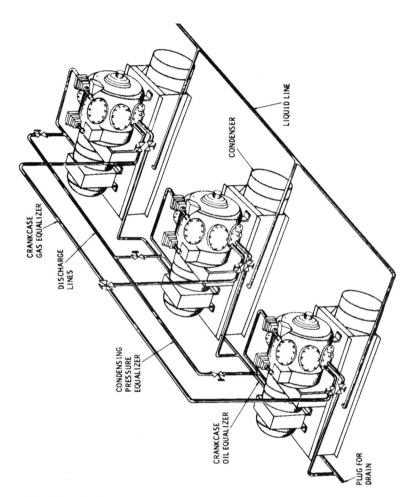

Fig. 25-7. Piping interconnection for multiple condensing units.

condensers to prevent hot gas from blowing through one of the condensers and into the liquid line. To do this, install a condensing-pressure equalizer line, as shown. If the piping is looped as shown, vibration should not be a problem. The bottoms of all condensers should be at the same level to prevent backing liquid into the lowest one.

621

The condensing pressure equalizer line between units should be the same size as the largest hot-gas line to which it connects. The liquid lines from the condensers should be piped as shown in Fig. 25-7 to ensure equal level in all condenser receivers.

Multiple-Compressor Controls

The control of reciprocating compressors must be such that excessive accumulation of liquid refrigerant in the crankcase during *off* cycles is prevented. In addition, the following pointers should be observed:

1. Step control by a modulating thermostat in the conditioned space should be provided to start and stop each compressor according to the load variation. No solenoid valves should be used.
2. A step controller operated by a modulating low-pressure switch should be provided to start and stop each individual compressor according to variations in the suction pressure. Solenoid valves should be used to control the refrigeration to coils and conditioners.
3. When evaporative condensers are interconnected, each condenser should be started and stopped along with its compressor.
4. In case condensers and compressors cannot be properly balanced, all compressors and condensers should be operated as a unit.

PIPE CONNECTIONS

Steel pipe has been used with success on many installations, particularly on large jobs where the cost of copper pipe or tubing prohibits their use. The inside of steel pipe or steel tubing should be free of scale and dirt. Do not use pipe with a mill-oil coating; it must be internally clean, to the bare metal. Refrigerant acts as a cleansing agent and will quickly loosen all scale and dirt from the interior of pipe and fittings. This dirt not only clogs the strainers in the system but may possibly damage the compressor.

As far as possible, all steel pipelines should be welded to

prevent leaks at the joints. Welded pipe connections should be carefully made and proper care exercised so that scale and welding particles may be removed from the lines before putting the plant into operation. Try to make the final welding joint near a flange connection so the line can be opened and cleaned after the welding has been completed. Be sure, however, to keep the welds at least 1 foot or more away from threaded joints because the welding heat on the pipe is apt to ruin the threaded connections.

Wherever threaded joints are used, the threads should be cut full length and should be of standard taper and size so that they will tighten gradually and not shoulder at the end of the threads. The threads should be cleaned with benzine, rubbed bright and dry, and all rust removed with steel wool. Make threaded joints of extra heavy pipe by welding a section of extra heavy weight to the standard weight, if necessary. On sizes over 2 inches, peen the inside of the pipe where flanges are made up. Make the flanges as tight as possible, and do this peening carefully. If extra heavy pipe is used, trouble from leaks will not occur. Standard-weight pipe, while suitable for pressures involved, very often loosens up due to vibration. Pipe intended for use should be kept dry, the ends being plugged at all times until the pipe is placed in the line.

Soldered Joints

The making of soldered joints is of particular importance. The surface of the tubing, especially the inside of the fittings being joined, must be absolutely clean and free from oxidation, dirt, moisture, or grease. Steel wool is considered best for cleaning the fittings and pipe; and immediately after they have been cleaned, the joints should be made up.

One should be especially careful of the flux used in making a soldered joint. It has been found that No-korode and Streamline solder pastes give better service than fluxes. *Warning: Under no circumstances use any of the commercial solders with an acid core since the fluxes in them are corrosive.* A thin, even coating of this flux should be distributed over all surfaces to be joined, the flux preferably being applied by means of a dry brush. When soldering paste is applied, the pipe should be warm (heated, if necessary, to

623

approximately 80°F in advance). The end of the pipe or tubing should be cut square in order that it may fit tightly against the ledge, which usually forms a part of the fitting. The tubing should be perfectly round, with as little clearance as possible between the fitting and the pipe or tube wall. Poor fits will give trouble. Use a miter vise when cutting tubing to be soldered.

To solder, heat the entire surface evenly. Do not apply the solder until the entire surface around the joint is at an even temperature and just above the melting point of the solder. Do not have the surface too hot, and be sure there is enough solder to fill the entire space. Soldered joints require skill and care on the part of the mechanic. Fewer defects can be seen in a soldered joint than can be observed in a welded joint. The type of solder to be used will depend upon local ordinances and suggestions of the tubing manufacturers.

Flared Joints

Flared joints must be made with care. If the flaring tool is used without proper oiling, it will thin out the tubing to a point where there may be danger of breakage. Observe whether or not the tubing is being thinned at the flare. If the tubing is hard and tends to crack, then anneal the end before making the flare. Be sure to use the nonfreezing or vented nuts on all frosted flare connections. All tubing for flared joints must be soft copper.

Hanging Pipe Lines

The Freons, being heavy, will cause vibration in pipelines unless care is taken to avoid sharp bends and proper hanging and bracing are done. Make the runs of pipe as short as possible. Make all hangers rigid, but avoid tying into floors or columns that will cause noise to be carried through the building. In such cases, pipes should be braced from the machine column, foundations, or one another; special supports can be devised to suit each condition. Make the hangers solid and firm. Wherever possible, set the condensers on concrete piers. Noise in condenser tubes invariably can be traced to vibration, either in pipelines or in the whole condenser.

Where compressors are installed at a distance of 20 feet or more from the condenser, particularly if the condenser is higher than the compressors, an antipulsation drum or vibration eliminator is necessary and should be installed in the compressor discharge line.

Strainers

Strainers should be placed in the line ahead of all expansion valves of the automatic type. Put the cleanout cap on the bottom. Suction traps are standard equipment with all compressors and are intended to remove the fine particles of dirt that may accumulate in the system. As a means of eliminating dirt, a so-called sump leg is quite effective. Dirt or scale will tend to separate whenever the velocity of the fluid decreases or whenever a change in direction takes place.

Pipeline Repairs

If at any time a leak in a pipeline, weld, or soldered fitting is to be repaired, be sure that the line has been completely emptied of Freon before applying any heat to it. Open this line at a point near a shutoff valve. After the repair has been made and the joint has completely cooled, blow out the air and torch gases in the line by admitting a little Freon from the opposite end. Always remember that high temperatures obtained when welding or soldering will tend to break down any Freon that is present, releasing dangerous fumes. It is very necessary to have the line cool before admitting any of the refrigerant to it and to get rid of any gases that may have been generated in the line during the repair procedure.

PIPING INSULATION

Insulation is required for refrigeration piping to prevent moisture condensation and prevent heat gain from the surrounding air. The desirable properties of insulation are that it should have a low coefficient of heat transmission, be easy to apply, have a high degree of permanency, and provide protection against air and moisture infiltration. Finally, it should have a reasonable installation cost.

The type and thickness of insulation used depends on the temperature difference between the surface of the pipe and the surrounding air and also on the relative humidity of the air. It should be clearly understood that although a system is designed to operate at a high suction temperature, it is quite difficult to prevent colder temperatures occurring from time to time. This may be due to a carrying over of some liquid from the evaporator or the operation of an evaporator pressure valve. Interchangers are preferable to insulation, in this case.

The safest pipe insulation available is molded cork or rock cork of the proper thickness. Hair-felt insulation may be used, but great care must be taken to have it properly sealed. For temperatures above 40°F, wool felt or a similar insulation may be used, but, here again, success depends on the proper seal against air and moisture infiltration.

Liquid refrigerant lines carry much higher temperature refrigerant than suction lines; and if this temperature is above the temperature of the space through which they pass, no insulation is usually necessary. However, if there is danger of the liquid lines going below the surrounding air temperatures and causing condensation, they should be insulated when condensation will be objectionable. If they must unavoidably pass through highly heated spaces, such as those adjacent to steam pipes, through boiler rooms, etc., then the liquid lines should also be insulated to ensure a solid column of liquid to the expansion valve.

There are four types of insulation available commercially and in general use for refrigerator piping, namely:

1. Cork
2. Rock cork
3. Wool felt with waterproof jacket
4. Hair felt with waterproof jacket

Cork Insulation

Cork pipe covering is prepared by pressing dried and granulated cork in metal molds. The natural resins in the cork bind the entire mass into its new shape. In the case of the cheaper cork, an artificial binder is used. The cork may be molded to fit pipe and fittings, or it may be made into flat boards of varying sizes and thicknesses. Cork has a low thermal conductivity. The natural binder in the material itself makes cork highly water-resistant, and its structure ensures a low capillarity. It can be made practically impervious to water by surfacing with an odorless asphalt.

All fittings in the piping, as well as the pipe itself, should be thoroughly insulated to prevent heat gain to protect the pipe insulation from moisture infiltration and deterioration and eliminate condensation problems. Molded cork covering made especially for this purpose is available for all common types of fittings. Each covering should be the same in every respect as the pipe insulation, with the exception of the shape, and should be formed so that it joins to the pipe insulation with a break. Typical cork fitting covers are furnished in three standard thicknesses for ice water, brine, and special brine.

To secure maximum efficiency and long life from cork covering, it must be correctly applied and serviced, as well as properly selected. Hence, it is essential that the manufacturer's recommendations and instructions are followed in detail. The following general information is a summary of the data that are of general interest.

All pipelines should be thoroughly cleaned, dried, and free from all leaks. It is also advisable to paint the piping with waterproof paint before applying the insulation, although this is not recommended by all manufacturers. All joints should be sealed with waterproof cement when applied. Fitting insulation should be applied in substantially the same manner, with the addition of a mixture of hot crude paraffin and granulated cork used to fill the space between the fittings, as shown in Fig. 25-8.

Rock-Cork Insulation

Rock-cork insulation is manufactured commercially by molding a mixture of rock wool and a waterproof binder into any shape or thickness desired. The rock wool is made from limestone melted at about 3000°F and then blown into fibers by high-pressure steam. It is mixed with an asphaltum binder and molded into the various commercial forms. The heat conductivity is about the same as cork, and the installed price may be less. Because of its mineral composition, it is odorless, vermin-proof, and free

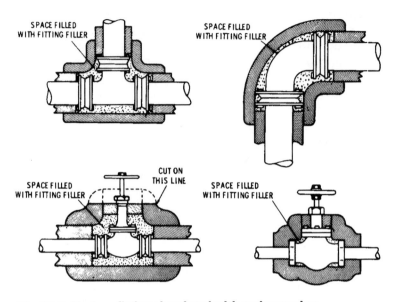

Fig. 25-8. Various fittings insulated with cork covering.

from decay. Like cork, it can be made completely waterproof by surfacing with an odorless asphalt. The pipe covering fabricated from rock wool and a binder is premolded in single-layer sections 36 inches long to fit all standard pipe sizes and is usually furnished with a factory-applied waterproof jacket.

When pipelines are insulated with rock-cork covering, the fittings are generally insulated with builtup rock wool impregnated with asphalt. This material is generally supplied in felted form, having a nominal thickness of about 1 inch and a width of about 18 inches. It can be readily adapted to any type of fitting and is efficient as an insulator when properly applied.

Before applying the formed rock-cork insulation (Fig. 25-9), it is first necessary to thoroughly clean and dry the piping and then paint it with a waterproof asphalt paint. The straight lengths of piping are next covered with the insulation, which has the two longitudinal joints and one end joint of each section coated with a plastic cement. The sections are butted tightly together with the longitudinal joints at the top and bottom and temporarily held in place by staples. That part of the exterior area of each section to be covered by the waterproof lap should be coated by the plastic cement and the lap pressed smoothly into it. The end joints should be sealed with a waterproof fabric embedded in a coat of the plastic cement. Each section should then be secured permanently in place with three to six loops of copper-plated annealed steel wire.

Fig. 25-9. Rock-cork pipe insulation.

Wool-Felt Insulation

Wool felt is a relatively inexpensive type of pipe insulation and is made up of successive layers of waterproof wool felt that are indented in the manufacturing process to form air spaces. The inner layer is a waterproof asphalt-saturated felt, while the outside layer is an integral waterproof jacket. This insulating material is satisfactory when it can be kept air- and moisture-tight. If air is allowed to penetrate, condensation will take place in the wool felt, and it will quickly deteriorate. Thus, it is advisable to use it only where temperatures above 40°F are encountered and when it is perfectly sealed. Under all conditions, it should carry the manufacturer's guarantee for the duty that it is to perform.

After all the piping is thoroughly cleaned and dried, the sectional covering is usually applied directly to the pipe with the outer layer slipped back and turned so that all joints are staggered. The joints should be sealed with plastic cement, and the flap of the waterproof jacket should be sealed in place with the same material. Staples and copper-clad steel wire should be provided to permanently hold the insulation in place, and then the circular joints should be covered with at least two layers of waterproof tape to which plastic cement is applied.

The pipe fittings should be insulated with at least two layers of hair felt (Fig. 25-10) built up to the thickness of the pipe covering; but before the felt is placed around the fittings, the exposed ends of the pipe insulation should be coated with plastic cement. After the felt is in place, two layers of waterproof tape and plastic cement should be applied for protection from moisture infiltration.

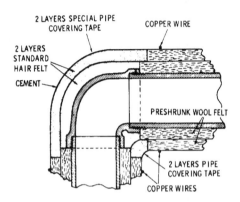

Fig. 25-10. Pipe fitting insulated with preshrunk wool felt.

Insulation of this type is designed for installation in buildings where it is normally protected against outside weather conditions. When outside pipes are to be insulated, one of the better types of pipe covering should be used. In all cases, the manufacturer's recommendations should be followed as to the application.

Hair-Felt Insulation

Hair-felt insulation is usually made from pure cattle hair that has been especially prepared and cleaned. It is a very good insulator against heat, having a low thermal conductivity. Its installed cost is somewhat lower than cork; but it is more difficult to install and seal properly, and hence its use must be considered a hazard with the average type of workmanship. Prior to installation, the piping should be cleaned and dried and then prepared by applying a thickness of waterproof paper or tape wound spirally, over which the hair felt of approximately 1-inch thickness is spirally wound for the desired length of pipe. It is then tightly bound with jute twine, wrapped with a sealing tape to make it entirely airtight, and finally painted with waterproof paint. If more than one thickness of hair felt is desired, it should be built up in layers with tar paper between. When it is necessary to make joints around fittings, the termination of the hair felt should be tapered down with sealing tape and the insulation applied to the fittings should overlap this taper, thus ensuring a permanently tight fit.

The important point to remember is that this type of insulation must be carefully sealed against any air or moisture infiltration, and even then difficulty may occur after it has been installed. At any point where air infiltration (or "breathing," as it is called) is permitted to occur, condensation will start and travel great distances along the pipe, even undermining the insulation that is properly sealed.

There are several other types of pipe insulation available, but they are not used extensively. These include various types of wrapped and felt insulation, but they are seldom applied with success. Whatever insulation is used, it should be critically examined to see whether it will provide the protection and permanency required of it; otherwise it should never be considered. Although all refrigerant piping, joints, and fittings should be covered, it is not advisable to so so until the system has been thoroughly leak tested and operated for a time.

SUMMARY

In this chapter, a detailed account of pipe selection and installation procedure has been treated. Pressure drop in the various parts of commercial refrigeration systems due to pipe friction and the proper dimensioning to obtain the best operating results are important items when installation of equipment is made. These have been fully covered. Tabulated data on pipe sizes for various refrigerants are given in the Appendix.

Multiunit installation of compressors, along with piping diagrams, provides information for the installer. Pipe-connection methods, treatment of pipe joints, soldering of joints, and support of pipes are additional features worthy of note. A part of this chapter deals with piping insulation, various types of insulation material, and methods of application.

By careful observation of the foregoing detailed description of refrigeration piping and methods of installation, the piping problem will be greatly simplified and result in proper system operation.

REVIEW QUESTIONS

1. What are the general precautions to employ in the selection of refrigeration piping?
2. Describe the causes of friction pressure drop in refrigeration lines.
3. What factors should be considered when installing liquid lines?
4. Describe the interconnection of suction lines in a typical refrigeration system.
5. Describe a typical water valve and its operation.
6. What precautions should be observed in a multiple-unit installation?
7. Describe the controls employed when installing multiple-unit compressors.
8. Why are steel pipes used in preference to copper pipes for refrigeration lines?
9. What precaution should be observed when hanging pipe lines?
10. What is the function of piping insulation?
11. Describe wool-felt insulation and its method of application.

CHAPTER 26

Commercial Absorption Systems*

Commercial absorption refrigeration is similar in operation to domestic absorption refrigeration, covered in Chapter 5. Each manufacturer of this equipment will have a different design, and the manufacturer of the unit should be consulted for service information, etc.

ABSORPTION-REFRIGERATION CYCLES

Absorption-refrigeration cycles are heat-operated cycles. A secondary fluid, the absorbent, is employed to absorb the primary fluid, a gaseous refrigerant. This refrigerant has been vaporized

*The authors wish to thank the publishers of *ASHRAE Handbook of Fundamentals*, Washington, D.C., for their assistance in preparing this chapter and for allowing some parts of their publication to be reprinted.

in the evaporator. Simplified diagrams for the two heat-operated methods of obtaining refrigeration are shown in Fig. 26-1 and Fig. 26-2.

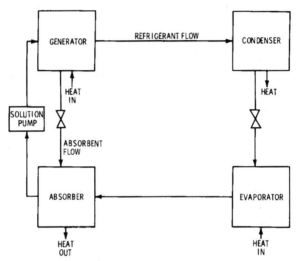

Fig. 26-1. Basic absorption-refrigeration cycle.

A similarity exists between the five main components of the absorption method and five of the components of the combination method.

1. *Generator and boiler*—Heat from a high-temperature source is transferred into the generator of the absorption method and into the boiler of the combination method. This is done to obtain relatively high-pressure vapor. Absorbent is also regenerated in the generator of the former method.

2. *Condenser and refrigerant condenser*—Heat is transferred out of the condenser of the absorption method and out of the refrigerant condenser of the combination method to an intermediate temperature sink in order to condense refrigerant at relatively high pressure.

3. *Evaporator and evaporator*—Heat from a low-temperature source is transferred into the evaporator of both methods. The result is the vaporization of refrigerant at relatively low pressure.

4. *Absorber and heat-engine condenser*—Heat is transferred out of the absorber of the absorption method and out of the heat-engine condenser of the combination method to an intermediate temperature sink. This is done to enable the conversion of relatively low-pressure vapor to a liquid state. Absorbent is added to the absorber of the former method to assist in this conversion.

5. *Solution pump and feed pump*—A small amount of work is put into the solution pump of the absorption method and into the feed pump of the combination method to pressurize liquid.

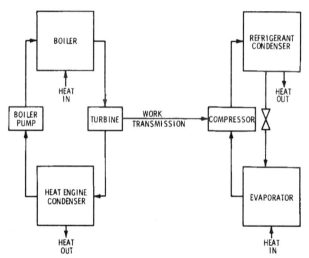

Fig. 26-2. Combination of heat-engine cycle and mechanical-refrigeration cycle.

The turbine (or expansion machine) and the compressor of the combination method do not have counterparts in the absorption method.

The relation between work and heat for an ideal heat engine operating in a Carnot cycle is given by the second law, as follows:

$$W = q_g \frac{T_h - T_s}{T_h}$$

where W = work, Btu/hr
q_g = heat input rate, Btu/hr
T_l = temperature of heat source, Rankine
T_s = temperature of heat sink, Rankine

The relation between work required and refrigeration load for an ideal mechanical refrigeration machine operating as a reverse Carnot cycle is as follows:

$$-W = q_e \frac{T_s - T_l}{T_l}$$

where $-W$ = work input rate, Btu/hr
q_e = refrigeration load, Btu/hr
T_l = temperature of refrigeration load, Rankine
T_s = temperature of heat sink (assumed to be same as for heat engine), Rankine

The coefficient of performance (COP) for the combination of the two ideal cycles is

$$\text{COP} = \frac{q_g T_h(T_h - T_s)}{q_e T_l(T_s - T_l)}$$

This equation also applies to the ideal absorption refrigeration process. It is plotted as a function of sink temperature, with the temperature of the heat source as the parameter for the case of refrigeration load at 40°F in Fig. 26-3.

There are many inefficiencies in both methods of refrigeration. Although the overall thermodynamic efficiency tends to be higher for the absorption method, this method lacks the versatility of the combination method and, by itself, is less capable of effectively utilizing high-temperature energy.

BASIC ABSORPTION-REFRIGERATION CYCLE

The distinctive parts of the basic absorption cycle are the absorber and generator. As has been described, there is a certain similarity between these components and those of the heat-engine condenser and boiler. However, there are differences in the internal processes and in the function of the components. The refrigerant condensation and evaporation steps are the same for the two cycles, however, and any difference in these components is related to the particular properties and purity of the refrigerant.

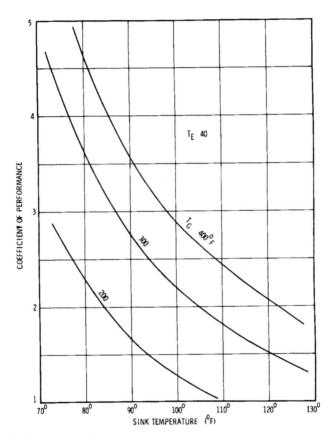

Fig. 26-3. Ideal coefficients of performance for reversible absorption of refrigeration cycles.

In the basic absorption cycle, low-pressure refrigerant vapor is converted to a liquid phase (solution) while still at the low pressure. The conversion is made possible by the vapor being absorbed by a secondary fluid, the absorbent. The absorption proceeds because of the mixing tendency of miscible substances and generally because of an affinity between the absorbent and refrigerant molecules. Thermal energy that is released during the absorption process must be disposed of to a sink. This energy arises from the heat of condensation, sensible heats, and, normally, heat of dilution.

The refrigerant-absorbent solution is pressurized in the solution pump. Since the volume is very much less than that of the refrigerant vapor, the work required by the solution pump is small in comparison with the work of compression required in a mechanical refrigeration cycle.

The refrigerant-absorbent solution is conveyed to the generator where the refrigerant and the absorbent are separated, i.e., regenerated, by a distillation process. A simple still is adequate for the separation when the pure absorbent material is nonvolatile, as in the water-lithium bromide system. However, fractional distillation equipment is required when the pure absorbent material is volatile, as in the ammonia-water system, because the refrigerant should be essentially free of absorbent. Otherwise, vaporization in the evaporator is hampered. The absorbent that is regenerated normally contains a substantial amount of refrigerant.

If the absorbent material tends to become solid, as in the water-lithium bromide system, it is necessary to have enough refrigerant present to keep the pure absorbent material in a dissolved state at all times. Certain practical considerations, particularly the avoidance of excessively high temperatures in the generator, make it generally desirable to leave a moderate amount of refrigerant in the regenerated absorbent. The high-temperature energy required for regeneration is approximately equal to the intermediate-temperature energy released in absorption.

As shown in Fig. 26-1, the refrigerant and absorbent have different patterns. The refrigerant goes from the generator to the condenser, to the evaporator, then to the absorber, to the solution pump, and back to the generator. The absorbent short-circuits from the generator to the absorber. The absorbent may be thought of as a carrier fluid that carries spent refrigerant from the low-pressure side of the cycle to the high-pressure side.

Practical Absorption-Refrigeration Cycles

The principal source of inefficiency in the basic absorption-refrigeration cycle is sensible heat effects. The conveying of hot absorbent from the generator into the absorber causes a considerable amount of thermal energy to be wasted. A liquid-to-liquid heat exchanger (generally called the *liquid heat exchanger*), which transfers energy from this stream to the refrigerant-absorbent solution being pumped back to the generator, saves a major portion of the energy for useful purposes. The use of this heat exchanger is shown in the flow diagram for a water-lithium bromide cycle, Fig. 26-4.

Conveying warm condensate from the condenser into the evaporator similarly causes a reduction in refrigerating effect. Although the loss here is not as great as described previously, it is

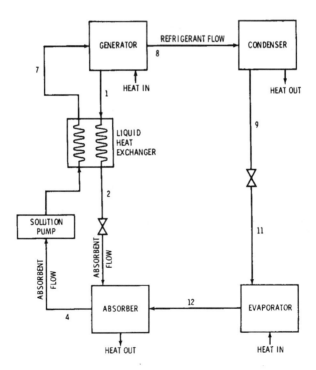

Fig. 26-4. Lithium bromide-water absorption-refrigeration cycle.

sometimes desirable to transfer energy from the condensate to the refrigerant vapor leaving the evaporator by means of a heat exchanger, commonly called a *precooler*.

There is further need of heat exchange in an ammonia-water machine. Absorbent reaches such a high temperature in the generator that it is desirable to transfer some heat by means of a coil in the analyzer or stripping section. Also, heat must be removed in a partial condenser, which forms reflux for the system. Means of accomplishing these two purposes, as well as the other changes, are shown in the flow diagram for an ammonia-water machine (Fig. 26-5).

In some modifications of the ammonia-water cycle, part of the heat of absorption is utilized for preheating the pressurized

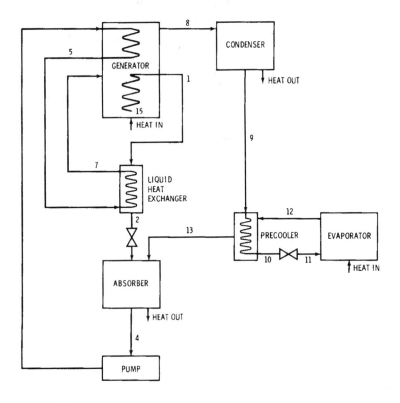

Fig. 26-5. Ammonia-water absorption-refrigeration cycle.

refrigerant-absorbent solution. The component that makes use of this heat of absorption is known as an *absorber heat exchanger* or a *solution cooled absorber*.

The heat-recovery modifications for the basic cycle that have been described do not bring the coefficient of performance over a threshold level of unity. In other words, the heat required to generate 1 lb of refrigerant is not less than the heat taken up when this 1 lb of refrigerant evaporates in the evaporator. A method of breaking through this threshold level is the use of a double-effect generator. It is possible with the water-lithium bromide pair to employ two generators: one at a high temperature and pressure, which is heated by an external source of thermal energy; and a second at a lower pressure and temperature, which is heated by heat of condensation of the vapor from the first generator. Condensate from both effects are conveyed to the evaporator. The coefficient of performance is nearly double that of an ordinary cycle.

CHARACTERISTICS OF THE REFRIGERANT-ABSORBENT PAIR

The two materials that make up the refrigerant-absorbent pair should meet nearly all of a number of requirements in order to be suitable for absorption refrigeration. Chief among these requirements are:

1. *Absence of solid phase*—The refrigerant-absorbent pair should not form a solid phase over the range of composition and temperature to which it might be subjected. In solid forms presumably it would stop flow and cause a shutdown of the equipment.

2. *Volatility ratio*—The refrigerant should be much more volatile than the absorbent in order that the two can be separated easily in the generator; otherwise the cost of the generator and the heat requirement for separation become large.

3. *Affinity*—It is commonly considered desirable that the absorbent have a strong affinity for the refrigerant under con-

ditions in which absorption takes place. This affinity causes a negative deviation from Raoult's law and results in an activity coefficient of less than unity for the refrigerant. It reduces the amount of absorbent that has to be circulated and consequently the waste of thermal energy due to sensible heat effects. Also, the size of the liquid heat exchanger that transfers heat from the absorbent to pressurized refrigerant-absorbent solution in practical cycles is reduced. Recently Jacob, Albright, and Tucker have made calculations that indicate that strong affinity does have disadvantages. With this affinity is associated a high heat of dilution, and consequently extra heat is required in the generator to separate the refrigerant from the absorbent.

4. *Pressure*—It is desirable that operating pressures, largely established by the physical properties of the refrigerant, be moderate. High pressures necessitate the use of heavy-walled equipment. Low pressures (vacuum) necessitate the use of large-volume equipment and special means of reducing pressure drop in the flow of refrigerant vapor.

5. *Stability*—Almost absolute chemical stability is required because the fluids are subjected to rather severe conditions over many years of service. Undesirable results of instability could be the formation of gas, a solid, or a corrosive substance.

6. *Corrosion*—It is especially important that the fluids themselves or any substance resulting from instability do not corrode the materials used in constructing the equipment.

7. *Safety*—The fluids must be substantially nontoxic and nonflammable if they are to be in an occupied dwelling.

8. *Viscosity*—Low viscosity for the fluids is desirable to promote heat and mass transfer and, to some extent, reduce pumping problems.

9. *Latent heat*—It is desirable for the latent heat of the refrigerant to be high so that the circulation rate of the absorbent can be kept at a minimum.

No known refrigerant-absorbent pair meets all the requirements that have been listed. Two pairs, ammonia-water and water-lithium bromide, are considered to come the nearest, and these are the only two that have found extensive commercial use. As has been indicated, the volatility ratio for ammonia-water is smaller than desired. Also, ammonia is in ANS B9 Safety Code Group 2, and thus its use inside a dwelling is restricted. The pressures encountered with this pair are somewhat high. Other requirements are quite well met.

The main problem with the water-lithium bromide pair is the possibility of solid formation. Since the refrigerant turns to ice at 32°F, the pair cannot be used for low-temperature refrigeration. Lithium bromide crystallizes at moderate concentrations. When the absorber is air-cooled, these concentrations tend to be reached, and thus the pair is usually limited to applications in which the absorber is water-cooled. There is a possibility that the use of a combination of salts as the absorbent will reduce the crystallizing tendency enough to permit air cooling. Other disadvantages of the water-lithium bromide pair are those associated with low pressure and with the high viscosity of the lithium bromide solution. These disadvantages are largely overcome by proper design of equipment. The combination does have the advantages of high safety, high volatility ratio, high affinity, high stability, and high latent heat.

Of the other refrigerant-absorbent pairs that have been investigated, the following types are of some promise:

1. Ammonia-salts
2. Methylamine-salts
3. Alcohols-salts
4. Ammonia-organic solvents
5. Sulfur dioxide-organic solvents
6. Halogenated hydrocarbons-organic solvents

Several of these types appear to be suitable for a relatively simple cycle and may not give as much crystallization problem as the water-lithium bromide pair does. However, stability and corrosion information on most of them is very sketchy. Also, the refrigerants, except for fluororefrigerants in the last type, are somewhat hazardous.

WATER-LITHIUM BROMIDE MACHINE

The cycle for a water-lithium bromide absorption refrigeration machine is shown in Fig. 26-4. This cycle includes a liquid heat exchanger for the absorbent and refrigerant-absorbent streams. Some of the limitations and details of the cycle are illustrated in the following Example.

Example—Make a first law analysis of a water-lithium bromide machine operating with the cycle shown in Fig. 26-4 and having the following requirements:

1. Refrigeration load—100 tons
2. Evaporator temperature—40°F
   This low temperature is desirable for achieving good dehumidification as well as cooling in air conditioning.
3. Absorber outlet temperature—90°F
   The absorber temperature should be kept at about 100°F or lower to reduce the danger of crystallization.
4. Condenser temperature—110°F
   This temperature is not critical and may be set higher than the absorber temperature to achieve better use of the cooling water.
5. Generator temperature—192°F
   This temperature is related to the condenser temperature in such a way that the absorbent will be in the feasible concentration range.

Solution—A number of assumptions are made. Among these is that the two phases are in equilibrium with each other at the points for which temperatures are given. Pressure drops, except at expansion devices, are negligible. The approach at the low-temperature end of the liquid heat exchanger is taken to be 10°F.

Table 26-1 is set up, and a start is made at filling in the values. Low- and high-side pressures are the water-vapor pressures for the evaporator and condenser temperature, respectively. Enthalpies for water and steam are found in the steam tables.

Relative flowrates are determined from the material balances, as follows:

$$\frac{w_a}{w_d} = \frac{x_b}{x_a - x_b}$$

where w_a = flowrate of absorbent, lb/hr
$\quad\quad\;\; w_d$ = flowrate of refrigerant, lb/hr
$\quad\quad\;\; x_a$ = LiBr concentration in absorbent, lb/lb solution
$\quad\quad\;\; x_b$ = LiBr concentration in refrigerant-absorbent solution, lb/lb solution

$$\frac{w_a}{w_d} = \frac{0.56}{0.61 - 0.56} = 11.2$$

$$\frac{w_b}{w_d} = \frac{w_a}{w_d} + 1$$

where w_b = flowrate of refrigerant-absorbent solution, lb/hr
$\quad\quad w_b/w_d$ = 11.2 + 1 = 12.2.

The enthalpy of the refrigerant-absorbent solution leaving the liquid heat exchanger is calculated from an energy balance, as follows:

$$h_7 = h_4 + [(h_1 - h_2) \times \frac{w_a}{w_b}]$$

$$= -75 + \left\{[-30 - (-70)] \times \frac{11.2}{12.2}\right\} = -38.3$$

Table 26-1. Conditions in Lithium Bromide Cycle

Point	Temp, °F	Pressure, mmHg	Weight Fraction, LiBr	Flow, lb/lb Refrigerant	Enthalpy, Btu/lb
1	192	66	0.61	11.2	-30
2	100	66	0.61	11.2	-70
4	90	6.3	0.56	12.2	-75
7	163	66	0.56	12.2	-38.3
8	192	66	0.0	1.0	1147
9	110	66	0.0	1.0	78
11	40	6.3	0.0	1.0	78
12	40	6.3	0.0	1.0	1079

The temperature corresponding to this enthalpy is found in the *ASHRAE Handbook of Fundamentals*.

The refrigerant flowrate is calculated from an energy balance, as follows:

$$w_d = \frac{q_e}{h_{12} - h_{11}}$$
$$= \frac{1,200,000}{1079 - 78}$$
$$= 1200 \text{ lb/hr}$$

The absorbent and refrigerant-absorbent solution rates are calculated next, as follows:

$$w_a = 11.2w_d = 11.2 \times 1200 = 13.400 \text{ lb/hr}$$
$$w_b = 12.2w_d = 12.2 \times 1200 = 14,640 \text{ lb/hr}$$

The net heat input to the generator is calculated from an energy balance, as follows:

$$
\begin{aligned}
q_g &= w_d h_8 + w_a h_1 - w_b h_7 \\
&= 1200 \times 1147 + 13,400(-30) - 14,640(-38.3) \\
&= 1,535,112 \text{ Btu/hr}
\end{aligned}
$$

The coefficient of performance on a net basis is

$$
\begin{aligned}
\text{COP} &= \frac{q_e}{q_g} \\
&= \frac{1,200,000}{1,535,112} \\
&= 0.781
\end{aligned}
$$

Heat-transfer rates for the other components are:

Liquid heat exchanger:

$$
\begin{aligned}
q_l &= w_a(h_1 - h_2) \\
&= 13,400(-30 - (-70))
\end{aligned}
$$

Condenser:

$$
\begin{aligned}
-q_c &= w_d(h_8 - h_9) \\
&= 1200(1147 - 78) \\
&= 1,282,800 \text{ Btu/hr}
\end{aligned}
$$

Absorber:

$$-q_a = q_g + q_e + q_c$$
$$= 1,535,112 = 1,200,000 - 1,280,000$$
$$= 1,455,112 \text{ Btu/hr}$$

In actual practice, somewhat lower concentrations of lithium bromide solutions than those given in the example are commonly used in small commercial units to ensure against crystallization of the salt, particularly on shutdown. These more typically are 54 and 58.5%. Higher concentrations, normally about 60 and 64.5%, are used in large commercial units in order to operate at higher absorber temperatures and thus save on heat-exchanger costs. Controls and a shutdown dilution cycle are employed with these large units to prevent crystallization.

AMMONIA-WATER CYCLE

As has already been indicated, the ammonia-water cycle must be more complex than the water-lithium bromide cycle to provide acceptable performance. More heat recovery means are required, and the simple still must be replaced by a complete distillation system. This system will be referred to as the generator, since it is housed in one component. Because of the complexity of the cycle, the design for optimum performance, based on a given set of design parameters, requires extensive calculations.

Characteristics and calculating methods for an ammonia-water machine are demonstrated in the following Example.

Example—Make a first law analysis of an ammonia-water machine operating with the cycle on which the following test data are available:

Refrigeration load	36,000 Btu/hr
Precooler liquid	80°F
Evaporator inlet temperature	38°F
Evaporator outlet temperature	50°F
Absorber outlet temperature	130°F
Condenser outlet temperature	123°F

647

Refrigerant-absorbent solution temperature	
leaving reflux condenser	168°F
Generator bottom temperature	364°F
Absorbent leaving generator temperature	270°F
Absorbent leaving heat-exchanger temperature	170°F
Vapor leaving reflux-condenser temperature	172°F
Condenser pressure	300 psia*
Absorber pressure	70 psia

Solution—It will be assumed that phases are in equilibrium with each other at the evaporator inlet, absorber outlet, bottom of generator, vapor outlet from reflux condenser, and condenser outlet. Pressure drops, except in expansion devices, are negligible.

Table 26-2 is set up, and values are filled in this table as they are determined. Enthalpies are likewise determined in either of these sources for the points for which sufficient physical data are now known.

Table 26-2. Conditions in Ammonia Cycle

Point	Temp, °F	Pressure, psia	Weight Fraction, NH₃	Flow, lb/lb Re- frigerant	Enthalpy, Btu/lb
1	270	300	0.100	1.97	215
2	170	300	0.100	1.97	110
4	130	70	0.400	2.97	0
5	168	300	0.400	2.97	43
7	230	300	0.400	2.97	113
8	172	300	0.991	1.00	599
9	123	300	0.991	1.00	102
10	80	300	0.991	1.00	51
11	38	70	0.991	1.00	51
12	50	70	0.991	1.00	527
13	92	70	0.991	1.00	578
15	364	300	0.100	1.97	315

The enthalpy at point 12 is the ordinate at which the 50°F, 70 psia tie line crosses the abscissa representing the refrigerant composition (0.991). The enthalpy at point 13 is calculated from an energy balance around the precooler:

°psia = pounds per square inch absolute.

$$h_{13} = h_{12} + H_g - h_{10}$$
$$= 527 + 102 - 51 = 578$$

The temperature at point 13 is determined by a tie-line procedure that is the reverse of the one used to find the enthalpy at point 12. Relative flowrates are determined from material balances as follows:

$$\frac{w_a}{w_d} = \frac{x_d - x_b}{x_b - x_a}$$

where x_d = concentration of ammonia in refrigerant, lb/lb

x_b = concentration of ammonia in refrigerant-absorbent solution, lb/lb

x_a = concentration of ammonia in absorbent, lb/lb

$$\frac{w_a}{w_d} = (0.991 - 0.400) \div (0.400 - 0.100) = 1.97$$

$$\frac{w_b}{w_d} = \frac{w_a}{w_d} + 1 = 1.97 + 1 = 2.97$$

The enthalpy at point 7 is calculated from an energy balance around the liquid heat exchanger:

$$h_7 = h_5 + (h_1 - h_2)\,\frac{w_a}{w_b}$$
$$= 43 + (215 - 110)\,\frac{1.97}{2.97}$$
$$= 43 + (105 \times 0.6632996633)$$
$$= 113$$

The refrigerant flowrate is calculated from an energy balance, as follows:

$$w_d = \frac{q_c}{h_{18} - h_{11}}$$
$$= \frac{36,000}{527 - 51}$$
$$= 75.63 \text{ lb/hr}$$

The absorbent and refrigerant-absorbent solution flowrates are calculated next, as follows:

$$w_a = 1.97w_d = 1.97 \times 75.63 = 149 \text{ lb/hr}$$
$$w_b = 2.97w_d = 2.97 \times 75.63 = 224.62 \text{ lb/hr}$$

The net heat input to the generator may now be calculated from an energy balance around the generator reflux condenser and liquid heat exchanger, as follows:

$$
\begin{aligned}
q_s &= w_d h_s + w_a h_2 - w_b h_4 \\
&= 75.63 \times 599 + (149 \times 110) - (224.5 \times 0) \\
&= 45,302.37 + (16390) - (0) \\
&= 61,692.37 \text{ Btu/hr}
\end{aligned}
$$

The coefficient of performance on a net basis is

$$
\begin{aligned}
\text{COP} &= \frac{q_c}{q_g} \\
&= \frac{36,000}{61,390} \\
&= 0.586
\end{aligned}
$$

Heat-transfer rates for other parts of the cycle are:

Liquid heat exchanger:

$$
\begin{aligned}
q_l &= w_a(h_1 - h_2) = 149(215 - 110) \\
&= 15,645 \text{ Btu/hr}
\end{aligned}
$$

Condenser:

$$
\begin{aligned}
-q_c &= w_d(h_s - h_g) \\
&= 75.5(599 - 102) \\
&= 37,523.5 \text{ Btu/hr}
\end{aligned}
$$

Reflux condenser:

$$
\begin{aligned}
q_r &= w_b(h_5 - h_4) \\
&= 224.5(43 - 0) \\
&= 9,653.5 \text{ Btu/hr}
\end{aligned}
$$

Absorber:

$$
\begin{aligned}
-q_a &= q_g + q_e + q_c \\
&= 61,390 + 36,000 \\
&= 59,890 \text{ Btu/hr}
\end{aligned}
$$

Precooler:

$$qp = w_d(h_9 - h_{10})$$
$$= 75.5(102 - 51)$$
$$= 3,850 \text{ Btu/hr}$$

SOLAR COOLING

Prior to 1972, little if any research had been done regarding the use of solar energy for cooling. Therefore, no air conditioning equipment specially designed for use with solar energy was available. Currently, however, there are cooling systems that lend themselves to easy modification for use with solar energy. One is the absorption-type system manufactured by Arkla Industries. The other is the Rankine cycle.

Of the two, only the Arkla absorption cooling equipment can use solar-heated water directly to produce cooling. The Rankine cycle needs an intermediate step. This involves replacing the electric motor in the conventional vapor compression refrigeration cycle with a turbine or using solar cells to produce electricity. In either case, making modifications for the Rankine cycle is more costly than making modifications of the absorption system.

Systems of Solar Cooling

There are two systems used in solar cooling: the direct and the indirect (Figs. 26-6 and 26-7).

The *direct system* of application uses the absorption cooling system. It provides higher firing water temperatures directly from the storage tanks to the unit's generator (Fig. 26-6).

The *indirect system* is a closed system in which a heat exchanger transfers the heat from the solar-heated water storage tanks. This allows the use of an antifreeze fluid (Fig. 26-7).

The Arkla-Solaire® unit operates on the absorption principle shown in Figure 26-8. It uses solar-heated water as the energy source. Lithium bromide and water are used as the absorbent/refrigerant solution. The refrigeration tonnage is delivered through a chilled water circuit that flows between the unit's evaporator and a standard fan-coil assembly located inside the conditioned

651

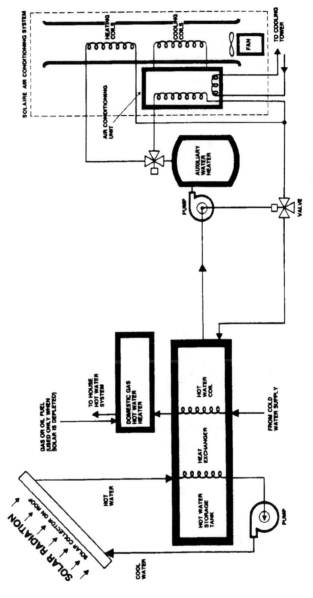

Arkla

Fig. 26-6. Solar energy air conditioning unit—direct system.

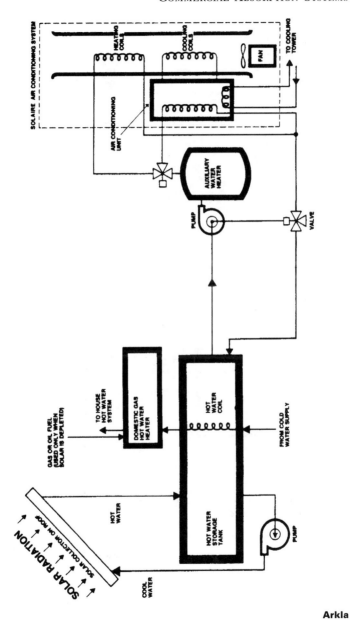

Arkla

Fig. 26-7. Solar energy air conditioning unit—indirect system.

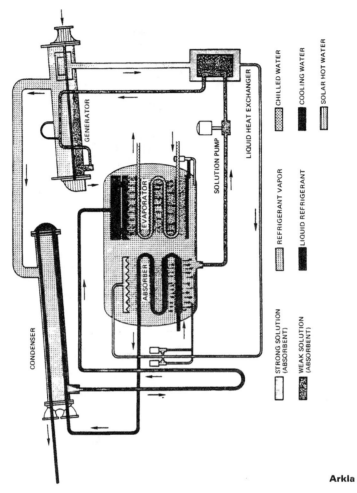

Arkla

Fig. 26-8. Solar air conditioning using lithium bromide and water absorption cycle.

space. The heat from the conditioned space is dissipated externally at the cooling tower.

Note how it is made up of four main components: the generator, condenser, evaporator and absorber. The only major difference between this and the other conventional systems is the method of obtaining heat—directly from the sun.

When the solar-heated water enters the tubes inside the generator, the heat from the hot water vaporizes the refrigerant (water), separating it from the absorbent (lithium bromide). The vaporized refrigerant vapor then flows to the absorber. There it again liquefies and combines with the circulating solution. The reunited lithium bromide and water solution then passes to the liquid heat exchanger. There, it is reheated before being returned to the generator.

SUMMARY

Absorption-refrigeration cycles are heat-operated cycles. A secondary fluid, the absorbent, is used to absorb the primary fluid, a gaseous refrigerant, which has been vaporized in the evaporator.

Five components of the absorption method are: generator and boiler; condenser; evaporator; absorber; and solution pump.

The distinctive parts of the basic absorption cycle are the absorber and generator. In the basic absorption cycle, low-pressure refrigerant vapor is converted to a liquid phase (solution) while still at the low pressure. The conversion is made possible by the vapor being absorbed by a secondary fluid, the absorber. The absorption proceeds because of the mixing tendency of miscible substances and generally because of an affinity between the absorbent and refrigerant molecules. Thermal energy, which is released during the absorption process, must be disposed of to a drain or sink. This energy arises from the heat of condensation, sensible heats, and, normally, heat of dilution.

The principal source of inefficiency in the basic absorption-refrigeration cycle is sensible heat effects. The conveying of hot absorbent from the generator into the absorber causes a considerable amount of heat energy to be wasted.

In some modifications of the ammonia-water cycle, part of the heat of absorption is utilized for preheating the pressurized refrigerant-absorbent solution. The component that makes use of this heat of absorption is known as an absorber heat exchanger, or a solution-cooled absorber.

REVIEW QUESTIONS

1. What type of cycles are present in the absorption-refrigeration process?
2. Name the five parts of the absorption-refrigeration system.
3. What is a heat-engine condenser?
4. Which system of refrigeration uses the solution pump?
5. What are the two distinct parts of the basic absorption cycle?
6. What is the principal source of inefficiency in the basic absorption-refrigeration cycle?
7. What is a heat exchanger used for?
8. What are the nine chief requirements for materials that make up the refrigerant-absorbent pairs?
9. Why is the ammonia-water cycle more complex than the water-lithium bromide cycle?
10. How do you calculate the absorbent and refrigerant-absorbent solution flowrates?

CHAPTER 27

Circulating Pumps

Circulating pumps are an important part of commercial and industrial refrigeration systems where brine is used because of the expensive equipment involved to refrigerate or freeze with the commonly used refrigerant. The refrigerant is contained in a comparatively small area and is used to cool the brine with refrigerant coils to a point where the brine will take over for further cooling or freezing. A loss of brine is economically inexpensive compared to the loss of the refrigerant. Pumps for this type of service may be classified according to service requirements and uses as rotary, centrifugal, or reciprocating.

ROTARY PUMPS

Rotary pumps are usually employed for pressure-lubrication service on ammonia compressors. They deliver a continuous liquid flow like the centrifugal type but operate at much slower speeds. Rotary pumps should be operated within the assigned speed limits to avoid rapid wear and depreciation.

Rotary pumps, sometimes called *gear pumps*, or *positive-displacement pumps*, depend for their operation on the principle that a rotating plunger, impeller, gear, or cam traps the liquid in the suction side of the pump casing and forces it to the discharge side. In the simple gear pump, there are two cams or gears meshed together and revolving in opposite direction, as shown in Fig. 27-1. The case surrounding these cams forms a closer fit on the ends of the teeth and also on the sides.

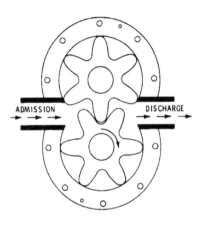

Fig. 27-1. Working principle of a rotary gear pump.

In operation, each pair of meshing teeth separates, a space forms with vacuum, and atmospheric pressure forces in the liquid to fill these spaces. Any liquid that fills the space between any two adjacent teeth must follow along as they revolve and be forced out of the discharge opening since the meshing of the teeth during rotation forms a seal separating the admission and discharge parts of the secondary chamber. Because of the number of teeth on the cams, this sequence of operation is practically continuous, some spaces carrying liquid to the discharge pipe and some spaces filling from the suction pipe. The action is positive. Each revolution transfers a quantity of liquid from suction to discharge because of the size and shape of the cam teeth. There are no valves in a rotary pump, and therefore it is better adapted to handling viscous liquid, such as lubricating oils, which would cause large frictional losses in passing through the restricted openings in valves of a reciprocating pump.

Another type of rotary pump is known as the *helical-screw-gear type*. This pump, shown in Fig. 27-2, consists of three rotors: one power rotor and two idle rotors. The convex surfaces of the power rotor mesh with the concave surfaces on the sealing rotors in such a way that a nearly fluidtight closure between the rotors is obtained. In these pumps, the standard direction of rotation is clockwise when the observer is standing at the driver and looking toward the coupling end of the pump.

The direction of rotation varies according to type of rotary pump. For instance, in the double-helical type, the direction of rotation is as follows: the standard direction of rotation is counterclockwise when facing the shaft extension end, indicated by an arrow on the pump body. The rotation of internal roller-bearing pumps may be reversed by removing the outside bearing cover and stuffing box and transferring the small plug in the side-plate casting to the opposite side. These plugs (one in each side plate) should be on the discharge side to induce circulation through the bearings to the inlet and maintain inlet pressure on the stuffing

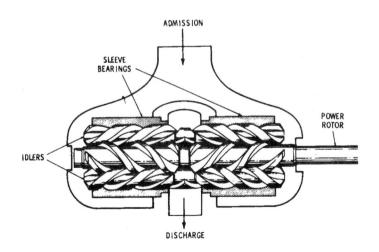

Fig. 27-2. Rotor arrangement of a helical-screw-gear pump.

box and ends of the drive shafts. Another pump, working on the internal-gear principle, makes note of the following directions:

1. In determining the desired rotation, the observer stands at the shaft end of the pump.
2. Note that the balancing groove in the shoe must be on the inlet side.
3. If a change in the direction of rotation is desired, it is necessary only to remove the cover, slip out both the top and bottom shoes, turn them end for end so that the grooves will be on the new inlet side, and reassemble the pump.

Another model of the same make is listed as an automatic reversing pump. It has a unidirectional flow regardless of the direction of shaft rotation and without the use of check valves. Finally, for a rotary pump of the helical-gear type, the instructions for determining the direction of rotation are as follows: to determine the direction of rotation, stand at the driving end facing the pump. If the shaft revolves from left to right, the rotation is clockwise; if the shaft revolves from right to left, the rotation is counterclockwise. The diagram in Fig. 27-3 is fundamental of the flow of liquid in this pump. Changing the direction of rotation of the pump drive shaft reverses the direction of the flow of liquid, causing the position of the inlet and discharge openings to be reversed. If the direction of rotation is not specified, the pump will be furnished for clockwise rotation.

The standard rotation of motors is in a counterclockwise direction. The direction of motor rotation is determined from a position at the end of the motor, which couples onto the pump.

CENTRIFUGAL PUMPS

A centrifugal pump (Fig. 27-4) is a pump in which the pressure is developed principally by the action of centrifugal force. Such pumps are commonly used where large quantities of water or brine are to be delivered at moderate pressures.

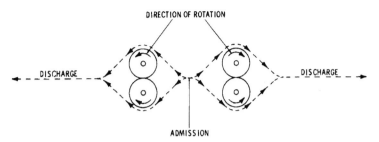

DIRECTION OF ROTATION

DISCHARGE

DISCHARGE

ADMISSION

Fig. 27-3. Liquid flow in a rotary gear pump.

The rotating member of a centrifugal pump gives a rapid rotary motion to a mass of water contained in a surrounding case; the centrifugal force forces the water out of the case through the discharge outlet. The vacuum thus created makes available atmospheric pressure to force in more water through the center. The process continues as long as motion is given to the rotor and there is a supply of water to draw upon. From this, a centrifugal pump may be defined as one in which vanes or impellers, rotating inside a close-fitting casing, draw in liquid at the center and, by centrifugal force, throw it out through an opening at the periphery of the casing.

Centrifugal pumps operate at high speeds and therefore may be connected directly to electric motors. Where noiseless opera-

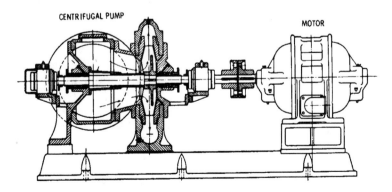

CENTRIFUGAL PUMP

MOTOR

Fig. 27-4. Drive arrangement for a typical centrifugal pump, with the motor and pump mounted on a large subbase.

tion is desired and little space is available, the centrifugal pump is usually the most logical choice.

After suction has been established by priming, the suction lift with a centrifugal pump is as high as with a reciprocating pump. It is absolutely necessary, however, to prime centrifugal pumps, which can be accomplished by submerging the unit or by equipping the inlet pipe with a foot or check valve so the suction pipe and casing may be filled with water or brine.

Single-stage centrifugal pumps are usually used for up to 100 feet total head. Above 100 feet, multiple-stage or turbine pumps are used. Multiple-stage or turbine pumps may also be used with advantage for heads less than 100 feet, particularly two-stage pumps, as they operate at about half the speed of the single-stage pump for the same capacity.

The design of impellers in centrifugal pumps varies greatly to meet the great variety of service conditions. The correct impeller for a given installation is of prime importance in order to secure economical and satisfactory operation. A high degree of efficiency can be obtained with open impellers, under certain conditions, by carefully proportioning the curvature of the blades and by reducing the side clearances to a minimum with accurate machining of the impeller edges and side plates. In general, however, more efficient results can be obtained by the use of enclosed, or shrouded, impellers. Figure 27-5 shows both an open and an enclosed impeller.

Most centrifugal pumps are fitted with ball bearings. A typical

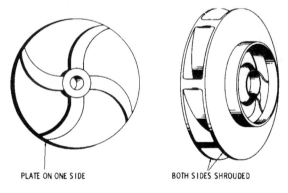

PLATE ON ONE SIDE   BOTH SIDES SHROUDED

Fig. 27-5. Semiopen and enclosed impellers.

construction consists of a single-row, deep-groove ball bearing of ample size to withstand all axial and radial loads. The bearing housings are of the cartridge type in order that the entire rotating element may be removed from the pump without disturbing the alignment or exposing the bearings to water or dirt. The bearing housings are positioned by means of dowel pins in the lower half of the casing and are securely clamped by covers split on the same horizontal plane as the pump casing. In this construction, the entire bearing may be removed from the shaft without damage by using the sleeve nut as a puller.

RECIPROCATING PUMPS

The reciprocating pump may be gear driven by electric motor or other prime mover. It may be single-acting, single-double-acting, duplex-double-acting, triple-single-acting, or triple-double-acting. The first two types should not be geared direct to an electric motor.

Steam pumps should not be considered for continuous duty because they are extremely wasteful and should be considered only where exhaust steam is required for heating purposes and where they are required intermittently for emergency purposes. Reciprocating pumps are generally employed where the amount of liquid (water or brine) is moderate and where the delivery pressure is high. They should be connected to secure a full and uniform supply where the suction lift seldom exceeds 22 feet. The less the suction lift, the better it will be for economy. To avoid air pockets, the suction line should have a uniform upward grade toward the pump of at least 6 inches in 100 feet. Every precaution should be taken to prevent foreign matter from entering the pump; if necessary, a strainer should be provided. Long-radius elbows and bends should be used on both suction and discharge lines, with valves provided near the pump.

POWER REQUIREMENTS

The following suggestions will be found useful for the selection of the proper motive power. For reciprocating pumps, the motor should have approximately 5 percent more power than that actually required. Manufacturers of these pumps give the best speed of the motor to be geared to them, and state the pump rpm in the event of a belt drive. For centrifugal and rotary pumps, the motor should have approximately 10 percent more power than actually required.

Shunt-wound dc motors are satisfactory for ordinary service, but a compound-wound motor should be furnished for single-stage pumps having a fluctuating load and for pumps frequently stopped and started under load. A compound-wound motor is preferred in all cases where there is a high starting torque and where a practically constant speed is desired. This is also one reason why compound motors are also preferable for centrifugal pumps.

For centrifugal pumps, either hand-started or automatic-control, two- or three-phase induction motors of the squirrel-cage or internal-resistance type are satisfactory. For single-phase current, any of the ordinary types of motors capable of starting the pump under load may be used.

PUMP MAINTENANCE

In general refrigeration practice, motor-driven centrifugal pumps are most generally used for handling water and brine in cold storage plants. Regardless of the type used, all will render satisfactory service if properly installed and maintained, assuming the pump has been properly selected and that it fits the actual operating requirements for any given situation. It is necessary, however, to make an occasional and thorough inspection of pumps of all types in order that defects or signs of trouble may be detected and corrected before either partial or complete breakdown of the equipment occurs.

As a guide for diagnosis of pump trouble, as well as for methods and a schedule of checks and tests, the following pointers will be of assistance:

No Liquid Delivered

1. Pump not primed
2. Suction lift too high (check with a vacuum gage at the pump suction)
3. Rotating in wrong direction

Not Enough Liquid Delivered

1. Air leaks in suction line or through stuffing box
2. Speed too slow
3. Suction lift too high
4. Too much suction lift (for hot liquids)
5. Foot valve or end of suction pipe not immersed deeply enough
6. Foot valve too small or obstructed
7. Piping improperly installed, permitting air or gas to pocket in pump
8. Mechanical defects:
   a. Pump damaged
   b. Pump badly worn
   c. Packing defective

Intermittent Delivery

1. Leaky suction lines
2. Suction lift too high
3. Air or grease in liquid
4. Pump scored by sand or by other abrasive in liquid

Pump Takes Too Much Power

1. Speed too high
2. Liquid heavier or more viscous than water
3. Suction or discharge line obstructed
4. Mechanical defects:
   a. Shaft bent
   b. Rotating elements bind
   c. Stuffing boxes too tight
   d. Misalignment due to improper connections to pipelines

or improper installation of foundation, causing spring in base

5. Misalignment of coupling (direct-connected units)

Periodic Checks and Tests on Rotary Pumps

1. Lubricate the speed-limiting, speed-regulating, and over-speed-trip governors at least once a week.
2. Run the pump under power.
3. Trip the emergency overspeed trip by hand.
4. Check the speed-limiting or speed-regulating governor to see that it maintains unit at proper speed.
5. Lift all relief valves by hand.
6. Check operation on discharge check valves.
7. Check lubricating oil for condition and presence of water.
8. Check thrust bearings, position of pump rotors, and turbine-blade clearance.

Periodic Checks and Tests on Centrifugal and Reciprocating Pumps

1. Inspect pump and motor shaft for alignment.
2. Examine the pump shaft for cuts and ridges on the section coming in contact with the stuffing-box packing.
3. Check impeller for wear.
4. Check pump and motor bearings for wear.
5. Determine whether or not the assembled unit is level.
6. Check pump speed and compare with that shown on pump nameplate.
7. Check suction lift to determine if total lift, including friction loss in suction pipe, is excessive (total suction lift, including friction loss, should not exceed 15 feet).
8. Check suction line for possible air leakage.
9. If foot valve is used, check the maximum free opening of same to determine whether or not it is at least equal to the area of the suction pipe.
10. Check pressure gage on pump discharge. If pressure gage is located on top of pump casing, its readings are worthless; it should be connected to pump discharge near discharge nozzle.

11. Check suction line to determine if same is correctly sloped so as not to contain air pockets.

SUMMARY

Circulating pumps are used in commercial and industrial refrigeration systems where brine is the refrigerant. Circulating pumps are classified as rotary, centrifugal or reciprocating.

Rotary pumps are usually employed for pressure-lubrication service on ammonia compressors and operate at a lower speed than the centrifugal type. Rotary pumps are sometimes called gear pumps, or positive displacement pumps.

The helical screw-gear type rotary pump consists of three rotors: one power rotor and two idle rotors.

Centrifugal pumps develop pressure principally by the action of centrifugal force. They are used where large quantities of water or brine are to be delivered at moderate pressures. Centrifugal pumps operate at high speeds and may be connected directly to electric motors. The design of impellers in centrifugal pumps varies greatly to meet the great variety of service conditions. The correct impeller for a given installation is of prime importance in order to secure economical and satisfactory operation.

Reciprocating pumps may be gear driven by electric motor or other prime mover. This may be single-acting, single-double acting, duplex-double acting, triple-single acting, or triple-double acting. The first two types should not be geared direct to an electric motor. Reciprocating pumps are generally employed where the amount of liquid (water or brine) is moderate and where the delivery pressure is high.

For centrifugal pumps, either hand-started or automatic control three-phase induction motors of the squirrel-cage or internal-resistance type are satisfactory. For single-phase current, any of the ordinary types of motors capable of starting the pump under load may be used.

Since the selection of pumps is of paramount importance, the manufacturer's recommendation should be procured and followed, but only after full details as to required capacity, speed and all other operating requirements have been furnished. The maintenance of pumps, including tabulated data on troubleshooting, has been described in considerable detail in order to assist the maintenance man or operating engineer in the correction of any operating difficulties that may occur.

REVIEW QUESTIONS

1. Describe the function of circulating pumps in a refrigeration system.
2. How are circulating pumps classified?
3. Describe the operation of a rotary pump.
4. What is the direction of rotation in a rotary pump?
5. Describe the standard rotation of an electric motor.
6. How does a centrifugal pump differ from that of a rotary pump?
7. What is the maximum total head for a single-stage centrifugal pump?
8. Describe a reciprocating pump and its use in a refrigeration system.
9. Give the power requirements for various types of pumps.
10. Where are motor-driven centrifugal pumps used?
11. What could cause no liquid to be delivered in a refrigeration system using motor-driven centrifugal pumps?
12. What trouble would be indicated by air leaks in the suction line or through the stuffing box?
13. What would possibly cause intermittent delivery of the brine?
14. If the pump motor starts heating up and takes too much power to operate, what would you check first?
15. List the 8 periodic checks and tests that need to be pulled on rotary pumps.
16. List the 11 periodic checks and tests that need to be pulled on reciprocating and centrifugal pumps.

CHAPTER 28

Installation and Operation

A good installation is the best insurance for troublefree operation. The refrigerating machinery should be located in a clean, dry, well-lighted, and well-ventilated area. Cleanliness and the absence of dampness will ensure long life to the motors and belts and will reduce the necessity of frequently painting the exposed piping to control corrosion. Good ventilation will make for greater safety of the operator during maintenance operation. In the case of air-cooled units, an adequate supply of outside air must be provided and a means of exhausting the air from the room to avoid recirculation.

When an installation is composed of two or more compressors, it is desirable to install the compressors on the same or parallel center lines both for appearance and so that the connections can be properly made. It should be noted, however, that in each case the compressors must be set to comply with the piping layout.

Leave enough room around the compressor and motor so that they can be easily serviced. Allow enough room at the oil-pump end of the compressor (approximately equal to the length of the compressor) so that the crankshaft may be removed if necessary. In locating the condenser, allow room equal to the length of the condenser for cleaning the tubes unless builtup cleaning rods are planned to be used. Be sure to protect water-cooled condensers, water lines, and other accessories from freezing damage during winter shutdown.

FOUNDATIONS

Foundations should be installed in accordance with the blueprints furnished with each machine. In the case of excavations running into quicksand or very marshy or soft ground, it will be necessary to make a wide footing under the foundation. In case of solid rock, a lesser foundation may be built. Foundations are preferably built of concrete in the proportion of one measure of Portland cement to three measures of sand and five measures of screened, crushed stone. Allow from 36 to 48 hours to set before the machine is put in place. Be sure the sand used in making the concrete is clean and has absolutely no soil mixed with it.

Foundations that are isolated so as to absorb vibration must be carefully engineered to suit the particular job. Simply purchasing some kind of springy supports generally will not prove satisfactory. When installing a refrigerating machine on the upper floor of a building, remember that, although the structure may be amply strong to support the weight, by chance it may have a frequency of vibration in some of its members that will cause them to vibrate in resonance with any equipment in rhythmic motion. This may call for some form of stiffening or for a foundation adapted to dampening the machine vibration until it is of a much lower intensity. A concrete base or a properly isolated spread mat will usually suffice. What is usually called noise is often really structural vibration.

If the compressor is to be located on a floor of medium or light construction, it should be supported directly over the joists or beams under the floor. When floors are weak or inadequate, supplementary floor supports must be provided. These supports should not contact the weaker floor between supports because this would tend to transmit vibrations to a larger surrounding area.

A concrete foundation used for a machine should be heavy enough to absorb operating vibrations. As a rule, the concrete slab should weigh from one to two times the total weight of the unit it supports. To ensure proper isolation, locate the concrete foundation not less than 6 feet from the footings of the building walls or columns. It is recommended that vibration isolators be used in mounting all units having steel bases when these units are not mounted on concrete foundations poured in the ground. In addition to providing more quiet operation, vibration isolators will make allowance for any slight irregularities in mounting that otherwise might result in distortion of the base and drive.

Anchor Bolts and Templates

Anchor bolts and templates are usually furnished by the compressor manufacturer. Prior to pouring the concrete foundation, the template should be braced, thus preventing its movement during the pouring operation. Nail the template to the form, being sure that it is in line and square with the foundation of the driving unit. The anchor bolts should be placed through the holes in the template, as shown in Fig. 28-1. Place ferrules made of tin spouting, pipe, or very thin lumber around the anchor bolts. These ferrules should have an inside diameter at least three times the diameter of the anchor bolt. This space is essential for moving the machine at the time of leveling and lining up before grouting.

The length of the ferrules and setting of the bolt can be determined by referring to Fig. 28-2 and by following these directions. The distance from the top of the foundation to the top of the bolt should be equal to the thickness of the motor- or compressor-base casting plus the thickness of the nut and washer, plus 1 inch for grouting. A spacing plug should be put at the top of each ferrule to keep it concentric about the bolt. When the motor and compressor are mounted on a steel base, it is advisable to set this steel base on a solid concrete foundation and level and grout it to give a firm footing. Sound isolation can be used where the noise of the compressor may be objectionable.

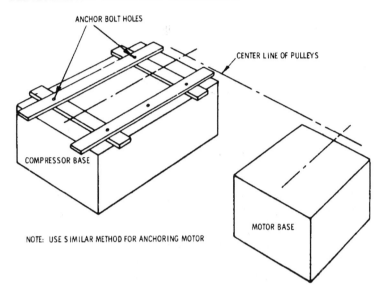

ANCHOR BOLT HOLES

CENTER LINE OF PULLEYS

COMPRESSOR BASE

MOTOR BASE

NOTE: USE SIMILAR METHOD FOR ANCHORING MOTOR

Fig. 28-1. Template arrangement to provide exact placement of anchor bolts in a belt-driven compressor installation.

Grouting

Build a frame around the top of the foundation that will hold at least 2 inches of grout. Mix two parts of clean sharp sand with one part of Portland cement, and add enough water to make a mixture that will flow freely. Wet the foundation top thoroughly, and pour the grout in from all sides, working it well with a thin strip of metal to ensure a solid grout over the entire base and in all the foundation bolt holes. Continue pouring and stirring until the grout rises to the top of the dam. If any low spots form after the dam is full, these should be promptly filled.

When the grout has hardened sufficiently to be self-supporting, remove the dam and trowel the grout down even with the bottom of the machine base, giving it a slight pitch to the outside so that any water or oil will drain off the foundation. To make a nice smooth job, trowel a thin layer of cement on the foundation.

After the grout is reasonably hard (which usually requires four or five days), draw the foundation bolt nuts uniformly all around

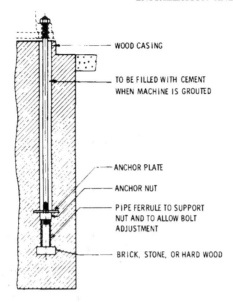

WOOD CASING

TO BE FILLED WITH CEMENT
WHEN MACHINE IS GROUTED

ANCHOR PLATE

ANCHOR NUT

PIPE FERRULE TO SUPPORT
NUT AND TO ALLOW BOLT
ADJUSTMENT

BRICK, STONE, OR HARD WOOD

Fig. 28-2. Anchor-bolt installation details.

until reasonably tight. Use care not to spring the base casting because this will throw the shaft out of line, cause heating of the bearings, and may even cause the crankshaft to break at some later date.

INSTALLING BELT-DRIVEN COMPRESSORS

When installing belt-driven compressors, it is necessary that the erection engineer check the foundations to see that they are built in accordance with certified blueprints and that the foundation tops are sufficiently rough to ensure proper grouting. The anchor bolts should be correctly placed, preferably by using a template such as shown in Fig. 28-1. Set the lower half of the belt wheel in place with the hub below the keyway on the shaft; then carefully place the compressor and motor, and lower them to within 1 inch of the tops of their respective foundations. Level the compressor by using an accurate level on the crankshaft and on the cross-leveling pads, and level by adjusting the leveling screws.

Leveling screws and iron plates are usually furnished with each compressor. If the compressor is not equipped with cross-leveling pads, it will be necessary to remove the top heads and cross level from the machined surface thus exposed. When the compressor is level, tighten the nuts on the foundation bolts slightly to prevent the compressor from shifting.

Set the key seat on the shaft, lower the top half of the belt wheel over the key, then raise the inner half and bolt the two halves of the wheel together. Heat the bolts for this purpose. Check to make sure the wheel is in the proper location as called for on the plans. Check the compressor to see that it has remained level, and place a jack under the shaft to relieve it of the weight of the wheel. Mount the motor pulley on its shaft with the projecting end of the hub toward the motor, and level and align it so that its belt grooves line up with the corresponding grooves in the flywheel. To do this, hold a long straightedge against the flywheel so that it touches the extreme edges of the flywheel and is in line with the centers of both the compressor and motor shafts. Make sure that the face of the motor pulley is parallel with the straightedge and that the center of the first groove on both the motor pulley and the flywheel are an equal distance from the straightedge. When the belt grooves are in line, tighten the set screws which hold the motor pulley key in place.

The compressor and motor can now be grouted. When the grout has set properly, the compressor leveling screws can be removed if desired. To apply the V belts, slide the motor forward sufficiently to place them in the grooves without stretching. When all belts are in their grooves, they should be tightened sufficiently to prevent slipping. Do not overtighten, or wear will be excessive. These belts require no lubrication or dressing.

INSTALLING DIRECT-CONNECTED, MOTOR-DRIVEN COMPRESSORS

With direct-connected, motor-driven units, the compressor should be placed on the foundation, leveled, and grouted. However, in this case, it is necessary to allow the grout to set before the motor is attached to the shaft. After the grout has set, place the stator frame and base plates in position over the anchor bolts, with the base-plate shims in place. The rotor is usually furnished with a split hub so that it can be easily slipped over the end of the shaft. Sometimes, however, it is a very tight fit, in which case it will be necessary to apply a light coating of oil and white lead to the shaft and force the rotor in place with a jack. In the installation of a large compressor, where the rotor is to be mounted on the shaft between the outboard bearing and the compressor, care must be used not to score the shaft. After the rotor has been keyed to the shaft, the stator should be moved over on the base plates until the stator laminations coincide with the field-pole laminations when the crankshaft is moved out so that the faces of the bearings take the thrust.

The outboard bearing should be leveled and aligned and the sole plate grouted before the motor is mounted on the shaft. To do this, remove the liners from between the pedestal and cap and pull the cap solid against the shaft. When the pedestal cap is drawn down tight on the shaft, the bottom of the box will be level providing the shaft is level. To level the pedestal the opposite way, place a level on the machined space on the sole plate and wedge underneath the sole plate until the pedestal is level. Also make sure the crankshaft has remained level. The bearing-cap liners must be replaced before the compressor is put into operation in order to give the bearing proper running clearance.

The air gap between the stator and rotor must be adjusted so that it is equal all around. To do this, first place the stator so that the gap is equal at both sides. Then wedge up or down on the base plates as required. During the process of checking the air gap, it is well to turn the rotor to several different positions and check the gap all around for each position. The air gap can be checked by means of a thin iron or wooden wedge, or, if desired, a gap gage can be obtained from the motor manufacturer. After the motor is thus aligned and with the stator shims still in place, bolt the stator, pedestals, and base plates securely. The base plates can then be grouted.

LEAK TESTING

When a refrigerant piping system is completed, the system should be checked very carefully for leaks. The most effective way of finding a leak in a Freon system is with a leak detector (Fig. 28-3). Testing with soap and water or with oil at the joints will only locate the larger leaks, and therefore these methods are unsatisfactory in determining the relative tightness of a system.

Fig. 28-3. Lightweight, portable, electronic, halogen leak detector.

When testing a new system for leaks, disconnect the lines from the water valve and pressurestat and cap the connections to the system in order to protect the bellows from injury by the high test pressure. Charge the system with a small amount of refrigerant (a couple of pounds for systems over 10 tons and less for smaller jobs). Replace the refrigerant drum with a cylinder of nitrogen, and build up the pressure to that required by local codes. *Caution: Do not use oxygen to build up pressure in the system. Serious explosions have resulted from using pure oxygen and oil under pressure.*

If no local codes are in effect, a pressure of 100 psi should be sufficient for locating any leaks. When a test pressure in excess of 150 psi is required, the compressor should be isolated to prevent damage to the seal. Disconnect the cylinder after the maximum test pressure has been reached. Test each joint and connection for leaks. The small amount of refrigerant will act as an indicator. Relieve the pressure and repair any leaks. Reconnect the lines to the water valve and pressurestat, and proceed to dehydrate the system.

DEHYDRATION AND EVACUATION

A dehydrator, as its name implies, is a device that removes moisture from the refrigerant in a system. Usually, it consists of a shell filled with a chemical moisture-absorbing agent. Its permanent installation in any field-connected installation is extremely desirable for both economy and proper operation. Its first cost is negligible compared to the service parts and labor it will save, and also the expansion-valve operation is improved by the elimination of moisture.

In systems that are field-connected, there is always a possibility of taking in moisture, especially where connections are hastily or carelessly made. Even carefully made connections may be loosened by continued vibration or possible abuse after installation. Obviously, these leaks should be repaired as soon as they are detected. Before detection, however, considerable moisture may be absorbed by the system, especially on low-temperature installations. Damage from such sources may be avoided if a dehydrator (Fig. 28-4) has been installed as a permanent fixture.

Fig. 28-4. Typical refrigeration dehydrator filled with moisture-absorbent silica gel.

Temporary dehydrator installation as a service procedure to overcome moisture is necessary, of course, but two things must be kept in mind. First, the condition probably would not have developed had a permanent dehydrator been installed; and, second, when the temporary dehydrator is removed, the system is necessarily opened up and moisture is given a chance to reenter and nullify all the good that had been accomplished. These points further emphasize the advisability of installing permanent dehydrators at the time of the original installation.

The dehydrating agent used is capable of absorbing only a certain amount of moisture. Should a greater quantity of moisture get into the system, the dehydrator will become saturated and inactive, and then must be replaced or refilled. Should dehydrators be opened or even not tightly sealed in the stock room, they will absorb moisture from the air, becoming saturated and inactive if put on an installation. Therefore, be sure the dehydrating agent is still active in any dehydrator that is to be installed.

It is extremely important that all refrigeration systems be absolutely free of moisture. Moisture in the system can cause sludging of the oil in the crankcase, resulting in corrosion and lack of proper lubrication. This contamination can be explained by two entirely different types of oil decomposition. The first type, reaction of the oil with oxygen in the noncondensables, usually is the first to take place. Even though the system is dry, oil breakdown products due to reaction of the oil with oxygen will start building up on the valve plate, particularly on the seat of the discharge valve at the high head temperatures with noncondensables present. In time, this can lead to wire drawing (forcing of the gas through a small leak on the discharge valve, creating very high temperatures—sometimes as high as 800 to 900°F) and initiate the second type of oil breakdown. The second type involves the oil reacting directly with the refrigerant to form hydrochloric acid, hydrofluoric acid, carbon, and water. This becomes a vicious cycle, snowballing until the system has a serious wet acidic condition, oil deterioration, and, with hermetic compressors, an eventual motor burnout.

If the system is properly evacuated by using a good vacuum pump that can pull down into the 50- to 100-micron range, the noncondensables are eliminated before the system is ever started. A good vacuum pump with electronic gage is desirable and highly recommended for this procedure. Actually, 25 to 28 inches of vacuum is simply not enough to accomplish any great amount of dehydration or noncondensable removal. The gage may be showing 50 microns at the pump, but if a long run is being pulled, it may be a completely different story at the other end of the line. The poorer the vacuum, the worse the situation becomes. Many times the situation is further complicated by using too small a connecting line between the vacuum pump and the system, which creates a high pressure drop that greatly reduces the efficiency of the vacuum pump.

Moisture Indicators

Moisture indicators are designed to show at a glance when moisture contents endanger a refrigeration or air-conditioning system and when the charge is low or a restriction is present. Elements change color when the moisture content in the system reaches a dangerous level. Bubbles appear in the sight glass when the pressure drops. The refrigerant 12 element, which works equally well for systems using refrigerants 11, 12, 13, 113, and 114, shows blue when the system is dry and pink when it is wet. The refrigerant 22 element, also used for refrigerants 500 and 502, shows green when the system is dry and pink when it is wet.

The moisture-indicator ring (or rings, in units that combine refrigeration 12 and 22 elements) and the sight glass are combined in a single unit, which screws out easily in one piece to permit worry-free soldering or solder connections. With the indicator assembly removed, there is no way to damage the elements when installing. Figure 28-5 shows a variety of moisture indicators, which have different means of being attached to the piping.

CHARGING THE SYSTEM

After the system has been completely evacuated and dehydrated (and with the compressor stopped), open the discharge and suction valves and connect the refrigerant drum to the charging valve on the system. Pay strict attention to the instruction tag attached to the drum by the manufacturer. When the connections to the drum have been made, open the valve on the drum slowly and make sure that there are no leaks between the drum and the charging valve. If no leaks are present, then open the drum valve

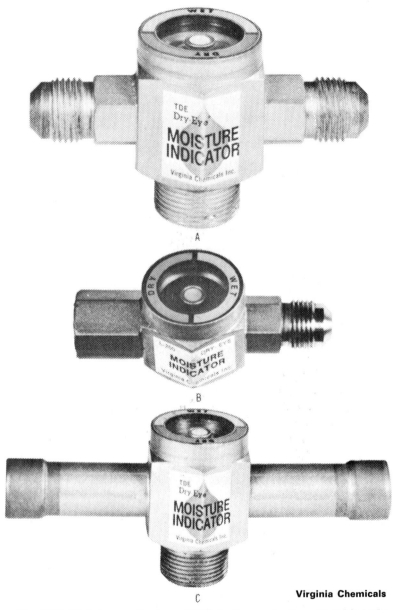

Virginia Chemicals

Fig. 28-5. Moisture indicators. (A) Male flare on both ends; (B) female flare and a male flare; (C) solder and solder ends.

two or three turns. Open the charging valve on the system slightly, and allow the refrigerant to enter the system, which has previously been pulled down to a vacuum while the evacuation was building up pressure throughout the system. Care should be exercised while charging and the drum valve should be closed immediately if any leakage is discovered in the system.

Open the main discharge valve on the manifold, and turn on the water to the condenser and compressor jackets, making sure that both are getting an ample supply. If the low side is a brine or water cooler, be sure the fluid is flowing so that freezing does not occur during charging. Start the compressor, and slowly open the main suction valve on the manifold. Close the valve on the liquid line from the receiver to the evaporator, and make sure that the valve in the liquid line from the condenser to the receiver is open.

As soon as the valve in the liquid line from the receiver is closed, the discharge pressure will begin to build up rapidly and suction pressure will drop correspondingly. Do not allow the suction temperature to operate below the freezing point of the cooling medium in the cooler. As the suction pressure drops, open the charging valve and drum valve, and regulate the flow from the drum to maintain normal operating suction pressure. This will allow the refrigerant to be taken into the system from the drum and a charge to be built up in the receiver. A drop in the suction pressure when charging the system from the drum and the frosting of the lower side of the drum near the valve end are indications that the drum is becoming empty. The drum valve should be opened wide immediately, and the suction should be pulled down to zero to completely exhaust the drum.

As soon as the drum is empty, the charging valve should be closed and the valve on the receiver opened and regulated in conjunction with the expansion valve to give the desired suction pressure or proper amount for the low-side design. The amount of refrigerant required will depend entirely upon the system, and if the initial charge proves insufficient, more may be added by connecting another drum to the charging connection, closing the valve on the receiver outlet, and allowing the refrigerant from the drum to maintain proper suction pressure. A close watch should be kept on the liquid-level gage on the receiver so that the proper level may be maintained. The receiver should contain enough liquid at all timers to keep the outlet sealed and prevent gas from entering the liquid line. It is advisable to weigh the drums before and after charging so that the amount of refrigerant added to the system may be recorded. This information is often of value for other operating or maintenance personnel.

SUMMARY

Installation of new refrigeration equipment is normally a function of various contracting companies and thus may not necessarily be of concern to the maintenance or operating engineer. It is felt, however, that a thorough review of installation methods will be of assistance since additional units at times may have to be installed without the assistance or guidance of experienced persons.

As noted in the first part of the chapter, a good installation is the best insurance for troublefree operation. Various details concerning this attainment have been presented, particularly dealing with the arrangement of the refrigerating equipment, foundations, and methods of securing compressors and accessories to a concrete foundation.

Various operating features, such as purging, leak testing, dehydration, and charging of typical refrigeration systems have been included, as well as the necessary precautions to be observed when accomplishing the foregoing maintenance details. The importance of vacuum pumps in evacuating a system and the method for determining proper vacuum ranges are discussed.

REVIEW QUESTIONS

1. What are the essential features in planning and installing refrigerating machinery?
2. What should be observed with respect to the installation of two or more compressors in one refrigeration plant?
3. Why is ventilation necessary in refrigeration plants?
4. What are the precautions to be observed when constructing the foundation?

5. What is the function of vibration insulators as installed in re-frigeration plants?
6. Give the recommended ratios between the weight of concrete foundation slabs and the machinery to be supported.
7. Describe the placement of anchor bolts and the methods used to secure the motor and compressor to the foundation.
8. What is meant by grouting?
9. Describe installation methods for belt-driven compressors and the various precautions to be observed when such installations are made.
10. How do belt-driven installations differ from direct-connected, motor-driven compressor installations?
11. Why is leak testing necessary, and what methods are used to leak-test a refrigeration system?
12. What is a dehydrator, and what is its purpose in a refrigeration system?
13. How does moisture affect compressor operation?

CHAPTER 29

Heat Leakage
Through Walls

Heat leakage is always given in Btu per hour per degrees Fahrenheit temperature difference per square foot of exposed surface (Btu/hr/°F/sq ft). Prior to designing a refrigeration system, an estimate must be made of the maximum probable heat loss of each room or space to be cooled. Therefore, before attempting to place even a small room cooler into service, check up on the walls and determine just what a certain sized unit will do, or is supposed to do, in the way of cooling and dehumidifying.

The heat losses may be divided into two groups: losses through confining walls, floors, ceilings, glass, or other surfaces; and infiltration losses due to leaks through cracks and crevices around doors and windows. The heat leakage through walls, floors, and ceilings can be determined by means of a formula and depends on the type and thickness of the insulating material used. The formula for heat leakage is:

$$H = KA(t_1 - t_2)$$

where   K   =   heat-transfer coefficient, Btu/hr/°F/sq ft
        A   =   area, sq ft
       $t_1 - t_2$ =   temperature gradient through wall, °F

Example—Calculate the heat leakage through an 8-inch brick wall having an area of 200 sq ft if the inside temperature is 10°F and the outside temperature is 70°F.

Solution—If it is assumed that the heat-transfer coefficient of a plain brick wall is 0.50, substitution of values in the foregoing equation will be as follows:

$$H = 0.50 \times 200(70 - 10) = 6000 \text{Btu/hr}$$

The heat leakage through floors, ceilings, and roofs may be estimated in the same manner, the heat-transfer coefficient, **K**, depending on the particular insulating material used. Tables 29-1 to 29-20 cover almost every type and combination of wall, floor, and ceiling encountered in the field. Partition walls are also included so that these may be estimated for heat leakage. Although the **K** values given in the tables do not agree entirely with the data given in various handbooks, they will give sufficiently close values for adaptation since in most instances only approximate values can be obtained.

In making use of the data, it must be remembered that the outside area is used as a basis for estimating. Ceiling and floor construction must also be determined and, in many instances, the four walls of the room may not be of the same construction since one or more of the walls may have partitions.

SUMMARY

Because the economy and efficiency of refrigeration depends largely on the insulation properties of walls, floors, and ceilings, an effort has been made to provide information on heat leakage in various structures and the mathematical methods used to determine the amount of this leakage. Tabulated data of the heat-transfer coefficient **K** for various insulating materials have been furnished.

Table 29-1. Concrete Wall (No Exterior Finish)

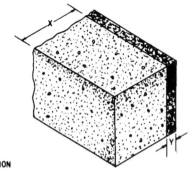

K=X+Y
X-WALL
Y-INTERIOR CONSTRUCTION

Values of *K* in Btu/hr/1 °F/sq ft

Wall Construction, *Y*	Thickness, *X*				
	6"	8"	10"	12"	16"
Plain wall—no interior finish	0.58	0.51	0.46	0.41	0.34
½" Plaster—direct on concrete	0.52	0.46	0.42	0.38	0.32
½" Plaster on wood lath, furred	0.31	0.29	0.26	0.24	0.22
¾" Plaster on metal lath, furred	0.34	0.32	0.29	0.27	0.24
½" Plaster on ⅜" plasterboard, furred	0.32	0.30	0.27	0.25	0.22
½" Plaster on ½" board insulation, furred	0.21	0.20	0.19	0.18	0.17
½" Plaster on 1" corkboard, set in ½" cement	0.16	0.15	0.14	0.14	0.13
½" Plaster on 1½" corkboard, set in ½" cement	0.15	0.14	0.14	0.13	0.12
½" Plaster on 2" corkboard, set in ½" cement	0.13	0.12	0.11	0.10	0.09
½" Plaster on wood lath on 2" furring, 1⅝" gypsum fill	0.20	0.19	0.18	0.17	0.16

Table 29-2. Concrete Wall (Exterior Stucco Finish)

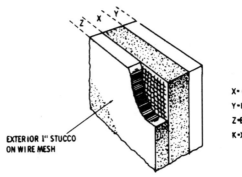

EXTERIOR 1" STUCCO
ON WIRE MESH

X = CONCRETE
Y = INSULATION
Z = EXTERIOR STUCCO 1"
K = X + Z + Y

Values of K in Btu/hr/1 °F/sq ft

Wall Construction, Y	Thickness of Concrete, X					
	6"	8"	10"	12"	16"	18"
Plain walls—no interior finish	0.54	0.48	0.43	0.39	0.33	0.28
½" Plaster direct on concrete	0.49	0.44	0.40	0.36	0.31	0.27
½" Plaster on wood lath, furred	0.31	0.29	0.27	0.25	0.23	0.22
¾" Plaster on metal lath, furred	0.32	0.30	0.28	0.26	0.24	0.23
½" Plaster on ⅜" plasterboard, furred	0.31	0.29	0.27	0.25	0.23	0.22
½" Plaster on wood lath on 2" furring strips with 1⅝" cellular gypsum fill	0.20	0.19	0.18	0.17	0.16	0.15
½" Plaster on ½" board insulation, furred	0.22	0.21	0.20	0.19	0.18	0.17
1"	0.17	0.16	0.15	0.145	0.140	0.13
½" Plaster on 1½" sheet cork set in cement	0.145	0.140	0.135	0.130	0.12	0.11
2"	0.12	0.118	0.115	0.110	0.105	0.100

Table 29-3. Concrete Wall (Brick Veneer)

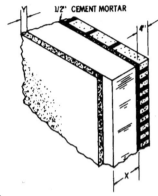

1/2" CEMENT MORTAR

X - CONCRETE

Y - INTERIOR FINISH

K = X + 1/2" CEMENT + 4" BRICK + Y

Wall Construction, Y	Concrete, X			
	6"	8"	10"	12"
Plain wall—no interior finish	0.39	0.36	0.33	0.30
½" Plaster direct on concrete	0.37	0.34	0.30	0.29
½" Plaster on wood lath, furred	0.24	0.23	0.21	0.20
¾" Plaster on metal lath, furred	0.27	0.25	0.23	0.22
½" Plaster on wood lath, on 2" furring strips, with 1⅝" cellular gypsum fill	0.18	0.17	0.16	0.15
½" Plaster on ⅜" plasterboard, furred	0.25	0.24	0.22	0.21
½" Plaster on ½" board insulation, furred	0.17	0.16	0.14	0.12
½" Plaster on 1" board insulation, furred	0.15	0.14	0.12	0.11
½" Plaster on 1½" board insulation, furred	0.13	0.12	0.11	0.10
½" Plaster on 2" board insulation, furred	0.11	0.10	0.09	0.08

Table 29-4. Concrete Wall (4-inch Cut Stone)

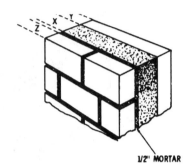

K= X + Y + Z
X = CONCRETE
Y = INSULATION
Z = 4' CUT STONE

1/2" MORTAR

Values of *K* in Btu/hr/1°F/sq ft

Wall Construction, *Y*	Thickness of Concrete, *X*				
	6″	8″	10″	12″	16″
Plain walls—no interior finish	0.46	0.42	0.38	0.34	0.30
½″ Plaster direct on concrete	0.42	0.38	0.34	0.32	0.28
¾″ Plaster on metal lath, furred	0.29	0.27	0.26	0.24	0.22
½″ Plaster on wood lath, furred	0.28	0.26	0.25	0.23	0.21
½″ Plaster on ⅜″ plasterboard, furred	0.28	0.26	0.25	0.23	0.21
½″ Plaster on wood lath on 2″ furring strips, filled with 1⅝″ gypsum	0.18	0.175	0.17	0.16	0.15
½″ Plaster on ½″ board insulation, furred	0.21	0.20	0.19	0.18	0.17
½″ Plaster on 1″ board insulation, furred	0.16	0.155	0.15	0.14	0.13
½″ Plaster on 1″ sheet cork, set in ½″ cement or mortar	0.15	0.145	0.140	0.135	0.130
1½″	0.138	0.135	0.130	0.125	0.12
2″	0.115	0.110	0.108	0.105	0.100

Table 29-5. Cinder and Concrete Block Wall

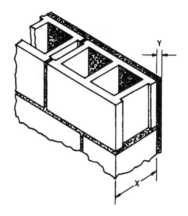

K = X + Y

X-WALL

Y-INTERIOR FINISH

Values of *K* in Btu/hr/1°F/sq ft

Wall Construction, *Y*	Thickness, *X*, and Kind of Blocks			
	Concrete		Cinder	
	8"	12"	8"	12"
Plain wall—no interior finish	0.46	0.34	0.31	0.23
½" Plaster—direct on blocks	0.42	0.32	0.29	0.21
½" Plaster on wood lath, furred	0.28	0.23	0.22	0.17
¾" Plaster on metal lath, furred	0.29	0.24	0.23	0.17
½" Plaster on ⅜" plasterboard, furred	0.28	0.23	0.22	0.17
½" Plaster on ½" board insulation, furred	0.21	0.18	0.17	0.14
½" Plaster on 1" board insulation, furred	0.16	0.14	0.14	0.11
½" Plaster on 1½" sheet cork set in ½" cement	0.13	0.12	0.12	0.10
½" Plaster on 2" sheet cork set in ½" cement	0.11	0.10	0.10	0.09
½" Plaster on wood lath, on 2" furring strips, 1⅝" gypsum fill	0.18	0.16	0.16	0.13

Table 29-6. Cinder and Concrete Block Wall (Brick Veneer)

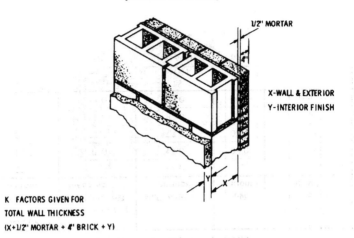

1/2" MORTAR

4"

X-WALL & EXTERIOR
Y-INTERIOR FINISH

K FACTORS GIVEN FOR
TOTAL WALL THICKNESS
(X+1/2" MORTAR + 4" BRICK + Y)

Values of K in Btu/hr/1°F/sq ft

Wall Construction, Y	Thickness, X, and Kind of Blocks			
	Concrete		Cinder	
	8"	12"	8"	12"
Plain wall—no interior finish	0.33	0.26	0.25	0.18
½" Plaster—direct on blocks	0.31	0.25	0.23	0.18
½" Plaster on wood lath, furred	0.23	0.19	0.18	0.15
¾" Plaster on metal lath, furred	0.24	0.20	0.19	0.15
½" Plaster on wood lath, on 2" furring strips, 1⅝" gypsum fill	0.16	0.14	0.14	0.12
½" Plaster on ½" board insulation, furred	0.18	0.16	0.15	0.13
½" Plaster on 1" board insulation, furred	0.14	0.13	0.12	0.10
½" Plaster on ⅜" plasterboard, furred	0.23	0.19	0.18	0.15
½" Plaster on 1½" sheet cork, set in ½" cement mortar	0.12	0.11	0.11	0.09
½" Plaster on 2" sheet cork, set in ½" cement mortar	0.10	0.09	0.09	0.08

692

Table 29-7. Brick Veneer on Hollow Tile Wall (Brick Veneer)

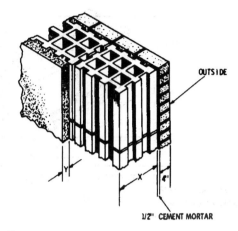

OUTSIDE

1/2" CEMENT MORTAR

K VALUES GIVEN FOR

TOTAL WALL THICKNESS AS:

X + 1/2" MORTAR + 4' BRICK + Y

X - WALL & EXTERIOR

Y - INTERIOR FINISH

Values of *K* in Btu/hr/1 °F/sq ft

Wall Construction, Y	Tile Thickness, X				
	4"	6"	8"	10"	12"
Plain wall—no interior finish	0.29	0.27	0.25	0.22	0.18
½" Plaster on hollow tile	0.28	0.25	0.24	0.21	0.17
½" Plaster on wood lath, furred	0.20	0.19	0.18	0.17	0.15
¾" Plaster on metal lath, furred	0.21	0.20	0.19	0.18	0.16
½" Plaster on ⅜" plasterboard, furred	0.20	0.19	0.18	0.17	0.15
½" Plaster on wood lath, on 2" furring strips, with 1⅝" cellular gypsum fill	0.14	0.13	0.12	0.11	0.10
½" Plaster on ½" board insulation, furred	0.16	0.15	0.14	0.13	0.12
½" Plaster on 1" board insulation, furred	0.14	0.13	0.12	0.11	0.10
½" Plaster on 1½" board insulation, furred	0.13	0.12	0.11	0.10	0.09
½" Plaster on 2" board insulation, furred	0.12	0.11	0.10	0.09	0.08

693

Table 29-8. Hollow Tile Wall (4-inch Cut Stone Veneer)

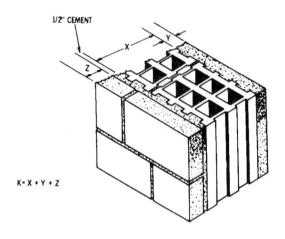

1/2' CEMENT

K = X + Y + Z

Values of *K* in Btu/hr/1°F/sq ft

Wall Construction, *Y*	Thickness of Tile, *X*			
	6"	8"	10"	12"
Plain walls—no interior finish	0.30	0.29	0.28	0.22
½" Plaster—direct on hollow tile	0.28	0.27	0.26	0.21
¾" Plaster on metal lath, furred	0.22	0.21	0.20	0.18
½" Plaster on wood lath, furred	0.21	0.20	0.19	0.17
½" Plaster on ⅜" plasterboard, furred	0.21	0.20	0.19	0.17
½" Plaster on ½" corkboard set in ½" cement	0.17	0.16	0.15	0.14
½" Plaster on 1" corkboard set in ½" cement	0.14	0.13	0.13	0.12
½" Plaster on 1½" corkboard set in ½" cement	0.13	0.12	0.11	0.10
½" Plaster on 2" corkboard set in ½" cement	0.11	0.10	0.09	0.08
½" Plaster on wood lath, 2" furring, 1⅝ gypsum fill	0.15	0.14	0.13	0.12
½" Plaster on ½" board insulation	0.18	0.17	0.16	0.15

Table 29-9. Hollow Tile Wall (Stucco Exterior)

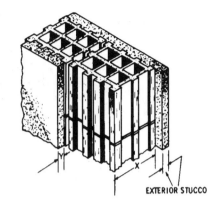

X - TILE THICKNESS
Y - INTERIOR FINISH

K VALUES GIVEN FOR
TOTAL WALL THICKNESS
X + Y + STUCCO FINISH

EXTERIOR STUCCO

Values of K in Btu/hr/1 °F/sq ft

Wall Construction, Y	Tile Thickness, X				
	6"	8"	10"	12"	16"
Plain wall with stucco—no interior finish	0.32	0.30	0.28	0.22	0.18
½" Plaster—direct on hollow tile	0.31	0.29	0.27	0.21	0.17
½" Plaster on wood lath, furred	0.22	0.20	0.18	0.16	0.12
¾" Plaster on metal lath, furred	0.23	0.21	0.19	0.17	0.13
½" Plaster on ⅜" plasterboard, furred	0.21	0.20	0.18	0.16	0.12
½" Plaster on wood lath, on 2" furring strips, with 1⅝" cellular gypsum fill	0.16	0.15	0.14	0.13	0.12
½" Plaster on ½" board insulation, furred	0.15	0.14	0.13	0.12	0.11
½" Plaster on 1" board insulation, furred	0.13	0.12	0.11	0.10	0.09
½" Plaster on 1½" board insulation, furred	0.12	0.11	0.10	0.09	0.08
½" Plaster on 2" board insulation, furred	0.11	0.10	0.09	0.08	0.07

695

Table 29-10. Limestone or Sandstone Wall

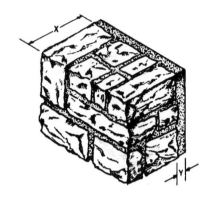

K = X + Y

X - WALL
Y - INTERIOR CONSTRUCTION

Values of *K* in Btu/hr/1°F/sq ft

Wall Construction, *Y*	Thickness, *X*			
	8"	10"	12"	16"
Plain wall—no interior finish	0.56	0.50	0.48	0.39
½" Plaster—direct on stone	0.50	0.45	0.42	0.36
¾" Plaster on metal lath, furred	0.33	0.31	0.29	0.26
½" Plaster on wood lath, furred	0.31	0.29	0.28	0.25
½" Plaster on ⅜" plasterboard, furred	0.31	0.30	0.28	0.25
½" Plaster on wood lath, on 2" furring strips, 1⅝" gypsum fill	0.20	0.19	0.18	0.17
½" Plaster on ½" board insulation, furred	0.23	0.22	0.21	0.19
½" Plaster on 1" board insulation, furred	0.17	0.16	0.15	0.14
½" Plaster on 1½" corkboard, set in ½" cement	0.14	0.14	0.13	0.12
½" Plaster on 2" corkboard, set in ½" cement	0.11	0.11	0.10	0.10
½" Plaster on wood lath, on 2" furring strips, 2" gypsum fill	0.19	0.18	0.17	0.16

Table 29-11. Brick Wall (4-inch Cut Stone)

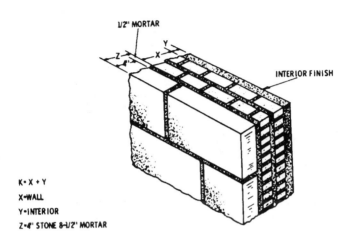

1/2" MORTAR

Y

Z · X

INTERIOR FINISH

K = X + Y
X = WALL
Y = INTERIOR
Z = 4' STONE 8-1/2' MORTAR

Values of *K* in Btu/hr/1 °F/sq ft

Wall Construction, Y	Thickness, X						
	8"	9"	12"	13"	16"	18"	24"
Plain wall—no interior finish	0.33	0.29	0.26	0.25	0.22	0.20	0.16
$\frac{1}{2}$" Plaster—direct on brick	0.31	0.28	0.25	0.24	0.21	0.19	0.15
$\frac{3}{4}$" Plaster on metal lath, furred	0.23	0.22	0.20	0.19	0.18	0.17	0.15
$\frac{1}{2}$" Plaster on wood lath, furred	0.22	0.21	0.19	0.18	0.17	0.15	0.12
$\frac{1}{2}$" Plaster on $\frac{3}{8}$" plasterboard, furred	0.23	0.22	0.20	0.19	0.18	0.17	0.14
$\frac{1}{2}$" Plaster, lath, 2" furring strips, 1 $\frac{5}{8}$" gypsum fill	0.16	0.15	0.14	0.13	0.12	0.11	0.10
$\frac{1}{2}$" Plaster on $\frac{1}{2}$" sheet cork	0.18	0.16	0.15	0.14	0.13	0.12	0.11
$\frac{1}{2}$" Plaster on 1" sheet cork	0.14	0.13	0.12	0.11	0.10	0.10	0.09
$\frac{1}{2}$" Plaster on 1 $\frac{1}{2}$" sheet cork	0.12	0.12	0.11	0.11	0.10	0.10	0.09
$\frac{1}{2}$" Plaster on 2" sheet cork	0.11	0.10	0.10	0.09	0.09	0.08	0.08

697

Table 29-12. Solid Brick Wall (No Exterior Finish)

K = X+Y

X = WALL

Y = INTERIOR CONSTRUCTION

Values of *K* in Btu/hr/1°F/sq ft

Wall Construction, Y	Thickness, X						
	8″	9″	12″	13″	16″	18″	24″
Plain wall—no interior finish	0.39	0.37	0.30	0.29	0.24	0.23	0.18
½″ Plaster—direct on brick	0.36	0.35	0.28	0.27	0.23	0.20	0.16
¾″ Plaster on metal lath, furred	0.26	0.25	0.23	0.21	0.19	0.18	0.15
½″ Plaster on wood lath, furred	0.25	0.24	0.21	0.19	0.18	0.17	0.14
½″ Plaster on ⅜″ plasterboard, furred	0.24	0.23	0.20	0.18	0.17	0.16	0.14
½″ Plaster on ½″ board insulation, furred	0.20	0.19	0.18	0.17	0.16	0.15	0.13
½″ Plaster on ½″ sheet cork, set in ½″ cement	0.18	0.17	0.16	0.15	0.14	0.13	0.12
½″ Plaster on 1″ sheet cork, set in ½″ cement	0.15	0.14	0.13	0.12	0.11	0.10	0.09
½″ Plaster on 1½″ sheet cork, set in ½″ cement	0.14	0.13	0.12	0.11	0.10	0.09	0.08
½″ Plaster on 2″ sheet cork, set in ½″ cement	0.13	0.12	0.11	0.10	0.09	0.08	0.07
½″ Plaster on 1½″ split furring, tiled	0.28	0.26	0.23	0.22	0.20	0.19	0.17

Table 29-13. Wood-Siding Clapboard Frame or Shingle Walls

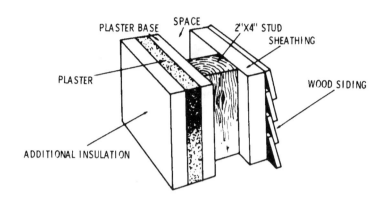

PLASTER BASE — SPACE — 2''X4'' STUD — SHEATHING — PLASTER — WOOD SIDING — ADDITIONAL INSULATION

Values of K in Btu/hr/1 °F/sq ft

Type of Sheathing	Insulation between Studding	Plaster Base						
		Wood lath	Metal lath	$\frac{3}{8}''$ Plaster-board	$\frac{1}{2}''$ Rigid insulation	1" Rigid insulation	$1\frac{1}{2}''$ Sheet cork	2" Sheet cork
½" Plaster-board	None	0.30	0.31	0.30	0.22	0.16	0.12	0.10
	Cellular gypsum, 18#	0.12	0.125	0.12	0.10	0.09	0.08	0.07
	Flaked gypsum, 24#	0.10	0.108	0.10	0.09	0.08	0.07	0.06
	½" Flexible insulation	0.16	0.17	0.16	0.14	0.11	0.09	0.08
1" Wood	None	0.26	0.28	0.26	0.20	0.15	0.12	0.10
	Cellular gypsum, 18#	0.11	0.118	0.11	0.10	0.09	0.08	0.07
	Flaked gypsum, 24#	0.10	0.108	0.10	0.09	0.08	0.07	0.06
	½" Flexible insulation	0.15	0.16	0.15	0.13	0.11	0.09	0.08
½" Rigid insulation (board form)	None	0.22	0.23	0.22	0.17	0.14	0.11	0.09
	Cellular gypsum, 18#	0.10	0.11	0.10	0.09	0.08	0.07	0.06
	Flaked gypsum, 24#	0.09	0.10	0.09	0.08	0.07	0.06	0.05
	½" Flexible insulation	0.14	0.148	0.14	0.12	0.10	0.08	0.07

Table 29-14. Interior Walls and Plastered Partitions (No Fill)

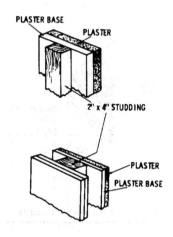

PLASTER BASE

PLASTER

2' x 4' STUDDING

PLASTER

PLASTER BASE

Values of *K* in Btu/hr/1°F/sq ft

Plaster Base	Single Partition, 1 Side Plastered	Double Partition, 2 Sides Plastered
Wood lath	0.60	0.33
Metal lath	0.67	0.37
³/₈" Plasterboard	0.56	0.32
¹/₂" Board (rigid) insulation	0.33	0.18
1" Board (rigid) insulation	0.22	0.12
1¹/₂" Corkboard	0.16	0.08
2" Corkboard	0.12	0.06

Table 29-15. Frame Floors and Ceilings (No Fill)

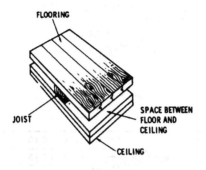

Values of K in Btu/hr/1 °F/sq ft

Type of Ceiling with ½" Plaster	No Flooring	1" Yellow-Pine Flooring	1" Yellow-Pine Flooring and ½" Fiberboard	1" Suboak or Maple on 1" Yellow-Pine Flooring
No ceiling		0.45	0.27	0.32
Metal lath	0.66	0.24	0.20	0.24
Wood lath	0.60	0.27	0.19	0.23
⅜" Plasterboard	0.59	0.28	0.20	0.24
½" Rigid insulation	0.34	0.20	0.15	0.17

Table 29-16. Masonry Partitions

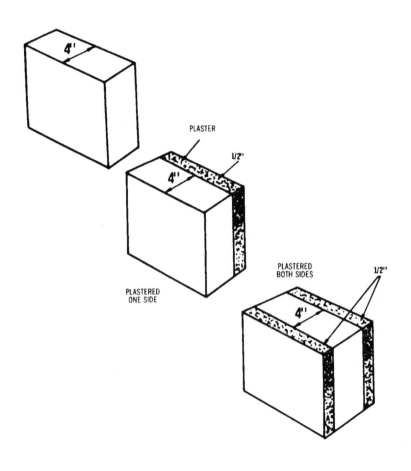

Values of *K* in Btu/hr/1°F/sq ft

Wall	Bare, No Plaster	One Side Plastered	Two Sides Plastered
4" Brick	0.48	0.46	0.41
4" Hollow gypsum tile	0.29	0.27	0.26
4" Hollow clay tile	0.44	0.41	0.39

Table 29-17. Concrete Floors (on Ground)

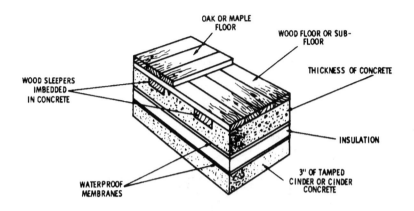

Values of K in Btu/hr/1 °F/sq ft

Type of Insulation	Thickness, Y	Concrete Thickness, X	No Flooring	Tile or Terrazzo Floor	1" Yellow Pine on Wood Sleepers	Oak or Maple on Yellow-Pine Subfloor on Sleepers
None and no cinder base, Z	0	4	1.020	0.930	0.510	0.370
	0	5	0.940	0.830	0.480	0.360
	0	6	0.860	0.790	0.460	0.340
	0	8	0.760	0.700	0.430	0.320
None with 3" cinder base, Z	0	4	0.580	0.550	0.510	0.310
	0	5	0.550	0.520	0.470	0.300
	0	6	0.520	0.490	0.440	0.290
	0	8	0.500	0.470	0.400	0.280
1" Rigid insulation with 3" cinder base, Z	1"	4	0.220	0.210	0.180	0.165
	1"	5	0.210	0.200	0.175	0.160
	1"	6	0.200	0.190	0.170	0.155
	1"	8	0.190	0.180	0.160	0.150
2" Corkboard insulation with 3" cinder base, Z	2"	4	0.125	0.120	0.115	0.105
	2"	5	0.120	0.115	0.110	0.100
	2"	6	0.115	0.110	0.105	0.095
	2"	8	0.110	0.105	0.100	0.090

Table 29-18. Concrete Floors and Ceilings

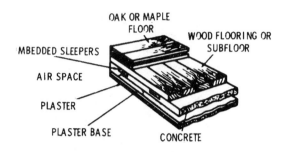

OAK OR MAPLE FLOOR

WOOD FLOORING OR SUBFLOOR

MBEDDED SLEEPERS

AIR SPACE

PLASTER

PLASTER BASE

CONCRETE

Values of *K* in Btu/hr/1°F/sq ft

Type of Ceiling and Base	Thick-ness	Floor Type			
		No Flooring	Tile or Terrazzo Floor	1" Yellow Pine on Wood Sleepers	Oak or Maple on Yellow-Pine Sub-floor on Sleeper
No ceiling	4	0.60	0.57	0.38	0.29
	6	0.54	0.52	0.35	0.28
	8	0.48	0.46	0.33	0.26
	10	0.46	0.44	0.31	0.25
½" Plaster direct on concrete	4	0.56	0.53	0.38	0.30
	6	0.51	0.49	0.35	0.29
	8	0.47	0.44	0.33	0.27
	10	0.42	0.41	0.31	0.26
Suspended or furred ceiling on ⅜" plasterboard, ½" plaster	4	0.33	0.32	0.24	0.20
	6	0.31	0.30	0.23	0.19
	8	0.29	0.28	0.22	0.18
	10	0.28	0.27	0.21	0.17
Suspended or furred metal lath, ¾" plaster	4	0.36	0.34	0.26	0.21
	6	0.34	0.32	0.24	0.20
	8	0.32	0.30	0.23	0.19
	10	0.31	0.29	0.22	0.18
Suspended or furred on ½" rigid insulation, ½" plaster	4	0.22	0.21	0.20	0.18
	6	0.21	0.20	0.19	0.17
	8	0.20	0.19	0.18	0.16
	10	0.19	0.18	0.17	0.15
1½" corkboard in ½" mortar, ½" plaster	4	0.16	0.15	0.14	0.13
	6	0.15	0.14	0.13	0.12
	8	0.14	0.13	0.12	0.11
	10	0.13	0.12	0.11	0.10

Table 29-19. Brick Veneer (Frame Walls)

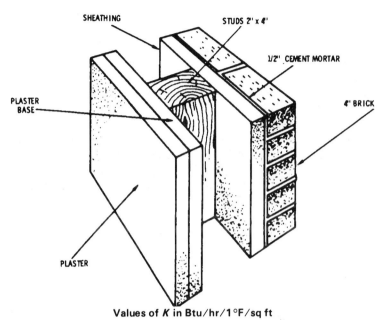

Values of *K* in Btu/hr/1°F/sq ft

Type of Sheathing	Insulation between Studding	Plaster Base							
		Wood Lath	Metal Lath	$\frac{1}{2}''$ Fiber board	$\frac{3}{8}''$ Plaster-board	$\frac{1}{2}''$ Cork-board	1″ Cork-board	$1\frac{1}{2}''$ Cork-board	2″ Cork-board
1″ Wood sheathing	None	0.260	0.270	0.20	0.27	0.18	0.15	0.12	0.095
	$\frac{1}{2}''$ Fiberboard	0.170	0.180	0.14	0.16	0.14	0.11	0.09	0.080
	Flaked gypsum 24#	0.095	0.100	0.09	0.10		0.08	0.07	0.060
	Cellular gypsum 18#	0.110	0.150	0.10	0.11		0.09	0.08	0.070
$\frac{1}{2}''$ Plaster-board sheathing	None	0.320	0.340	0.24	0.30		0.16	0.13	0.110
	$\frac{1}{2}''$ Fiberboard	0.180	0.190	0.14	0.17		0.11	0.09	0.080
	Flaked gypsum 24#	0.100	0.110	0.09	0.10		0.08	0.07	0.060
	Cellular gypsum 18#	0.120	0.120	0.10	0.12		0.09	0.08	0.070
$\frac{1}{2}''$ Fiber-board (rigid) insulation sheathing	None	0.250	0.260	0.19	0.26		0.15	0.12	0.100
	$\frac{1}{2}''$ Fiberboard	0.160	0.170	0.13	0.15		0.10	0.09	0.080
	Flaked gypsum 24#	0.090	0.090	0.08	0.09		0.07	0.07	0.060
	Cellular gypsum 18#	0.100	0.105	0.09	0.10		0.08	0.07	0.060

Table 29-20. Frame Floors and Ceilings (with Fill)

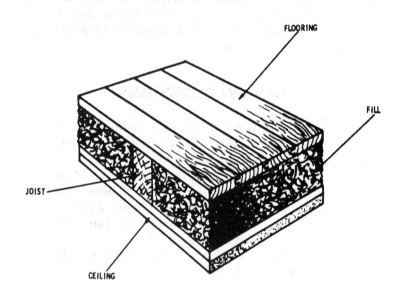

Values of *K* in Btu/hr/1°F/sq ft

Type of Ceiling with ½" Plaster	Insulation or Fill between Joists	No Flooring	1" Yellow-Pine Flooring	1" Yellow-Pine Flooring and ½" Fiberboard	1" Suboak, Maple on 1" Yellow-Pine Flooring
Wood lath	½" Flexible insulation	0.220	0.150	0.120	0.140
Wood lath	½" Rigid insulation	0.240	0.160	0.130	0.150
Wood lath	Flaked 2" gypsum	0.150	0.120	0.105	0.120
Wood lath	Rock wool, 2"	0.120	0.095	0.080	0.090
Corkboard ½"	None	0.155	0.115	0.100	0.105
Corkboard ½"	None	0.120	0.100	0.085	0.095

If it is recalled that 1 ton of refrigeration equals 12,000 Btu/hr, it is an easy matter to calculate the required size of plant equipment since all that is necessary is to divide the combined sum of the plant leakage and cooling loads in Btu/hr by 12,000 to obtain the condensing-unit capacity.

REVIEW QUESTIONS

1. Why is heat leakage through walls, ceilings, and floors an important consideration when erecting refrigeration plants?
2. Give the formula for leakage through walls.
3. What is meant by the heat-transfer coefficient, and why does this coefficient vary for different types of materials?
4. Calculate the heat leakage in Btu/hr for a plain 10-inch concrete wall (no exterior finish) having an area of 450 sq ft for a temperature difference of 50°F.
5. Is the inside or outside wall area used as a basis for estimating heat leakage?

CHAPTER 30

Refrigeration Load Calculations

A refrigeration load is measured by the heat units absorbed by the refrigerant while passing through the evaporator. With respect to a cold storage plant, this load consists of the following items:

1. Heat leakage through walls, floor, and roof
2. Heat emitted from the products to be cooled during the freezing process, usually termed *product cooling* and *freezing load*
3. Heat due to entrance of warm air when the doors are opened
4. Heat due to lights, motors, and people that may be in the rooms

If it is desired to calculate the load on a locker storage plant, it is practically impossible to estimate all of the foregoing items with any assurance of accuracy. Heat leakage, however, can be estimated with fair accuracy. The heat leakage through walls, floors, and ceilings can be determined by means of a formula and depends on the insulating material used and its thickness. The formula for heat leakage is

$$H = \mathbf{K}A(t_1 - t_2)$$

where $\mathbf{K}$ = heat transfer coefficient, Btu/hr/°F/sq ft
A = area, sq ft
$t_1 - t_2$ = temperature gradient through wall, °F

CALCULATING HEAT-TRANSFER COEFFICIENT

Assume that it is desired to find the heat-transfer coefficient $\mathbf{K}$ for a wall made up of 6 inches of concrete and 4 inches of cork. The outside of a concrete wall is presumably an outside wall exposed to outside air circulation. Experience has shown that for average conditions the heat transfer through the airfilm next to the wall is about 4.2 Btu/sq ft/hr for each degree difference in temperature between the inside and outside. The resistance is the reciprocal of this, or $\frac{1}{4.2}$ = 0.24. The coefficient for concrete is about 8 Btu/sq ft/hr per degree difference for each inch of thickness; the resistance is 0.125.

The resistance for a 6-inch wall is six times greater (6 × 0.125, or 0.75). The coefficient for cork is about 0.31 Btu/sq ft/hr per degree difference per inch of thickness, which is a resistance of $\frac{1}{0.31}$, or 3.23. The resistance for a 4-inch wall is therefore 4 × $\frac{1}{0.31}$ or 12.9. The inside of the wall is in contact with still air, and experience has shown that an average value of the coefficient for the film is about 1.4. The resistance then is $\frac{1}{1.4}$, or 0.71. The total resistance for this wall is then 0.24 + 0.75 + 12.9 + 0.71 = 14.60; and the overall coefficient is therefore $\frac{1}{14.60}$, or 0.0685. It is apparent that the principal resistance is offered by the cork. By following this same procedure, the overall coefficient for any type of wall can be obtained with a fair degree of accuracy.

CALCULATING HEAT LEAKAGE

After the heat-transfer coefficient is known, the heat leakage can readily be determined by the use of the formula. Assume that it is desired to compute the heat leakage of a room 25 × 25 feet × 10 feet high, with walls of 6-inch concrete and 4-inch cork. The perimeter of the room is 4 × 25, or 100 feet. The total wall surface is 10 × 100, or 1000 sq ft. With an outside temperature of 80°F and an inside temperature of 30°F, the heat leakage through the walls according to the formula is

$$H = 0.0685 \times 1000(80 - 30) = 3425 \text{ Btu/hr}$$

The leakage through the floor and ceiling can be determined it a similar manner to obtain the total heat leakage of the room. Any additional refrigeration needed to cool products in the room will have to be added to the heat leakage to obtain the total load.

Product Cooling Load

Heat emitted from the product to be cooled may also be calculated if the amount of product per locker is known. Most authorities assume that the average locker user will place an average of 2 to 2½ lb of produce per day in the storage compartment. Thus, in a chill room having a 300-locker installation, the load would be 300 × 2½, or 750 lb daily. The initial temperature may be as high as 95°F, and the final temperature can be taken as 36°F.

Because the specific heat of various kinds of produce will vary, an average specific heat of 0.7 is generally adopted as a good value for calculations. For a 300-locker installation, the daily product cooling load would be 0.7 × 750(95 − 36), or 30,975 Btu/day.

Air-Change Loads

Heat caused by the entrance of warm air when the doors are opened is termed *air-change load*. This load is difficult to estimate with any degree of accuracy since it is affected by such factors as usage of the room, interior volume, whether or not the room is entered through an outside door, size of the door, etc.

Product Freezing Load

When meat is removed from the chill room, it may have a temperature of about 35°F. During the process of preparation for the quick freezer, it may warm up to 40 or 50°F. It is evident that this meat will have to be cooled down to 32°F before it will begin to freeze. When freezing meat, the latent heat of fusion is removed and the average heat of fusion amounts to about 90 Btu/lb. After freezing, the meat is subcooled to the quick-freeze temperature, with an average specific heat of 0.4.

If it is required to calculate the amount of heat removed from 1 lb of meat (assuming a final temperature of the quick freeze to be −10°F), under the above conditions, we obtain:

Cooling meat to freezing point;
remove $0.7(45 − 28)$ = 11.9 Btu
Freezing meat (latent heat of fusion)= 90.9 Btu
Subcooling meat to −10°F
$= 0.4(28 + 10)$ = 15.2 Btu
              Total   118.0 Btu/lb

Again, allowing 2.5 lb of meat per locker, the quick freezer in a 300-locker installation should have a capacity to freeze 300×2.5, or 750 lb of meat daily. In this case, the product freezing load would be 750×118, or 88,500 Btu/day.

Miscellaneous Loads

Miscellaneous loads cannot readily be calculated. They consist of such items as heat from electric motors, lights, and people who may be in the room where the freezing process is in operation. With respect to this load (often referred to as the *service load*), it is assumed that ordinary precautions in handling of doors will be observed in order to keep this part of the total load to a minimum.

It has been found from experience that the average plant can be so handled that the service load experienced on the low temperature room will not be more that 15 to 20 percent of the leakage load.

Service Factor

Considering the foregoing, it is obviously difficult to arrive at a definite total value that will assist in a determination of the refrigeration capacity required by the compressor. To obtain a relatively close approximation of the load, it has been found by experience that these indeterminate factors can be closely compensated by allowing a certain percentage of the insulation loss. This percentage factor differs with the thickness of insulation used and temperature gradient between the walls.

CAPACITY OF MACHINES

The capacity of any refrigerating compressor depends largely on the number and size of its cylinders, its speed when running, the efficiency of compression, suction and discharge pressures, and number of hours of operation per day. The rated capacity is always based on continuous operation through 24 hours. Capacity rating is usually based on the conditions adopted as standard by the ASRE (American Society of Refrigerating Engineers), which are 5°F and 19.6 lb gage pressure for the suction, and 86°F and 154.5 lb for the discharge.

LOAD CALCULATION FOR A WALK-IN REFRIGERATOR

It is desired to estimate the size of the condensing unit for a walk-in refrigerator that is $8 \times 8 \times 10$ feet. The insulation consists of 3 inches of sheet cork. The cooling compartment is to be kept at 40°F in a 90°F room. The **K** factor for sheet cork insulation is 0.28 (Table 30-1).

Since the temperature difference is $90 - 40$, or 50°F, and the thickness of the insulation is 3 inches, the heat leakage per hour is obtained as follows:

$$H = \frac{A\mathbf{K}(t_1 - t_2)}{\text{Insulation Thickness}} = \frac{448 \times 0.28 \times 50}{3} = 2090.666 \text{ Btu / hr}$$

Because heat loads and equipment capacities are usually estimated on a 24-hour basis, the heat-leakage load in this cooler for 24 hours will be 2090.6666 × 24, or 50,176 Btu.

The product cooling load depends on the temperature of the products to be chilled. Assume that 200 lb of lean beef and 100 lb of poultry, both at room temperature, are brought into the 8 × 8 × 10 foot walk-in refrigerator daily. Assume, further, that the specific heat for beef and poultry is 0.77 and 0.80, respectively. The product cooling load is therefore obtained as follows:

> Beef: 200 × 0.77 × 50 = 7700 Btu
> Poultry: 100 × 0.80 × 50 = 4000

that is, a total 24-hour product load is 7700 + 4000, or 11,700 Btu.

Table 30-1. Conductivity Factors for Various Insulations

Insulation	K°
Sheet cork	0.28
Glass wool	0.26
Rock wool	0.28
Sawdust	0.41
Kapok	0.24
Hair felt	0.26

*Btu/sq ft/°F/hr/in. thickness.

It should be observed that when freezing ice cream, water, foods, or other products, large amounts of heat must be extracted because of the fact that the *latent heat of freezing or fusion* is very large and may add loads far in excess of the normal when merely storing these products prefrozen.

Air-change load is sometimes termed *service load* or *usage load,* and consists of heat introduced into the refrigerator products. This load varies a great deal and is affected by so many factors that it cannot be calculated with any degree of accuracy. Table 30-2 is supplied through the courtesy of the American Society of Refrigeration Engineers and is used for this purpose. To determine the air-change load, it is necessary first to calculate the inside cubic content of the walk-in refrigerator, and then multiply this value by the *usage factor* from Table 30-2.

Assuming a total wall thickness of 6 inches, the 8 × 8 × 10 foot refrigerator will have a net volume of 7 × 7 × 9 feet, or 441 cu ft.

Table 30-2. Usage Factors, Btu/24 hr/cu ft Interior Capacity

Inside Volume, cu ft	Type of Service	Temperature Difference, °F (Room temp. minus refrigerator temp.)								
		40	45	50	55	60	65	70	75	80
15	Normal	108	122	135	149	162	176	189	203	216
	Heavy	134	151	168	184	201	218	235	251	268
50	Normal	97	109	121	133	145	157	169	182	194
	Heavy	124	140	155	171	186	202	217	233	248
100	Normal	85	96	107	117	128	138	149	160	170
	Heavy	114	128	143	157	171	185	200	214	228
200	Normal	74	83	93	102	111	120	130	139	148
	Heavy	104	117	130	143	156	169	182	195	208
300	Normal	68	77	85	94	102	111	119	128	136
	Heavy	98	110	123	135	147	159	172	184	196
400	Normal	65	73	81	89	97	105	113	122	130
	Heavy	95	107	119	130	142	154	166	178	190

This puts the walk-in unit in the 400 cu ft class, since the total volume is well over 400 cu ft. If it is further assumed that the walk-in unit is subjected to *normal* rather than heavy usage, keeping in mind that a temperature difference of 50°F is required, it is found that the usage factor in this particular case is 81 per 24 cu ft of interior capacity. From the foregoing, it follows that the total air-change load is 81 × 441, or 35,721 Btu/24.

The miscellaneous load consists of heat from several sources, such as that generated by electrical equipment, motors, and lamps, in addition to heat from persons during servicing operations. Therefore, this load cannot readily be calculated with any degree of accuracy. With respect to the electrical equipment, it has been found that 1 W of electrical energy is equivalent to 3.415 Btu. Thus, a 25W lamp will generate 25 x 3.415, or 85.375 Btu. If it is assumed that three 25W lamps are used for lighting and are lit 12 hr/day, the additional load imposed on the system will be as follows:

Lamp load: $75 \times 3.415 \times 12 = 3073.5$ Btu
Motor load: $100 \times 3.415 \times 24 = 8196$ Btu

The total miscellaneous load will be 3073.5 + 8196, or 11,269.5 Btu.

From the foregoing calculations it follows that the total 24-hour load is obtained by adding together the heat-leakage load, product-cooling load, air-change load, and miscellaneous load, as follows:

Heat-leakage load	50,176.0	Btu
Product-cooling load	11,700.0	Btu
Air-change load	35,721.0	Btu
Miscellaneous load	11,269.5	Btu
Total	108,866.5	Btu

In order to compensate for the *on* and *off* periods of the condensing unit and to maintain the same refrigerator temperature despite variations in load (from summer to winter, in accordance with the amount of warm products chilled, or opening of the doors), it is necessary to use a condensing unit with a somewhat greater capacity per 24 hours than the total heat load. It is good practice to select a condensing unit with at least a 50 percent greater capacity per 24 hours than the total heat load. In this manner, the unit will be of sufficient capacity if it is assumed to run only two-thirds of the time, or 16 out of 24 hours. Thus, if the total load is 108,866.5 Btu and the condensing unit is run only 16 out of 24 hours, it must have an hourly capacity of 108,866.5/16, or 6804 Btu/hr.

At this point it is also necessary to determine the operating temperature of the evaporator. If an average temperature difference of 15°F in this 40°F refrigerator is assumed, the average temperature of the evaporator must be 25°F. The capacity of the condensing unit also varies with the evaporator temperature (and suction pressure, which is governed by the evaporator temperature). Therefore, it will be necessary to employ a condensing unit that has a capacity of 6800 Btu/hr at an evaporator temperature of 25°F, which corresponds to a Freon-12 suction pressure of 24.5 psi.

Table 30-3 shows the capacity of typical air-cooled units of several motor sizes operating on 25°F evaporators and in a 90°F room. It will be observed that a 1-hp condensing unit having a capacity of 8190 Btu/hr will be required since the ¾-hp unit has a capacity of only 5820 Btu/hr. With the calculated cooling load

Table 30-3. Capacity of Typical Air-Cooled Condensing Units

Condensing-Unit Horsepower	Capacity per Hour, Btu
⅓	2460
½	4010
¾	5820
1	8190
1½	12,050
2	16,150

required, it should be noted that since the chosen condensing unit is somewhat oversized, the actual operating period need be only 108,866.5 ÷ 8190, or approximately 13.29 out of 24 hours.

EVAPORATOR CAPACITY

After having selected a 1-hp condensing unit that, while in operation, refrigerates at the rate of 8190 Btu/hr, an evaporator that, at a 15°F temperature difference, will have a capacity of at least 8190 Btu/hr should be selected.

LOAD CALCULATIONS FOR A DAIRY PLANT

For a dairy handling 800 gallons of milk daily that must be cooled from a pasteurizing temperature of 145°F to 40°F, the following assumptions are made:

$$\text{Specific gravity of milk} = 1.03$$
$$\text{Weight of milk} = 8.6 \text{ lb/gal}$$
$$\text{Weight of water} = 8.34 \text{ lb/gal}$$
$$\text{Specific heat of milk} = 0.95$$

The equation for the heat load is

$$H = WS(t_1 - t_2)$$

where W = weight of the fluid being cooled
  S = specific heat, Btu/lb/°F
  t_1 = higher temperature
  t_2 = lower temperature

Substituting the values in the equation, we obtain

$$H = 800 \times 8.6 \times 0.95(145 - 40) = 686,280 \text{ Btu}$$

If it is assumed that ample water is available, it is entirely feasible to cool the milk with water to 80°F. Therefore, the heat to be removed by the water is

$$H = \frac{686,280 \times 65}{105} = 424,840 \text{ Btu}$$

If the water used for cooling is allowed to rise 10°F, the amount of water necessary is

$$\frac{424,840}{10 \times 8.34} = 5094 \text{ gallons}$$

If a 4-hour period is allowed for the cooling, the amount of water necessary per hour is 5094 ÷ 110. 4, or 1274 gallons. If a 25 percent loss of water is allowed, the gal/min requirement is:

$$\frac{1274 \times 1.25}{60} = 26.5 \text{ gal/min (approx.)}$$

To cool the milk from 80 to 40°F, the amount of heat to be removed is:

$$H = 686,280 - 424,840 = 261,440 \text{ Btu}$$

The cooling substance is usually brine, which for this purpose should enter the cooling coil at an initial temperature of approximately 18°F and leave at about 28°F. The specific heat of brine is dependent on the amount of salt in it, but 0.86 is a reasonable value. The weight of the brine required to remove 261,440 Btu of heat can be obtained if the foregoing equation is solved for W; that is,

$$W = \frac{H}{S(t_1 - t_2)} = \frac{261,440}{0.86(28 - 18)} = 30,400$$

If the weight per gallon of brine is assumed to be 9.1 lb, we obtain the number of gallons of brine required as 30,400/9.1, or approximately 3340 gallons. Again, if a 4-hour period is allowed for cooling, the requirement of brine in gpm is $^{3340}/_4 \div 60$, or 13.9 gal/min. Allowing for heat losses, a pump having a minimum capacity of 15 gal/min. should be provided.

If it is desired to calculate the cooling surface required for the water and brine cooler, the following procedure is recommended. For that part of the cooler in which water is the cooling medium, the amount of heat to be removed per hour is 424,840/4, or 106,210 Btu, assuming a 4-hour cooling period. If 20 percent is added to provide for losses, the additional heat to be removed is $1.2 \times 106,210$, or 127,452 Btu. The milk enters the cooler at 145°F and leaves the water-cooled section at 80°F.

Although the water temperature varies in different localities and with the season of the year, for the purpose of this calculation it may be assumed that water enters the cooler at 60°F and leaves it at 70°F. If the counterflow principal is employed, the milk enters the top of the cooler and leaves at the bottom, while the water enters the cooler at the bottom and leaves at the top. From this, it follows that the coolest milk is cooled by the coldest water and the warmest milk by the warmest water. The temperature difference at the warm end is $145 - 70 = 75$°F, and at the cold end, $80 - 60 = 20$°F. The formula for computation of the cooling surface is:

$$H = KAT_d$$

where H = heat transfer, Btu/hr
       K = coefficient of heat transfer
       A = surface area,
       T_d = mean temperature

If the formula is solved for A and the numerical values substituted, we obtain

$$A = \frac{H}{KT_d} = \frac{127,450}{60(75 + 20/2)} = 44.7 \text{ sq ft (approx.)}$$

The amount of surface required for the brine cooler may be calculated in a similar manner. The heat to be removed by the brine is 261,440 Btu. Allowing 20 percent additional heat losses, the total is $261,440 \times 1.2 = 313,728$ Btu. If it is assumed that the compressor is operating 8 hr/day and that, during the cooling period, the evaporating ammonia will remove half of the 313,728 Btu and the other half is absorbed by the brine to be removed by the evaporating ammonia later, then by applying the equation, we have:

$$H = WS(t_1 - t_2)$$

where H = heat to be absorbed by brine
W = weight of brine in system
S = 0.86 = specific heat of brine
t_1 = temperature of brine at end of milk-cooling period
t_2 = temperature of brine at beginning of milk cooling period

Allowing the brine to warm 5°F during the cooling period, then

$$t_1 - t_2 = 5$$

and

$$W = \frac{H}{S(t_1 - t_2)} = \frac{313,728}{2 \times 0.86 \times 5} = 36,480 \text{ lb}$$

At 9.1 lb/gal, this is 36,480/9.1, or about 4009 gallons of brine. On the basis of 8 hours operating time, the refrigerating equipment must remove heat at the rate of 313,728/8, or 39,246 Btu/hr. In this particular case, therefore, the tonnage capacity of the compressor must be 39,216/12,000, or 3.27 tons.

The evaporating-surface area in the brine tank may also be calculated. With an average brine temperature of 28°F, the temperature of the evaporating ammonia at 5°F, and an assumed coefficient of heat transfer of 15, we obtain:

$$A = \frac{H}{K \times T_d} = \frac{39,216}{15(28 - 5)} = 114 \text{ sq ft (approx.)}$$

It has been found by experience that it requires 2.3 lineal feet of 1 ¼-inch pipe to furnish 1 sq ft of surface. Since the number of square feet in this particular case is 114, the pipe requirement is 2.3 × 114, or 262 lineal feet.

Milk Storage

Most dairies have a walk-in type of cooler for storage of the milk after aerating, in addition to storage of other products such as cream and butter. Immediately after aerating, the milk is usually bottled, cased, and stored. During the process of bottling, temperature of the milk usually rises two or three degrees due to contact with the warm bottles. The amount of refrigeration required for such storage varies over a rather large range, depending on such factors as insulation, amount of products to be stored, etc. However, experience has shown that if a certain amount of refrigeration is provided per cubic foot of room space, satisfactory conditions can be maintained. If a room having dimensions of 18 × 12 × 10 feet (approximately 2200 cu ft) is chosen and the temperature is to be held at 35°F, about 1 ton of refrigeration should be provided. For a room 10 × 10 × 10 feet (1000 cu ft of space), about 0.6 ton of refrigeration should be provided.

A brine tank is often located in the milk storage room. When so located, it should be placed near the ceiling. The refrigeration stored in it during the time the machine is in operation will help to hold the room at the proper temperature during the shutdown period, and the fact that it is in an insulated room eliminates the necessity of insulating the tank itself. If it cannot be placed near the ceiling, it should not be placed in the room at all because it occupies valuable space, does not help to hold the room temperature very well, and renders it much more difficult to keep the room neat and sanitary.

In a dairy manufacturing plant, two compressors, each operating at a different suction pressure, are often desirable. The machine that cools the brine for freezing and cools the hardening room can be operated with an evaporating temperature of –20°F, which corresponds to a pressure of 18.3 psia, or 3.6 psig. The machine that serves the milk-storage room, the brine tank for milk cooling, and the mix-holding vat should have an evaporating temperature of about 5°F, which requires a suction pressure of 34.3 psia, or 19.6 psig. If ice is made, the ice tank should also be connected to this machine.

LOAD CALCULATIONS FOR SMALL FREEZERS

The so-called package units that are larger than the standard household refrigerator but smaller than the walk-in-type freezer and storage units, and that can be used for both freezing and storing after freezing as well as cooling products to about 35° to 40°F and hold at that temperature, are here considered. In order to evaluate some of the engineering aspects of such a unit, the following data are given: the cabinet consists of a compartment 16 × 19 × 35 inches, having a volume of about 6 cu ft, and a larger compartment 20 × 22 × 64, having a volume of about 16 cu ft. The smaller compartment is insulated with about 5 inches of a high-grade insulating material, and the larger one with about 3 inches. The small compartment is designed to be held at 5°F, and the larger compartment at 35°F. The amount of cooling surface in the low-temperature space is about 15.3 sq ft, and in the high-temperature compartment about 5.8 sq ft.

The refrigerating unit is a two-cylinder compressor driven by a ⅓-hp motor, and the refrigerant is Freon (F-12). The controls are set to start the compressor when the pressure in the evaporator is 12 psig, which corresponds to about 5°F, and to stop it when the pressure is reduced to 0 psig (atmospheric pressure), corresponding to −22°F. This avoids production of a partial vacuum in the system and the attendant evils resulting from air and water-vapor leakage into the unit. The average evaporating temperature is about −9°F. At this temperature, the rated capacity of the condensing unit is about 1580 Btu/hr, which is more, however, than the surface in the compartments can absorb with the temperature differences available. The result is that the compressor will continue to run until an equilibrium point, or balanced condition, is reached. As the following calculations will show, this equilibrium point is estimated to be about −20°F.

A reasonable heat-transfer coefficient for the freezer surface is

1.65 Btu/sq ft/hr/°F; for the cooler surface it is about 1.75. The temperature difference is 25°F in the freezer and 55°F in the cooler. The heat transfer is

$$H = AK\,(t_1 - t_2)$$

For the freezer:

$$H = 15.3 \times 1.65 \times 25 = 631\ \text{Btu/hr}$$

For the cooler:

$$H = 5.8 \times 1.75 \times 55 = 588\ \text{Btu/hr}$$

The total heat that the surface can absorb is the sum of these two heat transfers: 1189, or, in round figures, 1200 Btu/hr. At this evaporating temperature, the rated capacity of the condensing unit is substantially the same. As long as heat is supplied to the evaporator from the compartments at the rate of about 1200 Btu/hr, the compressor will continue to run; but if the heat supply is reduced, the evaporator temperature will begin to fall, and when it reaches the predetermined cutout point, the driving motor will be stopped automatically until the pressure has risen sufficiently to operate the starting mechanism.

The estimated heat leakage is about 160 Btu/hr, and the loss due to opening of the doors about 140 Btu/hr. These losses continue at a more or less uniform rate throughout the day and amount to about 300 Btu/hr. The necessary operating time required is

$$^{300}\!/_{1200} = 0.25\ \text{ or 25 percent}$$

The energy consumption is about 0.4 kW while the motor is running. The minimum power consumption is $0.4 \times 0.25 \times 24 = 2.4$ kWh/day, and the maximum is four times as much, or 9.6 kWh/day. The difference between 1200 and 300, or 900 Btu/hr, is the refrigeration available for cooling or freezing goods. It is sufficient to cool and freeze meat or poultry at the rate of about 6 lb/hr or berries at the rate of about 5 lb/hr.

SUMMARY

This chapter deals with detailed calculations of and capacity requirements for a variety of cooling applications. Thus, the heat leakage through the walls, floors, and roof of the structure is first determined, after which the product cooling and freezing loads, etc., are calculated by numerous practical examples.

Tabulated data include conductivity factors for various types of insulation, usage factors in Btu per 24 hours per cubic foot of space for various temperatures, and condensing capacity when the cooling requirement is known.

Refrigeration-load calculations for various structures have been made, such as for walk-in refrigerators and dairy plants. In this connection, it should be noted that other load problems may be calculated in a similar manner by substituting figures in the customary manner since no two problems in the field of refrigeration will be exactly alike.

REVIEW QUESTIONS

1. Enumerate the various cooling loads that must be considered in a typical cold-storage plant in order to obtain the required machine capacity.
2. What is meant by the product cooling load in a locker-plant installation?
3. What does the capacity of a refrigerating compressor depend on?
4. Of what do the so-called miscellaneous loads consist?
5. Give the formula for computing the cooling surface area of a brine cooler.
6. How does the calculation for smaller freezers differ from that for cold-storage plants?

Appendix

Temperature Requirements for Coolers and Refrigerants

Name of Product	Approximate Cooler Temperature	Approximate Operating Refrigerant Suction Temperature	Approximate Temperature Difference in Ordering Refrigerating Coils
Cheese, Eggs, Butter	Serving Cooler, 50 to 60°F Storage Cooler, 31 to 36°F	30°F 15°F	15 to 20°F 15°F
Poultry and Meats (Fresh)	Cooler, 34 to 38°F Cases, 38 to 42°F	20°F 20 to 25°F	12 to 15°F
Poultry and Meats (Frozen)	−5 to +5°F	−15 to −5°F	10°F
Vegetables, Fruits	35 to 45°F	25 to 30°F	10°F
Cream and Milk	Can Water Coiling, 36 to 38°F Cooler Storage, 36 to 40°F	15 to 20°F 20°F	15 to 20°F
Beer Storage	36 to 45°F	20°F	15 to 20°F
Fur Storage	Shock Temperature, 10 to 15°F Max. Temperature, 38°F	0°F 15°F	10°F 20 to 25°F
Ice Cream, Hardening	−20°F	−30°F	5 to 10°F
Ice Cream, Storing	−5 to −10°F	−15 to −20°F	10°F

Courtesy Dunham-Bush, Inc.

Solutions of Calcium Chloride (CaCl$_2$)

Percent of Pure Salt by Weight	Freezing Point, °F	Ammonia Gage Pressure lbs at Freezing Point	Specific Gravity* 60°F* 39°F	Baume Density 60°F	Specific Heat Btu/lb/°F at 60°F	Weight, lb/cu ft	
						CaCl	Brine
0	32.0	47.6	1.000	0.0	1.000	0.00	62.40
5	29.0	43.8	1.044	6.1	0.9246	3.26	65.15
6	28.0	42.6	1.050	7.0	0.9143	3.93	65.52
7	27.0	41.4	1.060	8.2	0.8984	4.63	66.14
8	25.5	39.0	1.069	9.3	0.8842	5.34	66.70
9	24.0	37.9	1.078	10.4	0.8699	6.05	67.27
10	23.0	36.8	1.087	11.6	0.8556	6.78	67.83
11	21.5	35.0	1.096	12.6	0.8429	7.52	68.33
12	19.0	32.5	1.105	13.8	0.8284	8.27	68.95
13	17.0	30.4	1.114	14.8	0.8166	9.04	69.51
14	14.5	28.0	1.124	15.9	0.8043	9.81	70.08
15	12.5	26.0	1.133	16.9	0.793	10.6	70.64
16	9.5	23.4	1.143	18.0	0.7798	11.4	71.26
17	6.5	20.8	1.152	19.1	0.7672	12.22	71.89
18	3.0	18.0	1.162	20.2	0.7566	13.05	72.51
19	0.0	15.7	1.172	21.3	0.746	13.9	73.13
20	−3.0	13.6	1.182	22.1	0.7375	14.73	73.63
21	−5.5	11.9	1.192	23.0	0.729	15.58	74.19
22	−10.5	8.8	1.202	24.4	0.7168	16.50	75.0
23	−15.5	5.9	1.212	25.5	0.7076	17.4	75.63
24	−20.5	3.4	1.223	26.4	0.6979	13.32	76.32
25	−25.0	1.3	1.233	27.4	0.6899	19.24	76.94
26	−30.0	1.6†	1.244	28.3	0.682	20.17	77.56
27	−36.0	6.1†	1.254	29.3	0.6735	21.13	78.25
28	−43.5	10.6†	1.265	30.4	0.6657	22.1	78.94
29	−53.0	15.7†	1.276	31.4	0.6584	23.09	79.62
29.5	−58.0	17.8†	1.280	31.7	0.6557	23.56	79.87

*Specific gravity and weights at 60°F referred to water at 39°F.
†Inches vacuum.
Note: To obtain the weight of commercial salt, divide the values given for anhydrous salt in the table by the percent purity of the commercial calcium salt. A calcium brine freezing at 0°F and requiring 13.9 lb of "pure" salt per cubic foot will require 0.73 into 13.9, or 19 lb of 73% salt. To obtain the weights of water for other than anhydrous salt, subtract the weight of that salt, as just determined, from the weight of the brine (73.13 − 19 = 54.13).

Courtesy Frick Company, Inc.

Solutions of Sodium Chloride (NaCl)
(Common Salt)

Percent of Pure Salt by Weight	Freezing Point, °F	Ammonia Gage Pressure lbs at Freezing Point	Specific Gravity* 60°F* 39°F	Baume Density 60°F	Specific Heat Btu/lb/°F at 60°F	Weight, lb/cu ft	
						NaCl	Brine
0	32.0	47.6	1.000	0.0	1.000	0.000	62.4
5	27.0	41.4	1.035	5.1	0.938	3.230	64.6
6	25.5	39.6	1.043	6.1	0.927	3.906	65.1
7	24.0	37.9	1.050	7.0	0.917	4.585	65.5
8	23.2	37.0	1.057	8.0	0.907	5.280	66.0
9	21.8	35.6	1.065	9.0	0.897	9.985	66.5
10	20.4	34.1	1.072	10.1	0.888	6.690	66.9
11	18.5	32.0	1.080	10.8	0.879	7.414	67.4
12	17.2	30.6	1.087	11.8	0.870	8.136	67.8
13	15.5	28.9	1.095	12.7	0.862	8.879	68.3
14	13.9	27.4	1.103	13.6	0.854	9.632	68.8
15	12.0	25.6	1.111	14.5	0.847	10.395	69.3
16	10.2	24.0	1.118	15.4	0.840	11.168	69.8
17	8.2	22.3	1.126	16.3	0.833	11.951	70.3
18	6.1	20.5	1.134	17.2	0.826	12.744	70.8
19	4.0	18.8	1.142	18.1	0.819	13.547	71.3
20	1.8	17.1	1.150	19.0	0.813	14.360	71.8
21	−0.8	15.1	1.158	19.9	0.807	15.183	72.3
22	−3.0	13.6	1.166	20.8	0.802	16.016	72.8
23	−6.0†	11.6	1.175	21.7	0.796	16.854	73.3
24	3.8	18.6	1.183	22.5	0.791	17.712	73.8
25	16.1	29.5	1.191	23.4	0.785	18.575	74.3
26	32.0	47.6	1.200	—	—	—	—

*Specific gravity and weights at 60° referred to water at 39°F.
†Eutectic point: addition of more salt raises the freezing point.
  Note: Weight given for pure NaCl salt can be used for commercial salt with a theoretical error of about 1%.

Courtesy Frick Company, Inc.

Suction Line* Sizes for R-12 (O.D.)

Suction Temperature

Total Equivalent Length of Suction Line (Feet)

Net Evaporator Capacity	40°F 25	40°F 50	40°F 100	40°F 150	20°F 25	20°F 50	20°F 100	20°F 150	0°F 25	0°F 50	0°F 100	0°F 150	-20°F 25	-20°F 50	-20°F 100	-20°F 150	-40°F 25	-40°F 50	-40°F 100	-40°F 150
3,000	3/8	1/2	1/2	1/2	1/2	1/2	1/2	5/8	5/8	5/8	3/4	3/4	5/8	3/4	7/8	7/8	3/4	7/8	1 1/8	1 1/8
6,000	1/2	1/2	5/8	5/8	1/2	5/8	3/4	3/4	3/4	7/8	7/8	1 1/8	7/8	1 1/8	1 1/8	1 1/8	1 1/8	1 1/8	1 3/8	1 3/8
9,000	5/8	5/8	5/8	3/4	5/8	3/4	7/8	7/8	7/8	7/8	1 1/8	1 1/8	7/8	1 1/8	1 1/8	1 3/8	1 1/8	1 3/8	1 3/8	1 5/8
12,000	5/8	3/4	3/4	7/8	3/4	3/4	7/8	1 1/8	7/8	1 1/8	1 1/8	1 3/8	1 1/8	1 1/8	1 3/8	1 3/8	1 3/8	1 5/8	1 5/8	1 5/8
18,000	3/4	3/4	7/8	1 1/8	3/4	7/8	1 1/8	1 1/8	1 1/8	1 1/8	1 3/8	1 3/8	1 1/8	1 3/8	1 5/8	1 5/8	1 3/8	1 5/8	2 1/8	2 1/8
24,000	3/4	7/8	1 1/8	1 1/8	7/8	1 1/8	1 1/8	1 3/8	1 1/8	1 3/8	1 5/8	1 5/8	1 3/8	1 5/8	1 5/8	2 1/8	1 5/8	2 1/8	2 1/8	2 5/8
30,000	7/8	1 1/8	1 1/8	1 1/8	1 1/8	1 1/8	1 1/8	1 3/8	1 3/8	1 3/8	1 5/8	2 1/8	1 3/8	1 5/8	2 1/8	2 1/8	2 1/8	2 1/8	2 5/8	2 5/8
36,000	7/8	1 1/8	1 1/8	1 3/8	1 1/8	1 1/8	1 3/8	1 3/8	1 3/8	1 5/8	1 5/8	2 1/8	1 5/8	1 5/8	2 1/8	2 1/8	2 1/8	2 1/8	2 5/8	2 5/8
42,000	1 1/8	1 1/8	1 3/8	1 3/8	1 1/8	1 3/8	1 3/8	1 5/8	1 3/8	1 5/8	2 1/8	2 1/8	1 5/8	2 1/8	2 1/8	2 5/8	2 1/8	2 5/8	2 5/8	2 5/8
48,000	1 1/8	1 1/8	1 3/8	1 3/8	1 1/8	1 3/8	1 5/8	1 5/8	1 5/8	1 5/8	2 1/8	2 1/8	1 5/8	2 1/8	2 5/8	2 5/8	2 1/8	2 5/8	2 5/8	3 1/8
54,000	1 1/8	1 3/8	1 3/8	1 3/8	1 1/8	1 3/8	1 5/8	1 5/8	1 5/8	2 1/8	2 1/8	2 1/8	2 1/8	2 1/8	2 5/8	2 5/8	2 1/8	2 5/8	2 5/8	3 1/8
60,000	1 1/8	1 3/8	1 3/8	1 5/8	1 3/8	1 3/8	1 5/8	1 5/8	1 5/8	2 1/8	2 1/8	2 5/8	2 1/8	2 5/8	2 5/8	2 5/8	2 5/8	2 5/8	3 1/8	3 1/8
72,000	1 1/8	1 3/8	1 5/8	1 5/8	1 3/8	1 5/8	1 5/8	2 1/8	1 5/8	2 1/8	2 5/8	2 5/8	2 1/8	2 5/8	2 5/8	2 5/8	2 5/8	2 5/8	3 1/8	3 1/8
90,000	1 3/8	1 3/8	1 5/8	2 1/8	1 3/8	1 5/8	2 1/8	2 1/8	2 1/8	2 5/8	2 5/8	2 5/8	2 5/8	2 5/8	2 5/8	3 1/8	3 1/8	3 1/8	3 1/8	3 5/8
120,000	1 3/8	1 5/8	2 1/8	2 1/8	1 5/8	2 1/8	2 1/8	2 1/8	2 1/8	2 5/8	2 5/8	3 1/8	2 5/8	3 1/8	3 1/8	3 1/8	3 1/8	3 1/8	3 5/8	4 1/8
150,000	1 5/8	1 5/8	2 1/8	2 1/8	1 5/8	2 1/8	2 1/8	2 5/8	2 5/8	2 5/8	3 1/8	3 1/8	2 5/8	3 1/8	3 5/8	3 5/8	3 1/8	3 5/8	4 1/8	5 1/8
180,000	1 5/8	2 1/8	2 1/8	2 5/8	2 1/8	2 1/8	2 5/8	2 5/8	2 5/8	2 5/8	3 1/8	3 1/8	2 5/8	3 1/8	3 5/8	3 5/8	3 1/8	3 5/8	4 1/8	5 1/8
210,000	1 5/8	2 1/8	2 1/8	2 5/8	2 1/8	2 1/8	2 5/8	2 5/8	2 5/8	3 1/8	3 1/8	3 5/8	3 1/8	3 1/8	3 5/8	4 1/8	3 5/8	4 1/8	5 1/8	5 1/8
240,000	2 1/8	2 1/8	2 5/8	2 5/8	2 1/8	2 5/8	2 5/8	3 1/8	3 1/8	3 1/8	3 5/8	3 5/8	3 5/8	3 5/8	4 1/8	4 1/8	3 5/8	4 1/8	5 1/8	5 1/8
300,000	2 1/8	2 1/8	2 5/8	2 5/8	2 1/8	2 5/8	3 1/8	3 1/8	3 1/8	3 1/8	3 5/8	4 1/8	3 1/8	3 5/8	4 1/8	5 1/8	4 1/8	5 1/8	5 1/8	6 1/8

*Type-L copper tubing.

Courtesy Dunham-Bush, Inc.

Suction Line* Sizes for R-22 (O.D.)

Net Evaporator Capacity	40°F				20°F				0°F				-20°F				-40°F			
	Suction Temperature — Total Equivalent Length of Suction Line (Feet)																			
	25	50	100	150	25	50	100	150	25	50	100	150	25	50	100	150	25	50	100	150
3,000	3/8	3/8	3/8	1/2	3/8	3/8	1/2	1/2	1/2	1/2	5/8	5/8	1/2	5/8	5/8	3/4	5/8	3/4	7/8	7/8
6,000	3/8	1/2	1/2	5/8	1/2	1/2	5/8	5/8	5/8	5/8	3/4	7/8	5/8	3/4	7/8	7/8	3/4	7/8	1 1/8	1 1/8
9,000	1/2	1/2	5/8	5/8	1/2	5/8	5/8	3/4	5/8	3/4	7/8	7/8	3/4	7/8	1 1/8	1 1/8	7/8	1 1/8	1 1/8	1 3/8
12,000	1/2	5/8	3/4	5/8	5/8	5/8	3/4	3/4	3/4	7/8	1 1/8	7/8	7/8	1 1/8	1 1/8	1 1/8	1 1/8	1 3/8	1 3/8	1 3/8
18,000	5/8	3/4	3/4	7/8	3/4	3/4	3/4	7/8	7/8	7/8	1 1/8	1 1/8	7/8	1 1/8	1 1/8	1 3/8	1 1/8	1 3/8	1 5/8	1 5/8
24,000	5/8	3/4	7/8	7/8	3/4	7/8	7/8	1 1/8	7/8	1 1/8	1 1/8	1 3/8	1 1/8	1 1/8	1 3/8	1 3/8	1 3/8	1 3/8	1 5/8	1 5/8
30,000	3/4	7/8	7/8	1 1/8	7/8	7/8	1 1/8	1 1/8	1 1/8	1 1/8	1 3/8	1 3/8	1 1/8	1 3/8	1 3/8	1 5/8	1 3/8	1 5/8	1 5/8	2 1/8
36,000	3/4	7/8	1 1/8	1 1/8	7/8	1 1/8	1 1/8	1 1/8	1 1/8	1 1/8	1 3/8	1 5/8	1 1/8	1 3/8	1 5/8	1 5/8	1 5/8	1 5/8	2 1/8	2 1/8
42,000	7/8	7/8	1 1/8	1 1/8	7/8	1 1/8	1 3/8	1 3/8	1 1/8	1 3/8	1 3/8	1 5/8	1 3/8	1 5/8	1 5/8	2 1/8	1 5/8	1 5/8	2 1/8	2 1/8
48,000	7/8	1 1/8	1 1/8	1 1/8	1 1/8	1 1/8	1 3/8	1 3/8	1 3/8	1 3/8	1 5/8	1 5/8	1 3/8	1 5/8	1 5/8	2 1/8	1 5/8	2 1/8	2 1/8	2 1/8
54,000	7/8	1 1/8	1 1/8	1 3/8	1 1/8	1 1/8	1 3/8	1 3/8	1 3/8	1 3/8	1 5/8	1 5/8	1 5/8	1 5/8	1 5/8	2 1/8	1 5/8	2 1/8	2 1/8	2 5/8
60,000	7/8	1 1/8	1 1/8	1 3/8	1 1/8	1 3/8	1 3/8	1 3/8	1 5/8	1 3/8	1 5/8	2 1/8	1 5/8	1 5/8	2 1/8	2 1/8	1 5/8	2 1/8	2 1/8	2 5/8
72,000	1 1/8	1 1/8	1 3/8	1 3/8	1 1/8	1 3/8	1 3/8	1 5/8	1 3/8	1 5/8	2 1/8	2 1/8	1 5/8	1 5/8	2 1/8	2 1/8	2 1/8	2 1/8	2 5/8	2 5/8
90,000	1 1/8	1 1/8	1 3/8	1 5/8	1 3/8	1 3/8	1 5/8	1 5/8	1 5/8	2 1/8	2 1/8	2 1/8	1 5/8	2 1/8	2 1/8	2 5/8	2 1/8	2 1/8	2 5/8	3 1/8
120,000	1 1/8	1 3/8	1 5/8	1 5/8	1 3/8	1 5/8	1 5/8	2 1/8	1 5/8	2 1/8	2 1/8	2 1/8	2 1/8	2 1/8	2 5/8	2 5/8	2 1/8	2 5/8	3 1/8	3 1/8
150,000	1 3/8	1 3/8	1 5/8	2 1/8	1 3/8	1 5/8	2 1/8	2 1/8	1 5/8	2 1/8	2 1/8	2 5/8	2 1/8	2 1/8	2 5/8	2 5/8	2 5/8	2 5/8	3 1/8	3 1/8
180,000	1 3/8	1 5/8	2 1/8	2 1/8	1 5/8	2 1/8	2 1/8	2 1/8	2 1/8	2 1/8	2 5/8	2 5/8	2 1/8	2 5/8	2 5/8	3 1/8	2 5/8	3 1/8	3 5/8	3 5/8
210,000	1 3/8	1 5/8	2 1/8	2 1/8	1 5/8	2 1/8	2 1/8	2 5/8	2 1/8	2 1/8	2 5/8	2 5/8	2 5/8	2 5/8	3 1/8	3 1/8	2 5/8	3 1/8	3 5/8	3 5/8
240,000	1 5/8	1 5/8	2 1/8	2 1/8	1 5/8	2 1/8	2 1/8	2 5/8	2 1/8	2 5/8	2 5/8	3 1/8	2 5/8	2 5/8	3 1/8	3 1/8	3 1/8	3 5/8	3 5/8	4 1/8
300,000	1 5/8	2 1/8	2 1/8	2 5/8	2 1/8	2 1/8	2 5/8	2 5/8	2 1/8	2 5/8	3 1/8	3 1/8	2 5/8	3 1/8	3 1/8	3 5/8	3 1/8	3 5/8	4 1/8	5 1/8

*Type-L copper tubing.

Courtesy Dunham-Bush, Inc.

Suction Line* Sizes for R-502 (O.D.)

Net Evaporator Capacity	40°F				20°F				0°F				-20°F				-40°F			
	Suction Temperature — Total Equivalent Length of Suction Line (Feet)																			
	25	50	100	150	25	50	100	150	25	50	100	150	25	50	100	150	25	50	100	150
3,000	3/8	3/8	3/8	1/2	3/8	3/8	3/8	1/2	1/2	1/2	5/8	5/8	1/2	5/8	5/8	3/4	5/8	3/4	3/4	7/8
6,000	3/8	1/2	1/2	5/8	1/2	1/2	1/2	5/8	1/2	5/8	3/4	3/4	5/8	3/4	3/4	7/8	3/4	7/8	7/8	1 1/8
9,000	1/2	1/2	5/8	5/8	1/2	5/8	5/8	3/4	5/8	3/4	7/8	7/8	3/4	7/8	1 1/8	1 1/8	7/8	1 1/8	1 1/8	1 3/8
12,000	1/2	5/8	3/4	3/4	5/8	5/8	3/4	3/4	3/4	7/8	7/8	1 1/8	7/8	7/8	1 1/8	1 1/8	1 1/8	1 1/8	1 3/8	1 3/8
18,000	5/8	3/4	3/4	7/8	3/4	3/4	7/8	7/8	7/8	7/8	1 1/8	1 3/8	7/8	1 1/8	1 1/8	1 3/8	1 1/8	1 3/8	1 5/8	1 5/8
24,000	3/4	3/4	7/8	7/8	3/4	7/8	1 1/8	1 1/8	7/8	1 1/8	1 1/8	1 3/8	1 1/8	1 1/8	1 3/8	1 3/8	1 3/8	1 3/8	1 5/8	1 5/8
30,000	3/4	7/8	7/8	1 1/8	7/8	7/8	1 1/8	1 1/8	1 1/8	1 1/8	1 3/8	1 3/8	1 1/8	1 1/8	1 3/8	1 5/8	1 3/8	1 5/8	1 5/8	2 1/8
36,000	3/4	7/8	1 1/8	1 1/8	7/8	1 1/8	1 1/8	1 3/8	1 1/8	1 3/8	1 3/8	1 5/8	1 1/8	1 3/8	1 5/8	1 5/8	1 3/8	1 5/8	2 1/8	2 1/8
42,000	7/8	7/8	1 1/8	1 1/8	7/8	1 1/8	1 3/8	1 3/8	1 1/8	1 3/8	1 5/8	1 5/8	1 3/8	1 3/8	1 5/8	2 1/8	1 5/8	1 5/8	2 1/8	2 1/8
48,000	7/8	1 1/8	1 1/8	1 1/8	1 1/8	1 1/8	1 3/8	1 3/8	1 3/8	1 3/8	1 5/8	1 5/8	1 3/8	1 5/8	1 5/8	2 1/8	1 5/8	2 1/8	2 1/8	2 1/8
54,000	7/8	1 1/8	1 1/8	1 3/8	1 1/8	1 1/8	1 3/8	1 3/8	1 3/8	1 3/8	1 5/8	1 5/8	1 5/8	1 5/8	2 1/8	2 1/8	1 5/8	2 1/8	2 1/8	2 5/8
60,000	7/8	1 1/8	1 1/8	1 3/8	1 1/8	1 1/8	1 3/8	1 3/8	1 3/8	1 3/8	1 5/8	2 1/8	1 5/8	1 5/8	2 1/8	2 1/8	1 5/8	2 1/8	2 1/8	2 5/8
72,000	1 1/8	1 1/8	1 3/8	1 3/8	1 1/8	1 3/8	1 3/8	1 5/8	1 3/8	1 5/8	1 5/8	2 1/8	1 5/8	1 5/8	2 1/8	2 1/8	2 1/8	2 1/8	2 5/8	2 5/8
90,000	1 1/8	1 3/8	1 3/8	1 5/8	1 1/8	1 3/8	1 5/8	1 5/8	1 5/8	1 5/8	2 1/8	2 1/8	1 5/8	2 1/8	2 1/8	2 5/8	2 1/8	2 1/8	2 5/8	3 1/8
120,000	1 1/8	1 3/8	1 5/8	1 5/8	1 3/8	1 5/8	1 5/8	2 1/8	1 5/8	2 1/8	2 1/8	2 5/8	2 1/8	2 1/8	2 5/8	2 5/8	2 1/8	2 5/8	3 1/8	3 1/8
150,000	1 3/8	1 3/8	1 5/8	2 1/8	1 3/8	1 5/8	2 1/8	2 1/8	1 5/8	2 1/8	2 1/8	2 5/8	2 1/8	2 1/8	2 5/8	2 5/8	2 5/8	2 5/8	3 1/8	3 1/8
180,000	1 3/8	1 5/8	2 1/8	2 1/8	1 3/8	1 5/8	2 1/8	2 1/8	2 1/8	2 1/8	2 5/8	2 5/8	2 1/8	2 5/8	2 5/8	3 1/8	2 5/8	3 1/8	3 5/8	3 5/8
210,000	1 3/8	1 5/8	2 1/8	2 1/8	1 5/8	2 1/8	2 1/8	2 5/8	2 1/8	2 1/8	2 5/8	2 5/8	2 5/8	2 5/8	3 1/8	3 1/8	2 5/8	3 1/8	3 5/8	3 5/8
240,000	1 5/8	1 5/8	2 1/8	2 1/8	1 5/8	2 1/8	2 5/8	2 5/8	2 1/8	2 5/8	2 5/8	3 1/8	2 5/8	3 1/8	3 1/8	3 1/8	3 1/8	3 5/8	3 5/8	4 1/8
300,000	1 5/8	2 1/8	2 1/8	2 5/8	2 1/8	2 1/8	2 5/8	2 5/8	2 1/8	2 5/8	3 1/8	3 1/8	2 5/8	3 1/8	3 1/8	3 5/8	3 1/8	3 5/8	4 1/8	5 1/8

*Type-L copper tubing.

Courtesy Dunham Bush, Inc.

730

Mechanical, Electrical, and Heat Equivalents

Unit	Equivalent Value in Other Units	
1 kilowatt-hour	1000	watt-hours
	1.34	horsepower-hours
	2,654,200	foot-pounds
	3,600,000	joules
	3412	heat units
	367,000	kilogram-meters
	0.235	pound carbon oxidized with perfect efficiency
	3.53	pounds water evaporated from and at 212° Fahrenheit
	22.75	pounds water raised from 62 to 212° Fahrenheit
1 horsepower-hour	0.746	kilowatt-hour
	1,980,000	foot-pounds
	2545	heat units
	273,740	kilogram-meters
	0.175	pound carbon oxidized with perfect efficiency
	2.64	pounds water evaporated from and at 212° Fahrenheit
	17.0	pounds water raised from 62 to 212° Fahrenheit
1 kilowatt	1000	watts
	1.34	horsepower
	2,654,200	foot-pounds per hour
	44,240	foot-pounds per minute
	737.3	foot-pounds per second
	3412	heat units per hour
	56.9	heat units per minute
	0.948	heat unit per second
	0.2275	pound carbon oxidized per hour
	3.53	pounds water evaporated per hour from and at 212° Fahrenheit
1 horsepower	746	watts
	0.746	kilowatt
	33,000	foot-pounds per minute
	550	foot-pounds per second
	2545	heat units per hour
	42.4	heat units per minute
	0.707	heat unit per second
	0.175	pound carbon oxidized per hour
	2.64	pounds water evaporated per hour from and at 212° Fahrenheit

Courtesy Calcium Chloride Institute

731

Properties of Standard Ammonia
(U.S. Bureau of Standards)

Temp., °F	Pressure in psi		Volume of Vapor, cu ft/lb	Density of Vapor, lb/cu ft	Heat Content		Latent Heat, Btu/lb	Entropy	
	Absolute	Gage			Liquid, Btu/lb	Vapor, Btu/lb		Liquid, Btu/lb/°F	Vapor, Btu/lb/°F
−40	10.41	8.7*	24.86	0.04022	0.0	597.6	597.6	0.0000	1.4242
−35	12.05	5.4*	21.68	0.04613	5.3	599.5	594.2	0.0126	1.4120
−30	13.90	1.6*	18.97	0.05271	10.7	601.4	590.7	0.0250	1.4001
−25	15.98	1.3	16.66	0.06003	16.0	603.2	587.2	0.0374	1.3886
−20	18.30	3.6	14.68	0.06813	21.4	605.0	583.6	0.0497	1.3774
−18	19.30	4.6	13.97	0.07161	23.5	605.7	582.2	0.0545	1.3729
−16	20.34	5.6	13.29	0.07522	25.6	606.4	580.8	0.0594	1.3686
−14	21.43	6.7	12.66	0.07898	27.8	607.1	579.3	0.0642	1.3643
−12	22.56	7.9	12.06	0.08289	30.0	607.8	577.8	0.0690	1.3600
−10	23.74	9.0	11.50	0.08695	32.1	608.5	576.4	0.0738	1.3558
−9	24.35	9.7	11.23	0.08904	33.2	608.8	575.6	0.0762	1.3537
−8	24.97	10.3	10.97	0.09117	34.3	609.2	574.9	0.0786	1.3516
−7	25.61	10.9	10.71	0.09334	35.4	609.5	574.1	0.0809	1.3495
−6	26.26	11.6	10.47	0.09555	36.4	609.8	573.4	0.0833	1.3474
−5	26.92	12.2	10.23	0.09780	37.5	610.1	572.6	0.0857	1.3454
−4	27.59	12.9	9.991	0.1001	38.6	610.5	571.9	0.0880	1.3433
−3	28.28	13.6	9.763	0.1024	39.7	610.8	571.1	0.0904	1.3413
−2	28.98	14.3	9.541	0.1048	40.7	611.1	570.4	0.0928	1.3393
−1	29.69	15.0	9.326	0.1072	41.8	611.4	569.6	0.0951	1.3372
0	30.42	15.7	9.116	0.1097	42.9	611.8	568.9	0.0975	1.3352
1	31.16	16.5	8.912	0.1122	44.0	612.1	568.1	0.0998	1.3332
2	31.92	17.2	8.714	0.1148	45.1	612.4	567.3	0.1022	1.3312
3	32.69	18.0	8.521	0.1174	46.2	612.7	566.5	0.1045	1.3292
4	33.47	18.8	8.333	0.1200	47.2	613.0	565.8	0.1069	1.3273
5	34.27	19.6	8.150	0.1227	48.3	613.3	565.0	0.1092	1.3253
6	35.09	20.4	7.971	0.1254	49.4	613.6	564.2	0.1115	1.3234
7	35.92	21.2	7.798	0.1282	50.5	613.9	563.4	0.1138	1.3214
8	36.77	22.1	7.629	0.1311	51.6	614.3	562.7	0.1162	1.3195
9	37.63	22.9	7.464	0.1340	52.7	614.6	561.9	0.1185	1.3176
10	38.51	23.8	7.304	0.1369	53.8	614.9	561.1	0.1208	1.3157
12	40.31	25.6	6.996	0.1429	56.0	615.5	559.5	0.1254	1.3118
14	42.18	27.5	6.703	0.1492	58.2	616.1	557.9	0.1300	1.3081
16	44.12	29.4	6.425	0.1556	60.3	616.6	556.3	0.1346	1.3043
18	46.13	31.4	6.161	0.1623	62.5	617.2	554.7	0.1392	1.3006
20	48.21	33.5	5.910	0.1692	64.7	617.8	553.1	0.1437	1.2969
22	50.36	35.7	5.671	0.1763	66.9	618.3	551.4	0.1483	1.2933
24	52.29	37.9	5.443	0.1837	69.1	618.9	549.8	0.1528	1.2897
26	54.90	40.0	5.227	0.1913	71.3	619.4	548.1	0.1573	1.2861
30	59.74	45.0	4.825	0.2073	75.7	620.5	544.5	0.1663	1.2790
35	66.26	51.6	4.373	0.2287	81.2	621.7	540.5	0.1775	1.2704

*Inches of mercury below one standard atmosphere (29.92 in.).

Properties of Standard Ammonia (Cont'd)
(U.S. Bureau of Standards)

Temp., °F	Pressure in psi		Volume of Vapor, cu ft/lb	Density of Vapor, lb/cu ft	Heat Content		Latent Heat, Btu/lb	Entropy	
	Absolute	Gage			Liquid, Btu/lb	Vapor, Btu/lb		Liquid, Btu/lb/°F	Vapor, Btu/lb/°F
40	73.32	58.6	3.971	0.2518	86.8	623.0	536.2	0.1885	1.2618
45	80.96	66.3	3.614	0.2767	92.3	624.1	531.8	0.1996	1.2535
50	89.19	74.5	3.294	0.3036	97.9	625.2	527.3	0.2105	1.2453
55	98.06	83.4	3.008	0.3325	103.5	626.3	522.8	0.2214	1.2373
60	107.6	92.9	2.751	0.3635	109.2	627.3	518.1	0.2322	1.2294
65	117.8	103.1	2.520	0.3968	114.8	628.2	513.4	0.2430	1.2216
70	128.8	114.1	2.312	0.4325	120.5	629.1	508.6	0.2537	1.2140
75	140.5	125.8	2.125	0.4707	126.2	629.9	503.7	0.2643	1.2065
78	147.9	133.2	2.021	0.4949	129.7	630.4	500.7	0.2706	1.2020
80	153.0	138.3	1.955	0.5115	132.0	630.7	498.7	0.2749	1.1991
82	158.3	143.6	1.892	0.5287	134.3	631.0	496.7	0.2791	1.1962
84	163.7	149.0	1.831	0.5462	136.6	631.3	494.7	0.3833	1.1933
86	169.2	154.5	1.772	0.5643	138.9	631.5	492.6	0.2875	1.1904
88	174.8	160.1	1.716	0.5828	141.2	631.8	490.6	0.2917	1.1875
90	180.6	165.9	1.661	0.6019	143.5	632.0	488.5	0.2958	1.1846
92	186.6	171.9	1.609	0.6214	145.8	632.2	486.4	0.3000	1.1818
94	192.7	178.0	1.559	0.6415	148.2	632.5	484.3	0.3041	1.1789
96	198.9	184.2	1.510	0.6620	150.5	632.6	482.1	0.3083	1.1761
98	205.3	190.6	1.464	0.6832	152.9	632.9	480.0	0.3125	1.1733
100	211.9	197.2	1.419	0.7048	155.2	633.0	477.8	0.3166	1.1705
105	228.9	214.2	1.313	0.7615	161.1	633.4	472.3	0.3269	1.1635
110	247.0	232.3	1.217	0.8219	167.0	633.7	466.7	0.3372	1.1566
115	266.2	251.5	1.128	0.8862	173.0	633.9	460.9	0.3474	1.1497
120	286.4	271.7	1.047	0.9549	179.0	634.0	455.0	0.3576	1.1427

Courtesy Calcium Chloride Institute

Hot-Gas Line Sizes*

Net Evaporator Capacity	Total Equivalent Length of Hot-Gas Line (Feet)							
	R12				R22 and R502			
	25	50	100	150	25	50	100	150
3000	1/2	1/2	5/8	5/8	3/8	3/8	1/2	1/2
4500	1/2	5/8	5/8	3/4	1/2	1/2	1/2	5/8
6500	5/8	5/8	3/4	3/4	1/2	1/2	5/8	5/8
8500	5/8	3/4	7/8	7/8	1/2	5/8	3/4	3/4
12,000	3/4	3/4	7/8	1 1/8	5/8	5/8	3/4	7/8
18,000	3/4	7/8	1 1/8	1 1/8	5/8	3/4	7/8	1 1/8
24,000	7/8	1 1/8	1 1/8	1 3/8	3/4	7/8	1 1/8	1 1/8
30,000	1 1/8	1 1/8	1 3/8	1 3/8	3/4	7/8	1 1/8	1 1/8
36,000	1 1/8	1 1/8	1 3/8	1 3/8	7/8	1 1/8	1 1/8	1 1/8
42,000	1 1/8	1 3/8	1 3/8	1 5/8	7/8	1 1/8	1 1/8	1 3/8
48,000	1 1/8	1 3/8	1 3/8	1 5/8	1 1/8	1 1/8	1 3/8	1 3/8
54,000	1 1/8	1 3/8	1 5/8	1 5/8	1 1/8	1 1/8	1 3/8	1 3/8
60,000	1 1/8	1 3/8	1 5/8	1 5/8	1 1/8	1 1/8	1 3/8	1 3/8
72,000	1 3/8	1 5/8	1 5/8	2 1/8	1 1/8	1 3/8	1 3/8	1 5/8
90,000	1 3/8	1 5/8	2 1/8	2 1/8	1 1/8	1 3/8	1 5/8	1 5/8
120,000	1 5/8	2 1/8	2 1/8	2 1/8	1 3/8	1 5/8	1 5/8	2 1/8
150,000	1 5/8	2 1/8	2 1/8	2 5/8	1 3/8	1 5/8	2 1/8	2 1/8
180,000	2 1/8	2 1/8	2 5/8	2 5/8	1 5/8	1 5/8	2 1/8	2 1/8

*Type-L copper tubing.
For suction temperatures less than –20°F, the next larger line size must be used.

Courtesy Dunham-Bush, Inc.

Liquid Line Sizes*

Net Evaporator Capacity	Total Equivalent Length of Liquid Line (Feet)							
	R12 and R502				R22			
	25	50	100†	150	25	50	100†	150
3000	1/4	1/4	1/4	3/8	1/4	1/4	1/4	1/4
4500	1/4	1/4	3/8	3/8	1/4	1/4	3/8	3/8
6500	1/4	3/8	3/8	3/8	1/4	1/4	3/8	3/8
8500	3/8	3/8	3/8	1/2	1/4	3/8	3/8	3/8
12,000	3/8	3/8	1/2	1/2	3/8	3/8	3/8	1/2
18,000	3/8	1/2	1/2	1/2	3/8	3/8	1/2	1/2
24,000	1/2	1/2	1/2	5/8	3/8	1/2	1/2	1/2
30,000	1/2	1/2	5/8	5/8	3/8	1/2	1/2	5/8
36,000	1/2	5/8	5/8	5/8	1/2	1/2	5/8	5/8
42,000	1/2	5/8	5/8	3/4	1/2	1/2	5/8	5/8
48,000	1/2	5/8	3/4	3/4	1/2	5/8	5/8	3/4
54,000	5/8	5/8	3/4	3/4	1/2	5/8	5/8	3/4
60,000	5/8	5/8	3/4	3/4	1/2	5/8	3/4	3/4
72,000	5/8	3/4	7/8	7/8	5/8	5/8	3/4	3/4
90,000	3/4	3/4	7/8	7/8	5/8	3/4	7/8	7/8
120,000	3/4	7/8	1 1/8	1 1/8	3/4	3/4	7/8	1 1/8
150,000	7/8	7/8	1 1/8	1 1/8	3/4	7/8	1 1/8	1 1/8
180,000	7/8	7/8	1 1/8	1 1/8	7/8	7/8	1 1/8	1 1/8
210,000	1 1/8	1 1/8	1 3/8	1 3/8	7/8	7/8	1 1/8	1 1/8
240,000	1 1/8	1 1/8	1 3/8	1 3/8	7/8	1 1/8	1 3/8	1 3/8
300,000	1 1/8	1 1/8	1 5/8	1 5/8	1 1/8	1 1/8	1 3/8	1 3/8

*Type-L copper tubing.
†Line sizes indicated are suitable for condenser-to-receiver applications.

Courtesy Dunham-Bush, Inc.

Discharge Line Sizes*

Net Evaporator Capacity	Total Equivalent Length of Hot-Gas Line (Feet)							
	R12				R22 and R502			
	25	50	100	150	25	50	100	150
3000	3/8	3/8	1/2	1/2	3/8	3/8	3/8	3/8
4500	3/8	1/2	1/2	1/2	3/8	3/8	1/2	1/2
6500	1/2	1/2	5/8	5/8	3/8	1/2	1/2	1/2
8500	1/2	1/2	5/8	5/8	3/8	1/2	1/2	5/8
12,000	1/2	5/8	3/4	3/4	1/2	1/2	5/8	5/8
18,000	5/8	3/4	3/4	7/8	1/2	5/8	3/4	3/4
24,000	5/8	3/4	7/8	1 1/8	5/8	5/8	3/4	7/8
30,000	3/4	7/8	1 1/8	1 1/8	5/8	3/4	7/8	7/8
36,000	3/4	7/8	1 1/8	1 1/8	5/8	3/4	7/8	7/8
42,000	7/8	7/8	1 1/8	1 1/8	3/4	3/4	7/8	1 1/8
48,000	7/8	1 1/8	1 1/8	1 3/8	3/4	7/8	1 1/8	1 1/8
54,000	7/8	1 1/8	1 1/8	1 3/8	3/4	7/8	1 1/8	1 1/8
60,000	7/8	1 1/8	1 3/8	1 3/8	3/4	7/8	1 1/8	1 1/8
72,000	1 1/8	1 1/8	1 3/8	1 3/8	7/8	1 1/8	1 1/8	1 1/8
90,000	1 1/8	1 3/8	1 3/8	1 5/8	7/8	1 1/8	1 1/8	1 3/8
120,000	1 1/8	1 3/8	1 5/8	1 5/8	1 1/8	1 1/8	1 3/8	1 3/8
150,000	1 3/8	1 3/8	1 5/8	2 1/8	1 1/8	1 3/8	1 3/8	1 5/8
180,000	1 3/8	1 5/8	2 1/8	2 1/8	1 1/8	1 3/8	1 5/8	1 5/8
210,000	1 3/8	1 5/8	2 1/8	2 1/8	1 3/8	1 3/8	1 5/8	1 5/8
240,000	1 5/8	1 5/8	2 1/8	2 1/8	1 3/8	1 5/8	1 5/8	2 1/8
300,000	1 5/8	2 1/8	2 1/8	2 5/8	1 3/8	1 5/8	2 1/8	2 1/8

*Type-L copper tubing.

Courtesy Dunham-Bush, Inc.

Glossary

Absolute Pressure—The sum, at any particular time, of the gage pressure and the atmospheric pressure. Thus, for example, if the gage reads 164.5 psi, the absolute pressure will be 164.5 + 14.7, or 179.2 psi.

Absolute Temperature—The temperature of a substance measured above absolute zero.

Absolute Zero (Pressure)—The pressure existing in a vessel that is entirely empty. The lowest possible pressure. Perfect vacuum.

Absolute Zero (Temperature)—The temperature at which the motion of a substance theoretically ceased. This temperature is equal to -459.6^0F and -273.1° Celsius. At this temperature, the volume of an ideal gas maintained at a constant pressure becomes zero.

Absorber—A device for absorbing a refrigerant. A low-side element in an absorption system.

Absorption—The adhesion of the molecules of gases or dissolved substances to the surface of solid bodies, resulting in a concentration of the gas or solution at the place of contact.

Absorption Machine—A refrigeration machine in which the gas involved in an evaporator is taken up by an absorber.

Accumulator—A shell placed in a suction line for separating the liquid entrained in the suction gas.

Acrolein—A warning agent often used with methyl chloride to call attention to the escape of refrigerant. The material has a compelling,pungent odor and causes irritation of the throat and eyes. Acrolein reacts with sulfur dioxide to form a sludge.

Activated Aluminum—A form of aluminum oxide (AL_2O_3) that absorbs moisture readily and is used as a drying agent.

Adiabatic—A change in gas conditions where no heat is added or removed except in the form of work.

Adiabatic Process—Any thermodynamic process taking place in a closed system without the addition or removal of heat.

Aeration—A term generally employed with reference to air circulation or ventilation. In milk cooling, it refers to a method where the milk flow over refrigerated surfaces is exposed to the atmosphere.

Agitation—A condition in which a device causes circulation in a tank containing fluid.

Air Conditioning—The simultaneous control of all (or at least the first three) of the following factors affecting the physical and chemical conditions of the atmosphere within a structure: temperature, humidity, motion, distribution, dust, bacteria, odors, toxic gases, and ionization, most of which affect human health or comfort.

Air Conditioning Unit—A piece of equipment designed as a specific air-treating combination, consisting of a means for ventilation, air circulation, air cleaning, and heat transfer, with a control means for maintaining temperature and humidity within prescribed limits.

Air Infiltration—The in-leakage of air through cracks, crevices, doors, windows, or other openings caused by wind pressure or temperature differences.

Air Washer—An enclosure in which air is forced through a spray of water in order to cleanse, humidify, or dehumidify the air.

Ambient Temperature—The temperature of the medium surrounding an object. In a domestic or commercial system having an air-cooled condenser, it is the temperature of the air entering this condenser.

Ammonia Machine—An abbreviation for a compression refrigerating machine using ammonia as a refrigerant. Similarly, Freon machine, sulfur dioxide machine, etc.

Analyzer—A device used in the high side of an absorption system for increasing the concentration of vapor entering the rectifier or condenser.

Anemometer—An instrument for measuring the velocity of air in motion.

Antifreeze Liquid—A substance added to the refrigerant to prevent formation of ice crystals at the expansion valve. Antifreeze agents in general do not prevent corrosion due to moisture. The use of liquid should be a temporary measure where large quantities of water are involved unless a drier is used to reduce the moisture content. Ice crystals may form when moisture is present below the corrosion limits, and, in such instances, a suitable noncorrosive antifreeze liquid is often of value. Materials such as alcohol are corrosive and, if used, should be allowed to remain in the machine for a limited time only.

Atmospheric Condenser—A condenser operated with water that is exposed to the atmosphere.

Atmospheric Pressure—The pressure exerted by the atmosphere in all directions as indicated by a barometer. Standard atmospheric pressure is considered to be 14.695 psi (usually written 14.7 psi), which is equivalent to 29.92 in. Hg.

Atomize—To reduce to a fine spray.

Automatic Expansion Valve—A pressure-actuated device that regulates the flow of refrigerant from the liquid line into the evaporator to maintain a constant evaporator pressure.

Automatic Refrigerating System—A system that regulates itself to maintain a definite set of conditions by means of automatic controls and valves usually responsive to temperature or pressure.

Baffle—A partition used to divert the flow of a fluid.

Balanced Pressure—The same pressure in a system or container that exists outside the system or container.

Barometer—An instrument for measuring atmospheric pressure.

Bleeder—A pipe sometimes attached to a condenser to lead off liquid refrigerant parallel to the main flow.

Boiler—A closed vessel in which steam is generated or in which water is heated.

Boiling Point—The temperature at which a liquid is vaporized upon the addition of heat, dependent upon the refrigerant and the absolute pressure at the surface of the liquid and vapor.

Brine—Any liquid cooled by the refrigerating system and used for the transmission of heat.

Brine System of Cooling—A system whereby brine, cooled by a refrigerant system, is circulated through pipes to the point where the refrigeration is needed.

British Thermal Unit (Btu)—The amount of heat required to raise the temperature of one pound of water one degree Fahrenheit. It is also the measure of the amount of heat removed in cooling one pound of water one degree Fahrenheit and is so used as a measure of refrigerating effect.

Butane—A flammable, hydrocarbon refrigerant used to a limited extent in small units.

Calcium Chloride—A chemical having the formula $CaCl_2$ that, in granular form, is used as a drier. This material is soluble in water, and in the presence of large quantities of moisture it may dissolve and plug up the drier unit or even pass into the system beyond the drier.

Calcium Sulfate—A solid chemical, with the formula $CaSO_4$, which may be used as a drying agent.

Calibration—The process of dividing and numbering the scale of an instrument; also of correcting and determining the error of an existing scale.

Calorie—The heat per unit of weight required to raise the temperature of water one degree Celsius. Thus, a gram calorie is the amount of heat required to raise the temperature of one gram of water one degree Celsius.

Capacity—In a refrigerating machine, the heat-absorbing capacity per unit time, usually measured in tons or Btu/per hour.

Capacity, Container—The ability of a container to hold a material. The quantity of material that may safely be contained in a container.

Capacity, Refrigerating—The ability of a refrigerating system, or part thereof, to remove heat. Expressed as a rate of heat removal, it is usually measured in Btu per hour or tons per hour.

Carbon Dioxide Ice—Compressed solid CO_2; dry ice.

Carbon Tetrachloride—A liquid having the formula CCL_4 (also known as carbona), which is a nonflammable solvent used for removing grease and oil and loosening sludges.

Celsius—A thermometric system in which the freezing point of water is called 0° and its boiling point 100° at normal barometric pressure. Old term: Centigrade.

Centipoise—A unit of viscosity used in figuring pressure drop, etc.

Centrifugal Machine—A compressor employing centrifugal force for compression.

Centrifuge—A device for separating liquids of different densities by centrifugal action.

Change of State—A change from one state to another, as from a liquid to a solid, from a liquid to a gas, etc.

Charge—The amount of refrigerant in a system.

Charging—Putting in a charge.

Coefficient of Expansion—The fractional increase in length or volume of a material per degree rise in temperature.

Coil—Any cooling element made of pipe or tubing.

Cold Storage—A trade or process of preserving perishables on a large scale by refrigeration.

Comfort Chart—A psychrometric chart. Strictly, a chart showing effective temperatures.

Compound Compressor—A compressor in which compression is accomplished by stages in two or more cylinders.

Compression System—A refrigerating system in which the pressure-imposing element is mechanically operated.

Compressor—That part of a mechanical refrigerating system which receives the refrigerant vapor at low pressure and compresses it into a smaller volume at higher pressure.

Condenser—A heat transfer device that receives high-pressure vapor at temperatures above that of the cooling mediums, such as air or water, to which the condenser passes latent heat from the refrigerant, causing the refrigerant vapor to liquefy.

Condensing—The process of giving up latent heat of vaporization in order to liquefy a vapor.

Conduction, Thermal—Passage of heat from one point to another by transmission of molecular energy from particle to particle through a conductor.

Conductivity, Thermal—The ability of a material to pass heat from one point to another, generally expressed in terms of Btu per hour per square foot of material per inch of thickness per degree temperature difference.

Conductor, Electrical—A material that will pass an electric current as part of an electrical system.

Constant-Pressure Valve—A valve of the throttling type, responsive to pressure, located in the suction line of an evaporator to maintain a desired constant pressure in the evaporator higher than the main suction-line pressure.

Constant-Temperature Valve—A valve of the throttling type, responsive to the temperature of a thermostatic bulb. This valve is located in the suction line of an evaporator to reduce the refrigerating effect on the coil to just maintain a desired minimum temperature.

Control—Any device for the regulation of a machine in normal operation, either manual or automatic. If automatic, the implication is that it is responsive to temperature, pressure, liquid level, or time.

Control, Low-Pressure—An electric switch, responsive to pressure, connected into the low-pressure part of a refrigerating system. Usually closes at high pressure and opens at low pressure.

Control, Temperature—An electric switch responsive to the temperature of a thermostatic bulb or element.

Cooling Unit—A specific air-treating combination consisting of a means for air circulation and cooling within prescribed temperature limits.

Cooling Water—Water used for condensation of refrigerant. Condenser water.

Convection—Passage of heat from one point to another by means of a gravity fluid circulation due to changes in density resulting from absorbing and giving up heat.

Copper Plating—Formation of a film of copper, usually on compressor walls, pistons, or discharge valves.

Counterflow—In the heat exchange between two fluids, the opposite direction of flow, the coldest portion of one meeting the coldest portion of the other.

Critical Pressure—The vapor pressure corresponding to the critical temperature.

Critical Temperature—The temperature above which a vapor cannot be liquefied, regardless of pressure.

Critical Velocity—The velocity above which fluid flow is turbulent.

Crohydrate—A eutectic brine mixture of water and any salt mixed in proportions to give the lowest freezing temperature.

Cycle of Refrigeration—A complete course of operation of a refrigerant back to the starting point, measured in thermodynamic terms. Also used in general for any repeated process for any system.

Defrosting—Removal of accumulated ice from the cooling unit.

Defrosting Cycle—A cycle that permits the cooling unit to defrost during an *off* period.

Degree Day—A unit based upon temperature difference and time used to specify the nominal heating load in winter. For one day, there exist as many degree days as there are degrees Fahrenheit difference in temperature between the average outside air temperature, taken over a 24-hour period, and a temperature of 65°F.

Dehumidify—To remove water vapor from the atmosphere. To remove water or liquid from stored goods.

Dehydrator—A device used to remove moisture from the refrigerant.

Density—The mass or weight per unit of volume.

Dew Point (of Air)—The temperature at which a specified sample of air, with no moisture added or removed, is completely saturated. The temperature at which the air, on being cooled, gives up moisture or dew.

Differential (of a Control)—The difference between the cut-in and cutout temperatures or pressures.

Direct Expansion—A system in which the evaporator is located in the material or space refrigerated or in the air-circulating passages communicating with such space.

Displacement, Actual—The volume of gas at the compressor inlet actually moved in a given time.

Displacement, Theoretical—The total volume displaced by all the pistons of a compressor for every stroke during a definite interval. Usually measured in cubic feet per minute.

Domestic Refrigerator—A refrigerator for home use.

Drier—Synonymous with dehydrator.

Dry-Bulb Temperature—The actual temperature of air, as opposed to wet-bulb temperature.

Dry-Type Evaporator—An evaporator of the continuous-tube type where the refrigerant from a pressure-reducing device is fed into one end and the suction line is connected to the outlet end.

Ebullator—A device inserted in flooded evaporator tubes to prevent the evaporator from becoming oil bound.

Efficiency, Mechanical—The ratio of the output of a machine to the input in equivalent units.

Efficiency, Volumetric—The ratio of the volume of gas actually pumped by a compressor or pump to the theoretical displacement of the compressor.

Ejector—A device that utilizes static pressure to build up a high fluid velocity in a restricted area to obtain a lower static pressure at that point so that fluid from another source may be drawn in.

Emulsification—Formation of an emulsion, i.e., a mixture of small droplets of two or more liquids that do not dissolve in each other.

Entropy—In thermodynamics, the base of a heat diagram, the area of which is heat units and the altitude of which is absolute temperature.

Eutectic Solution—A solution of such concentration as to have a constant freezing point at the lowest freezing temperature for the solution.

Evaporation—The change of state from liquid to vapor.

Evaporative Condenser—A refrigerant condenser utilizing the evaporation of water by air at the condenser surface as a means of dissipating heat.

Evaporative Cooling—The process of cooling by means of the evaporation of water in air.

Evaporator—A device in which the refrigerant evaporates while absorbing heat.

Expansion Valve, Automatic—A device that regulates the flow of refrigerant from the liquid line into the evaporator to maintain a constant evaporator pressure.

Expansion Valve, Thermostatic—A device to regulate the flow of refrigerant into an evaporator so as to maintain an evaporation temperature in a definite relationship to the temperature of a thermostatic bulb.

Extended Surface—The evaporator or condenser surface that is not a primary surface. Fins or other surfaces that transmit heat from or to a primary surface that is part of the refrigerant container.

External Equalizer—In a thermostatic expansion valve, a tube connection from the chamber containing the evaporation pressure-actuated element of the valve to the outlet of the evaporator coil. A device to compensate for excessive pressure drop through the coil.

Fahrenheit—A thermometric system in which 32° denotes the freezing point of water and 212° the boiling point under normal pressure.

Filter—A device to remove solid material from a fluid by a straining action.

Flammability—The ability of a material to burn.

Flare Fitting—A type of soft-tube connector that involves the flaring of the tube to provide a mechanical seal.

Flash Gas—The gas resulting from the instantaneous evaporation of the refrigerant in a pressure-reducing device to cool the refrigerant to the evaporation temperature obtained at the reduced pressure.

Float Valve—Valve actuated by a float immersed in a liquid container.

Flooded System—One in which the refrigerant enters into a header from a pressure-reducing valve and in which the evaporator maintains a liquid level. Opposed to dry evaporator.

Fluid—A gas or liquid.

Foaming—Formation of a foam or froth of oil refrigerant due to rapid boiling out of the refrigerant dissolved in the oil when the pressure is suddenly reduced. This occurs when the compressor operates; if large quantities of refrigerant have been dissolved, large quantities of oil may "boil" out and be carried through the refrigerant lines.

Freezeup—Failure of a refrigeration unit to operate normally due to formation of ice at the expansion valve. The valve may be frozen closed or open, causing improper refrigeration in either case.

Freezing Point—The temperature at which a liquid will solidify upon the removal of heat.

Freon-12—The common name for dichlorodifluoromethane (CCl_2F_2).

Frost Back—The flooding of liquid from an evaporator into the suction line, accompanied by frost formation on the suction line in most cases.

Fusible Plug—A safety plug used in vessels containing refrigerant. The plug is designed to melt at high temperatures (usually about 165°F) to prevent excessive pressure from bursting the vessel.

Gage—An instrument for measuring pressure or liquid level.

Gas—The vapor state of a material.

Glycerol—This material has been suggested as a lubricant but has found little favor because of its tendency to rapidly absorb moisture. A suitable oil is more satisfactory.

Head Pressure—The pressure in the compressor head. Discharge pressure.

Heat—Basic form of energy that may be partially converted into other forms and into which all other forms may be entirely converted.

Heat Content—*See* **Total Heat**.

Hermetically-Sealed Unit—A refrigerating unit having no mechanical connections and containing no shaft seal.

High-Pressure Cutout—A control device connected into the high-pressure part of a refrigerating system to stop the machine when the pressure becomes excessive.

High Side—That part of the refrigerating system containing the high-pressure refrigerant. Also used to refer to the condensing unit, consisting of the motor, compressor, condenser, and receiver mounted on a single base.

High-Side Float Valve—A float valve that floats in high-pressure liquid. Opens on an increase in liquid level.

Holdover—In an evaporator, the ability to stay cold after heat removal from the evaporator stops.

Horsepower—A unit of power. Work done at the rate of 33,000 foot-pounds per minute, or 550 foot-pounds per second.

Humidifier—A device to add moisture to the air.

Humidistat—A control device sensitive to relative humidity.

Humidity, Absolute—The definite amount of water contained in a definite quantity of air. Usually measured in grains of water per pound or per cubic foot of air.

Hydrocarbons—A series of chemicals of similar chemical nature ranging from methane (the main constituent of natural gas) through butane, octane, etc., to heavy lubricating oils. All are more or less flammable. Butane and isobutane have been used to a limited extent as refrigerants.

Hydrochloric Acid, Hydrogen Chloride—An acid commonly used in soldering operations. It is formed when appreciable quantities of moisture are present in Freon-12 or methyl chloride, or when these refrigerants come in contact with a flame or hot object. The acid vapor is pungent and warns of its presence.

Hydrofluoric Acid, Hydrogen Fluoride—The acid formed when Freon-12 passes through a flame. Although the vapor is pungent at moderate concentrations, lower harmful concentrations may be undetected.

Hydrolysis—Reaction of a material, such as Freon-12 or methyl chloride, with water. Acid materials in general are formed.

Hydrostatic Pressure—The pressure due to liquid in a container that contains no gas space.

Ice-Melting Equivalent—The amount of heat (144 Btu) absorbed by one pound of ice at 32°F in liquefying to water at 32°F.

Indirect System of Cooling—*See* **Brine System of Cooling**.

Infiltration—The leakage of air into a building or space.

Insulation (Heat)—Material of low heat conductivity.

Irritant Refrigerant—Any refrigerant that has an irritating effect on the eyes, nose, throat, or lungs.

Isobutane—A flammable hydrocarbon refrigerant used to a limited extent.

Kilowatt—Unit of electrical power equal to 1000 watts, or 1.34 horsepower, approximately.

Lag of Temperature Control—The delay in action of a temperature-responsive element due to the time required for the temperature of the element to reach the surrounding temperature.

Latent Heat—The quantity of heat which may be added to a substance during a change of state without causing a temperature change.

Latent Heat of Evaporation—The quantity of heat required to change one pound of liquid into a vapor with no change in temperature. Reversible.

Leak Detector—A device used to detect refrigerant leaks in a refrigerating system.

Liquid—The state of a material in which the top surface in a vessel will become horizontal. Distinguished from solid or vapor forms.

Liquid Line—The tube or pipe which carries the refrigerant liquid from the condenser or receiver of a refrigerating system to a pressure-reducing device.

Liquid Receiver—The part of the condensing unit which stores the liquid refrigerant.

Load—The required rate of heat removal.

Low-Pressure Control—An electric switch and pressure-responsive element connected into the suction side of a refrigerating unit to control the operation of the system.

Low Side—That part of a refrigerating system which normally operates under low pressure as opposed to the high side. Also used to refer to the evaporator.

Low-Side Float Valve—A valve operated by the low-pressure liquid. Opens at a low level and closes at a high level.

Manometer—A U-shaped tube for measuring pressure differences.

Mechanical Efficiency—The ratio of work done by a machine to the work done on it or energy used by it.

Methanol—*See* **Methyl Alcohol**.

Methyl Alcohol—This material, also called wood alcohol, has a formula of CH_3OH. It has been used as an antifreeze but is corrosive.

Methyl Chloride—A refrigerant having the chemical formula CH_3Cl.

Micron Measurement—Actually, a micron is a unit of measure in the metric system and is equivalent to one-thousandth of a millimeter. There are 25.4 millimeters to 1 inch. Therefore, 1 micron is equal to 0.00004 inch; 100 microns are equivalent to 0.004 inch, or 29.996 inches of vacuum. Vacuums of less than 29.0 inches of mercury are not satisfactory for evacuating compressors.

Mollier Chart—A graphical representation of thermal properties of fluids, with total heat and entropy as coordinates.

Motor—A device for transforming electrical energy into mechanical energy.

Noncondensables—Foreign gases mixed with a refrigerant thatcannot be condensed into the liquid form at the temperatures and pressures at which the refrigerant condenses.

Oil Trap—A device to separate oil from the high-pressure vapor from the compressor. Usually contains a float valve to return the oil to the compressor crankcase.

Overrun—In ice cream freezing, the ratio of the volume of ice cream to the volume of the mix used.

Ozone—The O_3 form of oxygen. Used in air conditioning as an odor eliminator. Harmful in high concentrations.

Packing—The stuffing around a shaft to prevent fluid leakage between the shaft and parts around the shaft.

Packless Valve—A valve that does not use packing to prevent leaks around the valve stem. Flexible material is usually used to seal against leaks and still permit valve movement.

Performance Factor—The ratio of the heat moved by a refrigerating system to the heat equivalent of the energy used. Varies with conditions.

Phosphorus Pentoxide—An efficient drier material that becomes gummy reacting with moisture and hence is not used alone as a drying agent.

Pour Point (of Oils)—The temperature below which the oil surface will not change when the oil container is tilted.

Power—The rate of doing work, measured in horsepower, watts, kilowatts, etc.

Power Factor (of an Electrical Device)—The ratio of watts to volt-amperes in an alternating-current circuit.

Pressure—The force exerted per unit of area.

Pressure Drop—Loss in pressure, as from one end of a refrigerant line to the other, due to friction, static head, etc.

Psychrometric Chart—A chart used to determine the specific volume, heat content, dew point, relative humidity, absolute humidity, and wet- and dry-bulb temperatures, knowing any two independent items of those mentioned.

Purging—The act of blowing out refrigerant gas from a refrigerant-containing vessel, usually for the purpose of removing noncondensables.

Pyrometer—An instrument for the measurement of high temperatures.

Radiation—The passage of heat from one object to another without warming the space between. The heat is passed by wave motion similar to light.

Refrigerant—The medium of heat transfer in a refrigerating system that picks up heat by evaporating at a low temperature and gives up heat by condensing at a higher temperature.

Refrigerating System—A combination of parts in which a refrigerant is circulated for the purpose of extracting heat.

Relative Humidity—The ratio of the water-vapor pressure of air compared to the vapor pressure it would have if saturated at its dry-bulb temperature. Very nearly the ratio of the amount of moisture contained in air compared to what it could hold at the existing temperature.

Relief Valve—A valve designed to open at excessively high pressures to allow the refrigerant to escape.

Resistance (Electrical)—The opposition to electric current flow, measured in ohms.

Resistance (Thermal)—The reciprocal of thermal conductivity.

Room Cooler—A cooling element for a room. In air conditioning, a device for conditioning small volumes of air for comfort.

Rotary Compressor—A compressor in which compression is attained in a cylinder by rotation of a semiradial member.

Running Time—Usually indicates percent of time a refrigerant compressor operates.

Saponifiable—Capable of reacting with an alkali (i.e., lye) to form a soap. Fats may be saponified.

Saturated Vapor—Vapor not superheated, but of 100 percent quality; i.e., containing no unvaporized liquid.

Seal Shaft—A mechanical system of parts for preventing gas leakage between a rotating shaft and a stationary crankcase.

Sealed Units—*See* **Hermetically-Sealed Units**.

Sensible Heat—Heat that raises temperature.

Shell and Tube—Pertaining to heat exchangers in which a coil of tubing or pipe is contained in a shell or container. The pipe is provided with openings to allow the passage of a fluid through it, while the shell is also provided with an inlet and outlet for a fluid flow.

Silica Gel—A drier material having the formula SiO_2.

Sludge—A decomposition product formed in a refrigerant due to impurities in the oil or due to moisture. Sludges may be gummy or hard.

Soda Lime—A material used for removing moisture. Not recommended for refrigeration use.

Solenoid Valve—A valve opened by the magnetic effect of an electric current through a solenoid coil.

Solid—The state of matter in which a force can be exerted in a downward direction only when not confined. As distinguished from fluids.

Solubility—The ability of one material to enter into solution with another.

Solution—The homogeneous liquid mixture of two or more materials.

Specific Gravity—The weight of a volume of a material compared to the weight of the same volume of water.

Specific Heat—The quantity of heat required to raise the temperature of a definite mass of a material a definite amount compared to that required to raise the temperature of the same mass of water the same amount. May be expressed as Btu per pound per degrees Fahrenheit.

Specific Volume—The volume of a definite weight of a material. Usually expressed in terms of cubic feet per pound. The reciprocal of density.

Standard Air—Air weighing 0.7488 pound per cubic foot, which is air at 68°F dry bulb and 50% relative humidity at a barometric pressure of 29.92 in. Hg, or approximately dry air at 70°F at the same pressure.

Static Pressure—Pressure against the walls of tubes, pipes, or ducts.

Subcooled—Cooled below the condensing temperature corresponding to the existing pressure.

Sublimation—The change from a solid to a vapor state without an intermediate liquid state.

Suction Line—The tube or pipe which carries the refrigerant vapor from the evaporator to the compressor inlet.

Suction Pressure—Pressure on the suction side of the compressor.

Sulfur Dioxide—A refrigerant having the formula SO_2. Forms sulfuric acid when mixed with water.

Superheat—The heat contained in a vapor above its heat content at the boiling point at the existing pressure.

Superheated Gas—A gas with a temperature higher than the evaporation temperature at the existing pressure.

Superheat Setting of Thermostatic Expansion Valve—The depression of the evaporation temperature at the valve feed pressure below the temperature of the thermostatic bulb.

Sweating—Condensation of moisture from the air on surfaces below the dew-point temperature.

Temperature—Heat level or pressure. The thermal state of a body with respect to its ability to pick up heat from or pass heat to another body.

Temperature-Control Lag—The delay in action of a temperature responsive element due to the time required for the element to reach the surrounding temperature.

Thermal Conductivity—The ability of a material to conduct heat from one point to another. Indicated in terms of Btu per hour per square foot per inch thickness per degrees Fahrenheit.

Thermocouple—A device consisting of two electrical conductors having two junctions, one at a point with a temperature to be measured, and the other at a known temperature. The temperature between the two junctions is determined by the material characteristics and the electrical potential set up.

Thermodynamic—The science of the mechanics of heat.

Thermometer—A device for indicating temperature.

Thermostat—A temperature-actuated switch.

Thermostatic Expansion Valve—A device to regulate the flow of refrigerant into an evaporator so as to maintain an evaporation temperature in a definite relationship to the temperature of a thermostatic bulb.

Ton of Refrigeration Capacity—Refrigeration equivalent to the melting of one ton of ice per 24 hours. 288,000 Btu/day, 12,000 Btu/hr, or 200 Btu/min.

Total Heat—The total heat added to a refrigerant above an arbitrary starting point to bring it to a given set of conditions (usually expressed in Btu per pound). For instance, in a superheated gas, the combined heat added to the liquid necessary to raise its temperature from an arbitrary starting point to the evaporation temperature to complete evaporation and to raise the temperature to the final temperature where the gas is superheated.

Total Pressure—In fluid flow, the sum of static pressure and velocity pressure.

Toxicity—The characteristic of a material to intoxicate or poison.

Turbulent Flow—Fluid flow in which the fluid moves transversely as well as in the direction of the tube or pipe axis, as opposed to streamline, or viscous, flow.

Unit System—A system that can be removed from the user's premises without disconnecting any refrigerant-containing parts, water connections, or fixed electrical connections.

Unloader—A device in a compressor for equalizing high- and low-side pressures when the compressor stops and for a brief period after it starts, so as to decrease the starting load on the motor.

Vacuum—A pressure below atmospheric, usually measured in inches of mercury below atmospheric pressure.

Vapor Pressure—The pressure existing at the liquid and vapor-surface of a refrigerant. Depends upon temperature.

Viscosity—The property of fluid to resist flow or change of shape.

Wax—A material that may separate when oil/refrigerant mixtures are cooled. Wax may plug the expansion valve and reduce heat transfer of the coil.

Wet Compression-A system of refrigeration in which some liquid refrigerant is mixed with vapor entering the compressor so as to cause the discharge vapors from the compressor to tend to be saturated rather than superheated.

Xylene-A flammable solvent, similar to kerosene, used for dissolving or loosening sludges and for cleaning compressors and lines.

Zero, Absolute (Pressure)—*See* **Absolute Zero (Pressure).** The lowest possible pressure. Perfect vacuum.

Zero, Absolute (Temperature)—*See* **Absolute Zero (Temperature).**

Index

A

Absolute pressure, 21, 69
Absolute zero, 33
Absorption refrigeration system, 153-190
   absorption cycle, 155
   burner, 161, 172, 177
   components, 159-161
   electrical accessories, 164
   flue system, 158
   gas
      burner
         components, 159
         thermostats, 163, 172
        control devices, 158
        line connection, 169
        pressure adjustment, 172
   maintenance, 175
   service
      air circulation, 167
      leveling, 168
   trouble charts, 183-188
Accumulators, 144
Acidity
   of brine, 505
   of lubricating oil, 376
Advantages and applications of
   thermoelectric cooling, 191
Aerator type, milk cooler, 584
Agitator unit, milk cooler, 586

Air
   change loads, 716
   circulation fans and thermostats, 165
   cooled condensers, 482
   piping, 510
   volume, 603
Alkalinity of brine, 505
Alternating-current motors, 428
   polyphase, 428
   single-phase,
      capacitor, 435
      repulsion-start, 437
      shaded-pole, 436
      split-phase, 434
      squirrel-cage, 429
Altitudes and refrigeration
   capacity, 61
Ammonia
   and sulfur dioxide leaks, 82
   as a refrigerant, 79, 89
   -water cycle, 648
Anchor bolts and templates, 671
   installation details, 673
Applications of refrigeration, 12
Arrangement of
   accumulator, connections in loose-can tank, 514
   components for group-lift plant, 517

Arrangement of (cont.)
  equipment, typical locker plant, 562
Atmospheric factors in refrigeration, 22
Atmospheric pressure, 69
Automatic
  can filler, 512
  defrosting, 181,291
  expansion valve, 384
  ice makers, 520
Axial fans, 603

B

Baffles, 540
Balancing the system, 487
Barometer, 52
Basic absorption refrigeration cycle, 634, 636
Basic systems of refrigeration, 64
Belt alignment, compressor, 116
Bimetallic thermostats, 446
Blanching vegetables, 595
Block system, ice making, 518
Boiling point, 465
Bottle coolers, 569, 584
Boyle's law, 21
Brake horsepower, 122
Brine
  calcium chloride, 499
  cell plate system, ice making, 518
  coil plate system, ice making, 518
  foaming, 502
  initial charges, 501
  preparing new, 507
  system corrosion, 502
    acidity, 505
    alkalinity, 505
    pH, 504
      color table, 504
      color values, 504
    testing for ammonia, 506
British thermal unit, 34
Bubble method, leak detection, 82
Burner-air adjustment, 172
Burner-orifice cleaning, 171

C

Cabinet maintenance and repairs, 315-323
Calcium chloride brine, 499

Calculating
  heat leakage, 686
  heat transfer coefficient, 710
Calculations, compressor, 122
Calibrating controls, 269
Can dumps, 515
Can ice, 509
Capacitor
  motors, 435
  test, 578
Capacity
  of evaporator, 717
  of machines, 713
Capillary tube, 142, 396
  restrictions, 399
  replacement, 400
  system, characteristics of, 327
Carbon dioxide, 45, 80
Care in handling refrigerants, 84
Celsius scale, 24
Centrifugal compressors, 105
Centrifugal
  fans, 603, 604
  pumps, 660
Change of state, 39
Characteristics of the refrigerant-absorbent pair, 641
Charging of sealed system, 213
Charging the system, 679
Charles' law, 22
Chemical properties of freon, 73
Chill room, locker plant, 558
Circulating pumps
  centrifugal, 661
  maintenance, 107, 664
  power requirements, 663
  reciprocating, 663
  rotary, 657
Classification of refrigerants, 72
Cleaning refrigerator, 355
Closed meat-display cases, 535
Combination gage set, 204
Combination heat-engine cycle and mechanical-refrigeration cycle, 635
Combined law of Boyle and Charles, 67
Components
  absorption system, 154
  gas burner, 161
Compression ratio, 118
Compression testing with direct power, 260

Compressor, 93-133, 140
  belt alignment, 116
  calculations, 123
  care, 109
  efficiency test, 113
  knocks, 114
  lubrication system, 95, 115, 201,
    371
    acidity, 376
    wax separation, 377
  maintenance, 107
    oil, 108
    shaft seal, 95
    suction strainer, 108
    valves, 100, 109
  motor, 450
    replacement, 451
  necessary precautions, 372
  removal, 115
  replacement, 116
  safety precautions, 129
  servicing data, 110
  troubleshooting chart, 492
  types
    centrifugal, 105
    hermetic, 101
    reciprocating, 94
    rotary, 100
  unit replacements, 116
  wiring, 403
Condensation, 57
Condensers, 141
  replacement, 274
Condensing point, 66
Condensing unit, milk cooler, 585
Conduction, 55
Congealing tanks, 65,498
Connecting rods and wrist pins, 98
Connections for suction and hot-gas
  headers, multiple compressor
  installation, 618
Contactors, starters and relays, 438
Containers, refrigerant, 86
Control devices, gas burner,158
Control methods, 150
Convection, 56
Cooling
  load, 686
  ponds, 473
  system components, 479-489
  towers, 474
    control of, 478
    dimensions of, 475

Cooling (cont.)
  water requirements, 473
Cooling-coil frosting, 355
Cooling-water-line installation, 362
Copper tubing, 240
  sizes and dimensions, 241
Cork insulation, 626
Correct and incorrect hot-gas
  connection, parallel compressors,
  619
Correct method of making a
  multiple installation, ice cream
  cabinets, 582
Corrosion retarders, 503
Cracking a valve, 204
Crankshafts, 99
Critical pressure, 67
Critical temperature, 67
Cross-sectional view of typical ice
  cream cabinet, 581
Cutaway view of a Sherer-Gillett
  metal walk-in cooler, 546
Cutting plate ice, 520
Cycle of operation
  household freezers, 325
  typical bottle water cooler, 574
  typical pressure water cooler, 574

D

Dalton's law of partial pressure, 157
Defective compressor, 257
Defrost controls, 447
  hot-gas defrosting, 449
  timer operation, 447
Defrost thermostat testing, 299
Defrosting
  automatic, 181, 291
    operation, 291
    sequence, 296
    timer operation, 297, 448
  display-case, 542
  hot water, 291
  manual, 180, 290
  timers, 164
Dehydrator, 538, 566
Design considerations, locker plant,
  556
Desirable properties of refrigerants,
  71
Details of evaporator installation,
  typical locker room, 559
Detection of leaks, 82

Direct current, 414
Direct expansion system, 64
Direct-expansion-plate system, 516
Discharge lines, 615
Display case
    arrangement, 528
    defrosting
        pressure method, 543
        straight-time method, 543
        temperature method, 544
    drainage requirements, 546
    installation
        baffles, 540
        dehydrator, 538
        oil separators, 539
        suction lines, 538
    lighting, 533
    maintenance, 537
    radiant heat effect, 533
Domestic electrical systems, 401
    compressor wiring, 403
    overload protector, 404
    starting relay, 403
    temperature control, 404
Domestic refrigerators, 135-151
Door
    openings, 351
    replacement, 317
Double-suction-riser arrangement
    providing oil return, 613, 614
Drain-line installation, 339
Drier-replacement, 273
Driers, 483
Drier-strainer, 142
Drier with piping connections, 484
Drive arrangement for centrifugal
    pump, 661
Driver diameter, calculation of, 426
Dry-bulb temperature, 61
Dry ice, 522

E

Eccentric shaft, 99
Effect
    of low pressure, 53
    of pressure on refrigerants, 466
Efficiency test, compressor, 113
Electric circuit, 413
Electrical
    accessories, absorption system,
        164
    controls, 438-449

Electrical (cont.)
    system, automatic-defrost
        freezer, 338
    system maintenance, 449
    system troubleshooting, 456
    testing
        overload protector, 259
        running winding, 259
        starting winding, 258
        units with capacitors, 260
    troubleshooting chart, 456
Electronic leak detector, 81
Energy, 27, 421
    consumption, 119
    kinetic, 421
    potential, 421
    radiated, 56
Equipment and controls, locker
    plant, 561
Essential parts of a refrigeration
    plant, 467
Ethyl chloride, 45
Evaporating and recharging sealed
    systems, 206
Evaporation, 46
Evaporative condensers, 481
Evaporator, 144
    capacity, 717
    cooling fan, 275
    replacement, 272
    units, 144
Expansion-valve units, 214
Exterior view
    air-cooled condensing unit, ice
        cream cabinet, 581
    on-demand louvers, air-cooled
        compressors, 471
    sectional walk-in cooler, 560
External equalizer application, 391

F

Fan
    centrifugal, 603, 604
    characteristics, 604, 605
    motor types, 606
    power requirements, 606
    propeller, 604
    selection, 607
Filling ice cans, 511
Fittings and valves, 484
Flame stability, 174

Floor arrangement, walk-in coolers, 547
Flue system, 158
Foaming of brine, 502
Food
  arrangement, 350
  preparation for quick freezing, 595
Foundations, refrigeration unit, 670
Freezing
  methods, ice, 519
  requirements table, 582
  tanks, 514
  time, ice making, 525
Freon refrigerants, 73, 75-89
  operating pressures, 74
  properties, 73, 74, 125-128
  various, 45, 75-77
Frequency and slip, 418
Fresh fruits and vegetables in
  open-faced refrigerated display, 532
Frozen meat storage, 551
Fruits, preparation, 596

G

Gas
  burners, 171
  control devices, 160
  -line connection, 169, 182
  -line installation, 355
  -pressure regulator, 159
Gage pressure, 19, 41
  adjustment, 172
Gravity, specific, 17
Grouting, 672

H

Hair-felt insulation, 629
Halide torch, 81, 224
Halocarbon refrigerant safety pre-
  cautions, 130
Halogen leak detector, 676
Hand valves, 566
Handling refrigerant cylinders, 84
Harvesting plate ice, 518
Head-pressure control, 479
Heat
  exchangers, 486
    replacement, 273
  latent, 39

Heat (cont.)
  leakage, 685
  losses, ice making, 524
  sensible, 37
  specific, 35
  theory of, 29
  transfer coefficient, 710
  transmission methods, 55-57
  units of, 34
Hermetic compressors, 101
High-pressure side, 138
Hinge replacement, 317
Holdover tanks, 65, 498
Horizontal and vertical remote water
  chillers, 580
Horsepower, 28
Hot-gas defrosting, 449
Household freezers,
  capillary tube system, 327
  cycle of operation, 325
  electrical system, 328, 338
  troubleshooting chart, 340

I

Ice
  cans, 511
    freezing methods, 519
  making,
    calculations, 523
    capacity, 523
    freezing time, 525
    heat losses, 524
  removal, 511
  storage, 525
Ice cream refrigeration, 579
Ideal coefficients of performance
  for reversible absorption, 637
Importance of gages, 205
Incorrect refrigerant charge, 255
Indirect refrigeration system, 64, 497
Inferior ice, 513
Input controls, 160
Installation
  and service, 166
  methods, 359
  of absorption-type refrigerators, 362
  of belt-driven compressors, 673
  of direct-connected motor-driven
    compressors, 674

Insulated concrete-floor arrangement for a walk-in cooler, 549
Insulation requirements, locker plant, 559
Interconnection of suction lines, 614
Interior light arrangement, 321
Isothermal change, 22

K

Kinetic energy, 422
Knocks, compressor, 114

L

Lagging power factor, 417
Latch, replacement, 318
Latch-strike adjustment, 319
Latent heat, 39
  of evaporation, 42
  of fusion, 41
Law(s)
  Boyle's, 21
  Charles', 22
  Dalton's, 157
  of conservation of
    energy, 66
    matter, 65
  of gases, 65
Leading power factor, 417
Leak detection, 81
  ammonia and sulfur dioxide leaks, 82, 222
  bubble method, 82, 224
    electronic detector, 81, 225, 676
    halide torch, 81, 224, 676
  test, 362
Leaks in system, condensing unit, 588
Leveling, 168
Line connection, gas burner, 169
Liquid refrigerant lines, 612
Lithium bromide water absorption-refrigeration cycle, 644
Load calculations
  for a dairy plant, 717
  for a walk-in refrigerator, 713
  for small freezers, 722
Locker-plant construction, 555
Locker room, 559

Low pressure, 51
  effects of, 53
Low-pressure side, 138
Lubricants, refrigeration system, 119
Lubrication
  compressor, 120
  methods, 371
  motor and fan bearings, 120
  precautions, 373
  system, compressor, 371
  types of, 121

M

Maintenance
  electrical system, 449
  refrigerator, 175
Manual defrosting, 180, 290
Mass, 16
Meat preparation, 551
Mechanical centers, 540
Methyl chloride, 79, 89
Method of cold-storage insulation
  masonry-wall construction, 561
  as applied to concrete ceilings, 562
Method of connecting
  evaporators, 564
  freezer plates, 565
Method of interconnecting
  hot-gas discharge lines, 617
  water valves, 618
Milk
  coolers, 584
  pasteurization, 590
  storage, 721
Miscellaneous loads, 712
Molecular theory, 29
Motor
  calculations
    driven diameter, 427
    driver
      diameter, 426
      rpm, 427
  compressor controls, 438
  efficiency, 422
  lubrication, 371
  overload protector, 439
  winding relays, 441
Mounting arrangement
  centrifugal fan, 603, 604
  propeller fan, 604

Multiple refrigeration machine connection, 532
Multiple-unit installation, 618

O

Odor prevention, 355
Oil pressure controls, 441
  solenoid valve, 443
Open display cases, 530
Open, multishelf cold-meats display case, 530
Operating levers and valves on a Frick can filler, 513
Operating pressures of freon refrigerants, 71
Operation, thermostatic expansion valve, 389
Operational features, automatic ice makers, 520
Outdoor design conditions for air conditioning, 62
Overload protector, 259, 404, 450

P

Partial restriction in evaporator, 256
Performance factor, 122
Performance of thermoelectric couples, 191
Phantom view of a typical bottle water cooler, 572
Phase relationship between current and voltage, 414
Physical properties of freon, 71
Physical units, 13
Pipe connections
  flared joints, 624
  hanging pipe lines, 624
  pipeline repairs, 625
  soldered joints, 623
  strainers, 625
Pipe cutters, 227
Piping arrangement, direct-drive ammonia refrigerating unit, 468
Piping diagram showing liquid riser connections, 613
Piping insulation,
  cork, 626
  hair felt, 630
  rock cork, 627
  wool felt, 629
Piping interconnection, multiple condensing units, 622

Pistons and rings of compressor, 97
Plate ice, 510
Plumbing connections, water cooler, 573
Point of liquefaction, 66
Polyphase motors, 428
  starting and controlling, 431
Post-loop-condenser design change, 145
Potential energy, 421
Power, 26
  factor, 417
  requirements, 606
Practical dimensions, refrigeration system, 13
Precooler, water cooler, 573
Preparation for freezing
  fruits, 596
  vegetables, 597
Preparing new brine, 507
Pressure, 48, 69
  absolute, 21, 69
  adjustment, gas burner, 176
  atmospheric, 19, 69
  critical, 67
  -drop considerations, 612
  gage, 21, 69
  method of defrosting, 543
  regulators, 159
  standard atmospheric, 19, 69
  static, 604
  velocity, 604
  water coolers, 569, 572
Pressure-temperature chart, 82
Pressure-temperature relationship, 50
Prewired
  condensing unit, 542
  mechanical center, 540
Processing room, locker plant, 558
Product
  cooling load, 686
  freezing load, 712
Prony-brake test, 423
Propeller fans, 604
Pulley speed and sizes, 425
Pump maintenance, 664

Q

Quick-freeze room, 558
Quick freezing, 596

R

Radiant heat effect, 533
Radiated energy, 56
Radiation, 56
Reciprocating
  compressor, 94
  pumps, 663
  connecting rods and wrist pins, 98
  crankshafts, 99
  lubrication, 95
  pistons and rings, 97
  shaft seal, 111
  valves, 109
Refrigerant pressure vs. temperature, 82, 83
Refrigerant
  classifications, 75-89
  containers, 86
  cylinder handling, 84
  desirable properties, 71
  types of, 75, 89
Refrigeration
  applications, 12
  atmospheric applications in, 61
  basic systems, 64
  by vaporization, 62
  control devices,
    automatic expansion valve, 384
    testing, 393, 395
  cycle, 136
  load calculations
    air change loads, 686, 711
    capacity of machines, 713
    cooling load, 686
    dairy plant, 717
    evaporator capacity, 717
    heat leakage, 686
    heat transfer coefficient, 686, 710
    milk storage, 721
    miscellaneous loads, 712
    product freezing loads, 712
    service factor, 713
    small freezers, 722
    walk-in refrigerator, 713
  system
    components, 135-151
    lubricants, 115, 120
  ton of, 46
  trouble charts, 277

Refrigerator-unit removal and replacement, 116, 174
Remote
  bubbler, 573
  water chillers, 575
Removal of
  compressor, 115
  oil sludges and solids, 379
Replacement of compressor, 116
  motor, 451
Replacing
  controls, 268
  motor-protector relay, 590
  system components, 262
Repulsion-start motors, 437
Restricted capillary tube, 250
Rock-cork pipe insulation, 627-629
Roof-top packaged cooling unit, 477
Rotor arrangement of a helical
  screw gear pump, 659
Rotary
  compressors, 100
  pumps, 657
Running-time check, 117
Running winding, 259

S

Safety precautions, 129
Schematic arrangement, multiple
  installation of solenoid valves, 567
Sealed system
  charging, 213
  evacuating, 206, 209
  operation, 249
Seebeck effect, 192
Semiopen and enclosed impellers, 662
Sensible heat, 37
Service
  equipment and tools, 201
  factor, 713
  operations
    adding oil, 214
    charging,
      through the high-side, 217
      through the low-side, 215
    evacuating, 218
    tools, 225
    valves, 202, 408
Servicing data, lubrication, 376
Shaded-pole motors, 436
Shaft seal, compressor, 95, 111

Silver brazing, 243
Simple electric circuit, 413
Simple load calculations, milk pasteurization, 590
Simple refrigeration system, 86
Single-can dumper, 517
Single phase
  current, 414
  motors, 431
Soap-bubble method, leak detection, 224
Specific
  gravity, 17
  heat, 35
Spray-cooling ponds, 473
Spray-nozzle arrangement, spray cooling pond, 474
Squirrel-cage motor, 429
Standard refrigerating unit using Freon-12 or -22, 470
Starting
  and controlling polyphase motors, 431
  capacitors, 435
  relay, 261, 403
  winding, 267
Static
  efficiency, 604
  pressure, 604
Steam, 43
Straight-time method of defrosting, 543
Structural arrangement, open cooling tower, 476
Stuck or tight compressor, 114
Sugar and salt packing, 595
Suction line interconnection, 614
Suction strainer, 108
Sulfur dioxide, 45, 78
Synchronous motors, 430
System precautions, 106

T

Tables
  average specific heats, 37
  capacity of typical air-cooled condensing units, 717
  comparative figures of operation for refrigerators, 350
  condensed list of storage temperatures, 599
  conditions in ammonia cycle, 648
  conditions in lithium bromide cycle, 645
  conductivity factors for various insulations, 714
  copper tubing sizes and dimensions, 241
  effects of altitude on atmospheric pressure, 22
  energy conversion factors, 66
  freezer requirements, 582
  Freon-12 properties of saturated vapors, 125-128
  operating pressures of refrigerants, 74
  pH color table, 505
  properties of calcium chloride brine, 500
  refrigerant pressures vs. temperature, 63
  specific heats of
    foods, 38
    water, 37
  standard welded and riveted ice cans, 511
  suggested methods for removal of refrigerations solids, 379
  temperature and pressure properties of refrigerants, 71
  typical outdoor design conditions for air conditioning, 62
  usage factors, Btu/24 hr/cu ft interior capacity, 715
  variations of Ohms law, 413
  wall heat leakage tables, 687-701
Temperature, 30
  control, 404, 444
    by suction pressure, 406
    methods, 149
    servicing, 276
  critical, 67
Template arrangement, compressor installation, 672
Test-gage set, 226
Testing with volt-ohmmeters, 258
Theoretical compressor capacities, 123
Theory of heat, 29
Thermoelectric cooling, 191
  advantages and applications, 193
Thermometers, 31
Thermostatic

control, 150
Thermostatic (cont.)
  expansion valve
    testing, 393
      precautions, 396
      procedure, 395
Thermostats, 163, 165
Three-phase current, 414, 419
Time/pressure defrost-control
  diagram, 543
Ton of refrigeration, 46
Tools, 225
Total pressure, 603
Transformer connections, 419
Transmission of heat, 55
Troubleshooting charts
  absorption refrigeration
    system, 183
  compressors, 492
  electrical system, 456
  household freezer, 340
  refrigeration, 277
Tube cooler, milk cooling, 585
Tubing cutters, 228
Typical
  belt arrangement for motor and
    compressor drive, 426
  nameplate data, 417
  refrigeration dehydrator, 677
  shell ice maker, 521
Typical installation
  one-unit remote chiller, 578
  remote water chillers, 577
  two-unit remote chiller, 579

U

Units of
  electricity, 412
  energy, 27
  heat, 34
  power, 26
Upright freezers, 334
Use of ice, 46

V

Valve selection, 389
Valves
  compressor, 100, 109
  how to crack, 204
Vaporization, refrigeration by, 62
Various fittings insulated with

  cork covering, 628
Vat type, milk cooler, 585
Vegetables, preparation, 597
Velocity of pressure, 604
Voltage and current in a wye trans-
  former, 420
Volume, 16
Volumetric efficiency, 122

W

Walk-in coolers
  floor arrangement, 547
  frozen meat storage, 551
  walls, 549
  wholesale storage, 550
Warm-weather effect, 353
Water
  -cooled condensers, 480
  coolers,
    bottle, 570
      condensing unit, 570
      controls, 572
      cooling chamber, 571
    -lithium bromide machine, 644
  pressure,
    cycle of operation, 574
    plumbing connections, 573
      coolers
        precooler, 573
        remote bubbler, 573
  regulating valves, 480
  system corresponding to electric
    circuit, 415
  valves, 616
Waveforms showing relationship in
  an ac current, 417
Wax separation, 377
Wet-bulb temperature, 61
Wiring diagrams of cooling tower
  control, 478, 479
Wool-felt insulation, 629
Working principle of a rotary
  gear pump, 658
Wound-rotor motor, 430

Z

Zero, absolute, 33

**Over a Century of Excellence
for the Professional
and
Vocational Trades and the Crafts**

Order now from your local bookstore
or use the convenient order form
at the back of this book.

AUDEL

These fully illustrated, up-to-date guides and manuals mean a better job done for mechanics, engineers, electricians, plumbers, carpenters, and all skilled workers.

CONTENTS

Electrical......................... II
Machine Shop and
   Mechanical Trades............... III
Plumbing.........................IV
HVAC V
Pneumatics and Hydraulics.......... V

Carpentry and Construction VI
Woodworking......................VI
Maintenance and Repair........... VII
Automotive and Engines........... VII
Drafting.........................VIII
Hobbies.........................VIII

ELECTRICAL

House Wiring (Sixth Edition)
ROLAND E. PALMQUIST

5 1/2 x 8 1/4   *150 Illus.*
ISBN:

NEW EDITION FOR 1991

The ... National
Elect ... apply to residential
wiring ... ailed with examples and illustrations.

Practical Electricity
(Fifth Edition)
ROBERT G. MIDDLETON;
revised by L. DONALD MEYERS

*5 1/2 x 8 1/4 Hardcover 512 pp. 335 Illus.
ISBN: 0-02-584561-6 $19.95*

The fundamentals of electricity for electrical workers, apprentices, and others requiring concise information about electric principles and their practical applications.

Guide to the 1990 National Electrical Code
ROLAND E. PALMQUIST;
revised by PAUL ROSENBERG

*5 1/2 x 8 1/4 Hardcover 664 pp. 230 Illus.
ISBN: 0-02-594565-3 $24.95*

The most authoritative guide available to interpreting the National Electrical Code for electricians, contractors, electrical inspectors, and homeowners. Examples and illustrations.

Mathematics for Electricians and Electronics Technicians
REX MILLER

*5 1/2 x 8 1/4 Hardcover 312 pp. 115 Illus.
ISBN: 0-8161-1700-4 $14.95*

Mathematical concepts, formulas, and problem-solving techniques utilized on-the-job by electricians and those in electronics and related fields.

Fractional-Horsepower Electric Motors
REX MILLER and
MARK RICHARD MILLER

*5 1/2 x 8 1/4 Hardcover 436 pp. 285 Illus.
ISBN: 0-672-23410-6 $15.95*

The installation, operation, maintenance, repair, and replacement of the small-to-moderate-size electric motors that power home appliances and industrial equipment.

Electric Motors (Fourth Edition)
EDWIN P. ANDERSON;
revised by REX ...

5 1/ ... *405 Illus.*
ISBI

NEW EDITION FOR 1991

Insta ... ce, and repair of all
types ... ric motors.

Home Appliance Servicing
(Fourth Edition)
EDWIN P. ANDERSON;
revised by REX MILLER
5 1/2 x 8 1/4 Hardcover 640 pp.   345 Illus.
ISBN: 0-672-23379-7 $22.50
The essentials of testing, maintaining, and repairing all types of home appliances.

Television Service Manual
(Fifth Edition)
ROBERT G. MIDDLETON;
revised by JOSEPH G. BARRILE
5 1/2 x 8 1/4 Hardcover 512 pp.   395 Illus.
ISBN: 0-672-23395-9 $16.95
A guide to all aspects of television transmission and reception, including the operating principles of black and white and color receivers. Step-by-step maintenance and repair procedures.

Electrical Course for Apprentices and Journeymen
(Third Edition)
ROLAND E. PALMQUIST
5 1/2 x 8 1/4 Hardcover 478 pp.   290 Illus.
ISBN: 0-02-594550-5 $19.95
This practical course in electricity for those in formal training programs or learning on their own provides a thorough understanding of operational theory and its applications on the job.

Questions and Answers for Electricians Examinations
(Tenth Edition)
Revised by PAUL ROSENBERG
5 1/2 x 8 1/4 Hardcover 316 pp.   110 Illus.
ISBN: 0-02-604955-4 $22.95
Based on the 1990 National Electrical Code, this book reviews the subjects included in the various electricians examinations—apprentice, journeyman, and master. Question and Answer format.

MACHINE SHOP AND
MECHANICAL TRADES

Machinists Library
(Fourth Edition, 3 Vols.)
REX MILLER
5 1/2 x 8 1/4 Hardcover 1352 pp. 1120 Illus.
ISBN: 0-672-23380-0 $52.95

An indispensable three-volume reference set for machinists, tool and die makers, machine operators, metal workers, and those with home workshops. The principles and methods of the entire field are covered in an up-to-date text, photographs, diagrams, and tables.

Volume I: Basic Machine Shop
REX MILLER
5 1/2 x 8 1/4 Hardcover 392 pp.   375 Illus.
ISBN: 0-672-23381-9 $17.95

Volume II: Machine Shop
REX MILLER
5 1/2 x 8 1/4 Hardcover 528 pp.   445 Illus.
ISBN: 0-672-23382-7 $19.95

Volume III: Toolmakers Handy Book
REX MILLER
5 1/2 x 8 1/4 Hardcover 432 pp.   300 Illus.
ISBN: 0-672-23383-5 $14.95

Mathematics for Mechanical Technicians and Technologists
JOHN D. BIES
5 1/2 x 8 1/4 Hardcover 342 pp.   190 Illus.
ISBN: 0-02-510620-1 $17.95
The mathematical concepts, formulas, and problem-solving techniques utilized on the job by engineers, technicians, and other workers in industrial and mechanical technology and related fields.

Millwrights and Mechanics Guide (Fourth Edition)
CARL A. NELSON
5 1/2 x 8 1/4 Hardcover 1,040 pp. 880 Illus.
ISBN: 0-02-588591-x $29.95
The most comprehensive and authoritative guide available for millwrights, mechanics, maintenance workers, riggers, shop workers, foremen, inspectors, and superintendents on plant installation, operation, and maintenance.

Welders Guide (Third Edition)
JAMES E. BRUMBAUGH
5 1/2 x 8 1/4 Hardcover 960 pp.   615 Illus.
ISBN: 0-672-23374-6 $23.95
The theory, operation, and maintenance of all welding machines. Covers gas welding equipment, supplies, and process; arc welding equipment, supplies, and process; TIG and MIG welding; and much more.

Welders/Fitters Guide
HARRY L. STEWART

8 1/2 x 11   Paperback   160 pp.   195 Illus.
ISBN: 0-672-23325-8 $7.95

Step-by-step instruction for those training to become welders/fitters who have some knowledge of welding and the ability to read blueprints.

Sheet Metal Work
JOHN D. BIES

5 1/2 x 8 1/4   Hardcover   456 pp.   215 Illus.
ISBN: 0-8161-1706-3 $19.95

An on-the-job guide for workers in the manufacturing and construction industries and for those with home workshops. All facets of sheet metal work detailed and illustrated by drawings, photographs, and tables.

Power Plant Engineers Guide
(Third Edition)
FRANK D. GRAHAM;
revised by CHARLIE BUFFINGTON

5 1/2 x 8 1/4   Hardcover   960 pp.   530 Illus.
ISBN: 0-672-23329-0 $27.50

This all-inclusive, one-volume guide is perfect for engineers, firemen, water tenders, oilers, operators of steam and diesel-power engines, and those applying for engineer's and firemen's licenses.

Mechanical Trades Pocket
Manual (Third Edition)
CARL A. NELSON

4 x 6   Paperback   364 pp.   255 Illus.
ISBN: 0-02-588665-7 $14.95

A handbook for workers in the industrial and mechanical trades on methods, tools, equipment, and procedures. Pocket-sized for easy reference and fully illustrated.

PLUMBING

Plumbers and Pipe Fitters
Library (Fourth Edition, 3 Vols.)
CHARLES N. McCONNELL

5 1/2 x 8 1/4   Hardcover   952 pp.   560 Illus.
ISBN: 0-02-582914-9 $68.45

This comprehensive three-volume set contains the most up-to-date information available for master plumbers, journeymen, apprentices, engineers, and those in the building trades. A detailed text and clear diagrams, photographs, and charts and tables treat all aspects of the plumbing, heating, and air conditioning trades.

Volume I: Materials, Tools, Roughing-In
CHARLES N. McCONNELL;
revised by TOM PHILBIN

5 1/2 x 8 1/4   Hardcover   304 pp.   240 Illus.
ISBN: 0-02-582911-4 $20.95

**Volume II: Welding, Heating,
Air Conditioning**
CHARLES N. McCONNELL;
revised by TOM PHILBIN

5 1/2 x 8 1/4   Hardcover   384 pp.   220 Illus.
ISBN: 0-02-582912-2 $22.95

**Volume III: Water Supply, Drainage,
Calculations**
CHARLES N. McCONNELL;
revised by TOM PHILBIN

5 1/2 x 8 1/4   Hardcover   264 pp.   100 Illus.
ISBN: 0-02-582913-0 $20.95

Home Plumbing Handbook
(Third Edition)
CHARLES N. McCONNELL

8 1/2 x 11   Paperback   200 pp.   100 Illus.
ISBN: 0-672-23413-0 $14.95

An up-to-date guide to home plumbing installation and repair.

The Plumbers Handbook
(Seventh Edition)
JOSEPH P. ALMOND, SR.

4 x 6   Paperback _____ *IS.*
ISBN: 0-67~ _____

A h_____ NEW EDITION _____ ipe fitters,
and _____ FOR 1991 _____ ades. It has a rugged b_____ _____ for use on the job, and fits in t_____ box or conveniently in the pocket.

Questions and Answers for
Plumbers Examinations (Second
Edition)
JULES ORAVETZ

5 1/2 x 8 1/4   Paperb_____ _____ *45 Illus.*
ISBN: 0-8161- _____

A stu_____ NEW EDITION _____ take a licensi_____ FOR 1991 _____ pprentice, journeyma_____ _____ plumber. Question and Answer_____ _____ al.

HVAC

Air Conditioning: Home and Commercial (Fourth Edition)
EDWIN P. ANDERSON;
revised by REX MILLER

5 1/2 x 8 1/4 Hardcover 528 pp. 180 Illus.
ISBN: 0-02-584885-2 $29.95

A guide to the construction, installation, operation, maintenance, and repair of home, commercial, and industrial air conditioning systems.

Heating, Ventilating, and Air Conditioning Library
(Second Edition, 3 Vols.)
JAMES E. BRUMBAUGH

5 1/2 x 8 1/4 Hardcover 1,840 pp. 1,275 Illus.
ISBN: 0-672-23388-6 $53.85

An authoritative three-volume reference library for those who install, operate, maintain, and repair HVAC equipment commercially, industrially, or at home.

Volume I: Heating Fundamentals, Furnaces, Boilers, Boiler Conversions
JAMES E. BRUMBAUGH

5 1/2 x 8 1/4 Hardcover 656 pp. 405 Illus.
ISBN: 0-672-23389-4 $17.95

Volume II: Oil, Gas and Coal Burners, Controls, Ducts, Piping, Valves
JAMES E. BRUMBAUGH

5 1/2 x 8 1/4 Hardcover 592 pp. 455 Illus.
ISBN: 0-672-23390-8 $17.95

Volume III: Radiant Heating, Water Heaters, Ventilation, Air Conditioning, Heat Pumps, Air Cleaners
JAMES E. BRUMBAUGH

5 1/2 x 8 1/4 Hardcover 592 pp. 415 Illus.
ISBN: 0-672-23391-6 $17.95

Oil Burners (Fifth Edition)
EDWIN M. FIELD

5 1/2 x 8 1/4 Hardcover 360 pp. 170 Illus.
ISBN: 0-02-537745-0 $29.95

An up-to-date sourcebook on the construction, installation, operation, testing, servicing, and repair of all types of oil burners, both industrial and domestic.

Refrigeration: Home and Commercial (Fourth Edition)
EDWIN P. ANDERSON;
revised by REX MILLER

5 1/2 x 8 1/4 Hardcover 768 pp. 285 Illus.
ISBN: 0-02-584875-5 $34.95

A reference for technicians, plant engineers, and the homeowner on the installation, operation, servicing, and repair of everything from single refrigeration units to commercial and industrial systems.

PNEUMATICS AND HYDRAULICS

Hydraulics for Off-the-Road Equipment (Second Edition)
HARRY L. STEWART;
revised by TOM PHILBIN

5 1/2 x 8 1/4 Hardcover 256 pp. 175 Illus.
ISBN: 0-8161-1701-2 $13.95

This complete reference manual on heavy equipment covers hydraulic pumps, accumulators, and motors; force components; hydraulic control components; filters and filtration, lines and fittings, and fluids; hydrostatic transmissions; maintenance; and troubleshooting.

Pneumatics and Hydraulics
(Fourth Edition)
HARRY L. STEWART;
revised by TOM STEWART

5 1/2 x 8 1/4 Hardcover 512 pp. 315 Illus.
ISBN: 0-672-23412-2 $19.95

The principles and applications of fluid power. Covers pressure, work, and power; general features of machines; hydraulic and pneumatic symbols; pressure boosters; air compressors and accessories; and much more.

Pumps (Fourth Edition)
HARRY STEWART;
revised by TOM P...

5 1/2 x ... 360 Illus.
ISBN ...

NEW EDITION FOR 1991

The ... ...o-day operation of pump... ...ontrols, and hydraulics are thorou...ly detailed and illustrated.

CARPENTRY AND CONSTRUCTION

Carpenters and Builders Library
(Fifth Edition, 4 Vols.)
JOHN E. BALL;
revised by TOM PHILBIN

*5 1/2 x 8 1/4 Hardcover 1,224 pp. 1,010 Illus.
ISBN: 0-02-506450-9 $43.95*

This comprehensive four-volume library has set the professional standard for decades for carpenters, joiners, and woodworkers.

Volume I: Tools, Steel Square, Joinery
JOHN E. BALL;
revised by TOM PHILBIN

*5 1/2 x 8 1/4 Hardcover 384 pp. 345 Illus.
ISBN: 0-672-23365-7 $10.95*

Volume II: Builders Math, Plans, Specifications
JOHN E. BALL;
revised by TOM PHILBIN

*5 1/2 x 8 1/4 Hardcover 304 pp. 205 Illus.
ISBN: 0-672-23366-5 $10.95*

Volume III: Layouts, Foundations, Framing
JOHN E. BALL;
revised by TOM PHILBIN

*5 1/2 x 8 1/4 Hardcover 272 pp. 215 Illus.
ISBN: 0-672-23367-3 $10.95*

Volume IV: Millwork, Power Tools, Painting
JOHN E. BALL;
revised by TOM PHILBIN

*5 1/2 x 8 1/4 Hardcover 344 pp. 245 Illus.
ISBN: 0-672-23368-1 $10.95*

Complete Building Construction
(Second Edition)
JOHN PHELPS;
revised by TOM PHILBIN

*5 1/2 x 8 1/4 Hardcover 744 pp. 645 Illus.
ISBN: 0-672-23377-0 $22.50*

Constructing a frame or brick building from the footings to the ridge. Whether the building project is a tool shed, garage, or a complete home, this single fully illustrated volume provides all the necessary information.

Complete Roofing Handbook
JAMES E. BRUMBAUGH

*5 1/2 x 8 1/4 Hardcover 536 pp. 510 Illus.
ISBN: 0-02-517850-4 $29.95*

Covers types of roofs; roofing and reroofing; roof and attic insulation and ventilation; skylights and roof openings; dormer construction; roof flashing details; and much more.

Complete Siding Handbook
JAMES E. BRUMBAUGH

*5 1/2 x 8 1/4 Hardcover 512 pp. 450 Illus.
ISBN: 0-02-517880-6 $24.95*

This companion volume to the *Complete Roofing Handbook* includes comprehensive step-by-step instructions and accompanying line drawings on every aspect of siding a building.

Masons and Builders Library
(Second Edition, 2 Vols.)
LOUIS M. DEZETTEL;
revised by TOM PHILBIN

*5 1/2 x 8 1/4 Hardcover 688 pp. 500 Illus.
ISBN: 0-672-23401-7 $27.95*

This two-volume set provides practical instruction in bricklaying and masonry. Covers brick; mortar; tools; bonding; corners, openings, and arches; chimneys and fireplaces; structural clay tile and glass block; brick walls; and much more.

Volume 1: Concrete, Block, Tile, Terrazzo
LOUIS M. DEZETTEL;
revised by TOM PHILBIN

*5 1/2 x 8 1/4 Hardcover 304 pp. 190 Illus.
ISBN: 0-672-23402-5 $14.95*

Volume 2: Bricklaying, Plastering, Rock Masonry, Clay Tile
LOUIS M. DEZETTEL;
revised by TOM PHILBIN

*5 1/2 x 8 1/4 Hardcover 384 pp. 310 Illus.
ISBN: 0-672-23403-3 $14.95*

WOODWORKING

Wood Furniture: Finishing, Refinishing, Repairing
(Second Edition)
JAMES E. BRUMBAUGH

*5 1/2 x 8 1/4 Hardcover 352 pp. 185 Illus.
ISBN: 0-672-23409-2 $12.95*

A fully illustrated guide to repairing furniture and finishing and refinishing wood surfaces. Covers tools and supplies; types of wood; veneering; inlaying; repairing, restoring, and stripping; wood preparation; and much more.

Woodworking and Cabinetmaking
F. RICHARD BOLLER

*5 1/2 x 8 1/4 Hardcover 360 pp. 455 Illus.
ISBN: 0-02-512800-0 $18.95*

Essential information on all aspects of working with wood. Step-by-step procedures for woodworking projects are accompanied by detailed drawings and photographs.

MAINTENANCE AND REPAIR

Building Maintenance
(Second Edition)
JULES ORAVETZ
5 1/2 x 8 1/4 Paperback 384 pp. 210 Illus.
ISBN: 0-672-23278-2 $11.95
Professional maintenance procedures used in office, educational, and commercial buildings. Covers painting and decorating; plumbing and pipe fitting; concrete and masonry; and much more.

Gardening, Landscaping and Grounds Maintenance
(Third Edition)
JULES ORAVETZ
5 1/2 x 8 1/4 Hardcover 424 pp. 340 Illus.
ISBN: 0-672-23417-3 $15.95
Maintaining lawns and gardens as well as industrial, municipal, and estate grounds.

Home Maintenance and Repair: Walls, Ceilings and Floors
GARY D. BRANSON
8 1/2 x 11 Paperback 80 pp. 80 Illus.
ISBN: 0-672-23281-2 $6.95
The do-it-yourselfer's guide to interior remodeling with professional results.

Painting and Decorating
REX MILLER and GLEN E. BAKER
5 1/2 x 8 1/4 Hardcover 464 pp. 325 Illus.
ISBN: 0-672-23405-x $18.95
A practical guide for painters, decorators, and homeowners to the most up-to-date materials and techniques in the field.

Tree Care (Second Edition)
JOHN M. HALLER
8 1/2 x 11 Paperback 224 pp. 305 Illus.
ISBN: 0-02-062870-6 $16.95
The standard in the field. A comprehensive guide for growers, nursery owners, foresters, landscapers, and homeowners to planting, nurturing and protecting trees.

Upholstering (Updated)
JAMES E. BRUMBAUGH
5 1/2 x 8 1/4 Hardcover 400 pp. 380 Illus.
ISBN: 0-672-23372-x $15.95
The esentials of upholstering fully explained and illustrated for the professional, the apprentice, and the hobbyist.

AUTOMOTIVE AND ENGINES

Diesel Engine Manual
(Fourth Edition)
PERRY O. BLACK;
revised by WILLIAM E. SCAHILL
5 1/2 x 8 1/4 Hardcover 512 pp. 255 Illus.
ISBN: 0-672-23371-1 $15.95
The principles, design, operation, and maintenance of today's diesel engines. All aspects of typical two- and four-cycle engines are thoroughly explained and illustrated by photographs, line drawings, and charts and tables.

Gas Engine Manual
(Third Edition)
EDWIN P. ANDERSON;
revised by CHARLES G. FACKLAM
5 1/2 x 8 1/4 Hardcover 424 pp. 225 Illus.
ISBN: 0-8161-1707-1 $12.95
How to operate, maintain, and repair gas engines of all types and sizes. All engine parts and step-by-step procedures are illustrated by photographs, diagrams, and troubleshooting charts.

Small Gasoline Engines
REX MILLER and
MARK RICHARD MILLER
5 1/2 x 8 1/4 Hardcover 640 pp. 525 Illus.
ISBN: 0-672-23414-9 $16.95
Practical information for those who repair, maintain, and overhaul two- and four-cycle engines—including lawn mowers, edgers, grass sweepers, snowblowers, emergency electrical generators, outboard motors, and other equipment with engines of up to ten horsepower.

Truck Guide Library (3 Vols.)
JAMES E. BRUMBAUGH
5 1/2 x 8 1/4 2,144 pp. 1,715 Illus.
ISBN: 0-672-23392-4 $50.95

This three-volume set provides the most comprehensive, profusely illustrated collection of information available on truck operation and maintenance.

Volume 1: Engines

JAMES E. BRUMBAUGH

*5 1/2 x 8 1/4 Hardcover 416 pp. 290 Illus.
ISBN: 0-672-23356-8 $16.95*

Volume 2: Engine Auxiliary Systems

JAMES E. BRUMBAUGH

*5 1/2 x 8 1/4 Hardcover 704 pp. 520 Illus.
ISBN: 0-672-23357-6 $16.95*

Volume 3: Transmissions, Steering, and Brakes

JAMES E. BRUMBAUGH

*5 1/2 x 8 1/4 Hardcover 1,024 pp. 905 Illus.
ISBN: 0-672-23406-8 $16.95*

DRAFTING

Industrial Drafting

JOHN D. BIES

*5 1/2 x 8 1/4 Hardcover 544 pp. Illus.
ISBN: 0-02-510610-4 $24.95*

Professional-level introductory guide for practicing drafters, engineers, managers, and technical workers in all industries who use or prepare working drawings.

Answers on Blueprint Reading
(Fourth Edition)

ROLAND PALMQUIST;
revised by THOMAS J. MORRISEY

*5 1/2 x 8 1/4 Hardcover 320 pp. 275 Illus.
ISBN: 0-8161-1704-7 $12.95*

Understanding blueprints of machines and tools, electrical systems, and architecture. Question and answer format.

HOBBIES

Complete Course in Stained Glass

PEPE MENDEZ

*8 1/2 x 11 Paperback 80 pp. 50 Illus.
ISBN: 0-672-23287-1 $8.95*

The tools, materials, and techniques of the art of working with stained glass.